The Palgrave Handbook of Environmental Restorative Justice

"At this crucial, and terrifying, time in Earth's trajectory, when human/non-human action engagements have been, and are, so consequential, and when the prospect of a sixth mass extinction looms larger every day, this volume, with its wealth of thoughtful and telling insights drawn from restorative justice, charts possibilities for planetary restoration. It does so at a moment when nothing is more important, and more urgent, than collective action that will reconstitute our planetary engagements. This volume constitutes an urgent call for action by criminologists, to do what everyone of us living today must do—namely, contribute, urgently and with all our might to realizing the possibility of a liveable tomorrow."
—Clifford Shearing, Professor of Law, *Universities of Cape Town, Griffith, Montreal, and New South Wales, Australia*

"You may think 'restorative justice' sounds abstract, utopian, narrow—if so, you are wrong. This truly engaging and wide-ranging collection of essays covers the theory, philosophy and application of restorative justice. Insights and analysis range across the past and the future, case-studies from around the world, and ideas of trusteeship, remedy and repair. It is the definitive guide."
—Nigel South, Emeritus Professor of Sociology, *University of Essex, UK*

"In the face of an unprecedented global environmental crisis that is eroding the very foundations of life on Earth, it is clear that we need new ideas, creative solutions, and a profound rethinking of the relationship between humans and the rest of nature. This handbook delivers a thought-provoking smorgasbord of innovative proposals that collectively have the potential to spark rapid, systemic and transformative changes in society."
—Dr. David Boyd, UN Special Rapporteur on *Human Rights and the Environment*

"The IPCC tells us that we have only three years left to limit climate change and ensure a 'liveable future'. IBPES tells us that we have entered a sixth mass extinction. It is important to do everything possible to force industrial activity to respect the Earth system, with the recognition of the crime of ecocide being the most universal legal solution, but it is also clear that we must move quickly towards a society capable of adapting to the new living conditions that await us. This path is that of resilience, and resilience cannot be achieved without Environmental Restorative Justice and without transitional justice that recognises the status of victims for populations depending on them for survival, in particular Indigenous people, future generations, and non-human beings. This book is therefore fundamental because it gives us the tools to design tomorrow's world with ecosystemic new rules, and beyond that to reintegrate the Earth community."

—Valerie Cabanes, member of the Independent Expert Panel
for the *Legal Definition of Ecocide*

Brunilda Pali • Miranda Forsyth
Felicity Tepper
Editors

The Palgrave Handbook of Environmental Restorative Justice

palgrave
macmillan

Editors
Brunilda Pali
Department of Social and Cultural
Anthropology, Faculty of Social Sciences
KU Leuven
Leuven, Belgium

Miranda Forsyth
RegNet School of Regulation and Global
Governance
Australian National University
Canberra, ACT, Australia

Felicity Tepper
RegNet School of Regulation and Global
Governance
Australian National University
Canberra, ACT, Australia

ISBN 978-3-031-04222-5   ISBN 978-3-031-04223-2   (eBook)
https://doi.org/10.1007/978-3-031-04223-2

Cover illustration: © photo by Mark Požlep

This Palgrave Macmillan imprint is published by the registered company Springer Nature Switzerland AG.
The registered company address is: Gewerbestrasse 11, 6330 Cham, Switzerland

Foreword

'But how do you apologise to a river?' Eliza Victoria asks as Chap. 19 opens (by Jennifer Amparo, Ana Bibal, Deborah Cleland, Ma. Catriona Devanadera, Aaron Lecciones, Maria Mendoza, and Emerson Sanchez). If we tread on our dog's foot, we apologise, give her a cuddle. A river, however, is neither a single animal nor an object. Chapters of this book discuss how Indigenous wisdom instructs us that a river is a flow of life, life of diverse forms, that adapts its life flows to seasons, flood, and drought. If we seek a closer approach to a capability for saying sorry to rivers, first embrace and empower Indigenous voices of river custodians into restorative dialogue. Not any Indigenous voice, of course, but a custodian of that particular river who feels that river as part of herself, her ancestors as part of it. The spirit of the river thence flows into her spirit as a custodian of that particular river. Inviting the apt elders into a circle is a way to make a conversation about protection of the environmental heritage of that flow of life more spiritually profound.

White settlers on Indigenous lands are not excluded from spiritual engagement with Country and rivers. Personal experience is that my spiritual engagement with nature is shallow compared to Indigenous leaders who mentor me on this question. I listened to them when they said that I can become a less spiritually shallow white man within healing and yarning circles by taking my mind to that patch of nature for which I have special affection. It is love for a particular spot that I share with an old Aboriginal friend on Yuin country. Near where he and I live stands a particular spotted gum tree, wounded by big winds centuries ago, still strong and wide. I confess to being a white bourgeois tree hugger with that tree. Most days, rather than hug, I gently pat it or whisper hello. I say daily hellos too with two particular sea eagles who nest nearby and a cockatoo who disrupts my writing with noisy speech from our verandah railing. In a Bougainville village where I lived for a while in 1969, I was made an honorary chief of the Naboin (eagle) clan in 2006. The elders told me I must care for my special relationship with eagles; I must never shoot one just because I am hungry in the way they said white men do. I feel that connection to the Naboin clan's part of nature's whole when I look up to admire our soaring Malua Bay eagles.

A famous episode in the history of peacebuilding occurred in 1992 between current South African President Cyril Ramaphosa and former National Party Defence Minister Roelf Meyer. I interviewed Meyer about the spiritual participation of nature in this encounter. During 1992, the peace negotiations and the personal relationship between President de Klerk and African National Congress leader Nelson Mandela had soured. Violence was escalating. Mandela and de Klerk understood that Meyer and Ramaphosa had negotiated constructively over the National Union of Mineworkers' strike when Ramaphosa was the union secretary. So, de Klerk and Mandela agreed to step back and see if they could craft a new formula for a constitutional settlement. They did; it paved a path to a powersharing transition followed by the election of Mandela. The two men went trout fishing together. The serenity of nature helped soften their hearts as they settled a riverbank reconciliation. Then the river bit back. Meyer spiked a hook into his hand in the manner intended for the trout. Ramaphosa administered a stiff whisky to dull Meyer's pain and then firmly extracted the embedded hook. That was just one of the things that happened to strengthen their relationship as they negotiated amongst the trees. I suspect, but do not know for a fact, that Meyer may have betrayed his white constituency in some ways. Meyer and Ramaphosa were the experts on mining who had the ear of their two leaders. The day he was released from prison, Mandela announced an early priority to nationalise mining companies. He recanted upon realising that his policy of requiring these firms to clean up the environmental mess they had created would mean that many mines that were almost mined out would have assets less than their environmental liabilities. So Mandela turned against the policy of socialising capitalism's environmental losses to the benefit of white owners and western investors. What seemed a peacemaking concession to white business power was actually smart environmental custodianship.

Negotiating armed conflicts amongst the trees is smart peacemaking. It is not just Indigenous wisdom that teaches how trees soften hearts. Western science likewise shows that when people live in communities with a lot of trees, this soothes anxiety and anger. Planting great numbers of urban trees simultaneously improves urban environments, sucks carbon, and reduces urban crime. These are now quite well-established findings. Without western science to support it, Australians also understand how the soft beauty of the desert ecosystem spoke peacefully into one of our country's great literary and political communions with the ancestors, the Uluru Statement from the Heart.[1]

It is surprising how often conservative, warlike men choose to negotiate peace close to nature. In the Middle East conflicts from the 1970s, we saw this with bitter enemies meeting amongst the trees of the US Presidential retreat at Camp David and on other occasions amongst trees near places like Oslo. Long before this, Camp David was used for summits by presidents of World War II fame, Roosevelt and Eisenhower, who held summits there with the likes of Churchill and Khrushchev. Vladimir Putin is one of many Russian leaders who have been hosted there. President Reagan restored the nature trails of Camp David that President Nixon had paved over. It was here that Reagan discussed global leadership for the Montreal Protocol on healing the ozone hole. Reagan also held the most important of the peacemaking summits ever held with the Soviet Union in Iceland close to nature, with Mikhail Gorbachev. Presidents of these great powers who still carry codes to ignite the nuclear

[1] Please do watch the video and listen to the Uluru Statement from the Heart here: https://ulurustatement.org/the-statement/ (last accessed 6 February 2022).

conflagrations that would destroy most species of the planet through nuclear winter and famine do well to talk amongst the very trees that would be destroyed by failure in their peacemaking.

In the Peacebuilding Compared project,[2] a 25-year project designed to follow all the major armed conflicts around the world until 2030, with the aim of understanding key ingredients that make for the success of sustainable peacebuilding, we have tried to allow nature to speak to little peace processes we attempted to facilitate amongst the trees, for example, between one spoiler faction of the Bougainville peace, the President and the next President of Bougainville and the Minister for Bougainville Affairs in 2008, listening to many ordinary women and men from affected communities.

This handbook takes us to journeys of healing between so many kinds of environmental destroyers and healers. It is an inspiring book, brilliantly curated by the editors. Some authors are an established cream of green criminology and environmental restorative justice. Others are evocative new voices putting up fresh green shoots of ideas about how to mediate cries from nature and proffer remorse to rivers. They gift us readers with a rich blend of theory and green restorative practice. This reaches far beyond the practice of set piece restorative justice conferences. It is art, theatre, film, poetry, healing rituals of everyday life, restorative jurisprudence, intergenerational and United Nations justice in ecosystem restoration, green decolonisation, environmental learning, restorative regulation, green markets, and green democratic innovation. Other authors simply help us understand the difference between everyday speech acts amongst earthlings that are less and more infused with kindness and healing. Other chapters still help us see that environmental restorative justice must embody a robust politics of resistance to ecocide. We might read all chapters as helping to displace a hegemonic order based on domination of nature with one grounded at least in an alternative hegemonic order that is greener, more restorative, more just in its regulation of human dominations of nature.

---

[2] Peacebuilding Compared is a collective and long-term research project funded by the the Australian Research Council since 2004. For more information on the project visit: http://johnbraithwaite.com/peacebuilding/ (last accessed 6 February 2022).

A really simple way this book inspires is that it reveals the richness and sheer level of engagement of so many fine scholars with environmental restorative justice as a field. This when it barely existed as a field of scholarship a few years ago. Long may it grow from germinal seeds planted in this beautiful book.

Canberra, ACT, Australia   John Braithwaite

Preface

This handbook offers insights into some of the most important dimensions of environmental restorative justice (ERJ), with the aim of extending our understanding and appreciation of the concept's significance within our contemporary world. ERJ is an ethos and set of values and practices that respond to environmental harm through focusing on healing the harm, repairing relationships, deep listening, participation of everyone involved, and ensuring accountability for harm caused in ways that prevent its re-occurrence.

The handbook aims to actively inform and engage scholars, practitioners, policymakers, and concerned citizens through forming a shared language and providing a set of actionable principles to repair past and ongoing environmental harm and prevent future harm. In seeking to further expand and develop the domain of ERJ, the handbook brings together an interdisciplinary group of scholars from across the world who have studied different facets of ERJ.

Many thanks go to all the authors of this volume who have laboured so diligently and tenaciously to draw out the multiple aspects of ERJ and without whom this volume would not have been possible. Special thanks go to John Braithwaite and Ivo Aertsen for having enabled and inspired some of the thinking and encounters that led to this handbook. We also want to acknowledge the community of the European Forum for Restorative Justice (EFRJ) which has brought together throughout the years so many practitioners, policy makers, researchers, artists, and activists and under whose friendly umbrella so many creative conversations have been generated, including about ERJ. We thank especially the members of the EFRJ's Working Group on ERJ. Our work has also benefitted from the opportunity that was granted to us by the Oñati International Institute for the Sociology of Law offered in the summer of 2021 to organise the international workshop 'Environmental restorative justice: A new justice framework for preventing and addressing environmental harms', at which many contributions of this handbook had their genesis. We also thank our publisher for the trust in our work and their vision in commissioning this. Finally, we extend many thanks to our friends and families who supported us throughout the process of editing this volume.

Leuven, Belgium
Canberra, ACT, Australia
Canberra, ACT, Australia
1 February 2022

Brunilda Pali
Miranda Forsyth
Felicity Tepper

Contents

Notes on Contributors

Ivo Aertsen is Emeritus Professor of Criminology at the University of Leuven (Belgium). He holds degrees of psychology, law and criminology from the same university. At the Leuven Institute of Criminology, he has been leading the Research Line on 'Restorative Justice and Victimology' from 2001 to 2019. He has coordinated a series of European research projects and international publications and has been active as expert in restorative justice-related work in various countries and for international institutions. He is Co-Editor of *The International Journal of Restorative Justice*.

Ben Almassi is Associate Professor of Philosophy at Governors State University and the author of *Reparative Environmental Justice in a World of Wounds*, published in 2020 by Lexington Books. He is also an affiliated faculty member in Gender & Sexuality Studies, Interdisciplinary Studies, Political & Social Justice Studies, and GSU's Certificate in Restorative Justice. Ben lives with his partner Negin and their daughter Zeydi in Chicago, located on the traditional homelands of the Council of the Three Fires (the Ojibwe, Odawa, and Potawatomi Nations) and today home to one of the largest American Indian communities in the United States.

Jennifer Marie S. Amparo is an Assistant Professor at the Department of Social Development Services, College of Human Ecology, University of the Philippines. She was the former Country Coordinator of Blacksmith

Institute (now Pure Earth) in the Philippines from 2009 to 2012, including working on the integrative project on *Clean the Marilao-Meycauayan-Obando River System MMORS* of Blacksmith Institute from 2008 to 2009 and later rehabilitation projects in the river system. She completed her PhD at the Fenner School of Environment and Society, Australian National University, under the Australia Awards Scholarship, where she explored the social-ecological trapped fishing systems in the Philippines like in MMORS.

Mike Batley is a Co-founder and Director of the Restorative Justice Centre (RJC), a vibrant, multi-cultural civil society organisation. Within this context, he played a pioneering role in bringing restorative justice into the criminal justice system and public discourse and in developing associated services in South Africa. He was recognised as an Ashoka Fellow for this work. He has written several book chapters and journal articles on restorative justice. He was part of the group of experts that reviewed the UN Basic Guidelines for Restorative Justice in November 2017. In 2019, he developed a strategy for strengthening child justice in Eswatini.

Ana Christina M. Bibal is a Development Worker and Researcher engaged with Indigenous peoples rights networks, agroecology and food sovereignty alliances and forest conservation organisations in the Philippines, South and South East Asia. She holds a Masters in Environmental Science from the University of the Philippines Los Banos. Her policy advocacy and research involvement related to indigenous peoples, FPIC, mining, oil palm plantations, shifting rural livelihoods and impacts studies of extractive industries on traditional land use systems have led to her academic interest in environmental justice and political ecology. She is engaged as community development specialist related to watershed conservation programs.

Daniela Bolívar is an Assistant Professor at the School of Social Work at the Pontifical Catholic University of Chile and a Research Collaborator at the Centre Justice and Society from the same University. Daniela has conducted her PhD at the Leuven Institute of Criminology, KU Leuven (Belgium). She is author of *Restoring Harm: A Psychosocial Approach to*

Victims and Restorative Justice (2019, Routledge), among other articles and book chapters on the topic.

Klaus Bosselmann is Professor of Environmental Law and Founding Director of the New Zealand Centre for Environmental Law at the University of Auckland. He has current leadership roles in the Ecological Law and Governance Association, the Earth Trusteeship Initiative, the Global Ecological Integrity Group, and Common Home of Humanity. Klaus has authored 15 books (a number of them translated into multiple languages) and over 150 articles in the areas of environmental ethics and law, international environmental law, global governance and eco-constitutionalism. For his pioneering work on ecological law and Earth jurisprudence, he received numerous awards including the Inaugural Senior Scholarship Prize of the IUCN Academy of Environmental Law.

Carlos Frederico Braga da Silva is a Postdoctoral Researcher associated to the Faculty of Sociology of the Federal University of Minas Gerais (UFMG), Brazil, and to the Canadian Chair of Legal Tradition and Penal Rationality (University of Ottawa). He obtained his Doctorate degree in Sociology at UFMG, in a cotutelle system with the University of Ottawa, Faculty of Social Sciences, Department of Criminology. His research interests concern institutionalising environmental restorative justice by assimilating its concepts, principles and goals into law. He has also worked as a State Trial Court Judge since 2000.

Gale Burford is Professor Emeritus of the University of Vermont. He has experience as a social work practitioner and manager and has held full-time university teaching appointments in Canada and the US and visiting appointments in Australia, New Zealand, and the United Kingdom. His main areas of research have focused on the use of a restorative approach with families experiencing family violence.

Marine Calmet is an Environmental Lawyer, involved in the defence of the rights of nature. She is advisor and coach with and for organisations and companies that are committed to the protection of the planetary boundaries through the construction of innovative socio-economic models, in harmony with the Living. As a field work lawyer, Marine leads campaigns against extractivism projects, gold mine, and offshore oil.

Calmet is at the service of projects, that wish to meet the challenge of ecological transition. Inspired by the richness and creativity of nature, Calmet believe that it is up to us to be attentive to the laws of the Earth to find new solutions to face the current crisis.

Xenia Chiaramonte is a Jurist and a Socio-legal Scholar. She is a fellow at the ICI Berlin where she develops a project on law and ecology. She has recently written her monograph *Governare il conflitto: La criminalizzazione del movimento No TAV [Governing conflict: The Criminalization of the No TAV Movement]* (2019), which analyses the criminalisation of one of the most long-standing and high-profile environmental movements in Western Europe. Her recent publications include a co-authored book on criminal justice and two co-edited books on political violence.

Deborah Cleland is an Interdisciplinary Social Scientist, with a background in human ecology and natural resource management. Most recently, she has been working on environmental restorative justice, which draws on her previous work on fisheries conservation, environmental impact assessment and workplace health and safety. Deb is particularly interested in how regulators can work with communities to improve safety, quality of life and citizen engagement in our democracy. Following knee surgery in 2019, rehab is still taking a lot of Deb's time, so she is thinking a lot about the crossovers between restoration and repair for people and planet.

Maria Lucia Cruz Correia is an Environmental Artist and Activist based in Belgium. Correia's cross-sectoral and hybrid practice speaks to her deep engagement with the ecological crises and climate emergency. She reacts to the environmental conflicts of our times by bringing audiences and communities into participatory laboratories that connect the artistic with the voices of scientists, activists and lawyers in long-term investigation processes. Correia's work expresses a sense of cosmopolitics, advocacy and kinship between humans and the more-than-human-world. Between 2013 and 2019, her work was supported by Vooruit kunstencentrum. In 2017 she received the Roel Verniers Prijs at the Theaterfestival with her first theatre piece *Voice of nature: The trial*. In 2019, her new project KINSTITUTE was shortlisted for COAL prize.

Her work has been supported by Imagine 2020 and is currently supported by Displacementjourneys (SW), Be Part (EU), TerraBatida (PT). In 2020 she started a trajectory as an artist with Wpzimmer (BE).

Ma. Catriona E. Devanadera is an Associate Professor at the Department of Community and Environmental Resource Planning, College of Human Ecology, University of the Philippines Los Baños. She has a PhD in Environmental Engineering from the University of the Philippines Diliman. Her expertise is in environmental engineering, specifically, biological wastewater, and membrane treatment systems. She also researches wetlands conservation and management and has conducted several climate and disaster risk assessments for municipalities in the Philippines. She is an active member of the Society for the Conservation of Philippine Wetlands, Inc. (SCPW), and the Society of Environmental Engineers of the Philippines (SEEP).

Anna Di Ronco is Senior Lecturer in Criminology at the Sociology Department of the University of Essex and Director of its Centre for Criminology. She holds a doctorate in criminology from Ghent University, Belgium. Anna's main research focuses on urban incivilities, local-level policing, urban resistance, criminalised environmental movements and social media protest.

Ashleigh Dore is the Manager of the Environmental Restorative Justice Pilot Project and the Wildlife and Law Project at the Endangered Wildlife Trust in South Africa. She is an environmental lawyer, holding a master of laws in environmental law from the University of Cape Town and is admitted as an attorney of the High Court of South Africa. Ashleigh has a keen interest in exploring legal approaches and solutions to complex conservation challenges. Her research focuses are environmental law (specifically law relating to biodiversity), environmental offences (with a focus on wildlife offences) and restorative justice. She also lectures past time at the University of Pretoria in the field of environmental law.

Miranda Forsyth is an Associate Professor in the School of Regulation and Global Governance (RegNet) at the Australian National University. She is also the Director of the Centre for Restorative Justice at RegNet and the Convenor of the ANU Pacific Institute. She works on plural justice in contexts of multiple legal and normative orders, such as the Pacific Islands region, and on environmental restorative justice.

Rachel Gehman is a graduate student in the University of Vermont Clinical Psychology PhD program with an interest in restorative justice research and practice. She has a research focus on the psychological processes underlying positive outcomes for responsible parties in restorative justice.

Liliana Guerra is an Associate Professor at the School of Social Work of the Pontifical Catholic University of Chile. She has a PhD in Latin American Studies (University of Chile), a Master in Social Work (Pontifical Catholic University of Chile), and Postgraduate degree in Family Studies (Pontifical Catholic University of Chile). She researches and writes on culture, social policies, youth, public health and family.

Mark Hamilton is a Lecturer in the Thomas More Law School, Australian Catholic University, teaching in the law and criminology programs. He was awarded his Doctor of Philosophy (Law) from the University of New South Wales in 2019. Mark has had considerable exposure to, and experience in, local government, planning and environmental law, having worked at the Land and Environment Court of New South Wales before practicing law in a local government and planning practice in a national mid-tier law firm in its Sydney office. Mark has a strong interest in green criminology, environmental victims, and restorative justice and is the author of *Environmental Crime and Restorative Justice: Justice as Meaningful Involvement*, which is part of the Palgrave Studies in Green Criminology series.

Jan Erik Henriksen (Sámi) is Professor in Social Work at Norwegian Arctic University (UiT). He is leader of Indigenous Voices, a research group at the UiT. Henriksen was main organiser of the fourth International Indigenous Voices in Social Work Conference (Iiviswc) in Alta in 2017 and is member of Iiviswc's international committee.

Annette Hübschle is Senior Research Fellow in Global Risk Governance Programme at the University of Cape Town, South Africa. She leads the Environmental Futures research group. Her research focuses on the governance of safety and security with a specific focus on illegal wildlife economies, environmental restorative justice and environmental futures, as well as the interface between licit and illicit economies and criminal networks.

Ida Hydle is an Adjunct Professor at the Norwegian Arctic University, UiT, has a PhD in medicine and in social anthropology, has served as a mediator in the Norwegian mediation service, and is a member of the research group Indigenous Voices at UiT.

Rachel Jolly is Director of the Burlington Community Justice Center, Vermont. Her degrees and work in environmental education combined with ten years at a non-profit working with incarcerated women and those transitioning into the community led her to restorative justice and meaningful community building.

Vinny Jones is a Scenographer who uses light as her primary material to shape the experience of space and the relationship between the audience and a performance. Her practice brings together work as a researcher, a light designer, a maker of immersive installations, and a teacher. As a light designer, her approach is highly collaborative. She uses light as an aesthetic, sensory and dramaturgical tool of theatre-making.

Tanya Jones is a PhD researcher at the University of Dundee, exploring the potential application of restorative justice principles to climate injustice. She has previously worked as a litigation lawyer, writer and teacher and has been active as an environmental campaigner, especially in relation to fracking. Before moving to Scotland, she was deputy leader of the Green Party in Northern Ireland. Her first degree was in English Literature at King's College, Cambridge, and she has an MA in Medieval Studies from the University of York and a LLM in Environmental Law from the University of Dundee.

Rachel Killean is a Senior Lecturer in the School of Law at Queen's University Belfast, and a Fellow of the Senator George J Mitchell Institute for Global Peace, Security and Justice. Her research focuses on transitional justice, with a particular focus on sexual and gender-based violence, cultural heritage, and the environmental legacies of atrocity. She is an investigator on a research project exploring the role of participatory filmmaking in conservation advocacy and is leading a project exploring the concept of 'human dignity' in Cambodia.

Orika Komatsubara holds a Research Fellowship for young scientists in Japan Society for the Promotion of Science. She has a PhD in Human Sciences from Osaka Prefecture University in Japan. Her latest work (in English) is *The Role of Literary Artists in Environmental Movements: Minamata Disease and Michiko Ishimure* (2021) published in the *International Journal for Crime, Justice and Social Democracy*.

Aaron M. Lecciones is an Assistant Professor at the College of Architecture, University of the Philippines, Diliman. After graduating as an architect, he undertook an MSc in Practising Sustainable Development from Royal Holloway, University of London. His research focuses on sustainability in the designed and built environment, including disaster risk management, land use planning, sustainable communities, active urban design, and resilient urban waterfronts. He is especially interested in wetland centre design where his current research studies the influence of community perception of ecosystem services on land use planning at the Las Piñas Parañaque Wetland Park, Metro Manila, Philippines.

Felipe Martínez is a Lawyer and LLM at the Pontifical Catholic University of Chile. He is Lecturer on negotiation and arbitration, at the Law Faculty, of the same university. He has a diploma in environmental and sustainable development and negotiation.

Margarida Mendes is a PhD candidate at the Centre for Research Architecture, Goldsmiths University of London. Her research explores the overlap between infrastructure, ecology, experimental film and sound practices—investigating environmental transformations and their impact on societal structures and cultural production. She has curated several exhibitions and was part of the curatorial team of the 11th Liverpool Biennale (2021); 4th Istanbul Design Biennial (2018), and; 11th Gwangju Biennale (2016). She consults for Sciaena environmental NGO working on marine policy and deep-sea mining and has co-directed several educational platforms, such as *escuelita*, an informal school at Centro de Arte Dos de Mayo—CA2M, Madrid and the ecological research platform *The World in Which We Occur/Matter in Flux*. Between 2009 and

2015, Mendes directed The Barber Shop, a project space in Lisbon dedicated to transdisciplinary research.

Maria Emilinda T. Mendoza is an Associate Professor at the Department of Social Services, College of Human Ecology, University of the Philippines Los Baños, and is affiliated with the Climate Risk Studies Center at the same institute. She is finishing her PhD Environmental Science from UP Los Baños. She is a sociologist whose research focuses on transdisciplinary approaches to the study of disasters and climate. She is also one of the Study Leaders on the integrative project 'Clean the Marilao-Meycauayan-Obando River System MMORS)' of Blacksmith Institute from 2008 to 2009 served as the basis for the MMORS case discussion.

Evanne Nowak works as a Program Maker, Curator and Moderator at the intersection of philosophy, ecology, art and justice. Evanne has a Masters in Humanistics at the University of Humanistics (2016); Masters in Theater & Global Development at the University of Leeds (2011); and Bachelor of Theater in Education at the Utrecht School of the Arts (2010). Recurring questions that she explores in her work are: How to live in a time of ecological disruption? How can we take our feelings of climate depression and eco-anxiety more seriously, and more specifically how can we live through them, investigate and convert them into political emotions: solidarity with non-human and with future, distant, unknown others?

Brunilda Pali is a Senior Researcher at the Department of Social and Cultural Anthropology, Faculty of Social Sciences, University of Leuven (Belgium), and board member of the European Forum for Restorative Justice. With Ivo Aertsen she has co-edited *Critical Restorative Justice* (2017) and *Restoring Justice and Security in Intercultural Europe* (2018). Brunilda has studied Psychology at the University of Bosphorus in Istanbul, Gender Studies at the Central European University in Budapest, Cultural Studies at Bilgi University in Istanbul, and Criminology at the University of Leuven. She researches and writes on gender and feminism, critical social theory, environmental and restorative justice, cultural and critical criminology, and arts.

Chiara Perini is Associate Professor of Criminal Law at the University of Insubria (Como, Italy), where she teaches Criminal Law, Restorative Justice and Penal Mediation, Corporate Criminal Law. She is co-chair of the Working Group on Environmental Restorative Justice established by the European Forum of Restorative Justice in 2020. Her researches focus on criminal law, environmental criminal law, criminal justice policy, restorative justice, environmental restorative justice.

Mark Požlep is an Artist who works mostly with video and photography. He has studied at the Academy of Fine Arts at the University of Arts in Ljubljana, Slovenia, and HISK, Belgium, where he lives and works. All of Mark Požlep's explorations can essentially be expressed with the question: what is it that gives us power and motivation for action after the end of the great meta-narratives?

Iokiñe Rodriguez is an Associate Professor at the School of International Development in the University of East Anglia, UK. She researches, teaches, and publishes on environmental conflict transformation in Latin America, focusing on issues of local history, local knowledge, power, environmental justice, equity, intercultural dialogue, and action research.

Emerson M. Sanchez is a Postdoctoral Research Fellow at the Crawford School of Public Policy, Australian National University. He researches environmental discourses, natural resource politics, social movements, and community engagement. He completed his PhD at the Centre for Deliberative Democracy and Global Governance, University of Canberra, where he studied environmental discourses surrounding mining disasters.

Ulrike Tabbert is a Senior Public Prosecutor (Oberamtsanwältin) in Germany. She holds a PhD in Linguistics from the University of Huddersfield, UK, where she is also a visiting research fellow.

Felicity Tepper is a Senior Research Officer at the RegNet School of Regulation and Global Governance, Australian National University. As well as her degrees (LLB/Hons, LLM/Env, MDPR/ Resilience&Sustainability, BA/Hons) and academic experience, Felicity has worked on environmental law and policy issues for many years in a variety of roles, including for national government, two national parliaments, the private sector, and NGOs. Her research interests include environmental restorative justice (localising the global; civic engagement; animals; recovery post-disaster); complexity and governance; and knowledge brokering.

Gema Varona is Lecturer in Victimology and Criminal Policy at the University of the Basque Country, Senior Researcher at the Basque Institute of Criminology (Donostia/San Sebastian, Spain), and Co-editor of the *Journal of Victimology/Revista de Victimología*. She has authored books and articles on migration and human rights, restorative justice, violence against women, victims of terrorism, and victims of sexual abuse.

Cristina Mihaela Vasilescu is a Public Policy Analysist and Evaluator. In the justice area, she has extensive experience in social research and evaluation of policies/projects focused on the organisation of justice institutions, restorative justice and the social and labour inclusion of (ex) offenders. She is coordinating the European Forum for Restorative Justice Working Group on Restorative Cities.

Lode Vranken is an Artist, Aarchitect and Philosopher. He works on international artistic projects within various public contexts. He has been the lead architect and philosopher of Futurefarmers since 2008. He has been teaching since 2005 as a Ned delegate at The Institute for Advanced Architecture of Catalonia, Barcelona, Spain, and from 1993 to 1994 at the Asian Institute for Technology in Bangkok, Thailand. Lode co-founded the architectural research coalition, De Bouwerij in Belgium that focuses on social living structures and passive houses, Cradle 2 Cradle buildings and zero energy construction. He is also a partner of Dear Pigs in Belgium and member of the Ghent School for Metaphysics.

Hercules Wessels is a qualified and practicing Attorney at Cullinan & Associates Inc., Cape Town, South Africa, where he specialises in environmental law. He studied at Stellenbosch University, Stellenbosch, South Africa, where he obtained his BA (Law) and LLB degrees. He is also an associate at the Wild Law Institute and a member of the Global Alliance for Rights of Nature.

Rob White is Emeritus Distinguished Professor of Criminology at the University of Tasmania, Australia. He has written extensively in the areas of youth studies, criminology and eco-justice. Among his recent books are *Climate Change Criminology* (2018), *The Extinction Curve* (with John van der Velden, 2021), *Theorising Green Criminology* (2022), *Critical Forensic Studies* (with Roberta Julian and Loene Howes, 2022) and *Advanced Introduction to Applied Green Criminology* (Edward Elgar, forthcoming).

Femke Wijdekop is the Legal Counsel and Content Manager of Stop Ecocide Netherlands. She studied at Vrije Universiteit, Amsterdam, the Netherlands, where she obtained her LLM degree. She is also an independent expert at the UN Harmony with Nature initiative, an associate of Restorative Justice Nederland and a member of the International Society for Therapeutic Jurisprudence.

Martin Wright is a widely known restorative justice scholar and advocate. He has been director of the Howard League for Penal Reform and policy officer for Victim Support. He was a founding member of the European Forum for Restorative Justice, the (UK) Restorative Justice Consortium, and Action for Bhopal. He is author of the book *Restoring Respect for Justice* and is joint editor (with David Cornwell and John Blad) of *Civilising Criminal Justice: An International Agenda for Penal Reform* (2013). In 2012 he received the European Forum's European Restorative Justice Award.

List of Figures

List of Images

1

Environmental Restorative Justice: An Introduction and an Invitation

Miranda Forsyth, Brunilda Pali, and Felicity Tepper

1 Beyond 'Justice as Usual'

The intricate webs of relationships that bind us closely and sustain and nurture strong communities and a healthy planet are too frequently invisible and neglected. Their true value only manifests when suddenly, shockingly, they are so broken that everything comes to an abrupt stop. It is at these times of crisis and damage that the connections joining us all, the human and the more-than-human, visibly emerge. Many of the contributions within this book open their stories at such moments of harm—glaciers melting, dam walls bursting, factories exploding and disgorging

M. Forsyth (✉) • F. Tepper
RegNet School of Regulation and Global Governance, Australian National University, Canberra, ACT, Australia
e-mail: miranda.forsyth@anu.edu.au; felicity.tepper@anu.edu.au

B. Pali
Department of Social and Cultural Anthropology, Faculty of Social Sciences, KU Leuven, Leuven, Belgium
e-mail: brunilda.pali@kuleuven.be

© The Author(s), under exclusive license to Springer Nature Switzerland AG 2022
B. Pali et al. (eds.), *The Palgrave Handbook of Environmental Restorative Justice*,
https://doi.org/10.1007/978-3-031-04223-2_1

chemical waste into the air—with resulting cascades of harm to humans, more-than-humans, and ecosystems. Immediate reactions to these disturbing events are all too familiar—we look for who to blame, who must pay compensation, what laws have been breached, what regulatory changes are needed. Our systems of justice require us to isolate individuals or entities, to limit our enquiries to fixed moments in time. Sometimes, these reactions have positive outcomes. Mostly though, they don't. And even if compensation for harm is finally paid (on the assumption that money somehow fixes dead fish, polluted land, or cancer), often it is delayed for years, fails to account for trauma to people, communities and other species, and lacks genuine apologies or evidence of a true understanding of the harm done. All too often, compensation or fines do not repair the damage to countless more-than-human species or provide for the long-term recovery needs of fragile ecosystems.

We are yet to develop meaningful accountability, justice and repair when it comes to environmental harms, crimes, injustices and conflicts.[1] Many of our current legal and regulatory approaches to environmental harm have become so normalised as the 'only way to do justice' that we seldom ever stop to ask whether we are really seeking the right things from our processes of justice for our environment. Our book challenges this 'justice as usual' approach towards the environment. It suggests that punitively focused reactions to harm form part of our current system that makes it more likely that as a species we will continue to cause tremendous damage, and compound this insult by failing to adequately repair it or prevent future environmental harms. Our reluctance to acknowledge the failings of our justice systems across the world to prevent and repair environmental harm ignores the fundamental relational connections between us and our planet that are at the heart of building better futures. 'Relationships are the essence and fabric of collective impact', state Milligan et al. (2022), further noting that although this is a simple truth, it is one we often lose sight of. Environmental harm is complex, its causes

---

[1] In our book, the term 'environmental harm' is used broadly to include such actions, impacts and outcomes as: environmental crime; damage to the environment; degradation and despoilation of ecosystems; injury, death and illnesses caused to human and more-than-human species; pollution; damaging atmospheric emissions; thoughtless and excess use of resources; dumping of toxic materials; loss of biodiversity and habitats; and other similar eco-destructive actions and impacts.

and effects are interrelated, and any justice response that is of a purely technical or reductive nature is inherently limited in its transformative potential.

2 Introducing Environmental Restorative Justice

To address this failure of justice systems to make environmental harm a core focus, our book proposes a rethink. Drawing on different legacies and responses to environmental harm across many different jurisdictions, our book develops the contours and explores the agenda of Environmental Restorative Justice (ERJ) as a response to the need to reinvigorate justice approaches to environmental harm.

The term 'environmental restorative justice' indicates both how an environmental engagement can and should contribute to restorative justice, and how restorative justice can be used in the context of environmental harms and crimes. We do acknowledge that it has its limitations, as no term is ever 'just right'. For example, White[2] states that the concept of 'environmental justice' is human focused and does not accommodate ideas about justice to ecosystems and other species; he suggests we would do better to use the term 'eco-justice'. Wessels, Wijdekop and Bosselmann also speak about Earth Restorative Justice as an alternative term. However, the term we've adopted for this volume is young and we humbly suggest that it does already incorporate each of these valuable meanings within its spirit, intent and application. To this end, our book goes beyond terminology, accommodating conceptions of justice that deal with humans (environmental justice), with ecosystems and biospheres (ecological justice), with more-than-human animals and plants (species justice), and with climate change and its associated injustices (climate justice).

ERJ, a new framework for environmental justice centred on restorative justice, offers a different approach to environmental harm. The ERJ approach is based on relationality, both as a matter of inter-existence and as aspiration and on the need to respect and listen to different

---

[2] When there is no year for the cited references, they refer to chapters of this handbook.

perspectives; and on an understanding of the need to value and support connections between humans, between humans and more-than-humans, and between humans and Nature—even, or perhaps especially, when these relationships are thriving. ERJ calls attention simultaneously to the past and the future. For past harms, it requires us to acknowledge and repair the harms that have been done to the environment, to its human and more-than-human inhabitants and to communities. For the future, ERJ concerns itself with the need to ensure non-occurrence or recurrence of harm and focuses on building or rebuilding relational and ethical systems that prevent future harm, to present and future generations.

Therefore, ERJ is both healing and reparative because it can be both preventative and proactive. Preventing harm, building strong relationships, developing trust in institutions and leaders, nourishing our ability to have difficult conversations and to draw upon each other for support, are all at the heart of this approach. Throughout our book, we acknowledge that environmental harms, crimes and injustices raise specific conceptual challenges; it is important to note that these challenges are either not present, or manifest differently, in other domains where restorative justice is applied. Nevertheless, ERJ draws widely on the lessons learned, the applications and the values from other domains' experiences within restorative justice.

As the chapters of this book will clearly show, a restorative justice response has also the potential and aim to affiliate itself with and therefore learn from different domains, frameworks and heritages. This includes such fields as: environmental regulation (Forsyth); participatory governance (Vasilescu); criminal justice and criminal law (Braga Da Silva; Hamilton; Perini; Varona; White); Indigenous justice, epistemic justice, and decolonisation (Amparo et al.; Bolivar et al.; Hydle & Henriksen; Rodriguez); critical transitional justice (Killean); intergenerational and climate justice (Almassi; Jones); Earth jurisprudence and trusteeship and the Rights of Nature (Bosselmann; Wessels; Wessels & Wijdekop); social and community-based movements, resistance, art and activism (Di Ronco & Chiaramonte; Komatsubara; Jolly et al.; Pali et al.; Wright & Tabbert); and conservation and restoration (Dore et al.; Tepper). As the contributions of this volume make clear, ERJ presents promises and possibilities that are complementary to other existing responses to environmental harm, but also offer fresh practices and approaches.

It must be further noted that responses to environmental harm are often viewed from single disciplinary viewpoints. This tends to situate them in traditions of thinking that remain confined within singular frames of reference, for example, law, rights, care, restoration, activism and regulation. This volume showcases a multidimensional approach instead, taking a strong interdisciplinary perspective that looks at how to respond to and prevent environmental harm from multiple disciplinary and practitioner perspectives. This interconnected approach is essential when addressing complex problems and is reflective of the restorative justice virtue of valuing pluralistic approaches to justice, but also to problem-solving.

In the next part of this introduction, we provide an overview of the promises of ERJ for responding to and preventing environmental harm, to give an initial glimpse as to how the authors in this volume have approached using ERJ to address environmental harm. Following this overview, we examine the challenges that remain to be addressed as this new domain is further explored and developed.

3   Promises and Possibilities for ERJ

ERJ can potentially assist in guiding responses to a whole spectrum of environmental crimes, harms, injustices, and conflicts, whether these have happened in the past, are currently ongoing, or will happen in the future. In this way, ERJ can help to bring about a change in the way in which justice systems approach environmental harms. Through a variety of case studies and thought experiments, the authors in our book illustrate how restorative values and practices can assist in crafting different responses to environmental harm. In particular, the authors demonstrate: (1) the utility of the flexible approach to harm that lies at the centre of restorative justice and that expands its application to situations currently not covered by other justice frameworks, or to cover victims that are often unseen; (2) some of the practices and processes of restorative justice that can be drawn upon in an environmental context; (3) the ways in which restorative justice can address epistemic injustice through engaging with multiple perspectives and valuing different systems of knowledge; (4) values and practices that enable the addressing of power imbalances in the context of environmental harms; (5) the value and importance of refocusing attention on repair rather than on punishment; (6) novel ways to ensure accountability of those who are responsible for harm in ways that are meaningful to those harmed and to Nature; and finally (7) how restorative approaches can support an agenda of reconnecting humans with the environment in ways that recognise our essential interdependence and seek to put limits on extractivist ideologies. We outline each of these contributions briefly below.

3.1   Responding to a Broad Spectrum of Crimes, Harms, Injustices, and Conflicts

A restorative justice ethos and praxis starts the search for justice based on who and/or what has been harmed, who is accountable for that harm, how to protect those harmed, how to repair that harm, and how to prevent its re-occurrence. Thus, a restorative approach has the potential to expand notions of harm and victimhood, as it is not bound by a legal or hierarchical and exclusionary understanding of harm and victimhood. From this perspective, all harms (e.g. epistemic, material, moral, relational), and victims of environmental harm (e.g. humans, more-than-humans, individuals and communities, future generations, the climate and atmosphere, the Earth and the environment), could potentially be taken into account within restorative processes.

The chapters in this volume address issues of harm and wrongdoing, ranging from climate injustices in the Peruvian Andes (Jones); mining tailing dam disasters (Braga Da Silva); damage to rivers, wetlands and waterways (Amparo et al.; Forsyth; Pali et al.); harms and injustices of extractivism (Bolivar et al.); epistemic injustices and harms to local knowledge (Hydle & Henriksen; Rodriguez); criminalisation of environmental movements (Di Ronco & Chiaramonte; Jolly et al.); harms to animals and wildlife offences (Varona; Dore et al.); and 'historical' industrial disasters (Komatsubara; Wright & Tabbert).

3.2   Safe and Structured Spaces and Processes for Participation and Dialogue

One highly valuable contribution restorative justice makes to the environmental field is through creating safe and structured spaces and processes for all stakeholders to exchange different views, tell their stories and hold difficult conversations around what has happened and what must happen in the future to ensure accountability, repair, and non-repetition (Forsyth et al., 2021). Such dialogue can create empathy, mutual learning, a better understanding of other perspectives, increase in trust, and direct and widened responsibility amongst and between stakeholders.

The emphasis on direct participation by those who have caused harm and those who experience harm is in line with participatory governance as applied to environmental policymaking (Vasilescu), with environmental regulation (Forsyth), and with participation as a key feature and principle of justice (Jones; White), including throughout environmental repair initiatives (Pali et al.; Tepper; White).

The form of such spaces and processes will depend upon the nature of the parties and the harm involved, the resources available (e.g. time, finances), the specificities of the relationship between the parties, and the kind of outcome sought (Jones). This can range from Restorative Circles or Assemblies, to Council of All Beings Sessions (Wessels & Wijdekop), to Global Truth and Reconciliation Commissions (Bolivar et al.; Jones) through to Restorative Conferences (Hamilton) and even Restorative Inspections and Licensing (Forsyth). Whatever the precise form of the process, ERJ seeks to provide a space for respectful and honest conversations between members of harmed communities and representatives of institutions and entities responsible for the harm.

3.3   Enabling Epistemic Cohabitation and Epistemic Repair

Nature and natural resources are often perceived in radically different ways by different stakeholders in environmental conflicts (Minguet, 2021). For a company, natural resources are the raw materials from which their products will be made. For the state, natural resources are national resources that can lead to the economic development and progress of the country. For Indigenous populations, land, waters and ecosystems are often intrinsic elements in their lives and cultures. These different perspectives draw upon multiple knowledges and may lead to contestation between knowledge systems (Goyes, 2019), thus creating fundamental divides.

This disconnection and plurality of understandings creates a need for systems that enable collaborative knowledge-making, knowledge sharing and epistemic dialogue and cohabitation (Almassi; Rodriguez; White). Our authors often address the need for connecting the plurality of understandings. For example, White suggests, it is essential to assess each perspective and contribution to the picture as a whole, whether this perspective is Indigenous, non-Indigenous, traditional, local community or scientific, so as to fully explore how different types of knowledge can be integrated to achieve outcomes that support, nurture and heal our environment. White's discussion reveals the complexities involved in seeking to determine who is advocating for Nature, who is giving Nature a voice, but reiterates the importance of acknowledging the breadth of knowledges each person and community has to bring to the table, enlightening each other from varied perspectives, so as to enable Nature's holistic voice to come to the fore.

As explored in our book, ERJ offers a range of values and practices to support the sharing of multiple perspectives and knowledges. Several chapters highlight, however, that there is often a need to go beyond epistemic cohabitation, to actively engage in processes that enable epistemic justice. As Almassi argues, epistemic injustices occur when people are wronged in their capacities as knowers. Historically marginalised and oppressed people in settler colonial, post-colonial, and neo-colonial contexts have routinely had their lived experiences and perspectives dismissed, devalued, misunderstood, misattributed, and appropriated by dominant knowers and knowledge systems (Almassi; Dore et al.; Hydle & Henriksen; Rodriguez). Lack of respect for the environment and for local knowledges and cosmologies from corporations and governments is therefore a running theme throughout many of the chapters (Bolivar et al.; Di Ronco & Chiaramonte; Hydle & Henriksen; Komatsubara;

Rodriguez). Considering the applicability of ERJ in transitional settings, Killean for example argues that ERJ challenges the neo-colonial tendencies of transitional justice, by facilitating the design of mechanisms that are more inclusive of Indigenous harms and understandings of justice.

For example, several authors show how the state overreaches into traditional practices that do not need regulating as they have long traditions of cultural regulation. Unnecessary intrusion of the state results in great harm to both the traditional peoples, ecosystems and animals involved (Amparo et al.; Hydle & Henriksen). Rodriguez argues that Indigenous people are not only in need of more just legal frameworks or enforcement of laws aimed at safeguarding their cultural and physical survival, but there is also a need for processes that can help them to rebuild, reconnect and revitalise their own identities, knowledge and sense of self that modernity and colonialism have profoundly shattered and devalued. Drawing on Community Action Research in Venezuela and Bolivia, she discusses how a decolonial environmental conflict transformation and restorative justice approach has been used to support Indigenous and local peoples seeking to restore their cultural and environmental knowledges, and dignities as a people.

3.4   Reducing Distance Through Sharing Power

The dialogic element of ERJ has further potential to reduce distance and domination and thereby help prevent environmental harm. By distance, we mean in part a distancing from traditional or community ways of being in and living with Nature and the environment, caused by power imbalances, temporal disruptions and spatial variances. We also refer to distances as created, amongst other things, by the differences in experience (i.e. those who profit from causing environmental harm are rarely directly victimised by its impacts), by lack of access to socio-economic capital (i.e. the disparity between corporations and marginalised communities' access to legal and financial resources), by ongoing or upcoming temporal impacts (e.g. the continued impacts of colonialist approaches, the impacts of past and current generations' environmental harms on future generations), or by space (i.e. those impacted by exploitative practices and extractivism are often situated differently in terms of geography).

When engaging in addressing, seeking accountability for and repairing environmental harm, we enter into a world of entrenched systemic injustices, extreme power imbalances and high victim vulnerability (Pali & Aersten, 2021). Often, as can be seen from some of the case studies in our chapters, affected communities are silenced, offered tokenistic consultation or processes, have their local autonomy removed by experts and are ultimately left out of important decisions that regard them. ERJ's dialogue and participation values and practices are a way to challenge these invisibilising and marginalising actions or inaction. These ERJ practices are in line with the communities' and public's right to participate in decision-making, gain access to official and clarifying information and gaining access to justice (Vasilescu). ERJ's justice processes can be conceived, therefore, as ways to overcome the "ing" effects and ensure power-sharing actually takes place.

As Jolly, Gehman and Burford show in their chapter, however, restorative justice can be challenged by structurally reproduced inequalities, especially where power is not ceded and the status quo remains. Drawing from multiple sources, the authors conclude that to overcome this, there is a need to widen the restorative vision, along with the menu and sequencing of restorative and responsive pathways on offer. Hearkening back to our call for justice systems to reimagine their role in environmental harm, there is potential for this widening of the restorative vision to occur with a high level, all embracing dialogic practice, such as something like a restorative enquiry (Tepper), to address the injustices of structural inequalities that continue to beset communities and peoples. Many of our chapters show that this wider restorative lens is urgently needed to engage competently and ethically with resolving pervasive inequalities.

3.5   Prioritising Repair

Even though it is not usually what makes the news, the world is much more dependent on maintenance and repair than on innovation and invention. Humans are also *Homo Reparans*. Feminist philosopher Elizabeth Spelman (2003) argues that it is critical to bring repair out of the shadows, especially in an era when the need for repair is unavoidable as a result of the accumulated impacts of humanity on the Earth and its systems. Wessels & Wijdekop argue that what is most in need of repair is humanity's relationship with the Earth, to take it out of exploitation and instrumentalism and restore it as a relationship recognisant of interconnectedness and cooperation.

The ERJ perspective is driven largely by the principles of harm reparation, restoration, and healing of communities, relationships, institutions, and ecosystems damaged by human action or inaction. In cases of environmental harm, focusing solely on punishing the perpetrator of the pollution or destruction of natural resources is meaningless if there is no consequent repair of the damage done, repair of the relationship with the Earth, and if there is no reassurance that this type of behaviour will not happen again (Braga Da Silva; Bolivar et al.; Dore et al.; Perini; Wessels & Wijdekop).

As a result of restorative processes, a restoratively-focused plan of action or a restorative contract which contains suggestions to prevent or repair harms and damages can be collaboratively drawn up and agreed to, and because it is inclusive and participatory, such a contract and plan of action has more potential to be sustainable and transformative (Braga Da Silva; Hamilton; Pali et al.; Wessels; Wessels & Wijdekop). Instead of thinking of reparation merely in terms of financial compensation and restitution, in ERJ, reparation is primarily moral and relational (Almassi; Jones; Perini). As Walker (2015, p. 217) puts it, reparations are 'a medium for the contentious yet hopeful negotiation in the present of proper recognition of the past and proper terms of relation in the future'. Repair benefits not only the harmed, but also the harmers, enabling them to recover self-respect and to be reintegrated 'without stigma' into their community (Walker, 2006, p. 383; Wessels & Wijdekop). ERJ might also be an impetus behind the formation of environmental restoration groups drawn from volunteers amongst community members, where such persons want to remain engaged with the restoration process long-term. This is a form of ERJ values and practices leading to actions that endure, in order to maintain relationships between people who wish to stay engaged in caring for the environment.

3.6   Centring Accountability and Restoring Response-Ability

According to the international environmental lawyer and activist Polly Higgins, who drafted the ecocide law and launched the 'Stop Ecocide' campaign, we are very different from other species in that we have the capacity to recognise and understand the consequences of our actions and take collective action to remedy them (Higgins, 2010). Her main concern was how we might take responsibility for the way that we engage with our ecosystems (Higgins, 2017).

The concept of responsibility within restorative justice as Jones argues is rich and deep, and applies to accountability for past acts or omissions, to present participation in the restorative process, and to future commitments to prevent harm and build better relationships (Wallis, 2014). Responsibility and accountability for creating harm and for healing harm is central to ERJ as we see in many of our chapters (Almassi; Bolivar et al.; Bosselmann; Jones, Pali et al.). The chapters analyse more specifically the role of the state (Amparo et al.; Braga Da Silva; Bolivar et al.; Bosselmann; Di Ronco & Chiaramonte), of the minority world (Jones), of different generations (Almassi), of global community (Bosselmann; Tepper; Wessels & Wijdekop), and of corporations (Braga Da Silva; White; Wright & Tabbert) in repairing harm and assuming responsibility. Demands for accountability are based, among other things, on legal and ethical concepts including the polluter pays principle, historical responsibility, intergenerational justice, the global commons, Earth Community and the right to sustainable development.

But as Pali et al. argue, there is also a need to think of the term response-ability quite literally: our ability to respond as we stay with the trouble of living in a wounded and vulnerable Earth (Haraway, 2016). This ability is profoundly endangered by demoralising trends such as catastrophising, which lead to despair, or the other side of the coin, ecological techno-utopianism, which is characterised by naïve hope and fantasies of unlimited technological progress (Varona, 2021). Both trends impede social and moral accountability in relation to environmental harms.

3.7   Restoring Limits to Hubris

Both accountability and repair are closely associated with virtues of *humility* and restraint, which recognise the necessity and value of limits (Varona, 2021). Environmental harms and crimes are often characterised by excessive confidence in the inevitability of technological progress and the need for profit at any cost. This so-called *hubris* syndrome impedes a respectful relationship with different forms of life and the Earth (Bolivar et al.; Wessels & Wijdekop). Even though the term itself is disputed (Jones), the hallmark of the Anthropocene is humankind's broken relationship with Nature or the Earth Community (Wessels & Wijdekop). What needs to be restored therefore is limits to *hubris*, which can take place through acknowledging and putting into practice limitations on individual and collective behaviours that damage ecosystems (Varona, 2021). Our book explores one possible way for achieving this through enacting the crime of ecocide, supported by accompanying restorative values and practices.

We have reached such a level of broken relationship that environmental activists and civil society organisations are advocating for the amendment of the Rome Statute of the International Criminal Court to include ecocide as a crime. This further includes seeking a fundamental paradigm shift towards Earth jurisprudence to move us away from current anthropocentric legal systems and interests (which prioritise protection of human capital and enables dominion, exploitation and irreversible destruction of Nature), to a world where legal systems are premised on the idea that humans are but one part of the Earth Community, not apart from Nature, therefore requiring that decisions and relationships are maintained by taking into consideration what is in the best interests of the whole Earth Community (Almassi; Bosselmann; Jones; Tepper; Wessels & Wijdekop; White). This potential evolution of ecocide and Earth jurisprudence has implications for justice system change, as legal processes and courts would be required to give standing to, hear and account for the interests of the environment, something several of our chapters reveal remains scarce (Killean; Wessels; Wessels & Wijdekop).

4   Challenges and Dilemmas

The applicability of restorative justice to environmental issues comes with its own distinct challenges, with some being due to the particularity of restorative justice ethos and praxis, and some due to the contexts in which environmental harms take place. The authors in this volume highlight a range of challenges that remain to be addressed as the practice of ERJ is developed. These challenges include: (1) the ability of restorative approaches to respond adequately to situations of entrenched power imbalances and historical inequalities and oppression; (2) identifying who should have a voice in restorative processes and how do we decide what kind of expertise or 'surrogate presence' is required to adequately speak for or about the more-than-human, the environment and future generations and who speaks with authenticity, credibility and legitimacy; (3) identifying how to include corporate actors in restorative processes, especially in situations where legal harm cannot be proven; (4) addressing the real or perceived risk that restorative justice minimises accountability for environmental harm as it does not focus on punitive outcomes for those who have caused harm; and (5) finally, questions about the criteria by which decisions around restoration are to be made, such as what does restoration mean in the context of ecosystem destruction. We address these different challenges briefly in turn below.

4.1   Responding to Complexity and Power Imbalance

Both at the conceptual and practical level, responding to complexity remains one of the main challenges for ERJ in addressing issues of environmental harms, injustices, crimes and conflicts. Environmental conflicts are often long term and complex, rooted in socio-economic inequalities, corporate-political entanglements, global injustices, colonial history, and extractivist and neo-colonial realities (Almassi; Amparo et al.;

Bolivar et al.; Dore et al.; Hydle & Henriksen; Killean; Jones; Rodriguez; Wright & Tabbert).

For example, Bolivar et al. discuss how environmental harm in the Global South takes place under an economic, political, legal and social context of extractivism, which is encouraged by governments of different political persuasions as a way to promote economic growth and social development, with the result that there is entrenched impunity for environmental harms and abuses. There is also a lack of responsibility across time and space for damage caused, and the problem of transnational corporations and their stand-off/faraway attitudes towards environment and locals (Amparo et al.; Bolivar et al.; Wright & Tabbert). The case studies in the Philippines discussed by Amparo et al. show that it can be tricky to identify what harm clearly and definitively can and should be repaired, and who are the victims, offenders, regulators and broader community actors, when these roles are blurred and shifting across time and space. The authors argue that ERJ is not possible wherever and whenever legal responsibility can be divorced from local consequences of environmental harm and moral responsibility holds no sway. Without cross-scale accountability enforced by states and judiciaries far from the original sites of damage, ERJ has limited reach.

The question is then how well prepared are restorative justice practitioners to navigate the multi-layered aspects of environmental harm which involve, among other things, complexities in legal procedures, in establishing harms, in compensation issues, in accommodating different interests and needs, but also in relation to structural, intersecting and systemic harms and the role of the state (Amparo et al.; Bolivar et al.; Braga da Silva; Hydle & Henriksen; Jolly et al.; White)?

Often in cases of environmental conflicts, especially if they are long lasting and harmful, there are ongoing issues of deep mistrust and asymmetrical power relationships in affected communities, including layers of harm caused further by the state, and the values and principles of restorative justice might not be sufficient to fully turn the tide (Di Ronco & Chiaramonte; Komatsubara; Wright & Tabbert). Some of the values and pillars of restorative justice, such as restoration and stakeholder participation, especially when originating from the state services can be perceived by marginalised communities or social movements as placatory and instrumental of further domination, instead of being politically mean-ingful or transformational. Sometimes, restorative intentions and meth-ods might even be in direct clash with other coexisting strategies in the resisting or affected communities, such as protests (Jolly et al.; Di Ronco & Chiaramonte).

4.2   The Perspectives of More-Than-Humans, Future Generations, and Nature

Instead of being self-evident, perceptions and representations of harm and victimisation are often socially constructed and contestable and their identification involves a range of individuals, communities and organisa-tions with different interests and perspectives (Di Ronco & Chiaramonte; Komatsubara; White). An additional issue is the failure by states, legal systems and corporations to 'see' and recognise the harms caused and their impact (Natali, 2016; Bolivar et al.; Di Ronco & Chiaramonte; Wright & Tabbert). In responding to environmental harm and victimisa-tion, inevitably a range of vested interests and discourses contribute to the shaping of perceptions and issues. This implies differences in perspec-tive and a certain contentiousness of knowledge about the nature of the harm or crime. How and to whom/what harm is perceived and conceived depends upon the yardstick by which worth is determined and the main perspective is taken (White). An important risk for ERJ is that the iden-tification and repairing of the harm will take place within an anthropo-centric framework.

Identifying the right stakeholders and giving voice or listening to more-than-human forms of life, to Nature or the environment and to future generations remains an important challenge in ERJ (Killean; Pali et al.; White). Who should have a voice in restorative processes? How do we decide what kind of expertise or 'surrogate presence' is required to adequately speak for or about the more-than-human, the environment and future generations and who speaks with authenticity, credibility and legitimacy? Is scientific evidence to be acknowledged as expert authority, or are Indigenous peoples, local community groups, environmental agen-cies or non-for-profit environmental organisations better placed to speak truth to power? We also have to ask as White puts it, 'how do we institutionalise who speaks for what, when and why?' ERJ may provide one way of at least getting everyone, from scientists and policymakers, Indigenous communities and peoples and local communities, academics and practitioners, etc., to come into the room together and listen to one another respectfully, and purposefully focus on ensuring that diverse perspectives are heard. In this way, no one person or authority speaks for the environment alone but rather a diversity of voices converges based on the restorative values of citizen empowerment, participation, healing and dialogue. We suggest that this can help us to sidestep the potential for inertia born of despair warned against above, and instead of seeking perfection or 'the right voice' as if there were such a reductionist subject, we can ensure at least that we try, that we base this trying on restorative values and that we focus most on building, rebuilding and maintaining relationships between each other and our environment.

4.3 Participation of Corporate Actors—Is Essential

In the event of a lack of social, political or judicial pressure, one major challenge ERJ faces is the necessity to engage businesses and corporations voluntarily in a reparation process and have them acknowledge their responsibility for harm to the environment. Many companies still fail to perceive themselves as offenders or refuse to do so, especially when their actions do not strictly speaking violate the law, and when they can budget for or pay their way out of offending.

How are we to identify the accountable stakeholder for ERJ when causality is not established in court, or when the harm caused is not covered by legislation (i.e. a legally defined environmental harm)? And how do we decide on different degrees of accountability? What degree of offender acknowledgement of responsibility should be a prerequisite for participation in restorative processes? What should occur in cases when the company disappears, either because it merges with another, or it withdraws from the country where the harm occurred (Wright & Tabbert; Bolivar et al.)? Once a restorative option is given to corporations within a criminal justice process and they participate, how can we be sure that their participation is genuine and not instrumental or motivated only by self-interest (Hamilton)?

Optimistic accounts will argue that most corporations have good reasons to maintain public trust, reputation and right relations with their consumers and their communities (Braithwaite, 2002). Corporate ethics, social licence to operate, fear of reputational damage, 'corporate social responsibility' (CSR), might offer windows to bring a corporation to the restorative table (Wright & Tabbert; Forsyth et al., 2021). Individual corporate champions can sometimes be positive examples for the rest of industry too. For example, Forsyth's chapter relates the story of a corporate operator who went above and beyond to comply with the law and saw herself as an industry champion. Yet, more pessimistic accounts will argue that corporates are inherently profit-driven and criminogenic, . unable to see their role as accountable members of society (Whyte, 2020). The latter commentators warn against things like automated apologies, pre-calculated regulatory sanctions/fines, greenwashing, mitigation or tokenistic exercises like consultation without real say.

The truth lies probably somewhere between these two perspectives. Corporations are different from one another, in terms of scale, behaviours and willingness to invest relationally with the community, so while one type of approach might be suitable for one, another approach is likely to be more suitable for another. This points to the importance of context and nuance in the debate, to the need for avoiding fixed positions when it comes to something called 'the corporate', and to develop a way of thinking and practice that revolves rather around a continuum of organisations.

4.4 Risk of 'Downgrading the Importance of the Harm'

For environmental and animal rights activists, restorative justice might entail the risk of downgrading the seriousness of victimisation of living beings and ecosystems by not focusing on criminalisation or by allowing too lenient a sanction. It is often the risk of impunity for corporations that leads many emancipatory movements to fall into the temptation of claiming the use of symbolic punitivism through criminal law, which runs counter to the restorative ethos (Varona; Wright & Tabbert). Amparo et al., Bolivar et al., and Di Ronco and Chiaramonte argue that ERJ cannot count on the altruism of the perpetrators of harm, particularly when state and corporate actors circumvent rather than acknowledge culpability.

For White it is a question of finding a balance between, and bridging the intentions of, eco-justice with restorative justice in leading to *good environmental outcomes*. For example, considering the example of wildlife poaching, often committed by Indigenous peoples, many of whom are poor and economically vulnerable, both Dore et al. and White illustrate clearly that not all people are situated equally, and that some offenders are themselves victims of unequal, unjust systems. In this context, a punitive approach does not make sense. When it comes to harmful actions of the powerful against the environment and the crime of ecocide, however, the problem is the opposite: we find an entrenched context of impunity (Bolivar et al.; Wright & Tabbert; White).

A creative, collaborative and willingness to 'stay with the trouble' (Haraway, 2016) perspective would continue to see a role for ERJ even—or perhaps especially—in contexts of corporate wrongdoing. As the example of the Bhopal disaster in Wright and Tabbert's chapter shows, we know that many of the biggest unpunished criminals have access to political power or have vast economic wealth, or both. With extremely serious crime, when punishment risks are draconian, incentives for cover-up are huge for sophisticated criminals who have competence in playing the game. Even the wealthiest states cannot afford to consistently prosecute the most common kinds of very serious crime. For poor societies, to do so is unthinkable. Braithwaite (2022) argues that for the crimes of powerful, sophisticated criminals that are prohibitively costly to investigate consistently, such as environmental crimes, we should use restorative justice more often, precisely because it acts as a delivery vehicle for effective prevention of future offending. As Braga Da Silva's chapter shows us, there are some glimmerings of what type of role ERJ can play in the context of environmental disasters from closely examining emergent legal practices.

4.5   Decisions Around Repairing Harm

Another challenge relates to the criteria by which decisions around restoration are to be made. Whether it is physical or emotional, whether it applies to humans, more-than-humans, or the environment, the process of repair can all too easily generate ambivalence when different perspectives fail to be taken into account and where collaborative involvement is avoided. Similar processes are applied in other fields such as reconstruction of historic heritage; remediation of human bodies; restoration of ecosystems. The reconciliation of relationships and reconfiguration of cultural artefacts are full of debates that hint at such ambivalences and ethical questions (McLaren, 2018) such as: should we repair, when to repair, how to repair?

Similarly, our authors raise several questions in relation to repair. How do we disentangle material from non-material priorities (Jones; Almassi; Komatsubara)? How do we repair individual and collective harm whose impact might only be seen in the long term (Bolivar et al.; Komatsubara; Wright & Tabbert)? When do we know we're done repairing? This isn't always evident with something as complex and potentially fragile as an ecosystem, where an 'end' to repairing may just not be possible, calling us instead to be active stewards. And eventually, how to repair the irreparable—can irreversible environmental degradation even be healed (Braga da Silva)? Another difficulty around decisions related to repair is balancing of conflicting interests. Can ERJ simultaneously safeguard the rights and interests of communities, species, the states, corporations and the environment (Di Ronco & Chiaramonte; Jolly et al.; White)?

What the chapters overall highlight, is that healing and repair are not static and fixed but dynamic processes, contextually based and relational. Intentions and actions of repair (of our Earth, of each other, and of more-than-humans) must take into account the particular histories, potentialities, and limitations of each context.

5   Our Hope for This Book

Our overarching aspiration for this book is to actively inform and engage scholars, practitioners, policymakers and concerned citizens through forming a shared language and providing a set of actionable principles applicable to restoring existing environmental harm and preventing future harm. In seeking to open a dialogue and debate about what form an agenda for ERJ might take, our book has brought together scholars from different parts of the world across different stages of their careers, each of whom are experts in their fields and committed to the topic. Many of our contributors are already part of an existing network on ERJ, and this volume is part of a deepening and widening of this cooperative engagement and mutual learning between existing and future scholars and practitioners. We hope to see some of our readers join us in this journey in the near future.

The chapters in this volume present a balance between exercising imagination and practical demonstrations of ERJ and implications for ERJ in practice. In Part I, the chapters envision a new and better way to respond to and prevent environmental harm. In Part II, the chapters work through concrete examples of ERJ or like practices, providing both information and suggestions about how such an approach can be actioned in practice.

We are aware that there is some risk that the attitudes, suggestions and values explored and promoted in this handbook could be dismissed as idealistic, as romantic visions of human nature and of our connections with the natural world. Yet, if there is anything the first few years of the second decade of the twenty-first century have taught us, it is that the problems we are facing are ultimately collective problems. Not a single government, company or community has sufficient resources or adequate knowledge to address the complexities and breadth of environmental problems we face on its own. To address collective problems, we need collective efficacy which can only come from authentic connections between people nourished by empathy and respect. The complexities of environmental harm require expertise and ideas from scientists, traditional users of land, environmental activists, lawyers, artists and citizens, among others. One of ERJ's greatest strengths in this respect is its ability to contribute the space and values needed to bring together everyone affected by and responsible for environmental harm. It is through dialogue and participation, that ERJ can help to ensure collaborative involvement, response-ability and restorative, relational outcomes that repair environmental harm.

The challenges for ERJ discussed in this volume above are not insuperable problems, but they do point to the need to continue developing a theoretical, research and practice basis for ERJ. This volume provides a strong foundation for this work. We invite the reader to dig in deeply to unearth the problems, mull them over and then take a good, long look at how our authors provide pathways, stories and hope for resolving them. In turn, we hope that this book inspires the reader to take up the baton of helping grow the potential of ERJ, including by pushing through the challenges explained above, trying out some of the suggested approaches, and getting involved in sharing the knowledge of what does, and does not work.

References

Braithwaite, J. (2002). *Restorative justice and responsive regulation.* New York: Oxford University Press.

Braithwaite, J. (2022). *Macrocriminology and freedom.* Canberra: ANU Press.

Forsyth, M., Cleland, D., Tepper, F., Hollingworth, D., Soares, M., Nairn, A., & Wilkinson, C. (2021). A future agenda for environmental restorative justice? *The International Journal of Restorative Justice,* 4(1), 17–40.

Goyes, D. (2019). *Southern green criminology: A science to end ecological discrimination.* London: Emerald Publishing.

Haraway, D. (2016). *Staying with the trouble: Making Kin in the Chthulucene.* Durham: Duke University Press.

Higgins, P. (2010). *Eradicating ecocide.* London: Shepheard Walwyn Publishers.

Higgins, P. (2017). Changing the ground rules. Interviewed by Huw Spanner. *High Profiles.* Retrieved from https://highprofiles.info/interview/polly-higgins/ (last accessed 28 November 2020).

McLaren, D.P. (2018). In a broken world: Towards an ethics of repair in the Anthropocene. *The Anthropocene Review,* 5(2), 136–154.

Milligan, K., Zerda, J., & Kania, J. (2022). The relational work of systems change. *Stanford Social Innovation Review*. https://doi.org/10.48558/MDBH-DA38

Minguet, A. (2021). Environmental justice movements and restorative justice. *The International Journal of Restorative Justice*, 4(1), 60–80.

Natali, L. (2016). *A visual approach for green criminology: Exploring the social perception of environmental harm*. London: Palgrave Macmillan.

Pali, B., & Aersten, I. (2021). Inhabiting a vulnerable and wounded earth: Restoring response-ability. *The International Journal of Restorative Justice*, 4(1), 3–16.

Spelman, E.V. (2003). *Repair: The impulse to restore in a fragile world*. Boston: Beacon Press.

Varona, G. (2021). Why an atmosphere of transhumanism undermines green restorative justice concepts and tenets. *The International Journal of Restorative Justice*, 4(1), 41–59.

Walker, M. (2015). Making reparations possible: Theorising reparative justice. In C. Corradetti, N. Eisikovits, & J. Rotondi (Eds.), *Theorising transitional justice* (pp. 211–223). London: Ashgate.

Walker, M.U. (2006). Restorative justice and reparations. *Journal of Social Philosophy*, 37(3), 377–395.

Wallis, P. (2014). *Understanding restorative justice: How empathy can close the gap created by crime*. Bristol: Policy Press.

Whyte, D. (2020). *Ecocide: Kill the corporation before it kills us*. Manchester, UK: Manchester University Press.

Part I

Theoretical and Legal Perspectives on Environmental Restorative Justice

2

Restorative Justice, Repairing the Harm and Environmental Outcomes

Rob White

1   Introduction

This chapter explores contentious issues pertaining to the theory and practice of environmental restorative justice (ERJ). The chapter argues that it is essential to frame ERJ in terms of situationally relevant strengths and limitations. The reasons for this include, that notions of justice pertain to different kinds of entities (including humans) whose interests may occasionally conflict; that significant power differentials intrude upon the justice process, and this requires acknowledgement and appropriate institutional responses; and that the participatory elements of ERJ are complicated by the inclusion of non-human environmental entities (such as animals, plants, rivers and mountains) in the criminal justice 'conversation'.

R. White (✉)
University of Tasmania, Hobart, TAS, Australia
e-mail: r.d.white@utas.edu.au

© The Author(s), under exclusive license to Springer Nature Switzerland AG 2022   **27**
B. Pali et al. (eds.), *The Palgrave Handbook of Environmental Restorative Justice*,
https://doi.org/10.1007/978-3-031-04223-2_2

To illustrate these points, the chapter examines how justice is conceptualised within a green criminology framework, and how this translates into specific justice principles. The chapter then considers the question of how to institutionalise ERJ, that is, how to embed it within criminal justice systems. This discussion includes the observation that 'repairing the harm' does not always depend upon restorative justice processes and procedures, particularly with regards to crimes of the powerful such as corporations, although responses can still reflect some of the overarching principles of restorative justice. However, in practice ERJ is also utilised in forums and ways that downplay the criminal nature of the harm, treating it more as a civil matter and hence, not taking environmental crime seriously enough. The chapter also examines issues pertaining to who should speak *about* Nature (experts) and *for* Nature (advocates) in the context of ERJ processes and the legal and social frameworks which circumscribe and/or facilitate this.

2   Conceptualising Harm, Justice and Victims

Green criminology is fundamentally about *eco-justice*, which has three key elements: conceptions of justice that deal with humans, with ecosystems and biospheres, and with non-human animals and plants (White, 2022). A vital question is 'who is justice for', particularly in relation to non-human environmental entities such as animals, plants, rivers and mountains. In considering this question, green criminologists also ask what eco-justice might actually mean in the context of criminal law. Legislation is designed to enable and protect, to permit and prohibit, and this involves a considered weighing up of interests, rights, costs and benefits. At the centre of criminal justice are the concepts of harm and morality (Lin, 2006; MacKenzie et al., 2010), although how these are interpreted and institutionalised depends upon the balance of social forces and class relations in any given historical moment.

Every social formation has rules that guide what is deemed acceptable and unacceptable behaviour, and serious or not serious transgressions.

These vary according to culture and context and they are overlaid by matters of power and interest. In effect, the ruling ideas of any society are those of the ruling classes, and so the specific content of what is deemed to be harmful and immoral is shaped by the structure of social relations. Law and law reform (including legal processes incorporating restorative justice) is a site of contestation that reflects divergent interests and power, that likewise revolve around social constructions of harm and morality. Criminal justice, as a system, may well be unjust, just as laws may be (the usual exemplar is Apartheid laws), but they are still constructed around these two central bases—harm and morality—just not necessarily 'ours'. The struggle for justice is basically a struggle over who gets to define which harms are criminalised, and who sets out the ethical basis for permitting/prohibiting certain activities and behaviours.

Within green criminology, the conceptualisations of harm at the centre of its aspirational reform agenda are construed in relation to notions of rights and justice, which vary according to their focus on humans, environments and/or plants and non-human animals (White, 2013, 2022). Justice within an eco-justice perspective is thus framed by the subject or natural object that is potentially harmed or those whose rights and interests need to be protected. *Environmental justice* refers to the equitable distribution of environments among people in terms of access to, and use of, specific natural resources in defined geographical areas, and the impacts of particular social practices and environmental hazards on specific populations (e.g., Indigenous communities). In other words, humans are at the centre of analysis, and it is they who matter most. *Ecological justice* refers to the relationship of humans generally to the rest of the natural world and includes concerns relating to the health of ecosystems, and the plants and animals that also inhabit these systems (Smith, 1998; Cullinan, 2003). *Species justice* incorporates non-human animal rights as relevant to eco-justice (Beirne, 2009, 2014) and takes into consideration the importance of biodiversity in regard to both plants and animals.

Harm is not necessarily the same as victimisation, especially if the latter is interpreted as applying strictly to humans (Hall, 2013; Flynn & Hall, 2017). For example, environmental victimisation has been defined as specific forms of harm which are caused by acts (e.g., dumping of toxic waste) or omissions (e.g., failure to provide safe drinking water) leading to the presence or absence of environmental agents (e.g., poisons, nutrients) which are associated with *human* injury (Williams, 1996). Management of these forms of victimisation is generally retrospective (after the fact) and involves a variety of legal and social responses. Importantly, the central actor in this definition is humans (not non-human animals, plants or ecosystems). The same can be said of re-conceptualisations of environmental rights as 'human rights' in that the concept is premised first and foremost upon notions of humanity (Council of Europe, 2012).

Laws also provide protection for the non-human as well as the human. For example, this is reflected in legislation pertaining to endangered species (e.g., animals such as tigers) and to conservation more generally (e.g., in the form of national parks). Harm is central to these forms of social regulation although it is usually measured by reference to the economic, aesthetic and/or cultural benefits of protection for humans (Lin, 2006). Natural objects (such as trees and forests) generally lack legal rights and so must rely upon humans to bring actions to protect them. Some commentators have argued that the inherent interests of 'natural objects' ought to be protected through legal actions by the objects themselves, with humans serving as their guardians or trustees (Stone, 1972). This is now, under specific circumstances, occurring in some jurisdictions—for example, some named rivers have legal standing in New Zealand and India (Pecharroman, 2018; O'Donnell & Talbot-Jones, 2018). Parallel developments around legal personhood are also occurring in some jurisdictions with respect to non-human species such as cetaceans (e.g., dolphins) and other primates (e.g., chimpanzees) (Nurse & Wyatt, 2020).

The definition of 'victim' is evolving and expanding, and this is likewise reflected in restorative justice processes. For example, a river was represented at a restorative justice conference in Aotearoa/New Zealand by the chairperson of the Waikato River Enhancement Society (Preston, 2011, p. 144). In a small but increasing number of cases, there are 'surrogate victims' who are recognised as representing the community affected (including harms to particular biotic [living creatures such as birds] groups and abiotic environs [non-living entities such as mountains]) for the purpose of the restorative process. In a similar vein, public interest law has been used to establish future generations as victims of environmental crime (Mehta, 2009), the victims including human as well as the environment and non-human animals, for which surrogate victims (such as parents or NGOs) have provided representation (UN Environment Programme, 2017).

Simultaneous to these legal developments, greater attention has been given to articulating what non-human rights might consist of. For example, Schlosberg (2007) identifies key dimensions of justice as including distribution, recognition, participation and capability. The notion of justice as fairness is associated with questions of *distribution*. Justice, according to this conception, defines how we distribute various rights, goods and liberties and how we define and regulate social and economic equality and inequality. These principles are deemed to be relevant to both human and the non-human, particularly in the context of extending the notion of rights to environments (the 'rights of Nature'), to non-human animals (via extensions of 'personhood' to include animals such as apes and dolphins) and to natural objects such as rivers and mountains (through legislation pertaining to specific sites) (Koons, 2009; Maloney & Burdon, 2014).

Recognition refers to the equal dignity accorded to all, as well as the politics of difference where each is recognised for their particular distinctiveness. It is observed that '[a] lack of recognition in the social and political realms demonstrated by various forms of insults, degradation and devaluation at both the individual and cultural level inflicts damage to oppressed individuals and communities in the political and cultural realms' (Schlosberg, 2007, p. 14). Derogatory language used in relation to non-human animals provides one illustration of subordinated and disrespected identity as this applies to the non-human (Beirne, 2009; Sollund, 2012, 2019). Contemporary practices of cultural domination are such that the rights, interests and needs of ecosystems and non-human animals tend to be rendered invisible in ordinary life, and accordingly fewer legal, social and economic resources are put into acknowledging, supporting and respecting the non-human.

The need for *participation* is also defined as an important component of justice. Participation generally refers to membership and engagement in the greater community and it is supported by the institutionalisation of democratic and participatory decision-making procedures. As applied to the non-human world, participation basically involves human advocacy, where the voice of non-human animals or trees or ecosystems is 'heard' via the human third party. Humans thus can speak 'on behalf of' that and those which cannot participate directly for themselves in human affairs that affect them. We shall return to this issue later, given the importance of participation to ERJ processes.

Capability is important to justice as well. This refers to the ability to achieve valuable functionings within the context of an entity's essential character and setting. For humans, capabilities are about a person's opportunities to do and to be what they choose in the context of a given society. Wellbeing is about 'doings' (activities) and 'beings' (states of existence), and enhancing capability means concentrating on the opportunity to be able to have combinations of functionings (Schlosberg, 2007). Translated into an eco-justice context, capability means that each entity should be able to flourish as the thing it is. It is argued, for example, that '[e]very component of the Earth Community has three rights: the right to be, the right to habitat, and the right to fulfil its role in the ever-renewing processes of the Earth Community' (Berry, 1999, quoted in Cullinan, 2003, p. 115). What this practically means can be difficult to determine, however, because nature is inherently complex, uncertain, interconnected and ever changing. Capabilities (as possibilities) are therefore changing, open-ended, dynamic and subject to ongoing deliberation. This, too, is potentially problematic from the point of view of ERJ.

Restorative justice is informed by concepts such as harm reparation, social restoration, community conflict resolution and problem-solving. A retributive system of justice is essentially punitive in nature, with the key focus on using punishment to deter future crime and to provide 'just deserts' for any harm committed. A restorative approach is concerned with promoting harmonious relationships by means of restitution, reparation, and reconciliation involving offenders, victims, and the wider community (White et al., 2017). The benefits of restorative justice—in its usual forms—are its emphasis on active agency (engagement of stakeholders), cost-effectiveness (compared with detention or imprisonment), recognition of the victim (often through face-to-face meetings with offenders), and community benefit (through participation and through community service). The central thread underlying restorative justice is the spirit within which 'justice' is undertaken—the intent and outcomes of the process are meant to be primarily oriented towards repairing the harm that has been caused by a crime, and this means working to heal victims, offenders, and communities that have been injured by the crime.

In the context of the *doing of justice*, it is important to identify principles of justice that will most likely lead to desired outcomes—in this case the protection of the needs, rights and interests of non-human environmental entities such as rivers, birds, trees and mountains, as well as humans. With respect to this, the republican theory of justice (Braithwaite & Pettit, 1990) is premised on the idea that if every act of crime represents damage to liberty and wellbeing, then the system's task is to promote positive liberty, by rectifying or remedying the damage caused by the crime. The theory incorporates many of the notions of justice discussed above in that it is interested in addressing harm evident at different levels, examining the institutional consequences of intervention, and supporting restorative practices that foster individual and overall health and wellbeing. While its original focus is on human liberty and human societies, the basic approach of republican justice can also be applied to the non-human. For example, when republican theory is translated into practice it usually takes the form of restorative justice in which action is informed by two main considerations: repairing the harm and ensuring widespread participation in justice decision-making. The obligation to repair the harm is a communal as well as an individual obligation.

Fundamentally, justice ought to be an active process (with an emphasis on participation and engagement); should aim to maximise liberty (that is, status, capacities and worth); deal with issues holistically (taking into account the interrelationship between humans, environments and diverse species); and acknowledge temporal and spatial dimensions (for example, harms have a past, present and future across geographical spaces) (White, 2013). Yet, there are problems with this formulation. For instance, it does not provide a blueprint for how to deal with conflicting rights in specific circumstances. While it is desirable to not kill kangaroos, for example, there are situations in which failure to 'cull' the kangaroo mob puts into jeopardy the local ecology. Or, it could be that species of rare plants may have to be saved from extinction through animal eradication and/or transfer schemes. In other words, a weighing up of rights and interests, involving humans, ecosystems and species, is inevitable (White, 2013). How best to do this is an ongoing conundrum, including for ERJ proponents. Moreover, in order to decide on how best to approximate desired eco-justice and restorative justice principles and approaches to justice requires us to examine how justice is and might be institutionalised within criminal justice and environmental regulatory systems. We turn now to discuss this in greater depth.

3   Institutionalising Restorative Justice

This section considers how ERJ might be institutionalised within criminal justice systems, in particular as part of court proceedings. It is one thing to acknowledge the harms to and rights of non-human environmental entities in addition to that of humans, but major challenges exist with respect to institutional responses that best capture the gravity of the harms, the differing resources and interests of the protagonists, and the remediation strategies needed to garner the best environmental outcomes.

The aim of criminal law is to prevent behaviour regarded as harmful or potentially harmful by lawmakers. A primary task of criminal law is to quantify the seriousness of criminal harm and to respond with appropriate sanctions or penalties. Importantly, there are extra-legal concepts and factors that also need to be included if one is to fully capture the nature of environmental harm, and this requires drawing upon disciplines such as ecology and plant and animal biology (White, 2018). Yet, what is defined as criminal harm, and what is the measure of its seriousness are contingent upon the social interests bound up with the definitional process. Matters of definition and degrees of harm are also central to discussions of where ERJ fits within broader criminal justice processes.

There are, for example, significant differences in regard to how restorative justice is translated into particular institutional systems. Jurisdictions have different rules, guidelines and operational procedures, which means that restorative justice varies greatly in practice. Specific laws set out the processes and sanctions available for particular criminal offences, adjudication processes may be adversarial or inquisitorial, systems may be based on common law or civil law codes, and penalties can be arbitrarily set or subject to judicial discretion. Likewise, the restorative justice emphasis on participation may be supported and fostered by domestic laws and international agreements (for example, the Aarhus Convention on Access to Information and Public Participation) or hindered by the lack of suitable directives or traditions which work against citizens playing a meaningful role in criminal justice.

The mission and processes of restorative justice in regard to environmental crime likewise vary according to jurisdiction (Voigt & Makuch, 2018; Hamilton, 2021; Wright, 2022). In some instances, the goal may be to achieve good environmental outcomes; in others, the focus is on addressing victim needs, changing offender practices and/or enabling community resolution of conflicts. The intended outcomes are in part determined by the institutions with which restorative justice is functionally attuned, for example, lower-level courts and tribunals or higher-level courts (or, indeed, across the spectrum of the court hierarchy). The position of restorative justice in the criminal justice system affects both perceptions of the seriousness of matters and the potential sanctions available via the restorative justice process. The relationship of restorative justice to the totality of possible sanctioning regimes and processes—that is, where it fits vis-à-vis overall criminal law processes—is of considerable importance. When restorative justice is reserved for medium to higher-level offences, it entrenches the notion that it is not a 'soft' option and reinforces the usefulness of restorative justice in unpacking and responding to complicated issues. If restricted to lower-level offences and first-time offenders, it reinforces the lower value and significance of the transgressions. Used exclusively in the latter way, restorative justice methods can also serve to legitimate the harsher and more punitive criminal justice sanctions which, in turn, often reflect profound social inequalities and historical injustices (see White, 2015a).

But how is ERJ utilised in actuality, especially by courts? To illustrate this, we can consider several developments in Australasia. For instance, the New South Wales Land and Environment Court [NSWLEC], from inception, has been conceptualised and constituted as a problem-solving court, with specific requirements to take heed of human interests, as well as those of natural objects and animals and plants. An emergent concern has been to repair environmental harm where possible and feasible. The emphasis on remediation and repairing harm evident in the decisions of the NSWLEC has prompted some commentators to describe it as engaging in forms of 'restorative justice' (Walters & Westerhuis, 2013). In a similar vein, 'naming and shaming' sanctions used by courts, including environment courts, have likewise been equated with restorative justice (Westerhuis, 2013; Bisschop & Vande Walle, 2013), presumably on the basis that it seems to mirror the moral reprobation associated with 'reintegrative shaming' (Braithwaite, 1989).

The label 'restorative justice' is, however, misleading. Typically, restorative justice centres on empowering victims and involves multiple stakeholders in decision-making who participate largely on a voluntary basis in processes that take place outside of mainstream forums, such as courts. They feature aspects such as restitution and reconciliation and attempts to change how offenders think about their crimes (Wormer & Walker, 2012). These are not central features of how the NSWLEC operates in regard to most cases, however, as in fact only a handful of cases have relied upon explicitly named processes of restorative justice (Hamilton, 2021). Indeed, scrutiny of the wider literature on environmental crime and its prosecution shows that, to date, there is scant actual application of restorative justice in dealing with environmental issues more generally, anywhere (Hall, 2015).

To take another example, Aotearoa/New Zealand has a reputation for innovative measures associated with restorative justice. This is especially the case in regard to juvenile justice, but it has also been noted with respect to environmental harms. Restorative justice conferences can be utilised across a wide variety of offences (e.g., pollution of air and water, breach of conditions of development consent, and destruction of trees); a wide variety of victims (e.g., individuals, communities, and the environment); and a wide variety of outcomes (e.g., a defendant apology, payment of costs, tree planting) (Hamilton, 2008, 2021). Yet, New Zealand's apparent exceptionalism aside, discussion of restorative justice in other jurisdictions has tended to consist mainly of abstract pronouncements by environmentally minded commentators (e.g., on the importance of listening to the voices of Nature) and extra-judicial comment (e.g., on taking into account non-human interests in court deliberations), rather than case law as such (Besthorn, 2004, 2012; Preston, 2011; White, 2016). Even within Aotearoa/New Zealand, the use of restorative justice has been far less extensive than outside perceptions may indicate. For example, Hamilton (2021, pp. 130–132) observes that restorative justice conferences were held in just 4 per cent of environmental offending prosecutions between 30 June 2002 and 20 September 2012, and only in 16 instances in the period from 1 October 2012 to 1 July 2020.

Perhaps more significant is the overall tendency for local authorities in Aotearoa/New Zealand, who are mainly responsible for the administration of the *Resource Management Act 1991* (RMA) in relation to criminal offences and their prosecution, to treat prosecutions more like they are civil proceedings (Wright, 2022). Under the *Act*, the configuration of environmental law and criminal law, and the varied purposes of environmental protection and resource development (as manifest in the notion of 'sustainable management'), have tended to reflect greater emphasis on development than environment as such (Wright, 2022). This overarching sense of what is important (and what is less so) provides the grounding for local authorities to act so as to protect and reinforce activity that views the environment first and foremost as a resource. Hence, the emphasis on gaining consent for resource related use, rather than compliance, monitoring and enforcement relating to environmental protection. Moreover, the judiciary involved in adjudicating cases involving the RMA tend to be mainly engaged in civil proceedings, and given their comments, primarily view the issues through the lens of regulation rather than criminal law (Wright, 2022). Indeed, for many RMA offences 'there is no victim' (Wright, 2022). This again reflects core anthropocentric philosophical leanings with regards to environmental harm generally.

Returning to the Australian context, the Aotearoa/New Zealand approach can be contrasted with the NSWLEC which is a specialist 'environmental court'. This court and its judicial officers are very much informed by eco-centric perspectives rather than anthropocentric conceptions of harm, and offences are treated with the seriousness that in the eyes of the criminal law they warrant (White, 2018). Interestingly, however, within the context of the criminal law and particular mandate of the NSWLEC, until recently there was no specific or explicit reference to 'restorative justice' per se as a method or remedy. The *Protection of the*

Environmental Operations Act 1997—S 250 made reference to an additional order (c) for the offender to carry out a specified project for the restoration or enhancement of the environment in a public place or for the public benefit; and subsection (1A) allows that without limiting subsection (1)(c), the court may order the offender to carry out any social or community activity for the benefit of the community or persons that are adversely affected by the offence (a 'restorative justice activity') that the offender has agreed to carry out. While clearly oriented towards 'repairing the harm', this did not necessarily include victim-offender interactions and exchanges characteristic of the restorative justice *process* more generally (Wormer & Walker, 2012). Recent practice initiatives, however, do recognise a more participatory restorative justice option as part of court proceedings. Nonetheless, to date, there have been relatively few instances in which restorative justice, involving processes of mediation and community conferencing, has been used, although opportunities to do so have occurred over time (Hamilton, 2021).

I turn now to a related but different issue relevant to the institutionalisation of restorative justice. Ultimately, one of the fault lines in considerations of ERJ is the issue of *power*. Ideas associated with restorative justice need careful critical consideration if powerful actors are involved. For instance, a non-human entity like a corporation perpetrates harm against non-human environment entities like rivers, non-human animals and plants, but it is very difficult to effectively sanction a corporation (Whyte, 2020). Optimistic accounts suggest that restorative justice approaches have a place in dealing with corporate wrongdoing (Braithwaite, 2002), including environmental crime (Nurse, 2016; Wijdekop, 2019). However, others are more sceptical and suggest that the threat of stopping a company from doing new business is likely to be more effective than 'shaming' as a restorative justice technique of control (Levi, 2003). Still others argue that the corporation is inherently criminogenic and that it is the politics surrounding social control and regulation that most counts in determining outcomes (Tombs & Whyte, 2015; Whyte, 2020). These divergent viewpoints are not necessarily mutually exclusive and ultimately, they point to the importance of specific context and particular circumstances in determining the effectiveness of restorative justice in relation to corporate behaviour.

An emergent issue here is how best to join up eco-justice with restorative justice. For example, some green criminologists advocate for stronger action against environmental crime given the gravity of environmental harm and the urgency for making effective institutional interventions in this area (White, 2022). Superficially, this appears to run counter to restorative justice positions that oppose widening the criminal justice net and/or increasing the severity of sanctions. However, the question that must be asked is, what is it that proponents are trying to achieve? What is the purpose of the intervention? One answer to these questions is that any marriage of eco-justice and restorative justice should lead to *good environmental outcomes*.

Accordingly, environmental harms should not be trivialised by outsourcing to restorative justice alternatives considered less 'punitive' and therefore a lesser sanction in popular and political opinion, nor should there be downgrading of criminal offences to civil proceedings. On the other hand, an emphasis on 'greater' sanctions should not necessarily translate into more punitiveness either; rather, there is a need for tailormade responses that address the essential contours of each crime committed. Restorative justice is and ought to be identified with non-punitive, but effective, social measures that orient towards changing behaviour, including towards good environmental outcomes. Where restorative justice measures are located within larger criminal justice systems has implications, both for its potential effectiveness (i.e., changing behaviour in a positive way), and how or if it presents as an alternative to punitive sanctions (rather than reinforcing the importance of the latter). There are also important social complexities that need to be unpacked here.

For example, as with victims, there are considerable variations in offenders. Much wildlife poaching committed globally is by Indigenous or historically long-established local peoples, many of whom are poor and economically vulnerable (White & Heckenberg, 2014). In general, the nature of wildlife crime varies greatly depending upon the specific crime (e.g., illegal fishing on the high seas; trade in lizards), the cultural context within which use of the natural resource occurs (i.e., Indigenous people and customary ways of life, local people and poaching as a traditional form of 'folk crime'), the character of the market vis-à-vis a particular kind of wildlife (i.e., personal use versus profit-making), and the relationship in any given situation between local people and criminal syndicates (i.e., chains and networks of connection). In other words, not all people are in equal circumstances, and some are more powerful than others. Frequently, the offenders are themselves victims of circumstance or victims of a system that is unequal and it just does not make sense therefore to adopt a punitive approach.

When it comes to crimes of the powerful, however, the problem is the opposite. Rather than being too punitive, the difficulty is securing meaningful penalties for state, corporate and criminally organised wrongdoing. As with other crimes, the key lies in developing sanctions that not only fit the crime and provide good environmental outcomes, but that also are best suited to the nature of the offender (for examples, see Whyte, 2020). This poses significant challenges to the use of ERJ, especially in regard to its conventional features of participation, repairing harm and communal objectives. While it makes sense to deal with individual offenders and small firms via restorative justice processes—since there is greater scope for offender change in conscience and understanding, as well as behaviour—this is less relevant in respect to large companies. More appropriate and effective is the use of 'reparative justice', a term used here to describe an emphasis on repairing harm within a generally more punitive context (White, 2017). Reparative justice means getting the powerful to repair the harm and making it serious enough in terms of the penalty or sanction so as to dissuade them from doing the harm in the future. Reparative justice draws on some elements of restorative justice, such as repairing the harm, but it addresses this key issue of power by deploying measures designed to hurt the reputation, economic bottom line and/or resource allocations of these entities, such as publication orders, enforced remediation plans, stop-work injunctions and fines scaled to the size of the organisation.

Often the worst perpetrator of environmental harm is the state itself. This has two harm aspects. One is by omission: not doing anything or enough to enforce the laws that are ostensibly there to protect the environment, humans and/or vulnerable species. The other is by commission. It is astonishing, for example, that subsidies are still being provided to fossil fuel industries worldwide (Watts et al., 2019; Campbell et al., 2021), a challenge the Paris Accords and the UN Climate Change

Conferences (COP) are aiming to reform, such as at the last Glasgow COP 26 meeting which agreed to phase out these subsidies. Dealing with corporate criminality, state crime and state-corporate crime is fundamental to environmental protection and eco-justice. How ERJ contributes to addressing these issues is an ongoing concern and challenge, although an obvious intervention is ERJ efforts to expand the scope of democratic participation in social and criminal justice debates and institutional processes.

4 Environmental Harm and Multiple Voices

This section considers the difficulties and complexities of deciding who speaks for and about 'nature' in the context of ERJ settings and processes. As part of this consideration, the notion of 'harm' is once again revisited. This is so because how the quantum of harm is measured and by whom is vital to how ERJ is institutionalised. The question of process is also central to this discussion insofar as who participates in decisions about specific kinds of environmental harm has concrete ramifications for criminal justice operations and systems.

A restorative approach seems to be ideally suited to dealing with environmental crimes insofar as they hold that promise that things will be done to rehabilitate or repair the harms that have occurred. For instance, a creative interpretation and implementation of restorative justice principles allows for recognition by courts of categories of victims that may not normally be considered (Preston, 2011). Identification of victims is only part of the restorative process, however. Typically, the voice of the victim needs to be heard as well as part of restorative justice proceedings. From the point of view of green criminology, this provides an opening for the extension of justice across the human, environment and species domains, particularly given that environmental harm can be distinguished according to whom or what precisely is being harmed or victimised (White, 2022).

Assessment of victimisation usually involves responding to interrelated questions such as how are 'harm' and 'risk' defined, by whom, at what point does 'risk' or 'harm' occur to an extent warranting action or intervention, and how do we stop the harm from occurring again? (White & Heckenberg, 2014). However, in responding to environmental harm and victimisation, there are inevitably a range of vested interests and discourses that contribute to the shaping of perceptions and issues. This implies differences in perspective and a certain contentiousness of knowledge about the nature of the harm or crime.

In evaluating human victims of environmental harm, for example, it is essential to explore the diverse and often conflicting discourses around 'risk' and 'harm' by different stakeholders (e.g., medical practitioners' consciousness of risk in relation to the health department; loss of livelihood in the case of farmers; limited perception that there is a problem from local miners). Consider, for example, the variety of players who might be associated with disputes over toxic landfill or stockpiled mining tailings in a residential community adjacent to a mining operation. Victimisation is a contestable social process that involves a range of individuals who have different interests (e.g., workers and jobs; residents and amenity). The language of crime and victimisation is reflective of how an environmental problem (in this case toxic landfill) is socially constructed depending upon how it is being considered and by whom, and who is potentially affected and how. Those harmed by environmental degradation may also be those who do not 'see' the harms because the harms are simultaneously a key source of economic security (Natali, 2010, 2016).

How harm is conceived very much depends upon the yardstick by which worth is determined. To assess the severity of harm requires criteria linked to value, scale and precise measure (White, 2013). Value is measured through quantitative assessments (the extent and type of harm) and moral or qualitative assessments (whether to include some types of activities as harm). This assessment of 'worth' is partly dependent upon the scale at which evaluation occurs. Is the focus on individual species or entire ecosystems? Should value also be applied to individual organisms, and if so, should this apply to all plants and animals? Ecosystems incorporate the biotic (plants, animals) and the abiotic (water, soil) that have value in their own right as self-maintaining and self-perpetuating systems. How does one determine the relative value of individual organisms, particular species and overarching biotic communities relative to each other? At the same time, harms to specific ecosystems threaten all within them, human and non-human alike. For example, melting glaciers have implications for future flows of fresh water, and thus affect many different biotic communities in diverse territories and climates. Interconnection and overlapping interests are as important to consider as discrete needs, rights and concepts of justice. The key questions here are who is doing the valuing and what tools can be utilised to assign value.

Critics argue there is a danger that if agencies such as courts adopt environmental restoration (repairing the harm) within an anthropocentric framework, then more harm could result. Katz (1997, quoted in Besthorn, 2004, p. 41) notes that restoration policy presents:

> the message that humanity should repair the damage that human intervention has caused the natural environment. The message is an optimistic one, for it implies that we recognise the harm we have caused in the natural environment and that we possess the means and will to correct these harms. These policies also make us feel good; the prospect of restoration relieves the guilt we feel about the destruction of nature. The wounds we have inflicted on the natural world are not permanent; nature can be made 'whole' again. Our natural resource base and foundation for survival can be saved by the appropriate policies of restoration, regeneration and redesign.

Part of what is being expressed here is the danger that the 'voice' that gets heard, when it comes to restoration, including via restorative justice proceedings, is that of the human, not that of the non-human. This raises the further dilemma of who, then, should speak on behalf of the voiceless, who are the legitimate proxies and spokespeople for entities that cannot otherwise articulate their claims?

One can agree with the sentiment that we need to 'hear' what the voiceless have to say, whether this refers to trees, soils, bees or orchids. This, in turn, should involve active listening, by humans, to the nonverbal communication from nature, the signals emanating from the natural world and its inhabitants, that denote things such as the impacts of climate change (e.g., oceans warming, insect eggs hatching earlier) (Schlosberg, 2007; Besthorn, 2012). There is much to learn by bringing the non-human into the dialogue about ecological health and wellbeing that affects all. But then things get very complicated.

For example, one way to approach the issue is to initially describe those who speak for nature as advocates and those who speak about nature as experts. There is an overlap between those two groups, and the composition of each is diverse. Abstractly, when we talk about speaking *for* nature, we can ask, why do we privilege one group over another and whose voices should be privileged? The environmental activist? The Indigenous person? The government? We also have to ask, 'how do we institutionalise who speaks for what, when and why?' Here we can point to concrete examples of how this might be achieved, for instance, where Indigenous rights and standing are embedded in legislation; this then provides a legal platform for recognition of their relationship with the land which thereby opens the door to official acknowledgement of their voice (White, 2015b).

And in regard to speaking *about* nature, we also have to recognise that there are many different knowledges. For example, a river is defined quite differently by an ecologist and by a geomorphologist and by an Indigenous person. They each have a very different construct of what the river means. And so, even when we talk about expertise, this is contentious. There are hunters and foresters who know the woods and who want to protect what they do in the woods, even though to others they are seen as part of the problem. There are fishers who want to protect the oceans and the fish, even though to others they are seen as part of the problem. But each group has intimate knowledge and they can be the experts for their particular environments and the species within them.

The complexities of environmental harm in terms of spatial and temporal dimensions, divergent opinions surrounding how to conceptualise injustice in regard to the human and the non-human, and the difficulties associated with measuring the extent, seriousness and scale of wrongdoing dictate that justice needs to be a free-flowing but grounded process. Something needs to be done in the here and now, but very often the 'what is to be done' requires a blend of expertise and ideas from many different quarters (including scientists, traditional users of land, environmental activists and laypeople amongst others). Increasing dialogue and deliberative democracy is essential, and this is one of the great strengths of the ERJ approach generally.

Assessment of environmental harm necessarily involves discussion of, and disputes over, the evidence to be drawn upon, the interpretations of impact, and the courses of action advocated. This is an ongoing process. A flexible approach to environmental harm would include elements such as documenting the uncertainties (i.e., being aware of chaotic unpredictability), examining a wide range of alternative courses of action (i.e., going beyond an either/or approach to consider a wider range of options), engaging in broad public deliberation (i.e., allow a plurality of voices and expertise), considering risks in the context of benefits (i.e., analyse the trade-offs), and instituting continuous monitoring and evaluation systems (i.e., evaluate and re-evaluate in the light so different perspectives and new evidence) (Scott, 2005). Justice as an active process does not have a defined endpoint, again something that ERJ would not dispute.

Acknowledgement of 'victim' status is crucial to understanding the ways in which environmental harm affects both human and the non-human. This means locating creatures and environments within their unique ecological niche and context. There are necessarily going to be multiple forums and multiple ways in which to express concern for the non-human environmental entity. There are also multiple knowledges and contestation between knowledge systems (Goyes, 2019). In light of this, there has to be some cross-checking of knowledge and 'facts' regardless of where these come from. For instance, there are some communities where people view calamities to nature (such as lead or asbestos contamination) as an act of God rather than related to the industrial complex down the road (Waldman, 2007). It is essential to assess each situation, each perspective, each contribution to the whole picture, whether this is Indigenous, non-indigenous, traditional or scientific. Exploring how different types of knowledge can be more productively integrated is the essence of intellectual labour. It is also the best pathway to good environmental outcomes.

5   Conclusion

There are a number of practical questions that need to be asked about restorative justice, whether it is in an environmental context or not. As this chapter has demonstrated, these include considerations pertaining to defining the elements and scope of 'justice', institutionalising preferred conceptions of justice (in this instance, ecocentrism) within criminal justice systems, dealing with crimes of the powerful effectively, and engaging with the voiceless in criminal justice forums. There is a conflation sometimes of the distinction between a restorative *process* that involves participants, actors and stakeholders, and a restorative *outcome* which sometimes is seen simply in terms of repairing the harm. It is important to build on both of these, but this too is highly contextual. It depends upon the nature of the stakeholders, including corporations and governments. In the end, though, we have to take into account the dual questions of *power* and *purpose*.

Also at the practical level, there are tensions within eco-justice and there are conflicting rights and conflicting social movements around and between the different spheres of environmental justice, ecological justice, and species justice. From the point of view of intervention, the task is to weigh up the harms in any given situation; rather than adopting a totalising or absolutist position (for example, humans come first, the Earth is most important, any harm to animals is bad); context is everything. Again, this applies to ERJ as it does to everything else, but in this instance, it is the approaches of ERJ—in particular, the impulse towards democratisation—that allows greater appreciation of the importance of context through engagement.

References

Beirne, P. (2009). *Confronting animal abuse: Law, criminology, and human-animal relationships.* New York: Rowman & Littlefield Publishers.

Beirne, P. (2014). Theriocide: Naming animal killing. *International Journal for Crime, Justice and Social Democracy,* 4(3), 50–67.

Berry, T. (1999). *The great work: Our way into the future.* New York: Harmony/Bell Tower.

Besthorn, F.H. (2004). Restorative justice in environmental restoration—The twin pillars of a just global environmental policy: Hearing the voice of the victim. *Journal of Societal and Social Policy,* 3(2), 33–48.

Besthorn, F.H. (2012). Speaking earth: Environmental restoration and restorative justice. In K. Wormer & L. Walker (Eds.), *Restorative justice today: Practical applications* (pp. 233–244). Los Angeles: Sage.

Bisschop, L., & Vande Walle, G. (2013). Environmental victimisation and conflict resolution: A case study of e-waste. In R. Walter, D. Westerhuis, & T. Wyatt (Eds.), Emerging issues *in green criminology: Exploring power, justice and harm* (pp. 34–54). Basingstoke: Palgrave Macmillan.

Braithwaite, J. (1989). *Crime, shame and reintegration*. Cambridge: Cambridge University Press.

Braithwaite, J. (2002). *Restorative justice and responsive regulation*. New York: Oxford University Press.

Braithwaite, J., & Pettit, P. (1990). *Not just desserts: A republican theory of criminal justice*. Oxford: Oxford University Press.

Campbell, R., Littleton, E., & Armistead, A. (2021). *Fossil fuel subsidies in Australia: Federal and state government assistance to fossil fuel producers and major users 2020–2021*. Canberra, ACT: The Australia Institute. Retrieved from https://apo.org.au/sites/default/files/resource-files//apo-nid311955.pdf (last accessed 30 June 2021).

Council of Europe (2012). *Manual on human rights and environment*. Strasbourg: Council of Europe Publishing.

Cullinan, C. (2003) *Wild law: A manifesto for earth justice*. London: Green Books in association with The Gaia Foundation.

Flynn, M., & Hall, M. (2017). The case for a victimology of nonhuman animal harms. *Contemporary Justice Review*, 20(3), 299–318.

Goyes, D. (2019) *Southern green criminology: A science to end ecological discrimination*. London: Emerald Publishing.

Hall, M. (2013). *Victims of environmental harm: Rights, recognition and redress under national and international law*. London: Routledge.

Hall, M. (2015). *Exploring green crime: Introducing the legal, social and criminological contexts of environmental harm*. Palgrave Macmillan.

Hamilton, M. (2008). Restorative justice intervention in an environmental law context: Garrett v Williams, prosecutions under the Resource Management Act 1991 (NZ), and beyond. *EPLJ*, 25, 263–271.

Hamilton, M. (2021). *Environmental crime and restorative justice: Justice as meaningful involvement*. London: Palgrave Macmillan.

Katz, E. (1997). *Nature as subject: Human obligation and natural community*. New York: Rowman & Littlefield.

Koons, J. (2009). What is earth jurisprudence? Key principles to transform law for the health of the planet. *Penn State Environmental Law Review*, 18, 47–68.

Levi, M. (2003). Suite justice or sweet charity? Some explorations of shaming and incapacitating business fraudsters. In E. McLaughlin, R. Fergusson, G. Hughes, & L. Westmarland (Eds.), *Restorative justice: Critical issues* (pp. 141–154). London: Sage.

Lin, A. (2006). The unifying role of harm in environmental law. *Wisconsin Law Review*, 3, 898–985.

Maloney, M., & Burdon, P. (Eds.) (2014). *Wild law: In practice*. London: Routledge.

MacKenzie, G., Stobbs, N., & O'Leary, J. (2010). *Principles of sentencing*. Sydney: Federation Press.

Mehta, M. (2009). *In the public interest: Landmark judgement & orders of the Supreme Court of India on environment & human rights, Vols 1–3*. New Delhi: Prakriti Publications.

Natali, L. (2010). The big grey elephants in the backyard of Huelva, Spain. In R. White (Ed.), *Global environmental harm: Criminological perspectives* (pp. 193–209). Uffculme: Willan Publishing.

Natali, L. (2016). *A visual approach for green criminology: Exploring the social perception of environmental harm*. London: Palgrave Macmillan.

Nurse, A. (2016). *An introduction to green criminology & environmental justice*. London: Sage.

Nurse, A., & Wyatt, T. (2020). *Wildlife criminology*. Bristol: Bristol University Press.

O'Donnell, E., & Talbot-Jones, J. (2018). Creating legal rights for rivers: Lessons from Australia, New Zealand, and India. *Ecology and Society*, 23(1), 7–13.

Pecharroman, L. (2018). Rights of nature: Rivers that can stand in court. *Resources*, 7(13). https://doi.org/10.3390/resources7010013.

Preston, B. (2011). The use of restorative justice for environmental crime. *Criminal Law Journal*, 35, 136–145.

Schlosberg, D. (2007). *Defining environmental justice: Theories, movements, and nature*. Oxford: Oxford University Press.

Scott, D. (2005). When precaution points two ways: Confronting "West Nile fever". *Canadian Journal of Law and Society*, 20(2), 27–65.

Smith, M. (1998). *Ecologism: Towards ecological citizenship*. Minneapolis: University of Minnesota Press.

Sollund, R. (2012). Speciesism as doxic practice versus valuing difference and plurality. In R. Ellefsen, R. Sollund, & G. Larsen (Eds.), *Eco-global crimes: Contemporary problems and future challenges* (pp. 91–115). Farnham, Surrey: Ashgate.

Sollund, R. (2019). *The crimes of wildlife trafficking: Issues of justice, legality and morality*. London: Routledge.

Stone, C. (1972). Should trees have standing? Toward legal rights for natural objects. *Southern California Law Review*, 45, 450–487.

Tombs, S., & Whyte, D. (2015). *The corporate criminal: Why corporations must be abolished*. London: Routledge.

United Nations Environment Programme (2017). *The status of climate change litigation: A global review*. Nairobi: UNEP.

Voigt, C., & Makuch, Z. (Eds.) (2018). *Courts and the environment*. Cheltenham, UK: Edward Elgar Publishing.

Waldman, L. (2007). When social movements bypass the poor: Asbestos pollution, international litigation and Griqua cultural identity. *Journal of Southern African Studies*, 33(3), 577–600.

Walters, R., & Westerhuis, D.S. (2013). Green crime and the role of environmental courts. *Crime Law and Social Change*, 59, 279–290.

Watts, N., et al. (2019). The 2019 report of The Lancet Countdown on health and climate change: Ensuring that the health of a child born today is not defined by a changing climate. *The Lancet*, 394(10211), 1836–1878.

Westerhuis, D. (2013). A harm analysis of environmental crime. In R. Walters, D.S. Westerhuis, & T. Wyatt (Eds.), *Emerging issues in green criminology: Exploring power, justice and harm* (pp. 197–217). Basingstoke: Palgrave Macmillan.

White, R. (2013). *Environmental harm: An eco-justice perspective*. Bristol: Policy Press.

White, R. (2015a). Indigenous young people and hyperincarceration in Australia. *Youth Justice*, 15(3), 256–270.

White, R. (2015b). Indigenous communities, environmental protection and restorative justice. *Australian Indigenous Law Review*, 18(2), 43–54.

White, R. (2016). Four problems facing environment courts in dealing with nonhuman environmental victims. In T. Spapens, R. White, & W. Huisman (Eds.), *Environmental crime in transnational context* (pp. 139–153). New York: Ashgate.

White, R. (2017). Reparative justice, environmental crime and penalties for the powerful. *Crime, Law and Social Change*, 67(2), 117–132.

White, R. (2018). Ecocentrism and criminal Justice. *Theoretical Criminology*, 22(3), 342–362.

White, R. (2022). *Theorising green criminology: Selected essays*. London: Routledge.

White, R., & Heckenberg, D. (2014). *Green criminology: An introduction to the study of environmental harm*. London: Routledge.

White, R., Haines, F., & Asquith, F. (2017). *Crime and criminology*. South Melbourne: Oxford University Press.

Wijdekop, F. (2019). *Restorative justice responses to environmental harm.* Amsterdam: IUCN.

Williams, C. (1996). An environmental victimology. *Social Justice, 23*(4), 16–40.

Wormer, K., & Walker, L. (Eds). (2012). *Restorative justice today: Practical applications.* Los Angeles: Sage.

Wright, M. (2022). *Responding to environmental crime: Lessons from New Zealand.* London: Palgrave Macmillan.

Whyte, D. (2020). *Ecocide: Kill the corporation before it kills us.* Manchester, UK: Manchester University Press.

3

Restorative Justice and Environmental Criminal Law: A Virtuous Interplay

Chiara Perini

1   Introduction

Taking the *European* legal space as a frame, this article would like to test the hypothesis that restorative justice—its ideal of justice, values and methods—could help criminal law to improve how it traditionally reacts to the commission of an offence against the environment. Specifically, restorative justice could manage the 'matters arising from the offence' (Council of Europe Recommendation CM/Rec(2018)8 concerning restorative justice in criminal matters, Rule 3) not only in a more careful, but also in a more effective way with respect to the purposes of criminal law itself: that is, safeguarding 'protected interests' and preventing offences.

There is a well-established body of scholarship arguing that there is a complementarity relationship between restorative justice and criminal law (Council of Europe, 2018b, p. 6; Mannozzi & Lodigiani, 2017, pp. 368–373). Many scholars also argue that there is benefit in spreading

C. Perini (✉)
University of Insubria, Varese, Italy
e-mail: chiara.perini@uninsubria.it

restorative justice principles throughout the criminal justice process (Marder, 2020, p. 396). Moreover, the aforementioned Recommendation CM/Rec(2018)8 (Council of Europe, 2018a) contains a reference to this complementary relationship in Rule 6, according to which 'Restorative justice may be used at any stage of the criminal justice process'.

In the field of environmental criminal law, therefore, it could be possible to establish a virtuous synergy between restorative justice and criminal law/proceedings with reference, in particular, to two aspects. On the one hand, restorative justice could support the criminal justice system in better understanding the *harm* (in its penal meaning), that is, the *disvalue* generated by the crime as a *harmful* fact for a *value* (the environment) that the community wants to protect through criminal law. On the other hand, restorative justice could help to identify the (possible) ways of repairing the harm, that is, what can be done to *remedy* the disvalue and *regenerate* the damaged value.

Harm and *reparation* are traditional concepts in criminal law (Cavalla & Todescan, 2000), but in certain sectors of criminality—such as, emblematically, that of environmental crimes—criminal law seems to encounter some difficulties in fully conceiving these elements and in making them effective in criminal proceedings. Given the undeniable value dimension inherent in them, a simple assessment of what happened objectively is not enough to appreciate harm and reparation. Although based on a necessary material and factual substratum, a correct understanding of these elements needs a preliminary contextualisation in relation to the 'charter of values' on which the community that has suffered the harm is founded. The space for a fruitful interlacing of restorative justice is right here: with its own methods, which mirror its own values, restorative justice would allow criminal law to establish effective contact with those (individuals or communities) who have suffered harm (both material and in value) from the crime (Marshall, 2019, pp. 180–182). This would enable a more accurate comprehension of the *offence*, not only on a quantitative but also on a *qualitative* level, that would enrich the legal representation of the facts in the criminal process and consequently illuminate possible paths of *reparation* in a truly *regenerative* direction.

The aim of this paper is therefore to clarify the complementary relationship between restorative justice and criminal law in environmental protection. To this end, the paper firstly intends to clarify the specificity of environmental protection implemented by criminal law in comparison with that provided by other sectors of the legal system (Sect. 2). Secondly, the paper seeks to reconstruct what the environment is as a value to be protected (also by criminal law) in the European legal area (Sect. 3). Finally, the paper highlights the points of contact between the European conception of the environment as a value to be protected and restorative justice, to specify the points of synergy between the latter and environmental criminal law (Sects. 4 and 5).

2   The Meaning of Environmental Protection Through Criminal Law

The analysis starts with the review of the European strategy on environmental protection through criminal law, outlined in Directive 2008/99/EC. In particular, Recital n. 3 suggests that not only the social disapproval associated with the crime, but also the *crime itself* is qualitatively different from offences provided for by other sectors of the legal system. When a community associates a criminal sanction to a fact, it does not consider that fact simply as an economically assessable damage to an asset (this is instead the nature of the *tort*). Nor does it consider that fact as a mere violation of a rule of conduct established by a public law to protect a certain good (this is instead the case of the *administrative offence*) (Pulitanò, 2021, pp. 6–7, 27–28, 563–564).

The *qualitative* difference, proper to the crime, goes *beyond* these elements (which are nevertheless present in the crime or in any case related to it). Moreover, it does not depend only on the importance of the value damaged by the crime: this is confirmed by the fact that, in the same legal system, *both* criminal sanctions *and*—at the same time—civil or administrative sanctions often protect the same values (such as, the environment).

The difference lies rather in the 'social' expectation, so to speak, regarding the observance of the precept (Luhmann, 1990). As Coffee (1992) has pointed out, when a community chooses (with its own laws) to sanction a fact through a penal norm, it becomes the object of an absolute prohibition (*prohibiting*): that is, the community expects that the rule of conduct established by the penal norm will not be violated under any circumstances. Likewise, an element in support of this is the fact that the threat of sanctions against goods of primary value for the offender (personal freedom, first of all) averts criminal rule violation with a well-known 'deterrence' effect. Other legal system sectors—*other* than criminal law—do not establish absolute prohibitions, but burdens: in the sense that one can violate the rule that prescribes a certain behaviour, keeping into account and 'budgeting' for the cost of non-compliance (*pricing*).

Applying the distinction between prohibiting and pricing to the analysis of the European strategy of environmental protection through criminal law, we can see that for the community gathered in the European Union, the environment represents, first of all, a value in itself that must be protected. In addition, this community traditionally recognises the need to protect the environment, thanks to the legal pact made at supranational level since the Treaty of Maastricht (1992), so much so that the environment could be seen as a founding factor of the very 'identity' of the united Europe (Sotis, 2007, pp. 86–87, 94–102).

Secondly, the prohibition of damaging the environment or endangering its integrity is *absolute*: this applies both to the environment as such (understood as the totality of ecological resources, fauna and flora) and to the negative effects that environmental offences could have on people's health and life in the light of the crime catalogue specified by Art. 3 Directive 2008/99/EC. So, it is not admissible to 'put a price' on an offence against the environment, to allow it to be committed. Consistently, Directive 2008/99/EC imposes on the EU Member States' legislators the obligation to provide in the corresponding legal system criminal sanctions and corporate criminal liability in case one of facts described in Art. 3 is committed.

3 The Environment as a Value in the European Legal Space

Especially if compared with other international sources or with the constitutional traditions of some EU States, Directive 2008/99/EC, however, proves to be only partial in grasping the actual extension of the concept of 'environment' as a value to be protected through criminal law. The legal-social awareness developed in the context considered seems to conceive this asset in a wider way.

The catalogue of offences described by Art. 3 Directive 2008/99/EC shows that, in this case, the European legislator has accepted a strictly ecological concept of 'environment', for which the latter is a purely objective and material asset. Environmental crimes cause, in fact, harm against soil, air, water, fauna and flora; or—in a mediated way—harm against people and physical safety. Even the revision process of Directive 2008/99/EC currently being undertaken by the European Union does not question such an idea of 'environment' and focuses on other aspects of the corresponding regulation. Generally, the Commission seems to follow a purely sanctioning perspective: 'Overall, the ECD [Directive on the protection of the environment through criminal law 2008/99/EC] did not have much effect in practice. In particular, the ECD did not affect the number of convictions or the level of imposed sanctions in the Member States' (European Commission, 2020, p. 1).

On the contrary, some factors have been present for a long time in the European legal space, demonstrating two relevant points. These are, on the one hand, the partiality of a purely material conception of the environment as a value protected through criminal law and, on the other hand, the opportunity to cultivate an approach, which is not only punitive, but also reparative with respect to environmental offences.

3.1 The 'Vision' Promoted by the European Landscape Convention

The first factor is the European Landscape Convention, signed by the Council of Europe in 2000, and the international sources related to it. As known, with regard to the landscape—an asset not coinciding, but physiologically connected to the environment (Giunta, 2008, p. 1151)—the Convention overcomes a purely material conception and welcomes a personalistic vision. What is valued and protected is the landscape as an essential element of a person's life or persons' lives.

The link between landscape, individual and community is clearly carved into Article 5(a) of the Convention, which commits the Parties to 'recognise landscapes in law as an *essential component* of *people's surroundings*, an *expression of the diversity* of their *shared cultural and natural heritage*, and a foundation of their *identity*,' (italics added).

As a protected good, therefore, the landscape undoubtedly requires the preservation of the composite naturalistic substrate on which it rests, but it derives its perimeter (also) from immaterial values, which express the relationship—inevitably dynamic—existing between landscape and population (Council of Europe, 2000, par. 42).

The point of view of people and communities runs through the entire discipline developed to protect the landscape (Drigo, 2012, pp. 4–6). First, in the Convention the perception gained by the population in relation to their own *landscape "of life"* is a fundamental element to reach the very definition of 'landscape' (Art. 1 lett. a) as an asset to be preserved.

The *'aspirations* of the *public* with regard to the landscape features of their surroundings' (Art. 1 lett. c, italics added) are also valued in the administrative procedure of identification of 'landscape quality objectives' (Art. 6 lett. D). Finally, the administrative procedure of identification and evaluation of landscapes to be protected in the national territory must take into account 'the particular values assigned to them [i.e. landscapes] by the *interested parties* and the *population concerned*' (Art. 6 lett. C.1.b), (italics added).

The notion of landscape elaborated by the Convention is clarified by the subsequent Recommendation CM/Rec(2008)3 of the Committee of Ministers to Member States on the guidelines for the implementation of the European Landscape Convention. It underlines how landscape is not only a material good that is part of a physical space, but necessarily includes people who live *in* the landscape and establish *relationships in it and with it* (Council of Europe, 2008, p. 4).

In the same spirit, Recommendation CM/Rec(2017)7 on the contribution of the European Landscape Convention to the exercise of human rights and democracy with a view to sustainable development ultimately draws a bridge between the notion of landscape embraced by the Convention and the most recent developments in scientific debate and policy in the environmental field in the direction of sustainable development (United Nations, 2015).

As is well known, the principle of sustainable development aims to reconcile economic progress and environmental protection, giving the weighing of environmental interests an essential and unfailing position in the definition of relevant policies and legislation (Grassi, 2007, p. 1126). In this frame, the Recommendation goes a step further in the direction already highlighted, inviting the States Parties to the European Landscape Convention:

* To consider 'the importance that quality and diversity of landscapes has for the minds and bodies of human beings, as well as for societies, in the reflections and work devoted to human rights and democracy, with a view to sustainable development' (lett. *a*);
* To ensure 'that landscape policies respond to the ideal of living together, especially in culturally diverse societies' (lett. *d*);
* To guarantee 'the right to participation by the general public, local and regional authorities, and other relevant parties including non-governmental organisations, with an interest in the definition, implementation and monitoring of landscape policies' (lett. *g*);
* And to include 'the "landscape", as defined by the Convention, in indicators of sustainable development relating to environmental, social, cultural and economic issues' (lett. *h*).

The European Landscape Convention and the consequent system of international sources conceive environmental interests searching for virtuous and synergistic relationships between material goods and people. They do not weigh down that idea with hierarchies of value, being on the contrary aware that the human species—as 'dominant'—has the *responsibility* (Jonas, 2009) for the preservation and promotion of 'dominated' goods also in the interest of future generations (*mutatis mutandis* Mancuso, 2019, pp. 13–14). For this reason, we may speak here of a *personalistic* conception of the environment, where the environment as a value to be protected is not reduced to a set of material goods, but is (also) in service of the people—current and future—who recognise it as their 'living environment' and to whose care it is responsibly entrusted. This is a third option (intermediate or nuanced) with respect to the traditional alternative between a purely 'anthropocentric' or 'ecocentric' conception of the environment (Giunta, 2008, p. 1153; Grassi, 2007, p. 1114): unlike the approach mentioned here, these terms imply that some interests take priority over others, which are to be understood as secondary or subordinate and, to a certain extent, expendable.

3.2   The Environment in the Jurisprudence of the European Court of Human Rights

In the European legal space, the jurisprudence of the European Court of Human Rights (ECtHR) confirms a broad conception of the good 'environment', not limited to its ecological-material side. As is well known, also for historical reasons linked to the relatively recent affirmation of a common awareness on environmental matters, the text of the European Convention on Human Rights (ECHR)—dating back to 1950—does not contain an explicit reference to the environment. However, through its interpretation of the Convention, the ECtHR has recognised the existence of the *fundamental human right* to live in a healthy environment based on Article 8 (Right to respect for private and family life), as well as—more rarely—Article 2 ECHR (Right to Life), moving in particular along the track of the 'positive obligations' doctrine, which requires States not only to not violate but also to actively protect human rights (Mazzanti, 2020, pp. 70–72; Perini, 2004, pp. 210–212).

The process of establishing the environment as a value to be protected began in the mid-1990s with the famous leading case *Lopez Ostra v Spain* (judgement of 9 December 1994). In that case, the ECtHR condemned Spain for violation of Article 8 ECHR, since a solid and liquid waste treatment plant had been built with public funds only 12 metres away from the applicant's home and, once it had been put into operation without authorisation, had caused serious pollution and health problems for the inhabitants. The link between the human right to environment and

Art. 8 ECHR was subsequently confirmed by the ECtHR on numerous occasions concerning cases of noise pollution, industrial pollution and illegal waste management (Fimiani, 2019, pp. 376–380).

With regard to Art. 2 ECHR, the leading case was the judgement of 30 November 2004, *Öneryildiz v Turkey*, in which the Court condemned Turkey for violation of Article 2 ECHR. The responsibility of the State consisted in the failure to take the necessary measures to prevent the accident (a methane explosion in a rubbish dump near the applicant's house), in which nine of the applicant's relatives lost their lives and his house was destroyed (Perini, 2004, pp. 210–212).

The fact that the ECtHR derives the human right to environment from Art. 8 ECHR in no way detracts from this recognition and is, indeed, particularly significant from the perspective followed here.

Through the link between the right to the environment and Art. 8 ECHR, in fact, the Court emphasises that the possibility of enjoying a healthy environment is an essential element, qualifying and, to a certain extent, a precondition of the fundamental right to respect for private and family life. In other words, '[e]veryone has the right to respect for his private and family life, his home and his correspondence' (Art. 8 par. 1 ECHR) without 'interference by a public authority while exercising this right' (Art. 8 par. 2 ECHR). However, this right requires an intact environment as a context: in the absence of such a context, the right in question can neither be conceived in the abstract, nor actually exercised.

Following an anthropocentric perspective (Mazzanti, 2020, p. 72), therefore, the ECtHR recognised the right to the environment as a component of the complex of values enshrined in Art. 8 ECHR. This norm protects unitarily 'the set of relations that revolve around the most intimate sphere of the person' (Fimiani, 2019, p. 376). These are, namely, the relationships that take place in the home, that is, in the place typically reserved for private life; the individual's right to see the autonomy and quality of his/her private life respected, without external interference that may alter his/her psycho-physical well-being; and family relationships that qualify and give substance to private life.

The framing of the human right to environment in Art. 8 ECHR thus focuses on the relations that the individual establishes with others in the physical space in which private life takes place. Even if this implication concerns precisely the home, the protection of such a right obviously cannot be limited to material and ecological aspects, but must also consider the fitness of the environment to meet the individuals' relational needs amongst themselves and in the 'social groups' corresponding to that sphere. In other words, the environment's physical and objective characteristics receive protection not so much in themselves, but insofar as they are suitable for creating a positive *context* for people's relations *with* and *in* the environment considered. Recognised in these terms, the human right to environment appears very promising, especially if one assumes to transpose it into a wider physical space, which goes beyond the boundaries of the home and private life in the direction of the community.

3.3   The Link Between Solidarity and Environment in the Charter of Fundamental Rights of the European Union

The existence of a fundamental right to the environment in the European legal space is made manifest and visible thanks to Art. 37 of the Charter of Fundamental Rights of the European Union (CFR, 2000) (Stern, 2014).

Since its first adoption, the Charter has been considered to be strongly influenced by other sources in the European legal space: amongst them, the European Convention on Human Rights of 1950, as interpreted by the European Court of Human Rights (Pace, 2001). For this reason, too, the case-law interpretation carried out with reference to articles 2 and 8 ECHR in relation to the human right to environment cannot be neglected in the reading of Art. 37 CFR.

The scope of this provision is, in my view, illuminated by its position within the Charter. It is in fact part of the rights that the EU brings together under the concept of 'solidarity', one of the four 'indivisible, universal values' (*human dignity, freedom, equality and solidarity*) on which the EU declares itself to be founded (CFR, Preamble, 2nd par.). The Chapter devoted to 'Solidarity' is considered to be varied in terms of content, as it encompasses very different rights, qualified by some as 'social rights' (Bonaventura, 2015), by others as rights aimed at realising the 'values of freedom and equality' (Menéndez, 2004, p. 98).

From my point of view, the link between the fundamental right to environment and solidarity is significant. In a first approximation, the term 'solidarity'—on an ethical-social level but also in the history of law—alludes to the relationship of *brotherhood* and *mutual support* that connects community members in the feeling of belonging to the same group of people, made firm by the sharing of common interests and goals. It is a concept closely related to that of 'fraternity' which appears in the famous revolutionary triad *'Liberté, Égalité, Fraternité'*. However, for a full juridical affirmation with regard to the constitutional basis of a State (specifically, the French State), the latter had to wait until the French Republic Constitution of 1848. In fact, political and cultural obstacles raised by 'proprietary individualism' and 'modern contractualism' based on a conception of humanity rooted in independence rather than in the 'constitutive *relationality* of human existence' (Comazzetto, 2021, pp. 259–260, italics added) have long stood in the way. Without claiming to examine here in depth the complexity of the concept from a sociological and philosophical angle (Camboni, 2018), solidarity describes a context of collective life, in which there are ties of interdependence amongst individuals within a general framework of social cohesion (Camboni, 2018, p. 78). Out of the different forms (or levels) of solidarity, a central role belongs to the so-called internal group solidarity, that is, to the 'social bond that implies coordination and cooperation between group members as members, oriented towards the promotion of the main objectives of the group' (Camboni, 2018, p. 90). 'Internal social solidarity' is in fact the precondition for the group to be able to show solidarity also towards external subjects, be they individual or collective. The semantic framework evoked by the concept of solidarity clarifies the conception of the fundamental right to environment proclaimed by the European legislator.

First, the recognition of solidarity as a *founding value* of the European Union is reflected in the environment, whose qualification as a primary value or good to be protected is beyond doubt. Moreover, the link between environmental protection and fundamental human rights has recently been reaffirmed in the 'European Parliament resolution of 20 May 2021 on the liability of companies for environmental damage', which recognises that 'the impact of environmental damages and crimes adversely affects not only biodiversity and the climate, but also human rights and human health' (lett. W).

Secondly, given the necessarily pluralistic framework connected to solidarity as a concept, also for EU law the value of the good 'environment' is fully grasped, not (only) with regard to the individual, but (also and above all) in the context of groups (or 'social groups'), in which each person expresses his/her own 'human personality'. This means that the environment as a protected value is not limited to the ecological-material substratum that defines its physical and objective structure, but necessarily includes social relations that the individual(s) and the community develop with a certain natural context and within that context.

4   Synergies Between Restorative Justice and Environmental Criminal Law

A clear convergence has emerged from the analysis carried out, which leads to the conclusion that in the European legal space the protection granted to the good 'environment' does not only include its physical and objective consistency, but also its suitability to serve as a context for the life of people, considered individually and gathered in collectives. More precisely, environmental protection is also protection of relationships (Higgins et al., 2013; Almassi, 2017, p. 30), in two senses.

On the one hand, there is the relationship that the individual develops within and in relation to a certain environment. This relationship is undoubtedly qualified by the presence of variously conceived cost/benefit calculations, which consider the environment as a place to be inhabited, a territory to be exploited, and so on. Nevertheless, it is also enriched by a subjective component (even if standardised in the general and abstract assessment inevitably underlying the law): an *emotional* component, so to speak, deriving from the *feelings* aroused in the individual by the existential link with his/her 'life environment'.

On the other hand, there are the social relations that develop in a certain environment, that is, relations between individuals and within a community. Even at the social and community level, the link with the environment does not have a purely material meaning, that is deriving from the mere utility of living in and exploiting a certain natural context. Similarly to what has been observed at the individual level, this relationship is also qualified by the emotional involvement that a certain environment arouses in those who live there as a 'group of people', developing a sense of belonging to that context and playing a founding role in the *identity of the group* as a 'community' (Aime, 2019).

The acceptance of such a broader notion of environment by criminal law is not straightforward. Since it is obviously conditioned by the principle of legality (Bricola, 1981), the criminal system protects the good 'environment' as established by the legislation in force. Therefore, if the personalistic conception is not yet textually accepted by each juridical system, the elements in favour of a paradigm change must first be included in the criminal-political debate at national and/or European level, with a view to promoting possible reforms.

This does not exclude, however, that *de lege lata* the documented presence of a personalistic conception of the environment can be taken into account by the various actors of the criminal process, to encourage therein the consideration of additional value profiles highlighted before, where this is allowed by the interpretation and in application of the existing penal legislation.

It is precisely on this second track that the premises for a synergy between criminal law and restorative justice seem to be laid. If the environment as a protected good is the 'living environment' and includes the relationships already mentioned, the criminal justice system should necessarily not only conceptualise environmental harm, but also respond to it with consistent tools. These tools should allow the criminal system to become aware of the crime's negative impact on these relationships too; and they should also contribute to the *restoring* of individual/community/environment links after the crime.

If we consider law in a formal way, the lack of the normative prevision of the possibility of using restorative justice instruments by criminal law and criminal process, that is judged to be an obstacle in other legal systems (Hamilton, 2021, p. 92), should not be regarded as discouraging. In the European legal space, in fact, the recommendation expressed by the Council of Europe in Rule 59 should be appropriately assessed.

This allows an evolutionary *restorative* interpretation of different legal notions present in criminal law and process, provided that they are compatible at the structural and functional level with the restorative justice principles.

In summary, with respect to both mentioned tasks (harm understanding and restorative paths development) restorative justice appears, as widely accepted, particularly appropriate in general. With special regard to environmental crimes, the restorative justice approach could allow, on one side, to fill the under-representation and the awareness gap that tends to characterise the victims (especially if the victims are not single persons but collectives). On the other side, it could help in the development of *regenerative* pathways to restore individual/community/environment relationships after the crime. In this sense, restorative justice could contribute also to the content clarification of restoring plans for environmental resources, traditionally designed only in a material and ecological sense (Boscolo, 2021, pp. 28–33). The support of restorative justice, finally, is promising also on the offender's side: environmental crimes are structurally 'contactless crime', being characterised by the fact that the offender tends not to know the victim, has no contact with him/her and does not develop any kind of emotional involvement towards him/her (Stevanović, 2020, p. 103). Precisely for this reason, restorative justice implementation, by placing the victim as a person at the centre of the response to the crime, could contribute to the process of self-responsibility of the offender in view of his/her re-education, also in order to prevent future recidivism.

In the case of environmental crimes, the synergy between restorative justice and criminal law works in two fundamental directions. On the one hand, restorative justice is useful for the correct understanding of the harm caused by the crime, which includes not only material elements (destruction or damage of natural resources), but also value elements (negative impacts on individual/community/environment relationships). On the other hand, restorative justice is worthwhile for effective harm repair, suitable to restore the environment both in an objective sense and in a value and relational sense.

As a preliminary to these aims, restorative justice allows criminal law to make progress on the long-standing problem of victim identification in environmental crimes (Stevanović, 2020, p. 103). This is because restorative justice is able to establish a direct, personal and concrete contact with those (at individual or community level) who have been harmed by the environmental crime.

Traditionally, criminal law considers the environment as a 'collective and diffused legal asset' (Marinucci & Dolcini, 2001, pp. 542–543). Such an expression would underline that each individual within the community is interested in environmental protection; and no one within the community can act against the environment. Compared to the personalistic notion of environment in the European legal space, the reference to the 'diffused' character of the environment ends up weakening not only the belonging bond of the environmental asset to an identified or identifiable subject, but also the identity bond between 'environment' and 'individual' and between 'environment' and 'community'. On the contrary, the characterisation as a 'diffused' good puts the emphasis on the fact that an indistinct group of subjects, who as such cannot dispose of the asset itself, takes interest in the protection of the environment.

In such hypotheses, greater clarity could be achieved if the criminal justice process used restorative justice tools at an early stage. In the alternative between *front-end* and *back-end* models (Al-Alosi & Hamilton, 2019), restorative justice inserts could be considered which, while not qualifying as diversion, would still allow for the identification of crime victims before the sentencing phase. For example, during the phase of the so-called preliminary investigations, the prosecution could benefit from the possibility to explore and get to know, through a restorative justice process, the community settled in the territory involved by the environmental crime. In this way, the prosecuting authority and, therefore, the judge could better ascertain who suffered a damage (i.e. which individuals or which groups of individuals), beyond the material and direct involvement in the fact that could be limited to a few subjects.

Secondly, as mentioned above, restorative justice can help the criminal justice process to take exact account of the harm caused by environmental crimes especially in a *qualitative* sense. It is worth noting that the focus on the non-material harm components is imposed by the definition of crime victim set by Directive 2012/29/EU, according to which the victim is first and foremost 'a natural person who has suffered harm, including physical, mental or emotional harm, or economic loss, which has been directly caused by a criminal offence' (Art. 2 par. 1 lett. a.i). With respect to the various constituent elements of the offence, legal systems are traditionally well equipped to appreciate economic or economically assessable items, while they are substantially unprepared to deal with the emotional ones (including negative impacts on individual/community/environment relations in the case of environmental crimes). This is also due to the marginal role often given to the victim in criminal proceedings. In this context, the voice of the victim is mainly used as a means of proof against the offender and not for providing the judge with elements on the 'damage, including physical, mental or emotional' suffered, which would instead be important to reconstruct *harm seriousness*.

Operating a synthesis between the level of victim identification and of offence understanding, restorative justice seems to come to the rescue in this respect. If used from the earliest stages of the criminal trial, in fact, even on the initiative of the prosecution alone, restorative justice methods could allow parties to correctly frame the offence, not only with respect to individuals materially involved by the crime's negative consequences, but also in its *relational* component. In this way, it would be possible to define unambiguously the extent of the collective perimeter of the environmental crime's victims, namely the community that considers the offended environment to be its own 'living environment' and that recognises itself in it. A plural subject would therefore be called upon to take part in the trial, obviously in a more circumscribed way than the population in general, but representative of the bond of belonging—on an identity level also—of the offended asset to an identified group of people.

5   'There Is No Planet B': Restorative Justice and Harm Healing in Environmental Crime

In case of environmental crimes, the partiality of a purely punitive approach to the crime and the need to repair the harm caused to the protected good appears with particular clarity. To borrow an impressive expression of the environmental movements recently engaged in the fight against climate change, 'there is no planet B' (Berners-Lee, 2020) and the urgency of restoring the capability of natural resources damaged by the crime to serve as a suitable context for people's lives is evident.

In general, Coffee's (1992) distinction between prohibiting and pricing does not allow the idea of repairing the harm associated with a crime by a simple monetary compensation mechanism. This would create a structural contradiction and distort the offence by downgrading it to a civil one. Reparation of the criminal offence necessarily has a value meaning as well: it requires to remedy the disvalue created by the crime and to regenerate the damaged value.

In order to understand how to repair (in this sense) and to verify whether a certain conduct has had an effect in such terms, it is essential to meet the victims and talk to them or, as suggested by the Recommendation CM/Rec(2018)8 (Council of Europe, 2018a, Rule 3), more broadly with the persons involved in the offence (also as perpetrators). That is, a judicial decision on the point should be based on cognitive elements that take into account the point of view of these people in the specific case. For this reason, it is essential to open the criminal process to restorative justice paths, when the judge is supposed to assess harm reparation activities (e.g. in view of a *mitigation* of criminal liability, or of a crime *extinction* declaration as a result of the trial without conviction).

The use of restorative justice methods is consistent with the environment personalistic conception documented in the European legal space. Even when it includes in the reparation agreement obligations to restore the environment in a material sense, restorative justice aims at establishing (or re-establishing) the—also moral—conditions for the relationship between people to exist, being fundamental in this direction the creation of a sharing between the parties of values and responsibilities with respect to the incident (Almassi, 2017, pp. 22, 36; Al-Alosi & Hamilton, 2019, p. 1469).

One of the most important outcomes of such a restorative process is a climate of trust and hope, which looks at the future relationships between the people in the community. Therefore, some issues traditionally linked to the theme of environmental restoration appear secondary (e.g. the identification of the original environmental conditions to be taken as a reference point for the recovery of damaged natural resources) and other aspects assume central importance.

These undoubtedly include the *participation* not only of the victims, but also more generally of the community in the reparation process and subsequent activities, so that the specific needs of current victims are taken into account in the restoration process (Almassi, 2017, pp. 29–31). After an environmental crime, victims require information, participation (effective and not merely symbolic) and assistance from experts in the technical-scientific field, as documented by concrete cases (Arantes Prata, 2020, p. 218). A possible restorative justice path should take into account the victims' search for specialised knowledge support, for example, by including also technical-scientific experts providing the relevant knowledge involved in the issues caused by and related to the crime.

Secondly, environmental conflicts are often characterised by an *imbalance of power*: this is mostly the case when the parties responsible for the damage are companies, on which the victims' community may have developed more or less intense forms of socio-economic dependence (Hübschle et al., 2021, p. 145; Kershen, 2021, pp. 163–164; Pali & Aertsen, 2021, p. 6). In these hypotheses, the increased victim vulnerability, as well as the risk that, due to the pressure exerted by the perpetrator, the reparation agreement has a purely abstract content (and therefore is not very binding for the offender), or is not respected in practice, should be carefully considered when choosing, designing and implementing a restorative justice path (Arantes Prata, 2020, p. 217).

Thirdly, another critical element, frequently reported, is represented by the heterogeneity of environmental crime victims, which the reparation path should involve: people, ecological matrices (flora, fauna, ecosystems, etc.), communities (with particular attention to Indigenous peoples, where present) (Hamilton, 2021), commercial operators, and so on (Al-Alosi & Hamilton, 2019, p. 1466).

On one side, the *assessment* of damage caused by the crime—especially for the components that go beyond the purely material level—cannot be standardised in a general and abstract way for all victims (Arantes Prata, 2020, p. 218), but—as restorative justice promises—must be carried out for every victim specifically. On the other side, there is a problem of representation within the restorative process with regard to each victim (Forsyth et al., 2021, pp. 33–35). This is also because, especially in case of collective victims, the appointment of a representative brings with it the further question concerning the legal capacity of this figure to commit, with his/her statements and his/her conduct, the wider community for which he/she intervenes (Kershen, 2021, p. 162). The identification of a representative may be easier for more socially organised categories or those coagulated around a defined set of interests (e.g. economic operators). Difficulties will be greater for population groups aggregated on a more spontaneous basis, often expressly by the occurrence of the environmental damage. In these cases, a community analysis from a sociological point of view seems to be a necessary preliminary step before starting the repair process. This may help in order to involve an authoritative and legitimated representative for each social group characterised by its own identity.

Finally, talking about restorative justice methods and always taking advantage of restorative justice's 'versatility and flexibility' (Kershen, 2021, p. 162), in case of environmental crimes one of the most suitable tools to establish synergy with the criminal process may be considered to be *conferencing*. As a method that generates a dialogue between the offender, the victim and, more generally, the community (i.e. representative figures of the latter) (UNODC, 2020, pp. 27–29), it can create a listening space for community expectations of reparation with regard to a specific environmental offence. Al-Alosi and Hamilton (2019, pp. 1466–1469) underline the advantages of the use of conferencing as a response to environmental crimes, as it is functional both to repair environmental damage and to prevent possible recidivism.

In the wide range of restorative justice tools, we may examine, secondly, the *circle* which could become the context to involve victims in the design of a path to repair the environment damaged by the crime. Thirdly, the *victim/community impact statement* deserves attention, understood here as the recording, by the institutional body in charge of evaluating or defining the reparation activity, of reports related to the negative consequences of the environmental crime experienced by those who—as individuals or as a community—consider themselves involved in it (Mannozzi & Lodigiani, 2017, pp. 296–301).

Even if the experience gained with regard to environmental crimes documents the use of restorative justice tools mainly as 'part of the sentencing process' (Al-Alosi & Hamilton, 2019, p. 1472) and, therefore, under the judicial control as to the proportion between criminal offence and restorative activity, such a model of justice may be looked at with confidence in a broader perspective of a *virtuous interplay* with criminal law and criminal trial. Even if only on a cognitive level—to be understood, however, in a functional projection, that is, as necessarily aimed at achieving the criminal justice system objectives—restorative justice's contribution appears valuable, whenever the legal system assigns not only to the judge, but also to the public prosecutor a space of discretionary evaluation, including elements such as criminal offence seriousness and reparation adequacy.

References

Aime, M. (2019). *Comunità*. Bologna, IT: il Mulino Editore.

Al-Alosi, H., & Hamilton, M. (2019). The ingredients of success for effective restorative justice conferencing in an environmental offending context. *University of New South Wales Law Journal, 42*(4), 1460–1488.

Almassi, B. (2017). Ecological restorations as practices of moral repair. *Ethics & the Environment, 22*(1), 19–40.

Arantes Prata, D. (2020). Corporate crime and environmental victimisation: Analysis of the Samarco Case. *Revue Internationale de Droit Pénal, 91*(1), 203–223.

Berners-Lee, M. (2020). *No Planet B. Guida pratica per salvare il nostro mondo*. Milano, IT: Il Saggiatore Editore.

Bonaventura, G. (2015). La tutela dei diritti fondamentali nella Piccola Europa. *La tutela dei diritti fondamentali in Europa*. Retrieved from: http://www.adir. unifi.it/rivista/2015/bonaventura/cap2.htm (last accessed 16 August 2021).

Boscolo, E. (2021). Bonifiche e risarcimento del danno ambientale: rapporti (incerti) entro la cornice della funzione di ripristino. *Rivista giuridica dell'edilizia*, LXIV(1), 3–33.

Bricola, F. (1981). Art. 25, 2° e 3° comma. In G. Branca (Ed.), *Commentario della Costituzione* (pp. 227–316). Bologna/Roma, IT: Zanichelli Editore/ Società Editrice del Foro Italiano.

Camboni, F. (2018). La solidarietà come concetto filosofico. *Biblioteca della libertà*, LIII(221), 73–98.

Cavalla, F., & Todescan, F. (2000). *Pena e riparazione.* Padova, IT: CEDAM Editore.

Coffee, J. C. (1992). Paradigms lost: The blurring of the criminal and civil law models—and what can be done about it. *Yale Law Journal,* 101, 1875–1894.

Comazzetto, G. (2021). La solidarietà nello spazio costituzionale europeo. Tracce per una ricerca. *Rivista AIC,* 3, 258–279.

Commissione di Studio presieduta dal Pres. Dott. G. Lattanzi, *Relazione finale e proposte di emendamenti al d.d.l. A.C. 2435,* 24 maggio 2021. Retrieved from: https://www.sistemapenale.it/it/documenti/relazione-commissione-lattanzi-riforma-giustizia-penale (last accessed 16 August 2021).

Council of Europe (2000). Explanatory report to the European Landscape Convention. Retrieved from: https://rm.coe.int/16800cce47 (last accessed 15 August 2021).

Council of Europe (2008). Recommendation CM/Rec(2008)3 of the Committee of Ministers to Member States on the guidelines for the implementation of the European Landscape Convention. Retrieved from: https://rm.coe.int/CoERMPublicCommonSearchServices/DisplayDCTMContent?documentId=09000016805d3e6c (last accessed 16 August 2021).

Council of Europe (2017). Recommendation CM/Rec(2017)7 of the Committee of Ministers to Member States on the contribution of the European Landscape Convention to the exercise of human rights and democracy with a view to sustainable development. Retrieved from: https://rm.coe.int/1680750d64 (last accessed 16 August 2021).

Council of Europe (2018a). Recommendation CM/Rec(2018)8 of the Committee of Ministers to Member States concerning restorative justice in criminal matters. Strasbourg: Council of Europe. Retrieved from: https://search.coe.int/cm/Pages/result_details.aspx?ObjectId=09000016808e35f3 (last accessed 16 August 2021).

Council of Europe (2018b). Commentary to Recommendation CM/Rec(2018)8 of the Committee of Ministers to Member States concerning restorative justice in criminal matters. Strasbourg: Council of Europe. Retrieved from: https://search.coe.int/cm/Pages/result_details.aspx?ObjectId=09000016808cdc8a (last accessed August 2021).

Disegno di legge "Delega al Governo per l'efficienza del processo penale nonché in materia di giustizia riparativa e disposizioni per la celere definizione dei procedimenti giudiziari", approvato dalla Camera dei Deputati del Parlamento italiano il 3 agosto 2021. Retrieved from: https://www.sistemapenale.it/pdf_

contenuti/1628186576_ddl-2353-362792-1.pdf (last accessed 16 August 2021).

Drigo, C. (2012). Tutela e valorizzazione del paesaggio. Il panorama europeo. *Consulta Online*, 1–11. Retrieved from: https://www.giurcost.org/ (last accessed 16 August 2021).

European Commission (2020). *Improving environmental protection through criminal law. Inception Impact Assessment*. Retrieved from: https://ec.europa. eu/info/law/better-regulation/have-your-say/initiatives/12779-Environmental-crime-improving-EU-rules-on-environmental-protection-through-criminal-law_en (last accessed 16 August 2021).

European Parliament (2021). *European Parliament resolution of 20 May 2021 on the liability of companies for environmental damage (2020/2027(INI))*. Retrieved from: https://www.europarl.europa.eu/doceo/document/TA-9-2021-0259_EN.pdf (last accessed August 2021).

Fimiani, P. (2019). Inquinamento ambientale e diritti umani. In F. Buffa & M. G. Civinini (Eds.), *La Corte di Strasburgo* (pp. 376–380). Special Issue of *Questione giustizia*. Retrieved from: https://www.questionegiustizia.it/speciale/articolo/inquinamento-ambientale-e-diritti-umani_81.php (last accessed 16 August 2021).

Forsyth, M., Cleland, D., Tepper, F., Hollingworth, D., Soares, M., Nairn, A., & Wilkinson, C. (2021). A future agenda for environmental restorative justice? *The International Journal of Restorative Justice*, 4(1), 17–40.

Giunta, F. (2008). Tutela dell'ambiente (diritto penale). *Enciclopedia del diritto. Annali*, II (2), 1151–1163.

Grassi, S. (2007). Tutela dell'ambiente (diritto amministrativo). *Enciclopedia del diritto. Annali*, I, 1114–1141.

Hamilton, M. (2021). Restorative justice conferencing in Australia and New Zealand: Application and potential in an environmental and Aboriginal cultural heritage protection context. *The International Journal of Restorative Justice*, 4(1), 81–97.

Higgins, P., Short, D., & South, N. (2013). Protecting the planet: A proposal for a law of ecocide. *Crime, Law and Social Change*, 59(1), 251–266.

Hübschle, A., Dore, A., & Davis-Mostert, H. (2021). Focus on victims and the community: Applying restorative justice principles to wildlife crime offences in South Africa. *The International Journal of Restorative Justice*, 4(1), 141–150.

Jonas, H. (2009). *Il principio responsabilità. Un'etica per la civiltà tecnologica*. Torino, IT: Einaudi Editore.

Kershen, L. (2021). Restorative approaches to environmental harm: Shifting the levers of power. *The International Journal of Restorative Justice*, 4(1), 157–165.

Luhmann, N. (1990). *La differenziazione dei diritti. Contributi alla sociologia e alla teoria del diritto.* Bologna, IT: il Mulino Editore.

Marder, I. D. (2020). The new international restorative justice framework: Reviewing three years of progress and efforts to promote access to services and cultural change. *The International Journal of Restorative Justice*, 3(3), 395–418.

Mancuso, S. (2019). *La nazione delle piante.* Bari-Roma, IT: Laterza Editore.

Mannozzi, G. & Lodigiani, G. A. (2017). *La giustizia riparativa. Formanti, parole e metodi.* Torino, IT: Giappichelli Editore.

Marinucci, G., & Dolcini, E. (2001). *Corso di diritti penale* (3rd ed.). Milano, IT: Giuffrè Editore.

Marshall, C. D. (2019). Justice as care. *The International Journal of Restorative Justice*, 2(2), 175–185.

Mazzanti, E. (2020). Environmental rights and criminal protection: The dialogue between EU and ECHR. *Revue Internationale de Droit Pénal*, 91(1), 67–84.

Menéndez, A. J. (2004). La linfa della pace: I diritti di solidarietà nella Carta dei Diritti dell'Unione Europea. *Diritto & Questioni pubbliche*, 4, 95–115.

Pace, A. (2001). A che serve la carta dei diritti fondamentali dell'Unione Europea? Appunti preliminari. *Giurisprudenza costituzionale*, 46(1), 193–207.

Pali, B., & Aertsen, I. (2021). Inhabiting a vulnerable and wounded earth: Restoring response-ability. *The International Journal of Restorative Justice*, 4(1), 3–16.

Perini, C. (2004). La mediazione dei conflitti nella società del rischio. In G. Mannozzi (Ed.), *Mediazione e diritto penale. Dalla punizione del reo alla composizione con la vittima* (pp. 201–244). Milano, IT: Giuffrè Editore.

Pulitanò, D. (2021). *Diritto penale* (9th ed.). Torino, IT: Giappichelli Editore.

Sotis, C. (2007). *Il diritto senza codice. Uno studio sul sistema penale europeo vigente.* Milano, IT: Giuffrè Editore.

Stern, K. (2014). La Carta dei diritti fondamentali dell'Unione Europea. Riflessioni sulla forza vincolante e l'ambito di applicazione dei diritti fondamentali codificati nella Carta. *Rivista Italiana di Diritto Pubblico Comunitario*, 6, 1235–1260.

Stevanović, A. (2020). Environmental crime: Criminological reflections. *Revue Internationale de Droit Pénal*, 91(1), 99–112.

United Nations (2015). Resolution adopted by the General Assembly on 25 September 2015. 70/1. *Transforming our world: the 2030 Agenda for Sustainable Development.* Retrieved from: https://www.un.org/ga/search/view_doc.asp?symbol=A/RES/70/1&Lang=E (last accessed 16 August 2021).

United Nations Office on Drugs and Crime (UNODC) (2020). *Handbook on restorative justice programmes* (2nd ed.). Vienna: AT. Retrieved from: https://www.unodc.org/documents/justice-and-prison-reform/20-01146_Handbook_on_Restorative_Justice_Programmes.pdf (last accessed 16 August 2021).

4

Restorative Justice and Earth Jurisprudence

Hercules Wessels and Femke Wijdekop

1 Introduction

One of the hallmark functions of restorative justice is that it serves as a vehicle for reintegration of the wrongdoer into the community within which the wrongdoer caused harm. The reintegration function is the embodiment of the notion that a crime does not only cause harm, but it also affects and violates a multiplicity of relationships within the community in which the crime has been committed (Hamilton, 2019, pp. 197–211). If a wrongdoer is successfully reintegrated into the community through the use of a restorative justice process, the reintegration function would have served its purpose by creating the possibility for the repair of the broken or fractured relationships within the community, which were caused by the crime. In turn, humankind's broken

H. Wessels (✉)
Cullinan & Associates Inc., Cape Town, South Africa
e-mail: hercules@greencounsel.co.za

F. Wijdekop
Stop Ecocide Netherlands, Amsterdam, The Netherlands

relationship with Nature[1] or the Earth Community[2] is a hallmark of the Anthropocene.

The extent of the harm caused by anthropogenic activities to the Earth's life systems is so egregious that environmental activists and civil society organisations are advocating for the amendment of the Rome Statute of the International Criminal Court to include ecocide[3] as a crime. The rampant rate at which human activities are breaking down the integrity of Earth's biosphere is underpinned by a broken relationship between humans and Nature. Clearly, there is a need for humans to be reintegrated within the rest of the Earth Community, in order for humans to become a member with good standing, once more, and order our way of life in a manner which is not detrimental to the rest of Earth.

Proponents of Earth jurisprudence argue that the inability of current legal systems to foster a harmonious governance system in which humans have a sustainable and mutually enhancing relationship with Nature, cannot be cured by creating more laws, regulations or international treaties which aim to limit the effects of harmful human conduct on the rest of the Earth Community. This is so, because current international and domestic environmental laws are dominated by the themes of human, corporate or state possession and exploitation of Nature and the dominant aim is to protect and regulate human, corporate or state interests or rights in relation to the environment (Cullinan, 2002). Rather, a fundamental paradigm shift is required which moves us away from the current anthropocentric legal systems and interests that prioritise protection of human capital and enable dominion, exploitation and irreversible destruction of Nature, to a world where legal systems are premised on the

---

[1] During this chapter 'Earth', 'Earth Community', 'Mother Earth' and 'Nature' are used as proper nouns to signify that the use of the terms refers to specific subject and also to a legal subject. The terms are also used synonymously.

[2] 'Earth Community' denotes that the whole of Earth is based on a multiplicity of relationships between sentient and non-sentient beings on Earth and emphasises the interconnectedness of all beings on Earth. Individual humans and the human species as a whole are but one of countless other individuals, species, rivers, forests, ecosystems which form part of the Earth community.

[3] Ecocide is defined as 'unlawful or wanton acts committed with knowledge that there is a substantial likelihood of severe and either widespread or long-term damage to the environment being caused by those acts (see Stop Ecocide Foundation, 2021).

idea that humans are but one part of the Earth Community and not apart from Nature, and decisions and relationships are maintained by taking into consideration what is in the best interests of the whole Earth Community. The desired paradigm would be obtainable by adopting a legal system based on Earth jurisprudence, through which legal systems would no longer be designed to govern only anthropocentric interests and relationships. Instead, adoption of Earth jurisprudence would enable an eco-centric approach to governance, and legal systems would regulate the interests and relationships between humans and all other beings within the Earth Community.

It appears that the time is ripe for legal processes, norms and philosophies which could help humankind restore the broken, but diverse relationships it has with Earth and reintegrate the species (or at least its members who are causing the most environmental harm) into the rest of the Earth Community. This chapter explores the question of whether restorative justice and Earth jurisprudence could potentially aid the necessary restoration of relationships between humans and the Earth Community and the reintegration of our own species into the Earth Community.

Section 2 will further expand on the philosophy and principles of Earth jurisprudence and the concept of rights of Nature, in addition to the legislation, case law and policy which embody Earth jurisprudence. Section 3 will explore whether Nature has been represented in restorative justice processes and whether legislation has formally recognised or endorsed the use of restorative processes in disputes that concern the rights of Nature. Section 4 looks at similarities between the philosophies of Earth jurisprudence and restorative justice. This exploration serves as the basis for a practical proposal for the use of restorative processes to inform remedies in cases of environmental harm or infringements to the rights of Nature. This chapter is concluded in Sect. 5 by exploring the inherent tension which arises when restorative justice is applied and human beings would be required to compromise their own interests and rights in favour of the best interests of the entire Earth Community.

2   Understanding Earth Jurisprudence and the Rights of Nature

2.1   Earth Jurisprudence and the Great Jurisprudence

The concept of Earth jurisprudence was introduced in 2002 by Cormac Cullinan in his seminal book, *Wild Law: Governing people for Earth*. The concept is derived from and premised on the ten principles for the development of a new jurisprudence proposed by the eco-theologian Father Thomas Berry. In *Wild Law*, Cullinan proposes the Great Jurisprudence, from which Earth jurisprudence is also derived. The Great Jurisprudence can be understood and referred to as follows:

> The Great Jurisprudence refers to the inherent characteristics of the universe which determine how it is structured or ordered, and how different aspects relate to one another and behave. It can be understood both as a term used to describe the set of fundamental relationships and principles that constitute the Earth Community and produce the order we observe in Nature, and as a logical premise for the development of Earth jurisprudences by human societies. (Cullinan, 2021, p. 234)

Earth jurisprudence postulates that the rule of law itself is derived from, and is to be understood in the context of, the fundamental laws and principles that govern the functioning of the universe (i.e. the Great Jurisprudence) (Cullinan, 2002). Another tenet of Earth jurisprudence first proposed by Berry (1999) is that human and non-human beings obtain their rights from the same source that humans get their rights, which is the universe that brought them into being. Earth jurisprudence should be considered as an echo of the Great Jurisprudence, but confined and applicable to life and relationships on Earth.[4]

---

[4] The Great Jurisprudence applies to life on Earth as well, but it is not to be understood in terms of confining its application only to Earth. This is because the fundamental laws and principles which govern the universe, by their very nature, stretch far beyond the confines of Earth. Whereas Earth jurisprudence, which is a human interpretation of the Great Jurisprudence, is refined and attentive to the complexities of life on Earth and the Earth Community's web of relationships, which might not have the same application or relevance to life existing beyond Earth.

Earth jurisprudence is a legal philosophy which is alive to the fact that the wellbeing of humans is dependent on the wellbeing of the planet and all its inhabitants (and vice versa). Decisions must be made, and disputes must be resolved from an eco-centric viewpoint, by taking into consideration, *inter alia*, what is in the best interest of the whole and not only based on the interests of a specific species. This philosophy of law has been described as:

> Earth jurisprudence is a philosophy of law and human governance that is based on the idea that humans are only one part of a wider community of beings and that the welfare of each member of that community is dependent on the welfare of the Earth as a whole. (Cullinan, 2011, p. 13)

2.2 The Principles of Earth Jurisprudence

Earth jurisprudence seeks to enable and foster a harmonious co-existence between humans and other aspects of Nature, by allowing the rights of humans to be limited by the rights of other beings, to the extent necessary to the ensure the health and integrity of whole Earth Community. If the harmonious co-existence is disrupted, such a disruption is resolved in a manner which aims to restore the balance of the dynamic relationships and rights within the Earth Community. If the rights of a member of the Earth Community are infringed by human conduct or a member suffers, for instance, physical harm (which will be inadvertently a form of environmental harm), the methods by which the infringement or harm is addressed will be guided by the principles of Earth jurisprudence. The principles of Earth jurisprudence hold that:

* the universe is the primary lawgiver, not human legal systems;
* the Earth Community and all the beings that constitute it have fundamental 'rights', including the right to exist, to habitat, and to participate in the evolution of the community;
* the rights of each being are limited by the rights of other beings to the extent necessary to maintain the integrity, balance and health of the communities within which it exists;

* acts by humans, governments and companies which infringe these fundamental rights violate the fundamental relationships and principles that constitute the Earth Community, and are illegitimate and unlawful;
* humans must adapt their legal, political, economic and social systems to be consistent with the Great Jurisprudence and to guide humans to live in accordance with it, which means that human governance systems at all times take account of the interests of the whole Earth Community and must:

    – determine the lawfulness of human, state or company conduct by whether or not it strengthens or weakens the relationships that constitute the Earth Community;
    – maintain a dynamic balance between the rights of humans and those of other members of the Earth Community on the basis of what is best for that community as a whole;
    – promote restorative justice rather than punishment; and
    – recognise all members of the Earth Community as subjects before the law who have a right to an effective remedy for human acts that violate their fundamental rights (Cullinan, 2011, p. 13).

The legal philosophy that is Earth jurisprudence resists uniformity and rigidity, as these qualities are not reflective of the dynamic and ever-evolving nature of the interests and relationships within the Earth Community. The principles of Earth jurisprudence are there to guide the formulation or transition and implementation of any legal or governance system which claims to be founded on or influenced by eco-centric worldviews and concerns. Even though the principles of Earth jurisprudence are important, they are not an attempt to lay down uniform rules which must be followed in all communities on Earth. Therefore, Earth jurisprudence will take diverse forms depending on and influenced by different cultures and different contexts, but once accepted, the principles will likely be found being practised or used as guidance within the relevant community.

On the basis of these principles, human members of the Earth Community are able to determine whether or not their actions strengthen or weaken the rest of the web of interdependent relationships within the Earth Community. It is with the help of these principles that disputes, and conflict of rights and interests can be resolved in a manner which aims to achieve and promote Earth justice. Earth justice can be explained as a 'concern for reciprocity and the maintenance of a dynamic equilibrium between all members of the Earth community determined by what is best for the whole' (Cullinan, 2002, p. 139).

2.3   Earth Jurisprudence and the Rights of Nature

Earth jurisprudence can be realised by various legal methods and mechanisms. One of these legal methods is the recognition of the rights of Nature.[5] This mechanism can be broadly understood as the recognition of Nature as a legal person and which has specific rights of Nature, such as the right to live and restore if damaged, within a legal system, by virtue of its existence and that these rights create legal duties for other legal persons, such as individual persons, corporations and governments, to respect the rights of Nature and avoid unlawfully infringing upon these rights of Nature.[6] As an entity with rights, Nature cannot be owned or used by other legal persons as their property. This approach is contrary to the predominant legal systems across the world. The legal tools which can be used to recognise the rights of Nature are:

(a) expanding the class of legal/juristic persons to include all members of the Earth Community which have come into being as a consequence of the creation and evolution of this planet (i.e. what may be termed as 'ecological beings');

---

[5] 'The rights of Nature' as a legal tool to give effect to Earth jurisprudence should not be confused with the legal movement or organisations who are advocating for the recognition of the Rights of Nature globally, as the two concepts can be conflated when read together, and commonly are. An example of an organisation that is advocating for the transformation of legal systems into systems which are based on the philosophy of Earth jurisprudence is the Global Alliance for the Rights of Nature ('GARN'). GARN is part of the 'Rights of Nature movement' and should not be conflated with the legal tool which is the 'rights of Nature'. The rights of Nature is therefore both a mechanism and a movement through which Earth jurisprudence is realised. See GARN website: https://www.therightsofnature.org/get-to-know-us/.

[6] See also: https://www.therightsofnature.org/what-is-rights-of-nature/.

(b) recognising that all ecological beings have inherent and inalienable fundamental rights analogous to human rights;
(c) prohibiting human beings and human institutions (including artificial juristic persons such as corporations and governments) from infringing upon the rights of other ecological beings without adequate justification;
(d) imposing duties on human beings to seek to live harmoniously within the Earth Community; and
(e) establishing enforcement mechanisms (Cullinan, 2021, p. 237).

2.4   Examples: Recognising the Rights of Nature

The Universal Declaration of the Rights of Mother Earth ('UDRME'), proclaimed at the World People's Conference on Climate Change and the Rights of Mother Earth, which took place during 19–22 April 2010, contains examples of the inherent rights which Nature could be recognised to possess within a legal system. The UDRME should be understood as an instrument akin to the 1948 Universal Declaration of Human Rights, with similar legal status with respect to legal enforceability, as it is non-binding, but has the same sense of moral suasion. Whereas the UDRME addresses the rights of human and non-human beings as an integral part of Mother Earth, the Universal Declaration of Human Rights is species specific and focuses solely on the rights of humans (Cullinan, 2021, p. 241). The proclamation of the UDRME is considered a pivotal moment for the Rights of Nature movement. Article 2 of the UDRME reads as follows:

Article 2. Inherent rights of Mother Earth

(1) Mother Earth and all beings of which she is composed have the following inherent rights:

  (a) the right to life and to exist;
  (b) the right to be respected;
  (c) the right to regenerate its bio-capacity and to continue its vital cycles and processes free from human disruptions;

(d) the right to maintain its identity and integrity as a distinct, self-regulating and interrelated being;

(e) the right to water as a source of life;

(f) the right to clean air;

(g) the right to integral health;

(h) the right to be free from contamination, pollution and toxic or radioactive waste;

(i) the right to not have its genetic structure modified or disrupted in a manner that threatens its integrity or vital and healthy functioning;

(j) the right to full and prompt restoration for violations of the rights recognized in this Declaration caused by human activities;

There are, however, jurisdictions that have formally recognised the rights of Nature in either their constitution or legislation, and in other instances, Nature's rights have been recognised by a court. In 2008, Ecuador became the first country to include the rights of Nature in its Constitution, containing similar rights as those defined in the UDRME (see Chapter 7, Article 71, Ecuadorian Constitution, 2008).[7] The rights of Nature have also been recognised through municipal regulations in, amongst others, the United States of America and are also being introduced through local ordinances in Argentina.[8] Courts have made decisions in which the rights of Nature are recognised by developing the specific jurisdiction's existing laws to offer protection to the interests of Nature, as opposed to, for instance, enacting new legislation (for example, this was done in Colombia and India) (Kaufmann & Martin, 2017, pp. 130–142). In *Center for Social Justice Studies et al. v Presidency of the Republic et al.*[9] the Columbian Constitutional Court recognised Nature

---

[7] See https://pdba.georgetown.edu/Constitutions/Ecuador/english08.html (last accessed 25 January 2022).

[8] See the United Nations' Harmony with Nature Programme for a comprehensive database of countries which have either recognised the rights of Nature through legislative measures or have announced measures through which the country purports to recognise the rights of Nature, http://www.harmonywithnatureun.org/rightsOfNature/ (last accessed 25 January 2022).

[9] *Judgment T-622/16 (The Atrato River Case)*, Constitutional Court of Colombia (2016), translated by Dignity Rights Project. Retrieved from: http://files.harmonywithnatureun.org/uploads/upload838.pdf (last accessed 25 January 2022).

as a legal person by expanding the application of the people of Columbia's constitutional biocultural rights. The Court recognised specifically that the Atrato River has the rights to protection, conservation and restoration.

Several organisations and institutions have been pivotal in contributing to the Rights of Nature movement by advocating for the recognition of Nature's rights and by serving as important global repositories of knowledge pertaining to the application of the rights of Nature and its recognition, by collecting together both legally binding and non-binding instruments and laws. The Global Alliance for the Rights of Nature ('GARN') which was formed in 2010 established the International Rights of Nature Tribunal.[10] The Tribunal makes recommendations on how to address infringements of the rights of Nature, and in this manner draws inspiration from the International War Crimes Tribunal and the Permanent People's Tribunal, and acts as an international forum for parties to speak and protest destruction on behalf of Nature. Furthermore, in 2009, the United Nations General Assembly adopted the first Resolution on Harmony with Nature and from this resolution the United Nations Harmony with Nature Programme was developed, which studies and issues reports to the General Assembly on the advancements made in the recognition of the rights of Nature.

3   Restorative Justice Processes and the Rights of Nature

3.1   Representation of Trees and Rivers in Restorative Conferences

Trees and rivers have been represented as victims in four New Zealand restorative justice conferences and one Canadian restorative justice conference. However, these cases do not concern trees and rivers that have officially been awarded rights through rights of Nature legislation or

---

[10] See the Global Alliance for the Rights of Nature at https://www.therightsofnature.org/ (last accessed 25 January 2022).

court decisions. Without the support of a formal rights of Nature framework, these natural entities were recognised as victims that merited representation at the following restorative conferences:

In *Auckland City Council v 12 Carlton Gore Road Ltd and Mary-Anne Catherine McKay Lowe*,[11] and in *Rodney District Council v Sam Wong and Josh Topou*,[12] trees were cut down without the required resource consent. The trees were considered to be victims in their own right and were represented at a restorative conference, with the local council responsible for administering the laws protecting vegetation in the area acting as their representative.

In the *Waikato Regional Council v Huntly Quarries Ltd* case,[13] a river was represented at a restorative justice conference as a victim in its own right. In this case, sediment laden storm water was illegally discharged from the offender's quarry affecting the water quality of the Waikato River, a river of particular cultural significance for the local *Maori Taiui* people. The river was represented at the conference by the chairperson of the Lower Waikato River Enhancement Society. The outcome of the restorative conference was that the offender had to make a donation to the Lower Waikato River Enhancement Society instead of paying a fine. The offender complied and was discharged without conviction.

In *Canterbury Regional Council v Interflow (NZ) Ltd*,[14] underground services repair firm Interflow inadvertently had polluted Walnut Stream in Akaroa, Canterbury through the discharge of contaminants, which resulted in the death of 79 eels, 12 bullies (a fish species) and 51 whitebaits (a fish species), while also victimising the local community and the local Maori people who could no longer collect watercress from the stream. At the restorative justice conference, Walnut Stream was represented by the Maori people. Interflow requested the conference, apologised for the offence and offered to put $80,000 in trust for the

---

[11] Auckland District Court (McElrea DCJ), 11 April 2005 (see Hamilton, 2015, p. 552).
[12] Auckland District Court (Judge JP Doogue), 28 February 2005 (see Fisher & Verry, 2005, p. 59).
[13] *Waikato Regional Council v Huntly Quarries Ltd and Ian Harrold Wedding*, Auckland District Court (McElrea DCJ), 30 July 2003 and 28 October 2003 (see McElrea, 2004, pp. 13–14).
[14] DC Auckland, CRN 20050040131612, 2 March 2006, Judge McElrea (see Porfido, 2021, pp. 116–118).

remediation and restoration of Walnut Stream and two other Akaroa streams, and also reassured the participants that it would change its processes so that this manner of pollution would never happen again. Interflows' offer was accepted by the other conference participants, and the court in its sentencing process incorporated the restorative outcome agreement and discharged Interflow with no further penalty (Clapshaw, 2009).

In the Canadian case *CopCan Contracting Ltd. and the District of Sparwood (2010)*,[15] a restorative conference was held in response to the killing of 29 fish caused by the dewatering of a side channel of Michel Creek in Sparwood, British Columbia, during construction of the Elk River Pedestrian Bridge in November 2009. During the conference, community members collectively represented the interests of the river. The community members, the offending company CopCan Contracting Ltd. and the District of Sparwood discussed and agreed upon restitution for the incident. Restitution included a habitat compensation plan, riparian improvements to increase the juvenile fish-rearing habitat and a letter of apology to the local community.

3.2   Restorative Remedy for Violation of Sacred Natural Sites in Western Uganda

Currently, there are no instances of legislation which both formally recognise the rights of Nature and make provision for the use of restorative justice mechanisms as a remedy in cases where the rights of Nature are infringed.[16] However, a recently adopted district ordinance in Uganda, based on Earth jurisprudence principles, comes close.

In 2019, Uganda became the first nation in Africa to recognise the rights of Nature in national legislation under Section 4 of the National Environment Act (2019). On 22 December 2020, the Council of the

---

[15] British Columbia, Natural Resource Compliance and Enforcement Database, https://bit.ly/3tFUfgj (last accessed 25 January 2022).

[16] The authors confined their research to the database of the United Nations' Harmony with Nature Program. Available at http://www.harmonywithnatureun.org/rightsOfNature/ (last accessed 25 January 2022).

Buliisa District in Western Uganda, in collaboration with the Indigenous Bagungu People, passed an ordinance that recognises the customary laws of the Bagungu, who live along the shores of Lake Albert. The ordinance provides for the protection of an interconnected network of sacred natural sites embedded within Bagungu ancestral territory—places of high spiritual, cultural and ecological significance. The ordinance recognises the rights of the custodians of the sacred natural sites to continue to access these sites, to carry out ceremonies and to protect the sites. Furthermore, it calls for restorative justice in case an offence is committed against the sacred natural sites. Those who violate customary laws are required to make amends in ways that restore the dignity and integrity of the sacred natural sites, such as restoring damaged areas, planting trees or offering seeds.[17]

Even though the ordinance does not recognise the sacred natural sites as rights bearing entities, the restorative provision in the ordinance affirms that the sites have intrinsic rather than utilitarian value, and are considered as legal subjects rather than 'lifeless' objects. It will be interesting to see how the restorative justice provision of the ordinance will be upheld and function in practice. If successful, it could be a template for other jurisdictions that would like to make local ordinances that seek to address violations of sacred natural sites and infringements of the rights of Nature by using restorative tools.

3.3   Restorative Justice at the Ecocide Mock Trial

While restorative justice processes have not yet been recognised in official rights of Nature legislation, there has been a promising experiment with restorative processes at the 2011 Ecocide Mock Trial, organised by the late Scottish barrister Polly Higgins in the Supreme Court of England and Wales to demonstrate the viability of a law of ecocide.

---

[17] Uganda recognises Rights of Nature, Customary Laws, Sacred Natural Sites, https://www.gaia-foundation.org/uganda-recognises-rights-of-nature-customary-laws-sacred-natural-sites/   (last accessed 25 January 2022).

Ecocide, according to the definition revealed in June 2021 by the Independent Expert Panel for the Legal Definition of Ecocide,[18] means 'unlawful or wanton acts committed with knowledge that there is a substantial likelihood of severe and either widespread or long-term damage to the environment being caused by those acts'.[19] Examples of ecocide include large scale deforestation of the Amazon, oil spills in the Niger Delta and the Athabasca Tar Sands. Higgins—who passed away on 20 April 2019, aged 50, but whose pioneering work to recognise ecocide as an international crime is continued by her foundation Stop Ecocide[20]—was very interested in the use of restorative justice as a dispute resolution mechanism:

> Restorative justice offers a safe space for a CEO or company director to accept responsibility for decisions they have made which lead to Ecocide, and then to step into a restorative justice circle. There, they come together with others who represent the beings who've been harmed, and collectively they decide what can be put in place to restore the land, to mend the damage. That's the radical part of Ecocide law, offering up the tools to allow those who have made decisions which cause harm to face that harm in a healing space. Yes, accountability is essential—but it's no use just locking people up or perpetuating a culture of blame. It's about finding ways of healing, and so changing things—and people—in a more meaningful and enduring way. (Blackie, 2016, p. 56)

In the 2011 Mock Ecocide Trial, two fictional Chief Executive Officers were put on trial for causing ecocide in the Athabasca tar sands and the Gulf of Mexico. As part of this Mock Trial, Queen's Counsel and former Chair of the UK Restorative Justice Council, Lawrence Kershen, facilitated the restorative circle that was part of the Athabasca Tar Sands case. He says:[21]

> The Ecocide Restorative Justice process was very successful in demonstrating how Restorative Justice could be used in cases of major environmental

---

[18] See Independent Expert Panel for the Legal Definition of Ecocide, https://ecocidelaw.com/independent-expert-drafting-panel/ (Stop Ecocide Foundation, 2021)

[19] Legal Definition of Ecocide Completed, https://www.stopecocide.earth/expert-drafting-panel (last accessed 25 January 2022).

[20] Stop Ecocide International, https://www.stopecocide.earth (last accessed 25 January 2022).

[21] Email conversation with Lawrence Kershen, 23 April 2019.

harm, within the limitations of it being a role play. One of the major lessons was the challenge of identifying who are appropriate parties and what interests could and should be represented. So, we had participants who spoke on behalf of the Earth, the birds, Future Generations, as well as those who were more immediately harmed by the Ecocide such as the Haisla People, a First Nations group that had in reality been profoundly affected by the Athabasca tragedy. Another lesson was how it was possible to distinguish in sentencing between the CEO who declined to take part in the restorative process (who received a sentence of four years imprisonment) and the other CEO who agreed to take part, whose sentence was deferred to allow him to demonstrate that he was willing to implement the Action Plan that had been agreed in the restorative process. And it was hugely gratifying how the judge was able to incorporate the points of the Action Plan into the conditions of the Deferred Sentence—which would be a very powerful incentive for a defendant to implement the necessary action arrived at in the restorative process.

The Action Plan for the Athabasca Tar Sands case stipulated among others that the CEO of Global Petroleum Company (GPC) would set up a working group that would study the possibility of establishing a Council of Legal Interests to oversee GPC's future projects, in which representatives of future generations and the Earth would have a seat. The outcome agreement of the restorative circle also included restoration orders to remove all tailing ponds and restore the affected area to the condition which predated the pollution and damage, to promote the restoration of birdlife, fauna and wildlife within the affected area, and to suspend operations and cease all tar sand extraction (Wijdekop, 2019, pp. 106–110).

4   Integrating Restorative Justice and Earth Jurisprudence

If the environment is recognised as being a victim of environmental crime and is represented in the restorative justice process, it becomes empowered. The environment is given a voice, validity and respect. This itself is a transformative act as it recognises the intrinsic value of the environment. Justice Brian Preston (2011, p. 152)

4.1   Common Ground Between Restorative Justice and Earth Jurisprudence

While 'restorative justice today is still a strongly anthropocentric approach, thanks to its underlying relational philosophy, it has the potential, more than any other justice approach, to incorporate eco-centric perspectives, indigenous justice approaches and perspectives from the rights of Nature movement' (Pali & Aertsen, 2021, p. 5). Restorative justice approaches crime as a violation of relationships and emphasises the need for relational healing. Our relationship with the other non-human members of the Earth community is in urgent need of healing and this calls for a questioning and correction of the dominant anthropocentric worldview. Restorative justice processes allow for a wide range of values and worldviews to be expressed and because of this 'open character', it might be uniquely positioned to facilitate Indigenous perspectives and rights of Nature approaches that can challenge this anthropocentric worldview. The fact that restorative justice has Indigenous roots and uses Indigenous processes such as peace-making circles, might further assist in creating a conducive and culturally appropriate environment for the expression of Indigenous cosmologies and justice approaches while strengthening the agency of Indigenous people as spokespersons for the harmed environment (Wijdekop, 2019, pp. 47–48). Furthermore, the purpose of recognising the rights of Nature is to give legal expression to the inherent value of Nature and to cultivate a culture of harmonious relationships between Nature and humans. Restorative justice aligns with this reconciling feature of Earth jurisprudence, as it is concerned with rehabilitation of offenders through reconciliation with victims and the community at large and with the cultivation of harmonious relationships.

Representation of the harmed environment in a restorative conference creates space for eco-centric viewpoints and voices and challenges the anthropocentric bias of our legal system. Furthermore, when other life-forms are recognised as victims in their own right and represented in a restorative justice conference, this can educate the offender about the harmful effects of their behaviour on the Earth Community.

4.2   Rights of Nature Restorative Circles

Given the commonalities and potential synergy between restorative justice and Earth jurisprudence, what kind of restorative process would be suitable to rehabilitate the offender for offences against the Earth Community? We suggest the development of 'Rights of Nature Restorative Circles' pilots, to be used as part of the resolution of disputes relating to smaller scale environmental harm caused by local corporations, considering that there is a track record for successful environmental restorative justice conferencing in such cases (Wijdekop, 2019, pp. 101–102). Rights of Nature Restorative Circles could also be used by organisations or tribunals that experiment with adjudicating violations of the rights of Nature, such as the International Rights of Nature Tribunal.

Rights of Nature Restorative Circles could take place either as an alternative to prosecution or civil litigation in response to environmental harm ('diversion'), or as part of prosecution or civil litigation, in which case the restorative agreement would reduce the sentencing outcome or fine (in the event of a criminal prosecution) or the relief claimed (in the event of civil litigation). The Circle would be a restorative process, led by a trained, impartial facilitator, in which the law enforcement or regulatory agency, the offending company, community-representatives and a 'Rights of Nature-guardian' would take part on a voluntary basis, and which would result in a formal agreement recording the collectively agreed remedies and restitution.

During the Circle, the participants would discuss the cause of the harm and its effects on Nature and the affected community. All participants are given an opportunity to be heard: the offender is asked to explain how and why the offence occurred and the community members describe the impact of the incident on the community and what the offender could do to restore the ecological harm. The Rights of Nature-guardian would speak about the impact of the environmental harm on other life-forms and give voice to the most appropriate remedies and restitution from the perspective of Earth jurisprudence. The guardian could be affiliated with a Rights of Nature organisation and prepare for their role as spokesperson for the harmed beings by doing a Council of

All Beings-session (see Sect. 5). Participation in a Rights of Nature Restorative Circle would enable the guardian to not only assert the rights of Nature and educate the offender about the intrinsic value of Nature, but to also influence the political and societal debate about the value of Nature, and the deficiencies and collateral damage of the dominant anthropocentric worldview. The remedies agreed upon in the restorative agreement could include:

* public apologies;
* restoration of the environmental harm and prevention of future harm through practical safety measures and ecological and Earth jurisprudence education of the offender; and
* compensatory restoration of environments elsewhere where the original environment cannot be restored and payment of compensation to community organisations dedicated to ecosystem regeneration and protection of the rights of Nature.

If we wish to achieve a lasting behavioural change in the offender and increase their understanding of the intrinsic value of Nature and 'ecological literacy', educating the offender about the rights of Nature and having the offender participate actively in the restoration of environmental harm would be the preferred outcome of a Rights of Nature Restorative Circle. According to insights from eco-psychology, engaging in environmental restoration work can foster in offenders a sense of belonging and connectedness to the natural world and help heal the emotional, cognitive and spiritual disconnect that caused them to experience themselves as separate from the natural world and act in exploitative rather than caring ways towards the Earth Community (Van Hoek & Wijdekop, 2019, p. 23).

In the case where the offender or wrongdoer does not comply with the terms of the restorative agreement, the affiliated Rights of Nature organisation could inform the responsible law enforcement or regulatory agency, which in turn could then decide to prosecute the offence or declare void the conditional reduction or mitigation in sentence.

5 Potential Success of Restorative Justice in an Earth Community: An Unavoidable Trade-Off?

Behaving in an ethical way always means behaving in a way that is appropriate to our community. There are do's and don'ts that arise from belonging to a family, a nation, or a cultural tradition. Ecoliteracy tells us that we all belong to oikos, the 'Earth household', and therefore we must behave accordingly.
(Capra & Mattei, 2015, p. 167)

Indeed, if human society can recognise that it is part of the Earth Community and not apart from the proverbial circle of life, we would have to strive for a way of being that is appropriate and harmonious with the interests and rights of the entire Earth household. Such a recognition would necessarily be precluded by a collective human ecological renaissance. Yet, sight should not be lost of the anthropocentric nature of the act of recognising the rights of Nature, in that Nature does not need rights to survive and the concepts of rights and duties are very much human. It goes without saying that no one has seen the occurrence of an actual right in Nature, in a forest or in the depths of the Mariana Trench. The anthropocentric tension within the act of recognising the rights of Nature could strengthen the impasses which might occur when humans are required to compromise their own interests. However, as humans are atop of the dominance hierarchy currently, in so far as their activities have the most impact on the rest of the Earth and are destroying Nature at a rampant pace, there is a dire need for a system in which Nature's inherent value is recognised and additionally its rights to live, regenerate and restore. If humans are to have a harmonious relationship with Nature, we would need Nature's rights to be recognised to curb the effects of our conduct and render them blameworthy in a legal sense. Stavru's (2016, p. 7) argument in this regard is relevant and compelling:

Nature does not need rights provided by people to continue its existence. It will survive after human extinction and is able to transform itself in various ways, many of which are not only hostile but totally incompatible with existence of mankind. Nature does not need the rights of people, but people need rights of [N]ature.

We are ultimately a part of Earth and very much dependent on it. Concurrently, humankind's wellbeing, its interest and rights, is influenced by the wellbeing and health of Mother Earth and all its constituents. If sceptics cannot find reason or value in recognising the rights of Nature because it would inadvertently lead to humans having to limit the exercise of their own rights or make decisions which do not advance their own interests, such as economic, political or social interests, they should be motivated by the need for their interests to be compromised to the benefit and wellbeing of the whole Earth Community (on which human wellbeing and interests unavoidably depend). In instances of conflicts over the rights and interests of Nature with those of humans, a restorative justice process in which there is substantive, community-driven engagement between wrongdoers and the victims of environmental harm creates the potential for a discourse where the perceived tensions could be resolved. Here, a restorative justice process could lead to an understanding that the compromise of a political or economic human right or interest would not necessarily be prejudicial to the interests or rights of humans, but most surely will be to the benefit of the wider Earth Community (and indirectly the human party which has had to compromise).

The impact of the anthropocentric tension in cases of conflict of human rights and interests with those of Nature could potentially be addressed by ensuring that Nature has appropriate representation. If Nature's interests were aligned with that of its representative, or if Nature could even be seen as a co-victim of the harm also experienced by the representative, then the argument can be made that the party which represents Nature will not necessarily have to compromise its interests in order to advance the rights or interests of Nature. Indigenous cultures already have ways of being which are in line with the principles of Earth jurisprudence and the culture of the Earth Community. Due to the close relationship with Nature experienced by many Indigenous peoples, their livelihoods, interests and rights are likely to more often be directly impacted when Nature's rights are infringed, whereas urbanised populations might not perceive a harm to Nature as harm done unto themselves, particularly where the harm occurs beyond urban life. When considering representation for Nature, social and cultural context is relevant and for this reason, Indigenous people should be preferred surrogate victims and representatives for Nature due to their intrinsic identification with the environment which they inhabit and their skilfulness in listening to the non-verbal communication of Nature.[22]

However, Westerners urgently need to take responsibility for becoming 'ecologically literate' themselves. Joanna Macy, founder of the Work That Reconnects,[23] calls such reconnecting with the Earth Community an act of moral imagination. To help strengthen this moral imagination, she developed, together with Australian deep ecologist John Seed, the 'Council of All Beings'. The Council of All Beings is a communal ritual which convenes people to speak for non-human members of the Earth community.[24] It is an experience in decentring humans in order to create space for perspectives that can help humans learn how to leave narrow self-interests behind and become present to the ways in which they can care for the Earth Community beyond the present-day's anthropocentric focus (Wijdekop, 2019, p. 46). The ritual helps participants to not just intellectually understand their interdependence with the Earth Community, but to embody it through awakening their senses to receive communication from the Earth. Ecologist Pella Thiel, co-founder of End Ecocide Sweden and the Swedish Network for Rights of Nature, explains this as follows:[25]

> From an ecopsychological point of view, doing a Council of All Beings-ritual helps us to step out of the assumption that the mind is something that you have in your own head, and instead see that the mind is a web of relationships. We can tap into that interconnected psyche—in deep ecology called our 'ecological self'- and speak from the perspective of another life-form.

---

[22] The signals emanating from the natural world that denote things such as the impacts of climate change—oceans warming, unusual periods of drought and insect eggs hatching earlier.

[23] The Work that Reconnects is a methodology developed by environmental activist and author Joanna Macey, which offers exercises to guide a process of reconnection with the land, with non-human beings and with future generations, https://workthatreconnects.org/ (last accessed 26 January 2022).

[24] Sanna Barrineau, A Council of All beings—communal rituals as transformative practice, https://www.slu.se/en/subweb/mistra-ec/news/blog-posts/a-council-of-all-beings%2D%2Dcommunal-rituals-as-transformative-practice/ (last accessed 25 January 2022).

[25] Zoom conversation with Pella Thiel on 30 April 2021.

The Council of All Beings starts with a ritual opening, after which participants become receptive to connect with another life-form, for whom they will speak in the Council. This preparation involves reflecting on the being of the life-form, often by making a mask to represent the being, or by practising moving and speaking as that life-form. After these preparations, the participants gather in a circle to speak of the grave threats faced by the life-forms that they represent.[26] Dr. Michelle Maloney, Director of the New Economy Network Australia and founder of the Australian Earth Laws Alliance, uses the Council of All Beings in her work with schools and does mock trials in which schoolchildren use their imagination and empathy to represent animals. She says:[27]

> Trying to speak for [N]ature should be a mix of deep love for something, imagination, but also factual knowledge of how systems work when left to their own devices, and of what Nature needs. Australian rivers for example need something quite different than most rivers. Most rivers always have water in them, so you talk about them having a 'right to flow, a right to have water, a right to be clean'. In Australia, I would speak for the river in terms that it has an ebb and flow, times of dry (because ecosystems and species have evolved and thrive specifically in dry conditions), and big floods that flush through the system. Scientists show that there are frogs that crawl into little creek beds for years and sit inside their sacks until the water comes, to have babies. If they're permanently in water, they don't thrive. So for me it's knowing—and caring—enough about a system to speak to other human beings about what they need.

Michelle spoke on behalf of the Great Barrier Reef, which is threatened by the development of large coal mines in Queensland, during the session of the International Tribunal for the Rights of Nature in Quito, 2014. She reflects on the experience as follows:[28]

---

[26] Work that Reconnects Network, Council of All Beings, https://workthatreconnects.org/resource/council-of-all-beings/ (last accessed 25 January 2022).

[27] Zoom conversation with Michelle Maloney on 19 April 2021.

[28] Zoom conversation with Michelle Maloney on 19 April 2021.

I'm a great advocate for speaking on behalf of Nature, because it changes how you express yourself and the words you choose. It really is just empathy: putting yourself into the viewpoint of another being. Empathy is the long-forgotten skill of the European descendant, but Aboriginals have a relational ethos. In Aboriginal philosophy, law is built on empathy. From birth, children are taught that empathy and care for each other are an obligation: everybody, all beings, have to be looked after. Empathy is taught, ingrained, in their decision-making process, governance process and social norms. Aboriginals literally have an obligation to care for their country, every day.

6   Conclusion

Developing empathy for the Earth and her beings is indeed essential if modernised humans who have lost their connection with Nature want to become ecologically literate, mature and rehabilitated members of the Earth Community. Rights of Nature Restorative Circles in which other life-forms are represented can help nurture this much needed interspecies empathy and eco-centric imagination. The harmed life-form can be represented in the Circle by a Rights of Nature guardian. Having a Rights of Nature guardian in a restorative mechanism when addressing conflict of rights between Nature and humans is, however, no guarantee that the tension of an anthropocentrically driven justice system will be completely eliminated. Notwithstanding, the development of mechanisms similar to the Council of All Beings, which can be an ancillary procedure to a restorative justice process, are important so that wrongdoers, facilitators and guardians of Nature can undergo the decentring experience and make an empathetic attempt at understanding the interests, needs and pain of Nature.

It is through understanding the interests, needs and pain of Nature that the human parties to the restorative proceeding, in whichever capacity they partake, can realise that both the harm and the rights and interests of members of the Earth Community must be remediated, addressed and limited in a manner which is in the best interest of the entire Earth Community, of which they as humans are part. Such a realisation will ultimately accelerate the restoration of relationships between humans and Nature. The processes, principles and norms found in Earth jurisprudence and restorative justice complement each other. Whilst the principles in Earth jurisprudence seek to foster a harmonious co-existence between humans and Nature, the potential for empathic learning, created by restorative justice processes, creates the possibility for humans to understand how a harmonious co-existence should be maintained within the Earth Community, and consequently, how we should conduct ourselves should we decide to reintegrate and once again become a member of good standing within the Earth Community.

References

Blackie, S. (2016). *If women rose rooted: A life-changing journey to authenticity and belonging.* London: September Publishing.

Berry, T. (1999). *The great work: Our way into the future.* New York: Bell Tower.

Capra, F., & Mattei, U. (2015). *The ecology of law: Toward a legal system in tune with nature and community.* Oakland: Berrett-Koehler.

Clapshaw, D. (2009). Restorative justice in resource management prosecutions: A facilitator's perspective. Retrieved from: http://www.deborahclapshaw.co.nz (last accessed 18 January 2022).

Constitution of the Republic of Ecuador. (2008). Retrieved from: https://pdba.georgetown.edu/Constitutions/Ecuador/english08.html (last accessed 2 July 2022)

Cullinan, C. (2002). *Wild law: Governing people for earth.* Cape Town: Siber Ink.

Cullinan, C. (2011). A history of wild law. In P. Burdon (Ed.), *Exploring wild law: The philosophy of earth jurisprudence* (pp. 12–23). Kent Town: Wakefield Press.

Cullinan, C. (2021). Earth jurisprudence. In *Oxford handbook of international environmental law*, 2nd edition (pp. 233–248). Oxford: Oxford University Press.

Fisher, R.M., & Verry, J.F. (2005). Use of restorative justice as an alternative approach in prosecution and diversion policy for environmental offences. *Local Government Law Journal*, 11(1), 48–59.

Hamilton, M. (2015). "Restorative justice activity" orders: Furthering restorative justice intervention in an environmental and planning law context? *Environmental and Planning Law Journal*, 32(6), 548–561.

Hamilton, M. (2019). Restorative justice intervention in an Aboriginal cultural heritage protection context: Chief Executive, Office of Environment and Heritage v Clarence Valley Council. *Environmental and Planning Law Journal*, 36(3), 197–211.

Kaufmann, G.M., & Martin, P.L. (2017). Can rights of nature make development more sustainable? Why some Ecuadorian lawsuits succeed and others fail. *World Development*, 92, 130–142.

McElrea, F.W.M. (2004). The role of restorative justice in RMA Prosecutions. *Resource Management Journal*, 3(7), 1–15.

Pali, B., & Aertsen, I. (2021). Inhabiting a vulnerable and wounded earth: Restoring response-ability. *The International Journal of Restorative Justice*, 4(1), 3–16.

Porfido, S. (2021). The use of restorative justice for environmental crimes in the European Union's legal framework. *Queen Mary Law Journal*, 1, 106–133.

Preston, B.J. (2011). The use of restorative justice for environmental crime. *Criminal Law Journal*, 35(3), 136–154.

Stavru, S. (2016). Rights of Nature: Is there a place for them in legal theory and practice?. *Sociological Problems*, 1–2, 146–166. (Translated from Bulgarian to English by Svilen Tashev).

Stop Ecocide Foundation (2021). Independent expert panel for the legal definition of ecocide: Commentary and core text. Retrieved from: https://static1.squarespace.com/static/5ca2608ab914493c64ef1f6d/t/60d1e6e604fae2201d03407f/1624368879048/SE+Foundation+Commentary+and+core+text+rev+6.pdf (last accessed 18 January 2022)

Wijdekop, F. (2019). *Restorative justice responses to environmental harm.* Amsterdam: IUCN.

5

Nature's Rights & Developing Remedies: Enabling Substantive and Restorative Relief in Civil Litigation

Hercules Wessels

1   Introduction

We have entered the United Nation's Decade on Ecosystem Restoration, and mainstream environmental litigation is dominated by climate litigation.[1] The focus on climate litigation is justified, in the sense that international efforts to redress the excessive emissions caused by anthropogenic activities has not yielded the desired change in human behaviour, nor slowed down the anthropogenic drivers of climate change. Litigation offers the potential to set legal precedents which influence and change societal behaviour. This potential is showcased by cases such as *Urgenda*[2]

---

[1] See   https://www.unenvironment.org/news-and-stories/press-release/new-un-decade-ecosystem-restoration-offers-unparalleled-opportunity (last accessed 25 January 2022).

[2] *The State of the Netherlands (Ministry of Economic Affairs and Climate Policy) v Stichting Urgenda*, 2018 (Netherlands) (see https://www.urgenda.nl/en/themas/climate-case/).

H. Wessels (✉)
Cullinan & Associates Inc., Cape Town, South Africa
e-mail: hercules@greencounsel.co.za

© The Author(s), under exclusive license to Springer Nature Switzerland AG 2022
B. Pali et al. (eds.), *The Palgrave Handbook of Environmental Restorative Justice*,
https://doi.org/10.1007/978-3-031-04223-2_5

and *Future Generations*,[3] in which cases civil society organisations used litigation to redress infringements to their rights by their respective governments, because of their governments' failure to take action to address the effects of climate change. In the case of and *Milieudefensie*[4] Shell, a private company, was sued by civil society groups in an effort to force the company to change its behaviour, which has and is still contributing to climate change.

The same anthropogenic activities that have caused climate-related harm to the human species and present the threat of future harm have already caused enormous harm to Nature.[5] The Intergovernmental Science-Policy Platform on Biodiversity and Ecosystem Services (IPBES) (2018), in their assessment of land degradation, found that there is widespread lack of awareness of the extent of land degradation and that this low awareness is a major barrier to action by policymakers (and potentially by civil society). IPBES also confirms that ecological degradation results in the loss of biodiversity. The World Wide Fund for Nature reported in 2018 a global decline of 60 per cent in the population size of vertebrate species between 1970 and 2014 (WWF, 2018). In their seminal work, Rockström et al. (2009) identified nine planetary thresholds or boundaries within which humanity can operate safely. Loss of biodiversity is identified as one such planetary boundary, but already in 2009, the authors' research indicated that the safe boundaries of biodiversity loss were being exceeded by at least one or two orders of magnitude.

If the reported findings of the World Wide Fund (WWF), IPBES and Rockström and colleagues are anything to go by, the scale of ecological destruction that has already occurred is huge and this harm already directly impacts the well-being of humans and other species of the Earth

---

[3] *José Daniel Rodríguez Peña, et al. v. Presidency of the Republic of Colombia, et al.* (see http://climate-casechart.com/non-us-case/future-generation-v-ministry-environment   others/#:~:text=The%20 plaintiffs%20allege%20that%20climate,%2D2018%2C%20threatens%20plaintiffs).

[4] *District Court of The Hague, Milieudefensie et al. v Royal Dutch Shell PLC (26 May 2021) C/09/571932/HA ZA 19-379, English Version (Milieudefensie v RDS) para 5.3.*

[5] See footnotes 1 and 2 in Chap. 4 for explanation of the terms 'Earth', 'Earth Community', 'Mother Earth', 'Nature' 'Earth Community'. These terms are afforded the same meaning in this chapter.

Community. The development of Earth jurisprudence[6] in countries such as Ecuador and Columbia is evidence that taking an ecocentric approach to litigation has the potential to influence and strengthen relationships between humans and Nature, and promote restorative justice and ecological restoration, when the overarching policies and executive agendas of governments fail. Yet, the majority of national legal systems do not recognise the rights of Nature and it is not clear how the recognition of the rights of Nature will change legal relationships in those jurisdictions or whether these legal systems could be developed or transformed to the extent necessary to provide for appropriate remedies when Nature's rights have been infringed.

What is certain, however, is that the recognition of the rights of Nature by legal systems has *de facto* legal implications for the horizontal application of rights[7] between legal subjects, and changes the behaviour of individuals and corporations towards Nature. The legal implications occur as a result of Nature being elevated to the status of a legal subject, on par with other legal subjects, such as natural humans, who have rights which must be respected and when infringed can be redressed by seeking legal remedies. When Nature's rights are recognised, its rights must be considered by other legal subjects when deciding on the manner in which they conduct their affairs and humans, communities and corporations therefore have to align their behaviour accordingly.[8]

Private law (in the case of common law jurisdictions), also known as the law of obligations in certain jurisdictions, is the cornerstone in legal systems which in part regulates horizontal rights between legal subjects, including corporations and humans, and provides for remedies to such

---

[6] See Sect. 2 in Chap. 4, in which the meaning of Earth jurisprudence and the rights of Nature are discussed.

[7] In the context of this chapter, horizontal application of rights refers to rights which are applied against legal subjects by one another, and which rights also carry obligations and duties, such as that the rights afforded to legal subjects (e.g. right to life) which may not be infringed upon by other legal subjects. Vertical applications of rights refers to rights which must be protected by and applied against public authorities, such as government.

[8] See Sect. 4 below for the discussion of how the Columbian Constitutional Court's recognition of the Atrato River confirmed that communities and the government must enable these rights of Nature, and consequently there was a change in the horizontal application of rights as a result of the recognition of the rights of Nature.

private actors in case of infringements of their rights.[9] Should Nature's rights be recognised, the recognition will affect the manner in which private law operates in a specific legal system, and so existing remedies in these systems would also need to be developed to ensure that infringements of Nature's rights would be substantively redressed.

For example, South Africa is one of the countries where the rights of Nature have not yet been recognised. Therefore, limited consideration has been given as to the extent to which the horizontal application of rights will be affected by the rights of Nature or how existing legal remedies could be used to redress any harm to Nature's rights. Part of private law, in South Africa, is the law of delict (also known as tort law in other jurisdictions).[10] Similar to tort law, it is a body of law concerned with the obligation of a wrongdoer to compensate a victim for harm suffered as a consequence of the wrongful act or omission. There are different forms of compensation or remedies which an injured party could pursue to redress the harm to its rights, and these remedies have different outcomes. If Nature's rights were recognised in South Africa and one of those rights were to be infringed by the wrongful action of another legal subject, what remedy would be the most appropriate to compensate Nature for the harm suffered?

This chapter explores the feasibility of existing delictual (or tort) remedies, in the context of civil or private litigation, to accommodate the principles of Earth jurisprudence, restorative justice, and ecological restoration, in order to meaningfully redress the harm suffered by Nature. The principles of Earth jurisprudence, restorative justice, and ecological restorative justice form an analytical and normative framework from which the development of existing delictual remedies are considered. Section 2 stipulates assumptions and limitations for the purposes of this chapter. Section 3, with reference to work covered in Chap. 4, discusses

---

[9] Legal systems can broadly be classified as either a common law system (which originated from England during the Middle Ages) or civil law systems (as developed in continental Europe). A few countries have a hybrid legal system, containing elements of both the common law and civil law systems.

[10] South Africa, due to being colonized first by the Dutch in the seventeenth century and subsequently by the English, has a hybrid legal system, in that it incorporates elements from both the common law and civil law systems.

the overlap between the principles of Earth jurisprudence, restorative justice and ecological restoration, and why these must be forged together, to create the bedrock which could support an interdisciplinary approach to redress harm to Nature within existing legal systems and without requiring radical transformation from these legal systems. Building on Sect. 3, Sect. 4 provides insight into how the rights of Nature could be recognised in South Africa, and whether there are existing concepts in South African law which complement those of Earth jurisprudence, restorative justice and ecological restoration. Whether there are remedies in the South African law of delict which could meaningfully redress harm to Nature is discussed in Sects. 5 and 6.

2   Important Caveats

The above-mentioned questions and considerations posed in relation to the South African law of delict are equally relevant to other common law or civil law jurisdictions in which the rights of Nature have not been recognised. The existing remedies in delict or tort in those jurisdictions also need to be scrutinised to determine whether the outcome of those remedies will be able to meaningfully redress the harm suffered by Nature, and what the practical and substantive outcome of the mainstream remedies would yield for Nature when perused through civil litigation. This chapter's analysis and case study of the South African delictual remedies, is therefore also of relevance to those jurisdictions and to the larger transnational movement of the rights of Nature and environmental restorative justice. It serves as a litmus test for determining whether infringements to the rights of Nature can be substantively redressed in those jurisdictions and accordingly, how the rights of Nature can be operationalised in like national legal systems.

The South African analysis serves a further transnational purpose, in that it could also offer guidance for identifying legal remedies in those jurisdictions where Nature's rights have been recognised, but where the relevant jurisdiction's tort law or law of delict has not been developed or used to redress an infringement to Nature's rights. This especially concerns infringements suffered as a result of the wrongful conduct of a corporation or private entity, as administrative law is used mostly as a basis for challenging decisions of governments that have infringed upon the rights of Nature.

As a point of departure for the analysis, it is assumed that if the rights of Nature were recognised by a legal system, the harm to Nature would be actionable. It is important to note this assumption because the analysis does not explore whether conduct by a wrongdoer in a specific scenario would constitute a delict or tort, or whether the requirements for a specific remedy could be proved by the legal representative of Nature. A detailed and theoretical analysis of whether certain conduct would constitute a delict or tort is beyond the scope of this chapter. Rather, the substantive outcome of a remedy, and whether or not the specific remedy and its effect would result in the harm to Nature being adequately redressed, considering the principles of Earth jurisprudence, restorative justice, and ecological restoration are given importance.

3   Identifying the Bedrock: Earth Jurisprudence, Restorative Justice and Ecological Restoration

Ultimately, the purpose of recognising the rights of Nature is to give legal expression or legitimacy to the inherent value of Nature and to impose duties on humans to act in ways that are conducive to harmonious relationships and coexistence within the *Aardsgemeenskap*.[11] If the inherent value of Nature is to be protected and credence in the legitimacy of the legal system which enables its protection is to be maintained, the importance of a remedy which could substantively redress harm suffered by Nature cannot be gainsaid. What after all are rights worth if they cannot be vindicated or protected through meaningful recourse? In order to ascertain what principles or values should underlie a meaningful remedy for Nature, one should investigate the legal philosophy and its principles which create the ecocentric paradigm within which it is recog-

---

[11] Afrikaans translation for 'Earth Community'.

nised, and therefore determine whether the remedy is reconcilable with Earth jurisprudence.

The law of delict is the area of South African law which creates an obligation on the wrongdoer to compensate the party that it has harmed. Given that the general principles of the modern South African law of delict are essentially derived from Roman law, would it even be possible for these remedies in the law of delict, and the substantive relief offered by these remedies to meaningfully redress the harm suffered by Nature according to the principles of Earth jurisprudence (Van der Walt & Midgley, 2016)? They are the legal principles of an ancient Western society which was concerned with retribution and penalising wrongdoers and are seemingly at odds with the principles of Earth jurisprudence, which champions an ecocentric and restorative approach to justice (Cullinan, 2002, 2011).[12] This is a transnational juxtaposition which legal practitioners or scholars from other jurisdictions might face when they try to reconcile their own 'Western' laws with the principles of Earth jurisprudence (or environmental restorative justice).

The impression left by the principles of Earth jurisprudence confirms that there is an appreciation that the well-being of all the members of the community, as a collective, is dependent on the relationships within the community and that if relationships have been damaged and an individual member has suffered harm, the damage is not remedied by focusing only on the relationship between the wrongdoer and the victim. Punishing or chastising the wrongdoer for their unlawful conduct and focusing on the harm of the individual, which was an approach followed by the Romans (Van der Walt & Midgley, 2016), or imposing a retributive form of punishment on the wrongdoer, as is the case with most criminal law sanctions, does not adequately take into account the damage done to the relationships within the community nor does it create the possibility for reconciliation between the wrongdoer, victim and community. Neither would the damage to the relationships within the Earth Community be adequately redressed if the wrongdoer only provided monetary compensation to the victim or multiple parties.

---

[12] See Chap. 4, Sect. 4.1 for a discussion on the principles of Earth jurisprudence, and how it overlaps with restorative justice.

A remedy which aims to punish, take revenge on, or deal a blow to the pockets of the transgressor for their actions is not in the spirit of Earth jurisprudence. By contrast, harm in a society governed according to the principles of Earth jurisprudence is remedied with an emphasis on restoring the damaged relationships between members of the community and maintaining the dynamic equilibrium within the Earth Community as a whole. The harm is not remedied by placing an emphasis on either retribution or compensation.

In seeking redress for harm to Nature, a different, non-conventional, or even 'Wild' approach should be followed to allow those harmed or those acting on behalf of environment/other species that have been harmed to pursue a remedy which both heals Nature's physical harm and restores the equilibrium within the entire community. Wild law seeks 'to foster passionate and intimate connections between people and Nature' (Cullinan, 2002, p. 10). The remedy must share the same focus as wild law, being a focus 'on relationships and the processes by which they can be strengthened, than on end-points and "things" like property' (Cullinan, 2002, p. 10). Are there ideas, principles or processes (applied in modern Western law) which have features that would enable the legal representative of Nature to seek a wild remedy, when instituting legal action, to redress the harm suffered by Nature? The concept of restorative justice appears to be the answer to this important question.[13]

A restorative justice approach to redressing harm to Nature makes sense, as it is concerned with reintegrating the offender into society as much as it is with repairing relationships between victim and offender (the one enforcing the other). Earth jurisprudence incorporates these concerns in its foundational principles, principles which serve as a framework for maintaining the dynamic web of relationships in the Earth Community, regardless of whether an offence was criminal or civil in nature. Restorative justice is therefore not only an apt way of practically implementing a remedy which is in line with Earth jurisprudence; it also buttresses the principles of Earth jurisprudence. The normative overlap between restorative justice and Earth jurisprudence is to be expected,

---

[13] See Chap. 4, Sect. 4 for a discussion of the similarities between Earth Jurisprudence and restorative justice and why remedies, which are embedded in restorative justice values and practices, enable meaningful recourse to the rights of Nature.

since they both are aimed at restoring relationships, rather than punishing an individual. Restorative justice is part of the answer as to whether remedies of Earth jurisprudence could find application in a Western legal system and redress such harm. Are there further theories or practices which could be relevant to forging a remedy for Nature in a Western legal system?

In practice, Nature's attorney would have to present a case to an adjudicator which shows evidence of the harm and explain how the harm should be redressed. Sight should therefore not be lost of the need to also incorporate evidence of the harm caused and scientific and biological processes of how to restore the harm. The Society for Ecological Restoration (2019, pp. 15–16) defines ecological restoration as the 'process of assisting the recovery of an ecosystem that has been degraded, damaged, or destroyed' and a discipline which 'aims to move a degraded ecosystem to a trajectory of recovery that allows adaption to local and global changes, as well as persistence and evolution of its component species'. Ecological restoration also focuses on repairing ecological damage and rebuilding a 'healthier relationship between people and the rest of nature' (ibid., p. 11). Ecological restoration's focus on restoring relationships is aligned with the principles of Earth jurisprudence and restorative justice. From an environmental restorative justice perspective, ecological restoration is a perfect addition as a legal remedy for redressing harm caused to Nature, to require the offender to partake in the restoration process where they were previously engaged in degrading the environment. This could assist in preventing recidivism.

By participating in, for instance, a restorative justice conference, the wrongdoer and the representative of Nature, in addition to other affected parties within the community, can explore with the assistance of ecological restoration experts what ecological restoration measures would be best suited to redress the harm to Nature. Based on the advice of the ecological restoration experts, it can be decided what must be done to restore the equilibrium between all the members of the *Aardsgemeenskap*.

As described in the introductory section, Nature has suffered and is currently suffering extensive harm, which is symptomatic of humans' relationship with Nature. When viewed together, the principles of Earth jurisprudence, restorative justice, and ecological restoration, or 'ERE',[14] could inform a remedy that not only substantively redresses harm to Nature, but could also reintegrate humans as a species into Earth's ecosystem. Such reintegration would be aimed at putting us on a trajectory of recovery, reconciling humans with the entire Earth Community, before our behaviour causes us to exceed all the planetary boundaries. ERE is therefore the bedrock from which a meaningful remedy for Nature can be developed.

4   Nature's Rights in South Africa, Restorative Justice and *Ubuntu*

The importance of recognising the rights of Nature at the appropriate level of government must not be underestimated, as this will determine whether the rights of Nature will be appropriately balanced or even limited by the rights of humans or executive actions by governments, especially so in the context of civil litigation. In Ecuador, constitutional lawsuits have often been based on a need to overturn laws or executive orders that place an unjustifiable limitation on the rights of Nature (Kaufmann & Martin, 2017). In Bolivia, the rights of Nature were recognised through legislation and it also included the specific right to the restoration of living systems affected by human activities directly or indirectly.[15] In Colombia, the Constitutional Court recognised Nature as a legal subject by expanding the application of its people's biocultural rights, a special category of rights recognised in the Colombian Constitution that unifies ethnic communities to a specific part of Nature, to the Atrato River.[16] The court order recognised the Atrato River as an entity and a subject with rights (i.e. a legal person), such as the rights to protection, conservation and restoration by the Colombian state and its

---

[14] 'ERE' is a preposition meaning before or in time but is also used in this chapter as an acronym for 'Earth justice, Restorative justice and Ecological restoration'.

[15] Article 7(6) of the Law of the Rights of Mother Earth, Law 071 of 2010.

[16] *Judgment T-622/16 (The Atrato River Case)*, Constitutional Court of Colombia (2016), translated by the Dignity Rights Project. Retrieved from: http://files.harmonywithnatureun.org/uploads/upload838.pdf (last accessed 25 January 2022).

ethnic communities. At this juncture, it is important to note that a recurring theme in the recognition of Nature's rights is that it has a right to be restored if harmed. This is a clear right which can be infringed. Despite there being a common approach amongst legal jurisdictions that have started to undergo their respective ecological renaissances, in that Nature has a right to restoration if it is harmed, there is no international consensus of what the best way is of redressing this harm and which remedies would allow Nature to restore its vital life cycles.

If Nature's intrinsic worth and rights were expressly recognised in South Africa, it would be appropriate to afford its rights constitutional protection, which is the highest form of protection. As was the case for Nature's rights in Ecuador, constitutional protection would allow the values which are behind such an act of recognition to take root in all other areas of the law (2008). The South African judiciary and its reasoning in case law have shown glimmers of a future where the values of Earth jurisprudence can become commonplace in South African law as well as placing desperately needed limitations on the rights of humans. For instance, in the case of *National Society for Prevention of Cruelty to Animals v Minister of Justice and Constitutional Development and another*,[17] the Constitutional Court considered cruelty to animals within the context of the constitutional values of South African society.[18] In addition to case law, there is also evidence that the concept of *ubuntu* in South African customary law and African cultures would be able to provide a foothold for the values of Earth jurisprudence and restorative justice.

Just like wild law, the concept of *ubuntu* resists a uniform definition in legal terms (Cullinan, 2002, p. 10; Himonga et al., 2013). *Ubuntu* has been described as a 'unifying vision of community built upon compassionate, respectful, interdependent relationships' and serves as 'a rule of conduct, a social ethic, the moral and spiritual foundation for African societies' (Swartz, 2006, p. 560). Similar to restorative justice, *ubuntu* strives for reconciliation and reformation instead of punishment or retribution. *Ubuntu* places an emphasis on the values of community, mutual

---

[17] 2017 (4) BCLR 517 (CC), para 56. Retrieved from: http://www.saflii.org/za/cases/ZACC/2016/46.html (last accessed 31 January 2022).
[18] Section 24 of the Constitution of the Republic of South Africa (1996).

enjoyment of rights by all members of the community and the ideas of sharing and co-responsibility.[19] It is easily discernible that there are overlaps between the values of *ubuntu*, Earth jurisprudence, and restorative justice.

In their study of the South African judicial views of *ubuntu*, Himonga et al. (2013) tracked the application of restorative justice and *ubuntu* in judicial decisions when a conflict of rights had occurred. They found that in many instances, the application of *ubuntu* was arguably applied under the guise of restorative justice, or vice versa. Traditionally, restorative justice was applied in the context of criminal law, but it has also found application in other areas of South African law where remedies are pursued through civil litigation. Specifically, the authors found that *ubuntu* and restorative justice influenced judicial decision making where conflicts of rights occurred as a result of an eviction or defamation.[20, 21]

In the case of *Dikoko v Mokhatla*, where the wrongdoer was found liable for defaming a party and causing injury to the plaintiff's dignity, the lower court ordered the defendant to pay damages as a form of compensation to the plaintiff (Himonga et al., 2013). The learned judges of South Africa's Constitutional Court heard the matter on appeal and upheld the High Court's decision that the party was defamed, but did not agree on the manner of compensation. The Constitutional Court invoked *ubuntu* and decided that the order for damages by the High Court was not appropriate. The honourable Judge Mokgoro found that in the context of compensation for defamation, the goal of the remedy should not be to 'enlarge the hole in the defendant's pocket' but rather that the compensatory damages are 'intended to restore the dignity of the plaintiff, rather than punish the defendant' (Himonga et al., 2013, p. 403).[22]

The honourable Judge Sachs viewed an apology to be the most appropriate way to redress the plaintiff's injured dignity in this instance. He

[19] *S v Makwanyane* 1995 3 SA 391 (CC) para 224. Retrieved from: https://lawlibrary.org.za/za/judgment/constitutional-court-south-africa/1995/3 (last accessed 31 January 2022).

[20] *Port Elizabeth Municipality v Various Occupiers* 2005 1 SA 217 (CC). Retrieved from: https://collections.concourt.org.za/handle/20.500.12144/2209 (last accessed 31 January 2022).

[21] *Dikoko v Mokhatla* 2006 SA 235 (CC). Retrieved from: https://collections.concourt.org.za/handle/20.500.12144/2489 (last accessed 31 January 2022).

[22] Ibid., para 68.

found that the 'post-apartheid constitutional ethos demands a move away from a preoccupation with monetary awards in the law of defamation to a flexible and "broadly-based" approach that promotes the restoration of social harmony and "interpersonal repair"' (Himonga et al., 2013, pp. 403, 105). Awarding remedies to plaintiffs who seek monetary compensation for their injured feelings is inappropriate, does not promote social harmony, and reinforces a community culture where dignity, reputation, and honour are commodified (ibid., p. 404). Such remedies can be argued to be counterintuitive to the spirit of *ubuntu*. Likewise, solely seeking monetary rewards to compensate Nature for the harm it has suffered would not be consistent with ERE.

5   Feasibility of a Remedy in the Law of Delict to Act as a Conduit for ERE

An amendment to South Africa's Constitution and the inclusion of Nature's rights in the Bill of Rights, would place a duty on the South African courts to apply and develop the common law to give effect to Nature's rights (to the extent that legislation does not do so) and also to limit other rights contained in the Bill of Rights (such as human rights), provided that the limitation is a justifiable one.[23] The scope of the definition of legal persons who are considered to be the holders of subjective rights,[24] as understood in the discipline of private law, will also have to be expanded to include Nature, so that private law is sensitive to the recognition of Nature as a subject with rights and that its subjective rights are aligned with its fundamental rights in the Constitution.

As an example of such a re-orientation, as a subject with rights, if Nature suffers harm, its representatives should be able to turn to the law of delict as a source of recourse and hold the wrongdoer accountable. The law of delict is the source of obligations, and when one legal subject

---

[23] Constitution of the Republic of South Africa (1996, article 8(3)).

[24] A subjective right is a legally supported claim by a legal subject to a legal object by virtue of which the subject acquires particular powers regarding the object. Additionally, any third party is legally obliged to respect the subject's powers and claim to the specific object. See du Plessis (1999, p. 137).

wrongfully causes another legal subject harm, the subject which caused the harm has an obligation to compensate the injured party, and the injured party has a personal right against that party for compensation for the harm inflicted on it. Delictual remedies and substantive relief for a wrongful harm, in the context of ERE, will be discussed further below.

5.1   Remedies in the Law of Delict

The two generic remedies available in the law of delict to an injured party are either an action for damages or an application for an interdict (or an injunction) restraining the wrongdoer from committing or continuing with the wrongful conduct (Midgley, 2016). The action for damages will vary depending on the form of harm that was incurred or the loss suffered. For instance, if the plaintiff suffered patrimonial loss, the plaintiff would be able to recover patrimonial damages, but the plaintiff must be able to quantify and prove the amount of damage. If a victim, however, suffered an injury to his or her personality, the plaintiff would also claim a monetary amount, but this would be awarded as sentimental damages. The claim for damages therefore results in monetary compensation being awarded to the aggrieved party for the harm it has suffered.

When seeking an interdict as a form of remedy, it can either be an interim or final interdict. The purpose of an interim interdict is to stop further harm from occurring, but only while the dispute between parties is being settled (Midgley, 2016).

Although interdicts are normally used to either compel the wrongdoer to take a certain action or restrain it from continuing with a form of harmful conduct, interdicts have also been used to simultaneously compel a wrongdoer to take certain actions and refrain from doing others whilst being overseen by the court which grants the order to this effect.[25] An interdict taking this form is called a structural interdict. As part of the structural interdict, the wrongdoer could be required to report back to

---

[25] *Hichange Investments (Pty) Ltd v Cape Produce Company (Pty) Ltd t/a Pelts Products and others* [2004] JOL 12538 (E). Retrieved from: https://www.informea.org/en/court-decision/hichange-investments-pty-ltd-v-cape-produce-co-pty-ltd-ta-pelts-products-and-others (last accessed 31 January 2022).

court at certain intervals. For instance, the wrongdoer could be required to report to court when it has been able to prevent the harm from continuing and at a subsequent stage, when the wrongdoer has been able to restore the harm it has caused. Theoretically, a structural interdict creates the possibility of the wrongdoer being instructed to complete multiple tasks.

The South African Constitution, being supreme, could be used to develop new remedies in the law of delict or to expand the application of existing remedies if the remedies were not appropriate to redress a certain harm. The Constitutional Court expressed and confirmed this view in their reasoning in *Dikoko v Mokhatla*.[26] Whether or not the existing remedies in the law of delict are appropriate to meaningfully redress physical harm to Nature and the impairment of its rights to restoration, will be determined by the remedy's compatibility with ERE.

5.2   Damages and ERE

Van der Walt and Midgley (2016) maintain that the purpose of delictual damages is to compensate the person who has suffered harm. When claiming damages for physical harm, the sole purpose of the damages is to restore the injured party's patrimony (or financial situation) through monetary compensation, to the position in which it would have been if the delict had not occurred.[27] Van der Walt and Midgley (2016) also state that damages awarded in this form 'do not serve to assuage wounded feelings or to compensate feelings or to compensate for mental suffering'. Indeed, that statement is correct and when viewed in isolation, it is possible to see that there are different actions for every form of harm which a legal subject might suffer, such as physical harm, pain and suffering, and injury to personality interests. But these traditional delictual actions are all aimed at paying money to the injured party as compensation.

---

[26] 2006 SA 235 (CC).

[27] *Union Government (Minister of Railways & Harbours) v Warneke* 1911 AD 657 662. Retrieved from: https://www.coursehero.com/file/71589066/Union-Government-v-Warnekepdf/ (last accessed 31 January 2022).

The fact that the law of delict allows a claimant to pursue damages as a remedy for all these forms of harm reflects that Western society, for the most part, viewed monetary compensation as adequate to assuage any grievance. We have been able to reduce different forms of conduct which cause harm into categories and view them in isolation, and only focus on remedying harm to the plaintiff, regardless of what the harm might have caused to the other relationships within the communities to which the party belongs. The non-physical injuries have been ascribed monetary values, even when money is not able to remedy the harm caused.

A remedy which would accord with ERE would need to create the possibility for reconciliation between Nature and the wrongdoer, and a damages claim potentially has the opposite effect, since the offender is not required to reconcile itself with the harm it has caused. On this count alone, damages as a remedy are at odds with the principles of ERE. Further, the remedy is sought for the individual, in isolation, and does not address the damaged relationship between the wrongdoer and the victim, or the greater community within which the wrong and harm occurred. Should damages only be claimed for a specific part of Nature and have no stakeholder engagement process to assess the harm to rest of the community, the real extent of the harm might not be redressed, and such substantive relief would not be aligned with ERE.

Consideration should also be given to the following: for repeat environmental offenders, paying damages is part and parcel of their way of business, be it in the form of large damages or monetary claim by aggrieved legal subjects or an administrative fine paid to an environmental authority. Yes, the hefty sum of money which a repeat environmental offender must pay to Nature might be able to sustain an ecological restoration campaign, but if the offender is not presented with an opportunity to engage with the harm which it has caused, there is little room to rebuild a healthier relationship between the people and Nature. Damages would therefore not present an opportunity for the wrongdoer to educate itself about its harmful conduct towards Nature, thereby failing to create change in the behaviour of the wrongdoer, which could help prevent future harm to Nature.

A remedy which only requires the wrongdoer to pay monetary compensation for the harm it has caused Nature, is arguably a reiteration of the polluter pays principle. Critics of the polluter pays principle argue that, for offenders, it is often less costly to pollute the environment and pay the fine than to implement the measures to monitor harm to the environment or to change their behaviour in ways which would help prevent future degradation. The principle fails because its focal point is on collecting damages instead of curing the behaviour that causes the degradation (Ingwani et al., 2010). If claiming damages amounts to an indirect form of the polluter pays principle, the transformative potential of restorative justice is lost.

The end result of a damages claim ending solely in monetary compensation risks meaning that behaviour which degrades the relationships between members of the Earth Community would be tolerated, and resulting in compensation that does not address damaged relationships. The damages paid might be utilised for restoration, but the degradation will continue. At its core, a sole claim for damages as a remedy to redress harm to Nature, is not reconcilable with the principles of ERE and the outcome of a successful damages claim would not meaningfully redress the harm to Nature within the Earth Community.

5.3   Could an Interdict be ERE?

In *Wild Law* (2002), Cullinan states that when human conduct does harm to a part of Nature or a non-human part of the Earth Community, requiring the payment of monetary compensation 'does not address the fundamental issue of how to restore healthy whole-maintaining relationships in that particular context' (Cullinan, 2002, p. 133). An adjudicator with a Western legal education, who has to decide on a conflict between the rights of Nature and a human or a corporation, might be ill-equipped to address this fundamental issue identified by Cullinan. The judge might avoid making any decision as to how the physical harm to Nature and damaged relationships between the parties and other stakeholders could be healed. A skilled litigator acting for Nature might help the overwhelmed judge by requesting a remedy in the form of an interdict, to not only stop the harm to Nature from further occurring, but that there also be a process followed, in accordance with the principles of ERE, to establish which is the best way to restore the physical harm inflicted on Nature and the relationships in the community. Such a process has the potential to provide appropriate substantive relief which will truly redress harm to Nature and its injured rights.

If the application for a structural interdict is granted, a judge could, for instance, order that one of the restorative justice processes be followed, and require the parties to report back to the court what the outcome of the restorative justice process was.[28] It has been shown that at restorative conferences, there are multiple outcomes and these outcomes are more widely accepted by the affected community, and regarded as legitimate, because of the engagement with a wider range of stakeholders, as opposed to just the injured party, the wrongdoer and the judge (as is the case for normal court proceedings). Outcomes of restorative justice conferences could include creating collaborative plans for the restoration of harm to the environment and prevention of future harm, apologies by the wrongdoer for the harm which it has caused, and compensatory restoration of parts of the environment elsewhere if the harmed part of the environment cannot be restored to its former condition (Wijdekop, 2019, p. 63). Another possibility is for the participants of a restorative justice conference to prepare a restorative justice agreement, and in cases where the wrongdoer is an organisation or the state, the agreement can include measures for the wrongdoer to raise cultural awareness and the skills of its people and educate them in the practices of ecological restoration.

If a structural interdict is imbued with the participatory nature of ERE, an opportunity for meaningful dialogue between all stakeholders is created. This dialogue is important in the process of healing the harm to Nature and the damaged relationships in the Earth Community, and is in accordance with the principles of ERE, because the harm is not viewed in isolation. A one-sided narrative of how the harm should be healed would not sit well in a dynamic Earth Community. Considering ERE, Nature's

---

[28] In Sect. 3 of Chap. 4, examples of restorative justice processes are provided, which could be implemented when Nature's rights are infringed.

harm will only be substantively redressed by a process based on broad stakeholder engagement, where collective decision making and co-responsibility is key, utilising both scientific and traditional knowledge. It would be for the court, as the final decision maker, to satisfy itself that the parties have abided by the terms of the structural interdict. Deviance from the process by wrongdoers can be disincentivised by fear of falling in contempt of court, and the fact that the court could hold parties accountable to perform in terms of the structural interdict, which gives the process legitimacy.[29]

6   The Final Verdict

Earth jurisprudence is concerned with maintaining and strengthening rela-
tionships between all members of the Earth Community and not just between
human beings. (Cullinan, 2002, p. 134)

Claiming damages as a method for redressing harm to Nature and repairing the damaged relationships caused by the unlawful conduct pales in comparison to an interdict, considering the potential which an interdict and specifically a structural interdict offer.

Practically, it would make sense for Nature's litigator to pursue an interdict as a remedy to redress harm to Nature, even an interim one, since this could stop the harm immediately until a final interdict is granted. By contrast, a claim for damages on its own could not halt the harm to Nature, the claim only assesses damage to Nature in monetary terms and from a certain point in time, resulting in a finite and isolated view of the harm which was caused and how it should be healed, which is at odds with ERE. For these reasons, a remedy in which only an interim interdict is requested, pending the outcome of a dispute in which solely monetary damages are claimed, will not be appropriate. Moreover, a conventional claim for damages has the potential to endure for years, especially when encountering an adversarial defendant who instructs their

---

[29] In South African law, failing to comply with an order of court is regarded as an offence. This offence is referred to as 'contempt of court'.

legal counsel to delay the litigation process at every opportunity, potentially resulting in the relief sought by Nature being granted far too late due to infliction of irreversible harm. In such a case, the dispute of rights will be left moot. A structural interdict would also allow public participation in the healing process and could be more effective for building best ecological practices targeted at remedying the breadth of injuries to Nature's constituent parts.

As mentioned, a claim for damages is finite and formulated at a particular point in time. It may be estimated, for example, that ecological restoration will cost €2 million, but when formulating the price of that restoration, the estimation is made at a certain point in time based on certain assumptions, causing the relief to be based on a linear trajectory of events. However, non-linear events (such as the COVID-19 pandemic) could interrupt the process of ecological restoration and the original amount claimed for the ecological restoration process might turn out to not be enough anymore. A process of ecological restoration that must be kept sacrosanct, as a result of a limited budget (derived from the monies claimed as damages), would not withstand the muster of the reality of non-linear environments within which ecological restoration must take place. Instead, a remedy which aims to heal harm to Nature and the interdependent relationships which are to be found in the Earth Community should be dynamic and robust, so that it can be resilient in the face of devastating non-linear events. Claiming damages at a fixed point in time cannot result in the harm to Nature and the relationships in the EarthCommunity being redressed, as it would not be possible to quantify the impact of such events, or the costs of implementing the ecological restoration, or the costs of restoring the damaged relationships.

It would also appear to be a rather misplaced, anthropocentric attempt to try to quantify the actual harm to Nature's person or its intrinsic value. The reasoning in *Dikoko v Mokhatla* confirms that even in the case of humans, not all harms and wrongs should be resolved by requiring the wrongdoer to pay monetary compensation. It is rather strange that when we humans defame one another we place a monetary value on the infringement of our right to dignity or a good name. Certainly, there is an argument to be made for when the defamation causes calculable loss in business revenue due to false information being spread, but why do humans and even, non-human corporations, in addition to monetary compensation require an apology from the alleged defamer? The social value of the apology vindicating the defamed person's name in public is simply not quantifiable. Apology repairs the person's dignity and his or her good name can now be maintained. Claiming payment of a hefty sum of compensation would not redress the harm felt or restore the relationship between the parties or their respective relationships with the rest of society. This specific demand highlights the shortcomings of placing a monetary value on the infringements of rights concerning personhood and dignity—there are some things money just can't fix.

Why then, in the case of a river having been polluted and harmed, would it seem most appropriate for the river (or Nature) to accept monetary compensation for the physical damage it had suffered and the resultant impairment of its riverhood? (Cullinan, 2002, p. 132) Would it be appropriate to define harm to the river's riverhood in monetary terms as we do, rather misguidedly on our own personhood? Would monetary compensation paid to adress the aggrieved river's rights be adequate? Unavoidably, some of the ecological restoration projects are going to be costly when employed at large scale. A restorative justice approach cannot be naïve to the monetary costs of large-scale ecological restoration projects. One of the world's most renowned large-scale ecological restoration process, the Loess Plateau restoration project, had an approximate budget of $500 billion (Commonland, 2016).

If we do try to restore relationships with Nature (when its rights have been infringed) by only paying monetary compensation, there will be less of an opportunity for those who are causing the harm to change their behaviour. Monetary compensation will only be able to redress harm in isolation from the damaged relationships around it. This would be counterintuitive to the purpose of addressing the interests of Nature through an ecocentric legal system and Earth governance approach. To achieve the envisioned behavioural and value change, we must pursue remedies which allow room for the restoration of relationships between members of the Earth Community, instead of only focusing on alienating offenders and polluters with the fear of financial retribution via a claim for damages.

Invariably, both remedies under consideration (damages and an inter-dicts), if granted by a judge, would have financial implications for the wrongdoer. However, with an action for damages the wrongdoer is required to *pay* to compensate for the harm (which compensation can be used to restore an ecosystem), but with an application for an interdict, the litigator is asking the wrongdoer to *do* something about the harm. When requesting the wrongdoer to stop the harm and take specific actions to restore the damage caused by the harm, by means of an inter-dict galvanised by the principles of ERE, it would be easier to ensure that the claimant's relief is integrated into what is in the best interests of the whole Earth Community. Nature's litigator should understand that the success of the remedy granted should be measured by its effect on the system as a whole, instead of only assessing the effect on the individual claimant and wrongdoer.

An interdict, preferably a structural interdict, influenced by ERE prin-ciples offers the potential to give effect to one of the central themes of a legal system influenced by Earth jurisprudence, which is 'a concern for reciprocity and the maintenance of a dynamic equilibrium between all the members of the Earth Community determined by what is the best for the system as a whole (Earth justice)' (Cullinan, 2002, p. 139). An inter-dict as applied in South Africa, and injunctions or interdicts in other legal systems, could act as a conduit for ERE. This legal approach is the most appropriate method of redressing harm to Nature and the damaged rela-tionships, because it does not necessarily purport to be the substantive relief which will directly heal the harm to Nature and the damaged rela-tionships, but instead makes the healing process possible.

There is enough evidence of the harm which Nature has suffered, caused by polluters like the Carbon Majors, which merits vindictive and retributive actions. However, it appears that such actions would be con-trary to the principles of the very philosophy which recognises the intrin-sic value of Nature in the first place. In so far as humans' reintegration into the Earth Community goes, we are at a crossroads. We can either paint all human and corporate environmental offenders as criminals, prosecute them and alienate them even further by trying to financially ruin them through claiming vast amounts of damages, or we can choose to include these offenders and have them take direct responsibility for the harm, in a true *ubuntu*-esque attitude, strengthening the relationships between all members of the Earth Community.

With the growing transnational support for the recognition of the rights of Nature, the time seems ripe for ecocentric perspectives to shape restorative responses to environmental crimes, and as this chapter's legal feasibility study has shown, these perspectives could find application in civil litigation as well. What is for certain is that we will not achieve the desired behavioural change and reintegration of humans into the Earth Community if we aim only to claim damages and recover pecuniary compensation from those who are complacent with being in the Anthropocene, instead of demanding their action and co-operation to break out of it.

In bringing wrongdoers to the restorative party, Nature's legal representatives should peruse a remedy in the form of an interdict or injunction to stop further harm from occurring while the Nature's legal representatives, affected members of the Earth Community and wrongdoers have a meaningful dialogue on how to redress the physical harm to Nature, through ecological restoration, but also on what is in the best interests of all members of the Earth Community. Such resilient ERE-inspired remedies implemented at the appropriate transnational scale, could help integrate humans back into the Earth Community ere an irreversible collapse of Earth's life systems occurs.

References

Commonland (2016). 4 returns from landscape restoration: A systemic approach to restore degraded landscapes. Retrieved from: https://www.iucn.org/sites/dev/files/import/downloads/commonlandpublicationv1lr_269496814.pdf (last accessed 31 January 2022).

Constitution of the Republic of Ecuador (2008). Retrieved from: https://pdba.georgetown.edu/Constitutions/Ecuador/english08.html (last accessed 25 January 2022).

Constitution of the Republic of South Africa (1996). Retrieved from: https://www.gov.za/documents/constitution-republic-south-africa-1996   (last accessed 31 January 2022).

Cullinan, C. (2002). *Wild law: Governing people for earth.* Cape Town: Siber Ink.

Cullinan, C. (2011). A history of wild law. In P. Burdon (Ed.), *Exploring wild law: The philosophy of earth jurisprudence* (pp. 12–23). Kent Town: Wakefield Press.

du Plessis, L.M. (1999). *An introduction to law* (3rd ed.). Cape Town: Juta & Co Ltd, Kenwyn.

Himonga, C., Taylor, M., & Pope, A. (2013). Reflections on judicial views of Ubuntu. *Potchefstroom Electronic Law Journal*, 16(5), 370–427.

Ingwani, E., Gondo, T., & Gumbo, T. (2010). The Polluter Pay Principle and the damage done: Controversies for sustainable development. *Economia. Seria Management*, Faculty of Management, Academy of Economic Studies, Bucharest, Romania, 13(1), 53–60.

IPBES (2018). The IPBES assessment report on land degradation and restoration. Montanarella, L., Scholes, R., & Brainich, A. (eds.). *Secretariat of the Intergovernmental science-policy platform on biodiversity and ecosystem services.* Bonn, Germany.

Kaufmann, G., & Martin, P. L. (2017). Can rights of nature make development more sustainable? Why some Ecuadorian lawsuits succeed and others fail? *World Development*, 92(C), 130–142.

Midgley, J. R. (2016). The law of South Africa (LAWSA), Delict (Volume 15—Third Edition). Retrieved from: https://www.mylexisnexis.co.za/Index.aspx?permalink=TEFXU0EgLSBWb2wgMTUoM2VkKSBQYXJhIDE4OCB mbiA5JDg0MDg0ODAkNyRMaWJJyYXJ5JEpEJExpYnJhcnk (last accessed 22 May 2020).

Rockström, J., Steffen, W., Noone, K., Persson, Å., Chapin, III, F. S. Lambin, E., Lenton, T. M., Scheffer, M., Folke, C., Schellnhuber, H., Nykvist, B., De Wit, C. A., Hughes, T., van der Leeuw, S., Rodhe, H., Sörlin, S., Snyder, P. K., Costanza, R., Svedin, U., Falkenmark, M., Karlberg, L., Corell, R. W., Fabry, V. J., Hansen, J., Walker, B., Liverman, D., Richardson, K., Crutzen, P., & Foley, J. (2009). Planetary boundaries: Exploring the safe operating space for humanity. *Ecology and Society*, 14(2), 32. Retrieved from: http://www.ecologyandsociety.org/vol14/iss2/art32/ (last accessed 26 January 2022).

Society for Ecological Restoration (2019). International principles and standards for the practice of ecological restoration. *Restoration Ecology*, 27(S1), pp. S1–S46.

Swartz, S. (2006). A long walk to citizenship: Morality, justice, and faith in the aftermath of apartheid. *Journal of Moral Education*, 35(4), 551–570.

Van der Walt, J. C., & Midgley, J. R. (2016). Principles of delict (LexisNexis South Africa, 4th ed.). Retrieved from: www.mylexisnexis.co.za/Index. aspx?permalink=emIvNmc3ZGUvN2c3ZGUvM3c5ZGUvcXg5ZGUvcng5 ZGUkLTEkNyRMaWJyYXJ5JGRwYXRooJExpYnJhcnkkv (last accessed 22 May 2020).

Wijdekop, F. (2019). *Restorative justice responses to environmental harm*. Amsterdam: IUCN Report.

WWF (2018). *Living planet report—2018: Aiming higher*. Grooten, M., & Almond, R. E. A. (Eds). Gland, Switzerland: WWF.

6

Earth Trusteeship and the Sovereign State

Klaus Bosselmann

1 Introduction

Environmental law is a generic term for legal concepts, statutes, institutions, and processes relevant to the protection of the natural environment (Bosselmann et al., 2013). This includes the concept of justice and institutions of governance in their functions to provide for fairness and equality. This chapter aims to show how justice, democracy, sovereignty, rights, and responsibilities are all relevant for a sustainable future.

For a start, sustainability is best understood as a form of justice in that it broadens the three dimensions of social justice: *space* to include all humans ('global' or 'intragenerational justice'), *time* to include humans living in the future ('intergenerational justice') and the human community as part of the larger community of life ('interspecies justice') (Bosselmann, 2017, pp. 8–10). This third dimension, the community of

K. Bosselmann (✉)
New Zealand Centre for Environmental Law, University of Auckland, Auckland, New Zealand
e-mail: k.bosselmann@auckland.ac.nz

B. Pali et al. (eds.), *The Palgrave Handbook of Environmental Restorative Justice*,
https://doi.org/10.1007/978-3-031-04223-2_6

life, is essential for understanding that all efforts for providing a fair, sustainable future will be in vain if they overlook our belonging to the community of life. We are not separate from nature as conventional Western philosophy and jurisprudence assumes, nor are we in any way superior to non-human species. We are just part and parcel of a single ecological system, called Earth.

For this reason, the notion of 'ecological justice' is preferable (Bosselmann, 2017, pp. 102–128) to the notion of 'environmental justice' which commonly refers to the social justice movement concerned with the 'fair' distribution of environmental benefits and burdens (Schlosberg, 2007). Ecological justice fully includes this social justice concern, but adds the concern for interspecies justice in reflection of ethical ecocentrism. Ecological justice is foundational to ecological or 'strong' sustainability as opposed to mere environmental or 'weak' sustainability (Hackett, 2011).

Restorative justice is an innovative approach to justice as it involves conversations between victims and offenders to rebuild mutual trust rather than merely punishing the offenders and separating them from their victims. Restorative justice aims for 'restoring' broken relationships and, in this way, helps to prevent future harm and injustice.

The same logic applies in an environmental context. If victims and offenders of environmental harm are in conversation with each other, chances are that relationships between them can be restored, so that future environmental harm is prevented. The key challenge for environmental restorative justice is, however, to bring non-human victims—fauna, flora and ecosystems—into the conversation (Wijdekop, 2019, p. 31). Earth, our home, does not distinguish between humans and non-humans, it may be more appropriate therefore to speak of 'Earth restorative justice'.[1]

If we can describe 'Earth restorative justice' as a social movement concerned with restoring the broken human-nature relationship, then it makes sense to think about the environment or *natürliche Mitwelt* (German for 'natural with-world') as our partner in dialogue. From an ecological point of view, the interrelations between the human and the

---

[1] See https://earthrestorativejustice.org/ (last accessed 26 January 2022).

non-human or beyond-human world are obvious and imply recognition of the intrinsic value of all beings. From a legal and practical point of view, the question arises how this ecological awareness can be best articulated. We can ask, for example, how traditional interpretations of human rights, legal procedures, and democratic processes can be advanced to include ecological relationships (Bosselmann, 1992). To this end, we need to ask who speaks for the beyond-human world and for the Earth (Wijdekop, 2019, p. 44).

This chapter aims at helping to answer this question. In law, the typical tools for allowing someone to speak on behalf, and in the interest of, those who cannot speak for themselves are the concepts of guardianship and trusteeship. In an ecological context, the challenge is to determine who should act as a trustee for Earth. The short answer is, each of us, at least potentially. We are all citizens of a country and, at the same time, of this planet, our home. As global citizens we are, or should be, acting as Earth trustees. Likewise, our political institutions should be acting as Earth trustees.

The chapter offers some ideas why and how the most powerful of our political institution, the sovereign state, could adopt Earth trusteeship responsibilities both, domestically and internationally. Earth governance (Bosselmann, 2015) has become a matter of great urgency and Earth system science and planetary boundaries modelling (Rockström et al., 2009) have put this challenge to governments and states. Traditionally, states have been resistant to accepting responsibility for areas outside their national jurisdictions, known as the global commons. Essentially, the focal-point of governing the global commons—such as the polar regions, the oceans, the atmosphere, and outer space—needs to shift from states with their often competing national interests to Earth as a whole.

The conceptual origins of Earth trusteeship can be traced to the mission of Judge Christopher Weeramantry, former Vice President of the International Court of Justice. In his Separate Opinion on the ICJ's 1997 *Gabcikovo-Nagymaros* case he described 'the principle of trusteeship of earth resources' as the 'first principle of modern environmental law' (Weeramantry, 1997, p. 105). In the observation of Weeramantry, the concept of trusteeship is, as a living example, rooted in traditional farming practices with respect to the land and to irrigation systems. Trusteeship resonates with the worldviews of Indigenous peoples and the teachings of world religions, as well as with secular philosophies (Chakrabarty, 2017).

Earth trusteeship proclaims a shift from state-centred governance to multi-actor governance. The shift involves states, but not exclusively. The new approach emphasises the role of the citizen rather than the nation-state as the source of governance. In a democracy, government is empowered by citizens and accountable to them. It is appropriate, therefore, to perceive of governments as fiduciaries or trustees acting on behalf of their citizens whose well-being depends on the well-being of all people and the planet. This calls for transnational processes of forming the collective will, at least potentially. For this reason, governments need to be concerned with areas outside their national jurisdiction, that is, the global commons.

2   The Global Commons and International Law

One prominent example of the global commons is the atmosphere. Intuitively, the atmosphere belongs to all living things, including humans, as our well-being and existence depends on a relatively stable climate system. The problem here is that currently, the atmosphere is being treated as an open access resource without any legal status; it is widely regarded as *ius nullius*, that is, a legal nullity or vacuum. This has worked to the advantage of property owners who filled the vacuum by exercising their property rights. Property rights may not include a right to pollute, but the absence of someone who could claim violation of their own rights means that actual pollution goes without any sanction.

In fact, it is free. You and I or the entire fossil fuel industry, for that matter, can freely emit greenhouse gases into the atmosphere as nobody's property rights are affected. This is the legal status quo. It will only be qualified if, and in so far as, the law sets rights limiting emissions standards. To date, this has been an uphill battle, which has not been made any easier merely by having the 2015 Paris Agreement on Climate Change requiring legally binding commitments, targets, and timelines, considering the slow and incomplete implementation practices of signatory states.

But we can win the battle, and it is quite simple. By asserting that all humanity living now and in the future owns the atmosphere, we begin to use the institutions of law and have them working in our favour. As legal owners we can charge for individual damage of our common property, provide rewards to those who protect it (e.g., producers and users of renewable energy), and in this way eliminate greenhouse gases. All we have to do is to declare the atmosphere a public common good, owned by all humanity and protected by law. This would provide an effective legal shield against the fossil fuel industry and its supporters (states, banks, corporations). The concepts of private property and state sovereignty would stay unchanged, but they would end where common property begins. The new rule for the Anthropocene is that common ownership prevails over private property.

This is a fairly simple legal construct. It could, for example, be supported by the well-established public trust doctrine. The public trust doctrine says that natural commons should be held in trust as assets to serve the public good. It is the responsibility of the government, as trustee, to protect these assets from harm and ensure their use for the public and future generations. So nationally, the government would act as an environmental trustee. Internationally, states would jointly act as trustees for the global commons such as the atmosphere. Considering that only about 90 companies are responsible for two thirds of carbon emissions emitted into the atmosphere, a global trusteeship institution could quickly fix the problem of climate change (Costanza, 2015).

The idea of global nature's trusts has been promoted by environmental lawyers such as Mary Wood (2013) or Peter Sand (2013) and economists such as Peter Barnes (2001) or Robert Costanza (2015). Trusteeship governance is also advocated by the rich literature on the commons (e.g., Bollier, 2014; Helfrich & Haas, 2009). The 'Reclaiming the Commons' movement has certainly found a momentum in recent times.

A strong case can be made to suggest that international law and the United Nations are not only in need, but practically ready to develop institutions of trusteeship governance. There is, for example, a tradition

of UN institutions with a trusteeship mandate including the (now defunct) UN Trusteeship Council, the World Health Organisation (WHO) with respect to public health and ironically also the World Trade Organisation (WTO) with respect to free trade (Bosselmann, 2015, pp. 198–232). A number of other UN or UN-related institutions with weaker trusteeship functions exist also. Quite obviously, states are capable of creating international trusteeship institutions, either expressly or implicitly. These developments prove that sovereignty of states can be transferred to international levels (as also the European Union demonstrates).

The underpinning motives are not so much of a moral nature but driven by political interests. More often than not, political interests are short-sighted and selfish, but they can change as morality can change to include responsibility for all people (human rights) and the entire planet (sustainability). It is important to note that citizens and social institutions are the agents of moral change, not the state per se. At present, states follow the outdated morality of national interests over global interests. What does it take for states to accept trusteeship responsibilities for the global commons?

We cannot expect the trusteeship governance to initiate from the 'top', that is, by the UN and its member states themselves, but rather by forces outside the system, in particular global civil society. To this end, we can build on many years of activism and proposals for institutional change. Nor should we envisage states to be in sole charge of running and controlling global trusteeship institutions such as a World Environment Organisation or a Global Atmospheric Trust. Rather, trusteeship governance should be seen as a joint effort of the UN, states, and civil society organisations, each with an equal say in decision-making.

3   Sovereignty and Trusteeship

There is, at present, an unholy alliance between politics (sovereignty) and private interests (property) severely undermining democratic process and the public concern for safeguarding the global commons. Furthermore, as Barnes points out, '[n]ot even seated at democracy's table—not organised, not propertied, and not enfranchised—are future generations, ecosystems, and non-human species' (Barnes, 2006, p. 38).

The practice of state governance in times of economic neoliberalism has affected the way that environmental policies and laws are being conceived, namely as entirely discretionary and located somewhere between resource exploitation and environmental protection. Mary Wood (2013, p. 592) calls this a 'discretionary frame', meaning that governments see themselves as perfectly entitled to favour short-term resource exploitation over long-term conservation.

State governance today is about a *quid pro quo*, symbiotic relationship between political institutions and corporations. The rewards include property rights, friendly regulators, subsidies, tax breaks, and free or cheap use of the commons. This leaves very little motivation for protecting the common good. In the words of Peter Barnes (2006, p. 45), 'we face a disheartening quandary here. Profit-maximising corporations dominate our economy. Their programming makes them enclose and diminish common wealth. The only obvious counterweight is government, yet government is dominated by these same corporations'.

Fundamentally, the legitimacy of the state rests on its function to act for, and on behalf of, its citizens. This requires consent with the governed as, for example, John Locke stated: '(G)overnment is not legitimate unless it is carried on with the consent of the governed' (quoted in Ashcraft, 1991, p. 524). Governmental duties can therefore be understood as fiduciary obligations towards its citizens (Fox-Decent, 2012). Such fiduciary obligations are recognised typically in public law, but generally exist in common law and civil law (although in varying forms and degrees) and are also known in international law. The fiduciary function of the state can also be described as a trusteeship function (Criddle et al., 2018; Finn, 1995, pp. 131–151).

In the following sections, I examine how state sovereignty can be reconciled with trusteeship. Prima facie, both seem to have different purposes; yet, as we will see, they are part of the same basic function of the state, which is to serve the citizens it depends on and is accountable to.

The environmental crisis and the state of the global commons give rise to the need for revisiting the relationship between sovereignty and trusteeship. Trusteeship must be pursued at both the international level and the domestic 'internal' level. As Eyal Benvenisti (2013, p. 301) notes, the private, self-contained concept of sovereignty is less compelling than it was in the past because of the 'glaring misfit between the scope of the sovereign's authority and the sphere of the affected stakeholders'. This misfit engenders inefficient, undemocratic, and unjust outcomes for under- or unrepresented affected stakeholders. Noncitizens, future generations, and the natural environment all fall into such a category of affected stakeholders.

There are two challenges to advancing the idea of trusteeship and both boil down to an outmoded understanding of sovereignty. On the one hand, to propose a system of international trusteeship is to directly challenge the principle of non-interference in states' domestic affairs. To propose that states become trustees themselves, in addition to an international system of trusteeship, is again, an intrusion into states' sovereign rights to determine their approach to the environment. However, without the latter we will not achieve the former. Regardless of what one thinks about the legitimacy of sovereignty and the entire makeup of international relations, the reality is that states are calling the shots. Leaving aside the theoretical possibility of radically reorganising global politics, we need to work within the context of nation-states and find practicable ways of incorporating trusteeship into sovereignty.

Trusteeship is an idea, which softens the blow of what would otherwise be seen as an unprecedented intrusion into sovereign state affairs. In the context of global governance, trusteeship serves the common interests of states. Yet, trustees are not states; a trust council might not even be an intergovernmental institution if it were comprised of individuals rather than drawn exclusively from 'states'. Arguably, this represents a less threatening intrusion into sovereignty. As Catherine Redgwell (2005, p. 179) explains, 'trust arrangements do not challenge sovereignty directly, for one of the advantages of trusteeship arrangements is the absence of sovereignty in the exercise of trusteeship functions—there is no transfer of sovereignty to the trust authority'.

Nevertheless, the many proposals of trusteeship arrangements at the level of the United Nations have been, more often than not, greeted with hostility (Bosselmann, 2015, pp. 233–244). States seem too attached to the principle of non-interference to appreciate cooperation of this kind.

Yet, the very origins of the concept of state sovereignty are closely linked with humanitarian concerns. The Peace of Westphalia, as the foundation of state sovereignty, was a key instrument for upholding humanitarian precepts related to freedom of conscience and religion. To the extent that it resolved a crisis of freedom of conscience and equality before the law and many pre-existing institutions had lost their legitimacy and ultimately collapsed, sovereignty has been and can be justified. But it should also be remembered that humanitarian concerns were at the root of the crisis that the new system of order resolved. Where new crises emerge that can only be solved through full cooperation, the principle of non-interference can be counterproductive and must be interpreted within a given context.

Likewise, with regard to the state itself as environmental trustee for those over whom it governs, it could hardly be refuted that a legitimate, typically democratically elected, government owes its citizens a duty to govern their natural wealth and resources in a sustainable way. The first step, then, is reminding ourselves as citizens and society that these rights and responsibilities rest with us, despite the state acting as our representative. The second step is convincing the consumer society of what these rights and responsibilities entail. This is no small feat.

Although there is a dedicated green movement, and ever more 'lite' green sentiment, convincing people to alter their ingrained, even unconscious neoliberal proclivities in favour of an ethic of stewardship and trusteeship is a very trying task. Yet, without a mobilised civil society—a demos—that is willing to hold governments to account, and demand that they represent their ecological interests internationally, states will continue to behave in the way they always have—reacting to the global environmental crisis according to the conflict model of international law that they are so used to and benefit from.

4   Trusteeship Functions of the State

The only way to turn things around and move international law from the Westphalian conflict model to a twenty-first century cooperation model is to redefine states as trusteeship organisations. Sovereignty and trusteeship must be seen as complementary and not mutually exclusive. The argument in favour of states as trustees goes as follows.

The state gains its legitimacy exclusively from the people who created it. While the legality of a state depends on recognition by other states, once in existence a state can only ever legitimise its continued existence through ongoing trust by its people. The core idea of the modern democratic state is that it acts through its people, by its people, and for its people. This implies a fiduciary relationship between people and state and was arguably the only legitimate basis for political authority in the English Civil War, American Revolution, and then again confirmed in the French Revolution. It is echoed in constitutional documents such as in Article IV of the Pennsylvanian Constitution of 1776: '[A]ll power being … derived from the people; therefore all officers of government, whether legislative or executive, are their trustees and servants, and at all times accountable to them'.[2] Likewise, John Locke had asserted that legislative power is 'only a fiduciary power to act for certain ends' and that 'there remains still in the people a supreme power to remove or alter the legislative, when they find the legislative act contrary to the trust reposed in them' (Locke, 1690, Sect. 149).

In the same vein, Immanuel Kant drew the moral basis of fiduciary obligations from the duty-bound relationship between parents and children. Kant claimed that children have an innate and legal right to their parents' care. In a similar sense, he believed that state legitimacy was the result of a contract that is necessarily created between people to form a Rousseauian 'general will'. Through this process, Kant claimed, we jointly authorise the state to announce and enforce law.

The idea that state sovereignty entails trusteeship obligations can be traced back not only to Middle Eastern origins, but also to Roman and Germanic law and equally to religious teachings. The idea is perhaps even more prevalent in non-Western societies than present-day Western societies because the former emphasise collective identities (e.g., family, clan, nation, religion) over individual freedom and dignity, imbuing implied fiduciary obligations into the structure of public and private legal institutions.

---

[2] See https://avalon.law.yale.edu/18th_century/pa08.asp (last accessed 26 January 2022).

Article 21(3) of the Universal Declaration of Human Rights states that, 'the will of the people shall be the basis of the authority of government'.[3] But as Ron Engel (2010, p. 28) has pointed out, 'democracy' can have differing interpretations. There is the 'thin' interpretation that includes procedural democracy, liberal democracy, representative democracy, or simply put 'the democratic process'. And there is 'thick' or material democracy where the government acts in trusteeship of the common good. As the Covid-19 pandemic has shown, in many countries democratic processes and commitments have failed to provide for effective health protection.

So, although we may have democracy (and many places do not) in a technical sense, we have lost sight of what duty the state owes to those whom it governs. At its most simplistic, the state's legitimacy to govern is based on its ability to serve the common interest. Benvenisti (2013, pp. 301–309) conceives of two normative bases according to which we should ascribe a trusteeship function to states' mandate to govern. Firstly, sovereignty should be viewed as a vehicle for the exercise of personal and collective self-determination. Collective self-determination embodies the freedom of a group to pursue its interests, further its political status, and 'freely dispose of [its] natural wealth and resources'. An outdated conception of sovereignty, which equates the voting constituency with only the affected stakeholders, can undermine communities' ability to exercise their right to self-determination.

Secondly, Benvenisti (2013) refers to a conception of sovereign states as agents of humanity as a whole. He bases this conception largely on the equal moral worth of all human beings and the corresponding foundation of international law in human rights. He argues that it is humanity at large that assigns to certain groups of citizens the power to form national governments. Accordingly, states can and should be viewed as agents of a global system that allocates competences and responsibilities for the promotion of the rights of all human beings and their interest in the sustainable utilisation of global resources. As such, the corollary of states' authority to manage public affairs within their domestic

---

[3] See https://www.un.org/sites/un2.un.org/files/udhr.pdf (last accessed 26 January 2022).

jurisdictions is an obligation to take account of external interests and balance internal against external interests.

Similarly, we could say that the privilege of territorial sovereignty can be legitimised only in so far as universal interests of humanity as a whole are not severely affected. This argument is based not only on ecological realities defying national state boundaries, but also on the observation that boundaries of states do not necessarily coincide with boundaries of nationalities, or more generally, with the boundaries of the groups whose members commonly share a distinct interest or conception of the good.

Benvenisti (2013) also refers to a conception of sovereignty as the power to exclude portions of global resources. He notes that both ownership and sovereignty are claims for the intervention in the state of nature by carving out valuable space for exclusive use. Such a perception of states as power-wielding property owners provides a solid normative foundation for the imposition of a positive obligation on states to take other-regarding considerations into account when managing the resources assigned to them. Property law theory can thus provide us a framework within which we can translate these moral grounds into legal obligations. Thus, we can and should conceptualise ownership of global resources as originating from a collective regulatory decision at the global level, rather than as an entitlement of sovereign states.

Although we might all agree a government has fiduciary duties, there is seemingly little to establish a precedent of state trusteeship for the environment (Bosselmann, 2015, pp. 155–197, 2017, pp. 145–174). However, this may be due to a lack of appreciation of just how closely connected rights and responsibilities have been in the past. The English 'commons' or German 'Allmende' were governance systems based on common property. States or local principalities exercised their powers in a way that access of the commoners to land and natural resources was assured. It is often forgotten, for example, that the Magna Carta of 1215 was accompanied by the Carta de Foresta (Charter of the Forest) of 1217 in an attempt to restrict exploitation of the forests by the king at the time. The Charter of the Forest provided a right of common access to (royal) private lands. Some of its clauses remained in force until the 1970s (Robinson, 2014).

For centuries, trusteeship responsibilities for the natural commons were part of ways of life around the world. Modern capitalism has dramatically turned things around, converting commons to commodities ('natural resources'). In the course of unfolding—and now unfettered—capitalism, governments and business had little regard to safeguarding natural cycles and ecological integrity. Yet, being vested in the people themselves, public sovereignty needs to protect the common good.

5   Framing Earth Trusteeship

Under international law, state sovereignty includes trusteeship responsibilities for common pool resources (Sand, 2014), but these are not well defined and are largely subject to specific arrangements in international agreements.

There is a general obligation of nation-states to cooperate in order to protect the integrity of the Earth's ecological system. More than 25 international soft and hard law agreements contain specific reference to this obligation. The first of such agreements was the Convention on the Conservation of Antarctic Marine Living Resources adopted in 1980, which recognised in its preamble, the importance of 'protecting the integrity of the ecosystem of the seas surrounding Antarctica'. Another example is the preamble of 1992 Rio Declaration on Environment and Development (Rio Declaration),[4] which calls for 'working towards international agreements which respect the interests of all and protect the integrity of the global environmental and developmental system'. Specifically, Principle 7 of the Rio Declaration stipulates, 'States shall cooperate in a spirit of global partnership to conserve, protect and restore the health and integrity of the Earth's ecosystem'. Similar obligations are contained in key documents such as Agenda 21, the 2002 Johannesburg Declaration, the 2012 United Nations Conference on Sustainable Development (Rio+20) outcome document *The Future We Want* or the 2015 Paris Agreement on Climate Change. Applying the usual standards

---

[4] See   https://www.un.org/en/development/desa/population/migration/generalassembly/docs/globalcompact/A_CONF.151_26_Vol.I_Declaration.pdf (last accessed 26 January 2022).

for the recognition of concepts as international law, we can say that the repeated and consistent references to ecological integrity amount to an emerging fundamental goal or *grundnorm* of international environmental law (Kim & Bosselmann, 2013, 2015).

Model frameworks for Earth trusteeship exist in the form of two agreements created by global civil society, that is, the Earth Charter (Earth Charter, 2000) and the Hague Principles (Hague Principles, 2018). The Earth Charter, launched in 2000, provides for an ethical framework for global governance based on the three principles of: (a) respect and care for the community of life; (b) ecological integrity, and; (c) social and economic justice. Eighteen years later, the Hague Principles for a Universal Declaration on Responsibilities for Human Rights and Earth Trusteeship (Hague Principles) were launched in the Peace Palace in The Hague. The three principles are: (a) responsibilities for the Earth (b) responsibilities for the community of life and (c) responsibilities for human rights. The Hague Principles are designed around the need for trusteeship responsibilities that both people and states have for human rights and the Earth.

Ongoing efforts towards implementing a Global Pact for the Environment (2018) provide an opportunity to include these three principles and make them legally binding. The current draft text includes important references to 'the Earth's community of life', 'the balance and integrity of Earth's ecosystem' (Preamble) and the state's duty to take care of 'conservation, protection and restoration of the integrity of the Earth's ecosystem' (Article 2). These notions are reflective of Earth Charter principles, the Hague Principles and of an Earth system approach to international environmental law. However, the text needs further strengthening to fully accommodate the Earth system approach and associated Earth trusteeship responsibilities (Bosselmann & Botrel, 2020; Kotzé & Kim, 2020).

The magnitude of the issues we are facing in times of dramatic climate change and global pandemics is truly immense. As yet, we are far from addressing the negative environmental and social outcomes as manifested in the various dimensions of the global ecological crisis as well as poverty, violence and war. The scale and complexity of our problems has pushed solutions beyond the grasp of current governance mechanisms.

We need integrated responses that are framed by the reality of Earth as a single ecological system and our common home demanding to live within planetary boundaries. The responses needed, therefore, are those of responsibility for the Earth. As noted in the Earth Charter:

> To realise these aspirations we must decide to live with a sense of universal responsibility, identifying ourselves with the whole Earth community as well as our local communities. We are at once citizens of different nations and of one world in which the local and global are linked. Everyone shares responsibility for the present and future well-being of the human family and the larger living world.

One way of institutionalising Earth trusteeship responsibilities at the international level is to create a new world organisation with direct environmental responsibilities, for example, a World Environment Organisation (WEO) (Bosselmann, 2015, pp. 257–267). The mandate of such a WEO would provide a trusteeship function over global common good, that is, those portions of the planet and its surrounding space which lie above and beyond the recognised territorial claims of any nation. Trusteeship duties would include:

* Global obligations for the integrity of planetary boundaries and the well-being of the greater community of life.
* Overseeing markets to ensure that they are protective of non-market common goods.
* Ensuring impartiality of all interests—individual, civil society, corporate, national—along with respect for human rights and ecological well-being.

The WEO could act in a way similar to that in which the now-defunct UN Trusteeship Council acted as a trustee of interests of states transitioning from colonisation to independence, that is, acting on behalf and in the interest of the global commons.

The legitimacy of such a powerful environmental institution will depend on its being widely democratic, representative, and participatory. As ongoing geopolitical events illustrate, Earth Charter principle 13 must rapidly become an imperative for all the cultures of the world: 'Strengthen democratic institutions at all levels, and provide transparency and accountability in governance, inclusive participation in decision making, and access to justice'. This democratic principle is particularly important with relation to the environment, as it is an issue which will affect everyone and often particularly those with the least power.

The concept of environmental or Earth restorative justice clearly demonstrates the importance of an institution such as a WEO. We need a truly democratic forum where the voice of the previously voiceless—the global environment and disenfranchised people and cultures—can be heard. The creation of a WEO at the present time of neo-liberalism and global capitalism may seem impossible to envisage. However, as we have entered the Anthropocene more, a global body of this nature could help to reassure people worldwide, particularly young people, that change is possible. To quote again the Earth Charter:

> The choice is ours: form a global partnership to care for Earth and one another or risk the destruction of ourselves and the diversity of life. Fundamental changes are needed in our values, institutions, and ways of living. We must realize that when basic needs have been met, human development is primarily about being more, not having more. We have the knowledge and technology to provide for all and to reduce our impacts on the environment. The emergence of a global civil society is creating new opportunities to build a democratic and humane world. Our environmental, economic, political, social, and spiritual challenges are interconnected, and together we can forge inclusive solutions.

6   Conclusion

Global commons governance reverses the traditional rule that international law and governance end where national borders begin. The dichotomy between national law and international law defies ecological reality and is counterproductive to genuine global governance. Hence, states need to exempt transnational ecological aspects from the concept of exclusive territorial sovereignty and in this way internalise global obligations into the meaning and scope of state sovereignty. The requirement for states, therefore, is to accept environmental trusteeship by restricting territorial sovereignty to environmental uses and impacts that are limited to their own territory (Bosselmann, 2004, pp. 293–313). This legal correction would remove the present paradox: instead of state sovereignty setting limits to environmental protection, environmental protection would set the limits to state sovereignty. It is the solution to the fundamental flaw of the current system of global governance: 'Limiting the self-interest of states by taking into account global concerns of humanity has become a fundamental aspect of international law' (Stec, 2010, p. 364).

Similar to international law, in political praxis states find themselves in a paradoxical situation. They cannot shake off the capitalistic logic of profits at all costs, even at social and environmental costs of suicidal dimensions. States may well want to avoid collective suicide, but they are as yet unable to resist the forces of global markets. These forces have heavily eroded state sovereignty—the same state sovereignty that is needed to resist complete dominance of global markets. The paradox of surrendering sovereignty to free trade and market forces, on the one hand, and on the other hand insisting on sovereignty when expected to protect the commons, has been described as the 'sovereignty paradox' (Kaul, 2013, pp. 33–58).

The way out of the sovereignty paradox is differentiation: more sovereignty where possible, less sovereignty where necessary. In a globalised world, this means protecting citizens and the environment from global economic forces (more sovereignty) and protecting the global commons through international rules controlling financial and economic markets (less sovereignty). The perspective of differentiated sovereignty, also referred to as 'responsible' or 'smart' sovereignty, inevitably calls for reforming and strengthening global institutions as outlined in this chapter.

The drivers for responsible and smart sovereignty are not states per se, of course, but real people, that is, citizens, activists, advocates, and decision makers in a bottom-up approach called democracy. To make democracy effective, the idea of citizenship needs to include a global dimension (global citizenship). Arguably, the concern for the global commons and for the Earth as a whole is a unifying feature of humanity. If we see ourselves as stewards of the Earth and states as trustees of the global commons, then this is a crucial step towards effective global governance.

References

Ashcraft, R. (Ed.) (1991). *John Locke: Critical assessments*. London: Routledge.

Barnes, P. (2001). *Capitalism 2.0: Who owns the sky? Our common assets and the future of capitalism*. Washington, DC: Island Press.

Barnes, P. (2006). *Capitalism 3.0: A guide to reclaiming the commons*. San Francisco: Berret-Koehler Publishing.

Benvenisti, E. (2013) Sovereigns as trustees of humanity: On the accountability of states to foreign stakeholders. *American Journal of International Law (AJIL)*, 107(2), 295–333.

Bollier, D. (2014). *Think like a commoner: A short introduction to the life of the commons*. Gabriola Island: New Society Publishing.

Bosselmann, K. (1992). *Im Namen der Natur—Der Weg zum ökologischen Rechtsstaat*. Munich: Scherz Publishing.

Bosselmann, K. (2004). Environmental governance: A new approach to territorial sovereignty. In R. J. Goldstein (Ed.) *Environmental ethics and law* (pp. 293–313). Burlington: Ashgate Publishing.

Bosselmann, K., Grinlinton, D. P., & Taylor, P. (Eds.) (2013). *Environmental law for a sustainable society*, 2nd ed. Auckland: New Zealand Centre for Environmental Law Monograph Series.

Bosselmann, K. (2015). *Earth governance: Trusteeship of the global commons*. Cheltenham: Edward Elgar.

Bosselmann, K. (2017). *The principle of sustainability: Transforming law and governance*, 2nd ed. London: Routledge.

Bosselmann, K., & Botrel, M. (2020). Constitutionalising international environmental law. *Sciences Po Law Review*, 18, 11–16.

Chakrabarty, B. (2017). Gandhi's doctrine of trusteeship. *Nabakrushna Choudhury Centre for Development Studies, Working Paper No. 67*.

Costanza, P. (2015). Claim the sky! *Solutions*, 6(1), 18–21.

Criddle, E., Fox-Decent, E., Gold, A., Kim, S. H., & Miller, P. (Eds.) (2018). *Fiduciary government*. Cambridge: Cambridge University Press.

Earth Charter (2000). Retrieved from: https://earthcharter.org/ (last accessed 21 January 2022).

Engel, R. (2010). Contesting democracy. In R. Engel, L. Westra, & K. Bosselmann (Eds.), *Democracy, ecological integrity and international law* (pp. 28–37). London: Cambridge Scholars Publications.

Finn, P. (1995). The forgotten 'trust': The people and the state. In M. Cope (Ed.), *Equity: Issues and trends* (pp. 151–155). Sydney: The Federation Press.

Fox-Decent, E. (2012). *Sovereignty's promise: The state as a fiduciary*. Oxford: Oxford University Press.

Global Pact for the Environment (2018). Retrieved from: https://globalpactenvironment.org/en/home/ (last accesses 21 January 2022).

Hackett, S. (2011). Weak vs strong sustainability debate. In K. Bosselmann, D. Fogel, & J. B. Ruhl (Eds.), *Berkshire encyclopedia of sustainability, Vol. 3: The law and politics of sustainability* (pp. 505–507). Great Barrington, MA: Berkshire Publishing.

Hague Principles (2018). Retrieved from: https://earthcharter.org/launch-of-the-hague-principles/ (last accessed 21 January 2022).

Helfrich, S., & Haas, J. (Eds.) (2009). *The commons: A new narrative for our time*. Berlin: Heinrich Böll Stiftung.

Kaul, I. (2013). Meeting global challenges: Assessing governance readiness. In Hertie School of Governance (Ed.), *Governance report 2013* (pp. 33–58). Oxford: Oxford University Press.

Kim, R., & Bosselmann, K. (2013). Towards a purposive system of multilateral environmental agreements. *Transnational Environmental Law, 2*, 285–309.

Kim, R., & Bosselmann, K. (2015). Operationalising sustainable development: Ecological integrity as a *grundnorm* in international law. *Review of European Community and International Environmental Law, 24*, 194–208.

Kotzé, L., & Kim, R. (2020). Planetary boundaries at the intersection of Earth System law, science and governance. *Review of European Community and International Environmental Law, 24*(2), 1–13.

Locke, J. (1690). *Second treatise of civil government* [republished by C. B. McPherson (Ed.) (1980), Hackett, Indianapolis and Cambridge].

Redgwell, C. (2005). Reforming the UN trusteeship council. In W. B. Chambers & J. F. Green (Eds.), *Reforming international environmental governance: From institutional limits to innovative* (pp. 178–203). New York: United Nations University Press.

Robinson, N. (2014). *The charter of the forest: Evolving human rights in nature*. Pace Law Faculty Publications. Retrieved from: https://digitalcommons.pace.edu/cgi/viewcontent.cgi?article=1988&context=lawfaculty (last accessed 26 January 2022).

Rockström, J., Steffen, W., Noone, K., Persson, Å., Chapin, III, F. S. Lambin, E., Lenton, T. M., Scheffer, M., Folke, C., Schellnhuber, H., Nykvist, B., De Wit, C. A., Hughes, T., van der Leeuw, S., Rodhe, H., Sörlin, S., Snyder, P. K., Costanza, R., Svedin, U., Falkenmark, M., Karlberg, L., Corell, R. W., Fabry, V. J., Hansen, J., Walker, B., Liverman, D., Richardson, K., Crutzen, P., & Foley, J. (2009). Planetary boundaries: Exploring the safe operating space for humanity. *Ecology and Society*, 14(2): 32. Retrieved from: http:// www.ecologyandsociety.org/vol14/iss2/art32/ (last accessed 26 January 2022).

Sand, P. (2014). Sovereignty bounded: Public trusteeship for common pool resources? *Global Environmental Politics*, 4(1), 47–71.

Sand, P. (2013). The rise of public trusteeship in international law. *Global Trust Working Paper Series 03/2013*. Retrieved from: http://globaltrust.tau.ac.il/ wp-content/uploads/2013/08/Peter-Sand-WPS-3-13-ISSN.pdf (last accessed 26 January 2022).

Schlosberg, D. (2007). *Defining environmental justice: Theories, movements, and nature*. Oxford: Oxford University Press.

Stec, S. (2010). Humanitarian limits to sovereignty: Common concern and common heritage approaches to natural resources and environment. *International and Comparative Law Review*, 12, 361–372.

Weeramantry, C. (1997). *Gabcikovo-Nagymoros Project: Separate Opinion of Vice-President Weeramantry* (pp. 88–116). Retrieved from: https://jusmundi. com/en/document/opinion/en-gabcikovo-nagymaros-project-hungary-slovakia-separate-opinion-of-vice-president-weeramantry-thursday-25th-september-1997 (last accessed 26 January 2022).

Wijdekop, F. (2019). *Restorative justice responses to environmental harm*. Gland: IUCN Report.

Wood, M. C. (2013). *Nature's trust: Environmental law for a new ecological age*. New York: Carolina University Press.

7

Turning Up the Restorative Dial in Environmental Regulation with an Adaptive Learning Loop

Miranda Forsyth

1 Introduction

Environmental Restorative Justice (ERJ) has much to offer environmental regulation. Environmental regulation, carried out principally by environmental protection authorities and other specialised environmental regulators or agencies across the world, is concerned with regulating pollution and waste, and associated impacts upon human health and the environment. Environmental regulation and regulators operate in a maelstrom of discontent. They are responsible for decisions in contexts of disagreements over who is responsible for polluting the environment; who is impacted that counts (communities, more-than-human, nature, the atmosphere); what sorts of pollution are and should be allowable; and whether the extent of pollution is just, fair and lawful (noting that these are three different standards, often leading to very different answers).

M. Forsyth (✉)
RegNet School of Regulation and Global Governance, Australian National University, Canberra, ACT, Australia
e-mail: miranda.forsyth@anu.edu.au

147
B. Pali et al. (eds.), *The Palgrave Handbook of Environmental Restorative Justice*,
https://doi.org/10.1007/978-3-031-04223-2_7

Environmental regulation is intensely place-based,[1] with communities, regulators, industry and environment often forced into decades-long or even generations-long relationships. Sometimes these relationships generate mutual support and sometimes there is ongoing mutual distrust, with parties feeling ignored, unheard or dismissed. It is within the context of these challenges that ERJ's orientation towards healing relationships, facilitating real dialogue, providing agency to all who are impacted and facilitating forums for mutual problem-solving has significant potential.

This chapter is concerned with identifying how restorative justice can form part of the regulatory ecosystem in which state and non-state regulators seek to achieve fair and equitable environmental relations for humans and nature. In thinking through how the potential of restorative justice could best be developed as a practical tool for environmental regulation, I draw upon the Adaptive Learning Loop, a framework I am currently developing in collaboration with my colleague Anthea Roberts as part of an ongoing project on Governing in Complexity.[2] The framework draws creatively from Peter Senge's (2006) and David Hurst's (2002) work on learning organisations, and John Boyd's (1995) OODA loop, which involves Observing, Orienting, Deciding and Acting. The overarching purpose of the Adaptive Learning Loop is to assist regulators, policymakers and other actors to better understand and navigate the complex system in which they are operating, and to adapt their policy interventions and regulatory approaches in response. The suggestion in this chapter is that regulators and practitioners using the Adaptive Learning Loop in the context of environmental regulation will benefit considerably from paying explicit attention to the possibilities presented

---

[1] The term 'place-based' is used by many Australian regulators to account for how regulatory actions and systems impact communities and populations and the subsequent need to account for the local population's perspective. For example, the Victorian Government states that 'place-based approaches target the specific circumstances of a place and engage local people as active participants in development and implementation, requiring government to share decision-making' (https://www.vic.gov.au/framework-place-based-approaches/).

[2] See Anthea Roberts, Risk, Reward and Resilience: The Triple-R Framework (manuscript on file with author, forthcoming), Governing in Complexity, https://regnet.anu.edu.au/research/research-projects/details/8153/uncaged-governing-complexity (last accessed 31 January 2022).

by restorative approaches. As such, the chapter focuses on providing examples and explanations to support such an exercise.

This chapter guides and illustrates the way actors using an Adaptive Learning Loop can include consideration of restorative justice through exploring three different domains of environmental regulation: (1) approvals and licensing; (2) inspections; and (3) responding to environmental harm. Frequently these domains of regulation are relatively small-scale, and each individual regulatory action viewed independently may not be of significance. Yet, we hypothesise that 'turning up the restorative dial'—that is, incrementally increasing the restorative dimensions—of regulatory actions across multiple individual regulatory actions, no matter how minor, may have a powerful cumulative effect.

In what follows, I first set out the background to the fieldwork drawn upon in the chapter and outline the methodology. I then describe the Adaptive Learning Loop Framework and how it can be used to imagine and implement more restorative regulatory approaches in relation to the three domains of environmental regulation.

2   Context for the Chapter

The chapter utilises examples from collaborative fieldwork with the Environment Protection Authority Victoria (EPA Victoria), the principal environmental regulator in the State of Victoria, Australia. EPA Victoria is an environmental regulator with which my colleagues and I from the Australian National University[3] have spent the past five years engaging as part of a participatory research project. EPA Victoria has itself been on a restorative journey for some time, a journey that has involved piloting restorative processes such as conferences, development of informal restorative practices in discrete pools across the organisation, and in 2021, the introduction of new explicitly restorative provisions into its legislation, as discussed below.

The primary source of data is 84 interviews with 188 EPA Victoria officers, plus observation of EPA officers in meetings where training

---

[3] John Braithwaite, Valerie Braithwaite, Deborah Cleland and Felicity Tepper.

occurred, policies were debated (including with cognate environmental agencies), and enforcement decisions were made. I and my research team further observed EPA interactions and consultations with community members and groups, Traditional Owner groups, and business actors (sometimes pursuant to formal EPA decision making, such as issuance of licences) and we observed court cases, field inspections, and meetings of an EPA Community of Restorative Practice. In addition, we observed and conducted 54 interviews of non-EPA actors from business, Traditional Owners and the community.

Our method was to search widely for incipient restorative justice, not to sample for estimation of its prevalence. We spent a lot of time hanging around EPA Victoria offices, central and regional, observed internal team meetings of disparate kinds, and joined in on countless casual and more serious conversations in lunchrooms and on long car trips. In the field, we participated in inspections, approval and licensing decision making conferences, Community Reference Group meetings and other community consultations.

3   The Adaptive Learning Loop framework

The Adaptive Learning Loop is based on four steps:

(1) Observe and Anticipate
(2) Envision and Assess
(3) Act and Communicate
(4) Monitor and Evaluate.

In step one, observation involves being aware of and understanding the current state of play, or the facts on-the-ground, whilst Anticipation involves projecting forward about what might arise in the future, and who may be impacted, how and when. Under step two, Envisioning involves imagining alternative options that could be adopted in response to the current state of play or anticipated situation, and Assessing means deciding which course or courses of action or regulatory approach should be pursued. For step three, Acting means deciding upon which course or courses of conduct to undertake and then working to give effect to those ideas on-the-ground. Communicating involves explaining those actions and the reasons for them so that others in the system can understand how and why decision makers have acted as they did. The fourth step is to Monitor what is happening and Evaluate it to determine whether, for instance, an experimental intervention has been a success and should be scaled up, or if it has been a failure and should be dampened down. It is also an important time to consider whether actions have had problematic unintended consequences that need to be ameliorated or modified. As in step two, any evaluation of an action needs to be comparative. Actions and options are neither good nor bad in the abstract; rather, they are better or worse than alternatives. As its name suggests, the Adaptive Learning Loop is an iterative process, ideally generating learning at all stages.[4]

The Adaptive Learning Loop is a general framework that any regulatory agency or policymaker may utilise to improve performance. Given the non-visibility of restorative justice in environmental regulation to date, it is unlikely that consideration of incorporating more restorative options would occur without an external prompt and clear guidance. In particular, a regulatory actor may struggle to Observe restorative processes already in operation, in Envisioning restorative alternatives and in Acting utilising restorative practices. This chapter therefore concentrates on these specific steps in the Adaptive Learning Loop, providing information, examples and illustrations for those seeking to guide and stimulate active consideration of the opportunities presented by restorative justice for environmental regulation.

In this chapter, I will demonstrate how each of the three steps helps to inspire, identify, develop and implement restorative practices in the environmental regulatory context. The Observation step can be utilised to identify existing or incipient restorative practices already being practised in a regulatory environment. Restorative practices may already be in use in a range of regulatory contexts, although those practising them may not be using the term 'restorative' or rendering restorative practices visible in any of the bureaucratic metrics being collected. Restorative practices are

---

[4] This description of the Adaptive Learning Loop draws upon Anthea Robert's forthcoming work on Risk, Reward and Resilience.

more often manifested in stories told by and between regulatory practitioners, including those about how the 'best' regulatory officials operate.

Second, the Envisioning step is, in many ways, the most challenging for those seeking to develop restorative practices in domains where it has not previously been implemented. It involves imagining alternative restorative approaches that could be adopted in place of or to enhance existing regulatory approaches. Envisioning involves what Braithwaite (2002) refers to as a firing of the 'restorative imagination'. This step needs to be grounded in understandings about which restorative values and principles are most relevant for the regulatory domain under question, and what possibilities have been already imagined elsewhere. It also involves a process of deliberate contrasting, of reflecting on differences between restoratively inflected approaches and those that are more punitively based, and seeking to understand what choices need to be made, including articulating and assessing relevant trade-offs.

The third step, Acting, involves implementing or operationalising the restorative approach. Below I describe some practices that can be used to act restoratively in particular environmental regulatory domains. Although not a focus here, it is also important to note the importance of Communicating, as it involves the explanation of what the restorative approach is and why it has been adopted. Communication about restorative approaches across an organisation and other relevant regulatory bodies can create a shared language and community of practice that can provide fertile ideas for Envisioning and support for responding to any pushbacks that may arise.

4   Environmental Regulation

4.1   Approvals and Licensing

The first domain of environmental regulation we consider is approvals and licensing—namely the approval of new, or extensions and modifications of old, proposals by those wishing to engage in development projects or other conduct that has the potential to impact on the environment or human health (or both). Decisions about what forms of environmental impact to allow are often fraught with trade-offs, competing needs and varying interests. A wide variety of perspectives often exists around what values matter, where the appropriate balance lies between human and environmental health, and how best to generate economic activity to maintain local jobs and provide essential public goods, such as energy or transport networks, waste removal or recycling. In today's increasingly interconnected world, it is becoming more obvious that inter-species co-existence is fiendishly complex, whilst at the same time our ability to hold informed and respectful dialogue about critical issues is eroding.

Envisioning a restorative approach for approvals and licensing requires us to consider the various principles and values such an approach would involve. The following inexhaustive list suggests some starting points:

* Providing a way for the community to play an active and meaningful role in deciding whether licences or other approvals are granted and maintained, such as through holding a restorative justice conference or similar forum;
* Asking the question 'on whose land is this happening?' (Connell, 2007) and where appropriate, seeking out Indigenous representatives of the land-holding groups for the affected area, and providing them with an opportunity to speak and be heard;
* Deep listening to and by all affected groups;
* Requiring senior management or company owners to be present at a restorative conference, to optimise the potential for developing relational trust and promoting genuine debate;
* Focusing on avoiding future harm to community and to environment through consciously seeking out answers to 'what could be better here?' in a collaborative manner that encourages co-design;
* Viewing the licensing or approvals process as an opportunity to build future relationships between all those who are involved;
* Enabling and valuing genuine debate and contestation through creation of appropriate forums and facilitation.

There are theoretical underpinnings supporting such an approach. According to procedural justice theory, stakeholders will accept and uphold decisions that they feel have been made fairly, where they have been heard, and the impact on them has been genuinely taken into consideration (Tyler, 1990), even if the outcome is not to their liking. One of the key principles of restorative justice theory is that better decisions are made collectively through widening the circle of participants to bring in new forms of knowledge and perspectives (Llewellyn, 2019). Regulatory theory also views trust as 'central to constructive relations' in regulation. Gunningham and Sinclair (2009, p. 167) state that '[w]hen those relations break down… then dialogue ceases, information is withheld rather than shared, in-firm accident investigation, prevention, and remedial action are inhibited and both sides retreat to a form of adversarialism that undermines regulatory effectiveness'. Incorporating restorative justice at the stage of awarding or modifying licences provides the opportunity to build and strengthen trust between the regulator, the regulated industry and the community.

So, how would such an approach be Actioned? There is a range of practices that could be useful that have been developed. In the state of Victoria, the *Environment Protection Act 2017* (EP Act) enables the regulator (EPA Victoria) to convene a Conference of Interested Parties for the express purpose of gathering multiple perspectives about the nature of the problem, or potential of the harm (section 236).[5] This can occur at the stage of industry seeking works approvals and licences. Within EPA Victoria, it is used for applications where there are community concerns, an example of which is discussed below. These conferences are facilitated by independent third parties and they seek to involve all affected parties in deciding whether the licence should be granted and, if so, what conditions should be imposed on it.

One important question in Envisioning a more restorative approach to approvals and licensing is what type of community participation is optimal, and to what extent there needs to be a real opportunity for the community to influence the outcome of a regulatory decision. Perhaps another way to ask the question is where restorative justice sits on Arnstein's (1969) 'ladder of citizen participation' that conceives of

---

[5] Under the now repealed *Environment Protection Act 1970*, the relevant section was s. 20B, which will be referred to in the case study below.

participation as sitting on different rungs of a ladder—ranging from forms of nonparticipation at the bottom through to tokenism in the middle and citizen control at the top. Of course, it should be noted that 'citizen control' on the ladder that becomes business control through front groups orchestrated by business interests can theoretically be worse than total citizen exclusion. Engaging with these issues in depth is beyond the scope of the chapter, but fieldwork with the EPA has suggested that a ladder approach, or having a focus solely on bureaucratic procedures, may not be the most productive way to view the potential of restorative processes to have meaningful impact on regulatory decisions. Instead, viewing the regulatory landscape as a non-linear and interconnected ecosystem enables a more expansive view of the possibilities of influence. Through such a lens, we can see that real influence can be exercised in multiple ways, often through relational processes that are not prescribed in regulations or practice notes and may be indirect. In fact, community voice can have a powerful catalytic effect, triggering regulatory responses through respectful dialogue with regulators or making connections between like-minded individuals within and outside of the regulatory institution, industry and community.

An example of the potential of the involvement of restorative approaches in a licensing decision can be seen through analysing an actual case involving Wannon Regional Water Corporation's (the Corporation) licence application in 2020. The Corporation was seeking a licence from EPA Victoria to upgrade their sewerage treatment plant, an ageing facility located on the coast in the State of Victoria. The plant treats a combination of domestic and industrial wastes and receives a high proportion of industrial wastes from industries such as abattoirs and dairy processing. The Corporation's operations, and this application in particular, generated considerable community concerns. These largely centred on the current and potential impacts of their operations on both human and marine health, due to the proximity of the mixing zone (where the effluent is discharged into the sea after treatment) to popular beaches and diving areas. Local residents described walking the beaches and gathering up large piles of micro-plastics, grey sewage sludge and balls of fat.

As a result of these concerns, after receiving the application EPA Victoria hosted a community conference in accordance with section 20B of the former *Environment Protection Act 1970*. The conference was held online, due to COVID-19 restrictions and was chaired by an independent facilitator. In the lead-up to the conference, an online forum was held in which the public could submit questions in preparation for the conference. At the conference, 57 community members had the chance to speak about their concerns, and to respond to the question: 'If the proposed development were to proceed, what could be done to minimise your concerns? (e.g. ideas for design or operation of the site)'. In reflecting on the conference, the independent facilitator stated:

> My experience with WRWC [the Corporation] over the 20B Conference process indicates a genuine desire to engage with community members authentically. The Information Sheet produced in response to community submissions is an example of this desire, as was a genuine effort by WRWC staff and consultants to answer as many community questions as possible during the Conference. WRWC's Community Engagement Framework and its Regional Advisory Forum are evidence of a good strategic approach to building community relationships. (Lilburn 2020)

However, she also noted that in order to build a social licence, a company must first do the right thing, and then be seen doing the right thing. In other words, actions to remedy the concerns of community are needed in addition to authentic involvement in processes to engage with the community. Thus, actions must accompany words. She further noted that during the conference the company had made claims that the plastics found on the local beaches came from other sources, and that this was perceived by some participants to be a failure to take responsibility for operational failings. I observed high levels of scepticism from community members that the Corporation would change its operations significantly, and also about EPA Victoria's willingness to require the Corporation to operate in ways that adequately protected the ecosystem and community health.

The licence was granted, but only after a tightening of the licence limits for the upgraded plant. The final report noted that the community's participation had resulted in the Corporation being required to answer two additional sets of questions, and that 'through the approval process', there had been a tightening of the licence limits and an improved model of the outfall had been developed.

Observing this case allowed us to identify a number of restorative elements already incorporated into the approvals and licensing process. In particular, the conference was an opportunity for all those who had a genuine stake in the operations of the Corporation to come together to express and discuss problems with current operations—and a potentially extended operation—with a view to contributing to the decision about whether the extension was granted, and, if so, under what conditions. Another restorative element was the funding by the regulator of the independent facilitator, who worked with participants beforehand to establish trust in the process, address any misinformation and document community concerns in a way that fairly represented them.

As part of the observation, I also identified a number of opportunities for 'turning up the restorative dial' on the existing process. These dial-adjusting opportunities present examples of mechanisms that could be used in a broader context to Act in more restorative ways in this environmental regulatory domain. Suggestions include:

* Where appropriate, include local Traditional Owner groups (who represent the Indigenous land-holding groups or custodians of the relevant lands) in the process and explicitly ask questions about any cultural heritage or values that are relevant;
* Where appropriate, resource dialogue-based processes sufficiently to allow community and Indigenous participants to be remunerated for their time, rather than relying on voluntarism;
* Actively seek to include specific representative groups for local wildlife/ecosystems and also broader environmental groups;
* Create explicit opportunities for those who have caused harm to acknowledge the impact of that harm and take responsibility for it;
* Create, expand or use the ability to appeal the decision (e.g. on the basis that certain voices and perspectives and evidence were excluded, that crucial environmental facts were wrongly identified or measured);
* Ensure that the relational development initiated through works approvals flows through to establishment of ongoing forums, such as

community reference groups, around the identified environmental risks by making this a condition of licences or approvals.

This section has identified several ways that environmental approvals and licensing can be modified to become more restorative and, in turn, increase the likelihood that decision making about developments with potential to impact the environment and community is based on a broader range of perspectives and understandings about harm than is currently the case. The information, suggestions and illustrations provided above have been intended to inform actors engaged in Observing current regulatory practices, Envisioning alternatives and Acting differently.

4.2   Inspecting and Holding to Account

The second domain of environmental regulation considered here is the role of inspections and minor forms of compliance and enforcement activity by regulators. Of course, this is not to say that restorative justice does not have a role to play in more serious enforcement. Rather, this chapter seeks to highlight the less obvious places where restorative justice can play a role in environmental regulation, as part of the overall argument that there are many different possible entrances for restorative justice. Inspections are a regular form of regulatory activity, involving the physical presence of regulatory officers at such places as factories, industrial sites, farms, landfills, recycling depots, and in other places with potential to harm the environment or community. Research by Schell-Busey et al. (2016, pp. 1–2) highlights the importance of inspections in regulatory efficacy, noting '[o]ur results suggest that regulatory policies that involve consistent inspections and include a cooperative or educational component aimed at the industry may have a substantial impact on corporate offending'.

Performance metrics of many regulators require the collection of data on how many sites are visited annually, including the numbers of notices, fines and other sanctions issued. This emphasis on sanction, and a focus on rules that have been broken, is in keeping with a common perception that the best way to do environmental regulation is to 'freak them out to keep them complying', as one of my interviewees put it.

This focus on easily counted regulatory *outputs* risks undermining the advantages that come from a focus on environmental *outcomes*, such as a pollution event being averted or a potential polluter learning about the consequences of their pollution and working out how to address it so that it never happens again. Often such 'non-events' are not counted, recognised or rewarded in bureaucracies. This lack of recognition cumulatively impacts on the enforcement culture, so that the regulator tends to negate or overlook practices that lead to preventive outcomes. In Envisioning how inspections can be made more restorative, this issue of bureaucratic metrics is an important one in the Assessment phase. In order to Envision what is involved in moving towards a restorative approach for inspections, the following restorative values and principles should be borne in mind by the regulator:

* Value the relationship between the regulator and those being regulated
* See those being regulated as potentially having the capacity for learning and empathy, and seek to work with those capabilities first before resorting to more punitive approaches
* Approach an inspection in a situated way, namely with a consciousness of the surrounding environment, communities and networks of relationships
* View achieving the prevention of harm and repair of harm as important goals in addition to sanctioning proscribed behaviour
* Look for opportunities for poor environmental stewards to re-story themselves into guardians of environmental and community health.
* Encourage those who perceive themselves as environmental champions who can spread the message of valuing the environment to their business peers.

The theoretical basis to support these values and principles comes from a few different places. The core insight of the International Institute of Restorative Practices' (IIRP) social regulation window is that doing things TO people or FOR them is less effective than doing things WITH people.[6] IIRP argue that 'human beings are happier, more cooperative and productive, and more likely to make positive changes in their behavior when those in positions of authority do things with them, rather than to them or for them' (see Wachtel, 2013). This suggests that industries and businesses that sustainably improve their environmental stewardship over time are more likely to have had supportive regulators who did things *with* them. Further support comes from insights from the regulation scholar, Malcolm Sparrow (2000), who writes that a regulatory encounter should start with a conversation that involves the regulator seeking to understand the problem. It should elicit the point of view of the person who is being regulated, rather than assuming instant understanding or failing to appreciate the value of multiple perspectives. This approach was also explained as best practice on the basis of lived experience by one EPA Victoria officer, who stated:

> I'll usually say 'this is why I am here' but I think it's best to approach that from I'm here not necessarily to crack the whip, although there will be circumstances where that is more than appropriate to go there and take a no-nonsense approach. But mostly, you want to have a conversation and get them talking about what they do. I usually start most visits with: 'What is it you do here? Take me through your process'. And I think that's a very good way of understanding it because you can't really understand what action you want them to take or what you're going to do, or even what they might have done wrong, unless you understand what it is they do. So, you've really got to take that time to understand their process, so that later on, when you're actually going round the place and looking at things, you get an idea of what those sorts of things they do are.

Braithwaite (1989) has also argued that when offenders understand and assimilate the consequences of their actions, then they are more likely to take responsibility for repairing harm. Braithwaite (2020, p. 70) further observes that 'responsive regulatory theory argues these days that regulators should be trained in motivational interviewing to help businesses find their own idiosyncratic motivations to comply and to

---

[6] See https://www.iirp.edu/defining-restorative/social-discipline-window (last accessed 31 January 2022).

transform their corporate culture to a more ethical one. A suite of meta-analyses on the effectiveness of motivational interviewing with other forms of compliance show impressive effectiveness'. The years of experience of the many regulators within EPA Victoria whom we interviewed reiterated these core understandings of a relational approach. One officer we interviewed stated:

> And [regulatees] understand that, this person's interested in what I'm doing, and we're here together working out how to fix the issue that's impacting on the community together…Sometimes it is hard to get the company in that mindset, sometimes it does involve court cases and all that, and punishment, but I like to approach it from the start that no-one's deliberately trying to do anything wrong, which most of the time they're not …Unless, of course, you go out there and they're just horrible and then you go, well! That's when you do have to exercise that judgement, going yep, nope, this person's not interested, they're not going to respond until I really start to apply pressure and use these powers that I've got. I suppose that's the thing, having powers is one thing, knowing when to use them [is another].

Turning now to how a restorative approach to inspections may be Actioned, the following serve as starting suggestions:

* Inspections can start with a 'greeting' as the relational entry point that sets the scene, an acknowledgement of respect. Renowned feminist scholar Iris Marion Young (2002) also regards the greeting as a valuable feminist practice for eliciting social justice conversations.
* It is important to seek to understand in a holistic way the problem that has given rise to the inspection (or has been discovered), rather than just in terms of what amounts to infractions of legislation. In Victoria, this approach has recently been made significantly easier for regulators due to the new EP Act that creates a General Environmental Duty not to cause environmental harm (Part 3.2 of the EPA Act 2017). This is particularly useful in environmental regulation contexts where 'the rules run out' (Zacka, 2017) and the disjunct between legal harm and actual harm widens. It is applicable to all citizens, not merely those obviously engaged in activities with a risk of harm, and allows enquiry

into what reasonably practicable steps ought to have been taken to avoid harm, what systems are in place to prevent harm, and whether there was a failure in setting up prevention mechanisms.

* Communicate praise when environmental improvement is accomplished. A quantitative study of aged care home inspection teams showed that teams with a propensity to praise improvement achieved markedly greater improvement on compliance with quality of care standards in the two years after inspections of 410 aged care homes (Makkai & Braithwaite, 1993). In recognition of this, EPA Victoria integrated praise into the 'supports and sanctions' prevention strategy of its 2018–2019 Regulatory Work Plan.

Consider conducting an exit conference at the end of major inspections attended by management and worker representatives, community and advocacy group representatives. This is done, for example, in many aged care cases (Braithwaite, 2002, pp. 17–18, 63) where the regulators, managers, workers and affected persons sit in a circle around a room at the end of an inspection to agree in a participatory way how to repair harms detected during the inspection.

During fieldwork, I had occasion to observe an example of an incipient restorative approach to inspections. I went with an EPA Victoria officer to a site where waste transportation trucks were being housed in regional Victoria. The Country Fire Authority (Victoria) had called EPA Victoria because one truck was leaking oil down the main street of a local town, and the EPA had traced the truck to these premises. There, officers uncovered many problems, including storage of industrial waste on site when not authorised to do so. Inadequate oil storage was causing leakage from the property into the neighbouring environment. At a first inspection, the EPA officer issued a number of notices—minor works notices, a requirement to remove the contaminated soil and to provide better storage for waste oil. Several months passed and we arrived for a follow-up inspection and we were met by the business owner; she announced she had just sent an email through to the EPA detailing all the work she had done. During our walk around the property, she received lots of praise from the EPA officer—'Awesome, you've done lots of work, that's really good' and, 'You've done the right thing in talking to Council'.

The owner appeared really pleased with all the cleaning up and transformation of the site. She had done things she had not even been required to do, such as install spill bins. She said, 'I said to you we'd got it done … It's been a wake-up call for us. I honestly didn't know beforehand about what was required. Now I explain to the boys, this is how we have to store things, so we can walk around them. … You've done me a bit of a favour'. This shows the possibility of using relational regulatory approaches to persuade more generalised compliance beyond the specific regulatory infractions identified, particularly through attempting to encourage broader considerations about what avoiding environmental harm entails.

When we were leaving, the owner asked, 'What else do you want us to do?' She seemed to be genuinely inquiring, but also wanted to know if she was likely to get a fine or had managed to avoid it by engaging so proactively. This illustrates that relational and restorative strategies operate together with more traditional sanctioning approaches, rather than one always replacing the other, and that whilst this can cause tensions at times, some of these may be productive, as was the case in this incident.

A final observation that the owner made as we departed was the need for others in the area to also lift their game—and she noted that she had been acting as a bit of an advocate in this regard since transforming her site. This illustrated the way in which the inspection had been able to provide an opportunity for her to re-story herself as an environmental champion, and avoid the more obvious (and possibly more immediately appealing in the context of her business peers) alternative positioning as a recalcitrant 'victim' of regulatory authorities. In the words of Braithwaite (2008), it proved to be an opportunity for turning a laggard into a leader, to flip from game playing to an engaged and committed motivational posture. Such a re-storying was enabled by the EPA officer's approach of working with her, providing praise, and listening deeply to what she was saying and acknowledging her work.

As with approvals and licensing, inspections are part of the bread and butter of regulatory life that are not often considered in the context of restorative justice. Yet, as shown above, they too provide a fertile space to imagine a more restorative approach that is alert to the relational opportunities they provide.

4.3   Responding to Environmental Harm

The third domain explored in this chapter is that of responding to environmental harm. A very large domain, responding to environmental harm includes many different areas and scales of regulatory action, ranging from disaster response through to harm remediation and achieving justice. Ideally, responding to environmental harm will also always incorporate an element of preventing future environmental harm. A restorative approach in this domain is characterised by the following guiding principles (amongst others):

* Focuses on healing harm and repairing or building relationships impacted by the harm
* Creates opportunities for stories of impact to be narrated by those harmed and heard by those who were responsible for the harm and who have power to change the harmful practices
* Involves those who have been impacted in co-creating responses to the harm and designing the pathway forward
* Holds those responsible for causing harm accountable in ways that are meaningful to those who have been impacted.

The values that inform this orientation are all well-established principles of restorative justice (Forsyth et al., 2021). As responding to harm is the essence of much restorative justice practice in other regulatory domains, it is less difficult to Envision restorative alternatives than in other domains of environmental regulation. Some of the most widely applicable Actions to be considered in an environmental harm context are:

* Always ask the question of whether there is anyone whom someone should apologise to over the incident. There is often considerable fear over the legal repercussions of an apology, but an apology over harm arising from the work of an organisation is not an admission of criminal responsibility but is an important relational way to acknowledge the affected person's or group's feelings and experience;

* Enquire into which humans might most appropriately speak for non-human victims of the incident: Indigenous custodians, river-keepers, Friends of X River, Landcare volunteers, bushwalking clubs, sometimes farmers or fishers, environmental advocacy organisations, and so on;
* When assessing impact and harm, create opportunities for different forms of knowledge and multiple perspectives about what constitutes harm to be shared and valued. This may include Indigenous knowledge about the land, scientific knowledge, local and place-based knowledge;
* Create opportunities for those impacted to be involved in determining how to repair the harm, whether through a formal restorative justice conference or a small-scale convening of affected people. This may include Actioning activities they can participate in directly, such as citizen science, monitoring, weeding and planting, ecological education, and the like.

A few illustrations from our fieldwork show the potential of this approach, providing concrete examples about restorative initiatives already being utilised in Australia that may inspire regulators across the globe.

First, I relate a seemingly small example that was raised as a thought experiment by a regulator in the course of a discussion about opportunities she saw for restorative justice. She discussed a case she had been involved with concerning a site that had been used for a long time as an unauthorised rubbish dump and which EPA Victoria had determined needed to be addressed. She explained that she had issued a clean-up notice and then thought she should check there were no cultural heritage values that might be hurt by the clean-up. In enquiring with the body responsible for the site about whether there had been such a check, she found they had not even considered the issue. She further noted that no plan had been developed by that body about what to do with the site once it had been cleaned up, a process that involved reducing it to bare earth. She said that she envisaged a restorative approach whereby they could use the fine money (several thousands of dollars) to do some plantings, involve local Indigenous groups, and use this as an opportunity for creating publicity on the issue of illegal dumping. She thought this would have flow-on impacts for prevention of future harm, noting that if the land remained unrestored, it would be likely someone else would use it as a dumping ground. This was an excellent example of a regulator Envisioning a more restorative approach, but she noted that for it to occur in practice [Action], it would require champions from within various regulatory bodies.

The restorative imaginations of environmental regulators in Victoria have recently been bolstered by a suite of recent legislative changes that support restorative responses to harm in a variety of ways. Section 332 (1) of the EP Act provides for Restorative Project Orders, whereby 'The Court may order the person to carry out a project for the restoration or enhancement of the environment in a public place or for the public benefit, whether or not the project is related to the offence or contravention'. It also provides for an order to make payments into a 'Restorative Project Account' so that Indigenous groups or environmental NGOs might be funded to do the work of the Restorative Project Order. Section 331 provides for 'General Restoration and Prevention Orders' that can enable, and explicitly provide for, education and training in environmental stewardship or environmental compliance, corporate restructuring and diverse other measures suitable for environmental restoration and prevention.

An example of the application of these provisions was a case in 2021 where a company was charged after a spill that sent a black oily substance into the adjacent Paisley Challis wetlands.[7] The wetlands consisted of a series of ponds that filter out urban pollutants from storm water and provide bird habitat amongst grasslands, coastal saltmarsh and mangroves. The company pleaded guilty to the two charges laid by EPA Victoria, and the EPA then sought out expressions of interest for undertaking environmental projects, asking local groups to apply for a Restorative Project Order. Two proposals were presented to the court, one led by the local Land Council Aboriginal Corporation to investigate the Aboriginal cultural values of the Bunurong People in relation to the land and waters surrounding the wetlands. The other was by the local

---

[7] See https://www.epa.vic.gov.au/about-epa/news-media-and-updates/media-releases-and-news/court-orders-cash-to-wetland-projects-instead-of-fine (last accessed 31 January 2022).

Friends of Williamstown Wetlands to do habitat fencing to keep traffic and litter out of the wetlands, do indigenous revegetation, and remove weeds. The regulator and the company put forward a proposal for $80,000 to be given to the two projects in place of a conventional penalty, and the magistrate ultimately ordered that it was appropriate for the company to contribute $100,000.

Another enabling new provision for restorative justice deliberation in the new EP Act is the 'Declaration of an issue of environmental concern' (section 160). This, in turn, can trigger citizen proposals or EPA demands for a 'Better Environment Plan' (section 180). As the Act is still new, it remains an open question how regulatory conversations around Better Environment Plans will be conducted. However, it appears to be an invitation to restorative innovation. Local Indigenous leaders we interviewed made it clear that they are ready to use such plans. They see Better Environment Plans as an opportunity to work more holistically. 'Who decides what is a better environment?', they enquired. 'We do, on our land'. These Indigenous leaders used the example of the need for funding dingo (native wild dog) reintroduction programmes. 'Restorative justice should deliver that healing of country', but they feel it is not currently doing so.

A final important update the new EP Act brings into play is that it explicitly provides for adjournment of criminal prosecutions before the court, so that judges can empower stakeholders to convene a restorative justice conference. After holding the conference, the parties can then return to the court with a conference agreement for the judge to decide whether to endorse, reject or amend. This legislative provision means that Victoria will be more easily able to engage in the types of judicial-directed environmental restorative justice as in the neighbouring Australian state of New South Wales and in New Zealand. Hamilton (this volume) terms this a 'back-end model' of restorative justice and it is a common use of restorative justice conferences in regard to many other domains. Like other legislative reforms, its impact will ultimately depend on the degree to which it is activated (Forsyth, 2018) through its use and championing by key individuals within the justice system such as judges, prosecutors and other judicial officers.

5   Conclusion

Whilst many of the other chapters in this Handbook reveal the hope that ERJ will increasingly manifest itself in relation to large-scale environmental harms, this chapter's main normative argument has been that there is also an important role for it in the context of less visible, smaller scale, and day-to-day regulatory actions and encounters. Ideally, these small changes in multiple regulatory domains will cumulatively lead to significant change through collectively leading the regulatory encounter to be more regularly characterised by a genuinely collaborative ethos. Achieving environmental regulation that places greater emphasis on restorative approaches must be a journey of adaptive learning about the possibilities of what a restorative approach suggests, trying out multiple ways to better heal the harm done to ecosystems and communities. I have suggested how regulators might use frameworks of reflection and reform, such as the Adaptive Learning Loop, as a constructive and effective way to incorporate restorative approaches. In particular, this chapter has offered information, suggestions and illustrations about how restorative considerations can be included when Observing current practices, Envisioning better alternatives, and Acting to prevent and heal environmental harm.

References

Arnstein, S. (1969). A ladder of citizen participation. *Journal of the American Planning Association* 35(4), 216–224.

Braithwaite, J. (2002). *Restorative justice and responsive regulation.* New York: Oxford University Press.

Braithwaite, J. (1989). *Crime, shame and reintegration.* New York: Cambridge University Press.

Braithwaite J. (2020). Regulatory mix, collective efficacy, and crimes of the powerful. *Journal of White Collar and Corporate Crime,* 1(1), 62–71.

Braithwaite, J. (2008). *Regulatory capitalism: How it works, ideas for making it work better.* Cheltenham: Edward Elgar.

Boyd, J. (1995). *The essence of winning and losing.* Retrieved January 31, 2022, from https://fasttransients.files.wordpress.com/2010/03/essence_of_winning_losing.pdf

Connell, R. (2007). *Southern theory: The global dynamics of knowledge in social science*. Cambridge: Polity Press.

Forsyth, M., Cleland, D., Tepper, F., Hollingworth, D. Soares, M., Nairn, A., & Wilkinson, C. (2021). A future agenda for environmental restorative justice?. *The International Journal of Restorative Justice*, 4(1), 17–40.

Forsyth, M. (2018). 'Activating' the powers of law in the South Pacific. *Development Bulletin*, 80, 140–144.

Gunningham, N., & Sinclair, D. (2009). Regulation and the role of trust: Reflections from the mining industry. *Journal of Law and Society*, 36(2), 167–194.

Hurst, D. (2002). *Crisis & renewal: Meeting the challenge of organizational change*. Harvard: Harvard Business School Press.

Lilburn, J. (2020). *Environment Protection Act 1970 Section 20B Conference Report Application: Upgrade of sewage treatment plant in Elliott St, Warrnambool. Applicant: Wannon Region Water Corporation*. Retrieved January 30, 2022, from https://s3.ap-southeast-2.amazonaws.com/hdp.au.prod.app.vic-engage.files/5816/2761/4515/Appendix_3_20B_Conference_Report_August_2020.pdf

Llewellyn, J. (2019). Responding restoratively to student misconduct and professional regulation—the case of Dalhousie dentistry. In G. Burford, J. Braithwaite & V. Braithwaite (Eds.), *Restorative and responsive human services* (pp. 127–142). London: Routledge.

Makkai, T., & Braithwaite, J. (1993). Praise, pride and corporate compliance. *International Journal of the Sociology of Law*, 21, 73–91.

Schell-Busey, N., Simpson, S. S., Rorie, M., & Alper, M. (2016). What works? A systematic review of corporate crime deterrence. *Criminology & Public Policy*, 15, 387–416.

Senge, P.M. (2006). *The fifth discipline: The art and practice of the learning organization*. New York: Doubleday.

Sparrow, M. (2000). *The regulatory craft: Controlling risks, solving problems & managing compliance*. Washington DC: Brookings Press.

Tyler, T.R. (1990). *Why people obey the law*. New Haven: Yale University Press.

Wachtel, T. (2013). *Dreaming of a new reality: How restorative practices reduce crime and violence, improve relationships and strengthen civil society*. Bethlehem: The Piper's Press.

Young, I. M. (2002). *Inclusion and democracy*, Oxford: Oxford University Press.

Zacka, B. (2017). *When the state meets the street public service and moral agency* Cambridge: Belknap Press.

8

Participatory Governance and Restorative Justice: What Potential Blending in Environmental Policymaking?

Cristina Mihaela Vasilescu

1   Introduction

Climate change is currently a major threat for both the Earth and humanity. As reckoned by experts and international organisations 'we are in a climate and environmental emergency' (European Commission Executive vice-president, 2021)[1] and 'if we continue on our current path, we will face the collapse of everything that gives us security' (Attenborough, cited in United Nations, 2021).[2] Climate change threatens not only our ecosystems, but also our well-being, societal cohesion, economic development, political stability and global security. Indeed,

---

[1] See https://ec.europa.eu/commission/presscorner/detail/e%20n/ip_19_6691 (last accessed 20 February 2022).
[2] See https://reliefweb.int/report/world/climate-change-water-scarcity-and-security#:~:text=In%20 February%20this%20year%2C%20in,%2C%20and%20ocean%20food%20 chains.%E2%80%9D (last accessed 20 February 2022).

C. M. Vasilescu (✉)
Association Comunità Il Gabbiano, Milan, Italy
e-mail: cristina.vasilescu@gabbianoonlus.it

where climate change hits hardest, it triggers conflicts over access to and use of resources, resulting in significant migration fluxes. While climate change affects us all, it impacts vulnerable groups and poorest societies the most. Societies across the world have embarked on a process of ecological transition to turn around the growth-driven economic model guiding our societies. Such a transition implies a shift from an extractive economy to a climate-neutral economy in an equitable way, 'redressing past harms and creating new relationships of power for the future through reparations'.[3] The ecological transition questions not only countries' economic models, but also people's daily behaviour. Ecological transition must be just and equitable, otherwise its outcomes will never be.[4] To be effective, it requires 'unparalleled levels of global cooperation' (United Nations, 2021) from the international level to the local levels, from states to citizens. Hence, participatory governance becomes a prerequisite of the ecological transition, as acknowledged by European (i.e. Green Deal) and international (i.e. Agenda 2030) policies aimed at ensuring the ecological transition.

However, while participatory governance is generally depicted as positive and beneficial for all, its ability to ensure a truly just and shared decision-making process, including in the environmental field, has been questioned in the literature (Lee, 2020; Groves, 2017; Lee et al., 2015; Carpentier & Jenkins, 2013; Teles, 2010; Cornwall, 2000). In this context, turning our attention to other approaches that may overcome participatory governance gaps and complement it on the ground seems paramount for ensuring not only an effective ecological transition, but also a just one. Restorative justice emerges as such a potential approach, through its focus on harm to the environment and communities, just relations and citizens' active engagement in preventing and dealing with social injustices, including with regards to the environment.

While citizen participation is one of the values of restorative justice, restorative justice and participatory governance are often tackled and practised as distinct areas. Taking an interdisciplinary view, this chapter aims to unveil linkages between the two approaches, with the objective

---

[3] See https://climatejusticealliance.org/just-transition/ (last accessed 20 February 2022).
[4] Ibid.

being to enhance their blending for an effective and just ecological transition. The chapter starts with an overview of participatory governance and restorative justice, continues with exploring the linkages between the two approaches and ends by discussing the lessons to be extracted from the two approaches that can be combined to ensure a just, effective and smooth ecological transition.

2 Introducing Participatory Governance

2.1 Definitions and Development

Participatory governance is grounded in citizens' and other stakeholders' participation in policymaking, namely the process through which public institutions turn their political vision into concrete actions aimed at providing an answer to a collective problem that tackles citizens' unsatisfied needs or opportunities for public intervention (Dunn, 1981).

While several definitions of public participation exist (see Scottish Parliament, 2004; Involve, 2005; Creighton, 2005), what these definitions have in common is citizens' active role in shaping the social and economic development of their communities through participation in the policy arena (Wampler & McNulty, 2011). Hence, participatory governance foresees the possibility that citizens, as individuals, community or represented people (i.e. stakeholders) can influence policymaking and its outcomes. Creighton (2005) defines public participation as the embedment of citizens' concerns, values and needs into the governmental and corporate decision-making process in order to make better decisions that are backed by the public. Thus, public participation applies to decisions taken by public administrations and sometimes also private, where the decision-making is based on the interaction between the decision-makers and the participants engaged in the respective decision, and where, to a certain level, participants' input has an impact on the outcomes of the decision taken.

As Nabatchi and Leighninger (2015) point out, citizens' engagement in public policymaking is hardly new. Indeed, the relevance of participation for decision-making processes goes back in time to the Enlightenment period. It regained significant attention in the 1960s with the increasing mistrust towards the elites'[5] (i.e. group of people who hold a significant amount of political and economic power or skills in a society) capacity to take effective decisions, especially in the American and Western European contexts. Several authors questioned the capacity of bureaucratic elites (i.e. public administration officials) to act for the public good, pointing out their interest in maintaining their authority at the expense of public input (Wright Mills, 2000; Lee et al., 2015). Concerns over the elites' capacity to act for the public good also stem from worries about the social and ecological effects of their decisions (Groves, 2017) or the pursuit of specific ideological positions, despite the known potential harm. Since the 1960s, citizens have started voicing their demand for more power in policy decisions that impact their daily life and future well-being in contrast to the elites' claims for authority in public policy-making in areas such as the environment, health and urban planning. Citizens' opposition to public decisions in spatial and urban planning (also known as Not in My Backyard, commonly referred to by the acronym of NIMBY since the 1980s) has also started to emerge more or less in the same period (Lee et al., 2015; Groves, 2017). Citizens' engagement in policymaking has been acknowledged as an effective way to ensure that social, economic and ecological progress is achieved for all citizens and, hence, as a good objective worth developing and institutionalising (Lee et al., 2015). Since the 1980s, citizens' engagement in policymaking has developed in the public policy arena as a valuable way to deal with the so-called democratisation deficit[6] and as a management (e.g. participation in the context of total quality management) and socialisation (e.g. promoting social solidarity) tool.

---

[5] In this chapter, the understanding of 'élite' draws on Wright Mill's concept of 'power élites', according to whom: 'the power élite is composed of political, economic, and military men (…) involved in virtually all widely ramifying decisions' (Wright Mills, 2000, pp. 276–277), with that proposed by Sartori (1987), who considers élites as a 'value' reference group of people who possess capacity and excellence.

[6] Council of Europe defines 'democratisation deficit' as the failure to comply with democracy principles such as, individual autonomy (referring to the fact that people should be able to exert control over their own lives within reason), or equality, (everyone should be in the position to influence decision-making processes that affect societies).

The increase in the relevance of public participation corresponds with an evolution of its intensification. If, at the beginning, public participation was mostly confined to information and consultation, nowadays it is mostly associated with community active engagement, namely partnering with citizens in all policy phases to define alternatives and select potential policy solutions (co-design), to deliver public policies (co-production), and to monitor and evaluate them (co-evaluation). In this model, citizens are not seen as passive beneficiaries, but as active actors who take over a part of the decision-making responsibilities in the policy process.

The growth in the intensity and breadth of public participation has made it a standard practice nowadays in democratic societies. Pimbert and Wakeford (2001, p. 23) consider that 'democracy without citizens' deliberation and participation is ultimately an empty and meaningless concept'. Public participation is now extensively practised in various policy fields (e.g. from territorial planning to innovation policies), and this includes policymaking for environmental protection.

2.2 Participatory Governance in the Environmental Field

Recourse to participatory governance in the environmental field gained wide attention following growing awareness that scientific knowledge does not manage to fully interpret and respond to socio-ecological challenges (Groves, 2017) and that socio-ecological transformations cannot be achieved without citizens' active contribution. Thorough responsiveness requires collaborative relations between all actors involved in the environmental policy arena. Several authors (Boehmer-Christiansen, 1995; Graham & Healey, 1999) have pointed out that collaborative relations are more relevant than technical solutions or technological innovation for the achievement of sustainability targets. This is due, on the one hand, to the fact that territories (be they cities/towns or rural areas) are governed by social-ecological relations and, on the other hand, to the presence of multiple actors with competing needs and objectives in the environmental policy arena. Often the relationship between the social-ecological features of territories and actors' interests in the environmental policy arena clashes, leading to conflicts. Management or prevention of conflict requires use of an approach that allows for collective construction of responses to competing needs and objectives. The management of the relationship between social-ecological aspects of territories and actors' competing interests in the policy arena also implies a change in the policy process, from top-down policymaking to bottom-up, involving all actors in the construction and delivery of public policies in the environmental field (i.e. participatory governance).

Participatory governance in the environmental field has also spread as a response to prevent or to deal with the NIMBY phenomenon (Suna et al., 2016). Adopting a participatory governance approach to environmental policymaking allows public administrations to understand citizens' concerns and to take them into consideration in the decision-making process, as well as being a way of improving citizens' knowledge and competence on the respective policy. In turn, citizens' engagement in policymaking increases the social legitimacy of the respective public policy and helps in reducing or preventing a potential NIMBY conflict.

In the environmental field, participatory governance has been framed both as a social process, that is, collective debates about and construction of alternative solutions and identification of means to environmental challenges, and as an administrative tool, that is, definition of strategies to achieve the goals of public institutions (Castro, 2007). However, as discussed later in this chapter, reliance on the latter frame to characterise participatory governance neglects the multiplicity of social values and the needs beneath it and conceives of environmental processes as an efficiency issue rather than as one of social justice (Castro, 2007; Paavola, 2007).

The relevance of participatory governance in environmental protection has resulted in its institutionalisation through its mainstreaming in European Union and national environmental legal frameworks: examples include the EU Water Framework Directive (2000), the Directive on Access to Environmental Information (2003), the EU Floods Directive (2007) and the Marine Strategy Framework Directive (2020). Participatory governance is also mainstreamed in the main international policy frameworks for ensuring a just ecological transition: the 2030 Agenda for Sustainable Development (2015) and the EU Green Deal

(2019b). Moreover, participatory governance is embedded in tools used in environmental decision-making processes, such as the Environmental Impact Assessment and the Strategic Environmental Assessment.

3   Introducing Restorative Justice

3.1   Definitions and Development

In the literature, restorative justice is often framed taking into consideration two distinct but interrelated perspectives: the process activated and the desired outcomes. While some of the definitions focus more on one or another dimension of restorative justice, others (e.g. European Forum for Restorative Justice) encompass both dimensions of restorative justice. For example, according to the European Forum for Restorative Justice (EFRJ), restorative justice is an 'approach of addressing harm or the risk of harm through engaging all those affected in coming to a common understanding and agreement on how the harm or wrongdoing can be repaired and justice achieved. (…) Restorative processes restore safety, security through bringing people together to undo injustice, repair harm and alleviate suffering' (EFRJ, 2018, pp. 3, 7).

UNDOC (2006) provides a restorative justice framework that encompasses both its ability to restore relations after an offence/injustice as well as its preventive nature. In this framework, restorative justice is conceived of as transformative, as 'it not only embraces restorative processes and steps to repair the harm, but it also focuses attention on structural and individual injustice (…) by identifying and attempting to resolve underlying causes of crime (poverty, idleness, etc.)' (UNDOC, 2006, p. 104). Furthermore, it challenges people to apply restorative principles in their daily lives in order to prevent crimes/wrongdoings.

Irrespective of whether the focus is on the process or on the outcomes of restorative justice and on the management of crimes/injustices or on their prevention, what these definitions have in common is the focus of restorative justice on: injustices or crimes; attention to restoring social relations, not only between victims and wrongdoers, but also between them and the society at large; and the active engagement of all parties (wrongdoers, victims, society) in this restoration process.

From this latter point of view, restorative justice shares with participatory governance a key element, that citizens' engagement as stakeholders with resources can be mobilised to enable a restorative process in the community. The role of community's active participation in the restorative process has been widely acknowledged in the literature (Zehr & Mika, 1998; Sullivan & Tifft, 2001; McCold, 2004).

As with participatory governance, restorative justice has developed as a response to concerns regarding, on the one hand, the limited effectiveness of public administrations (i.e. public justice system) in preventing crimes, reintegrating offenders, reconciling conflicts and supporting victims (Gal, 2016) and, on the other hand, increasing penal populism,[7] which often results in a strong punitive approach to law breaking (Pratt & Miao, 2017). Penal populism has been growing in the context of several trends, such as the decline in deference towards and trust in elites (e.g. politicians, bureaucratic officials, experts), increase in global uncertainties and complexity of public problems, and changes in how crimes are framed in mass media and social media, which depict crimes as episodes that could easily occur to anyone, resulting in increased public demand for greater punishment of crimes (Pratt & Miao, 2017).

Some of the reasons (such as mistrust of elites, increase in uncertainties, and complexity of public problems) that have contributed to the rise of penal populism and the subsequent identification of innovative solutions (such as restorative justice) are the same ones that explain the rise in the recourse to participatory governance.

3.2   Restorative Justice in the Environmental Field

Restorative justice has entered the environmental agenda, particularly in connection with environmental damage produced by heavy industries (McCauley & Heffron, 2018), protection of Indigenous people's rights

---

[7] Penal populism refers to criminal responses driven by their popularity amongst voters rather than by their effectiveness in dealing with crimes/injustices (Dobrynina, 2017).

and conservation of their culture, as well as ensuring a just and inclusive ecological transition (McCauley & Heffron, 2018; Pali & Aertsen, 2021).

By shifting the perspective adopted to injustices in the environmental field away from the laws broken and criminals to be punished, and instead to the harm done, restoration of relations between the wrongdoer, people harmed and the community at large, and the reparation of that harm restorative justice, this has the potential to deal with several complexities of traditional justice systems in the environmental field, such as:

* low reporting, prosecution and sentencing (Pali & Aertsen, 2021)
* hard-to-identify victims (Stark, 2016)
* present and future communities victimised through damage to natural resources, public property or to the general environment (Stark, 2016)
* 'non-traditional' victims, such as nature (natural landscape, animals, etc.) or heritage resources (Skinnider, 2011; Preston, 2011)
* exposing poor countries or Indigenous populations to environmental damage through the exploitation of their natural resources by foreign companies (Gibbs, 2009)
* unforeseen harms that will be managed throughout the transition to a post carbon world (McCauley & Heffron, 2018)
* systemic injustices, relevant power imbalances and significant victim vulnerability (Pali & Aertsen, 2021).

In what follows, the chapter explores potential linkages and differences between participatory governance and restorative justice in goals, values and expected outcomes, through a compare-and-contrast analysis, to understand better, how the two approaches can complement each other in pursuing a just ecological transition.

4   Goals of Participatory Governance and Restorative Justice

4.1   Goals of Participatory Governance

The main goals of participatory governance consist of (a) overcoming shortages of representative democracy; (b) improving the quality of public policymaking; (c) preventing and managing conflicts that can hinder the decision-making process; (d) empowering and educating the community.

Overcoming shortages of representative democracy. By placing power in the hands of citizens, in particular of those affected by the decision at stake, by engaging them as co-determinants of a decision-making/delivery policy process, by balancing élites' power and by building trust in institutions, participatory governance is expected to overcome the democratic deficit of representative democracy (Webler, 1995; Depoe et al., 2004; Creighton, 2005; Wampler & McNulty, 2011). In turn, this is expected to increase citizens' ownership of the public policies process.

Improving the quality of public policymaking. Societies today face complex social, economic, environmental and justice problems and related ethical concerns. Management of these problems requires a large amount of resources, which a sole actor does not possess. Moreover, the engagement of various actors in all phases of the policymaking process is essential to ensure effective circulation of resources. Through mobilising resources owned by citizens, and the promotion of a collaborative relationship between policymakers and citizens, participatory governance is expected to shape better public policies (Bobbio & Zeppetella, 1999). Citizens' engagement in the process allows for the inclusion of their values, rationality, concerns and onsite information in the policymaking process, from start to finish. In turn, this is expected to trigger a reframing of problems and the identification of new policy directions, shifting beyond existing policy frames towards more citizen-driven outcomes (Roniotes et al., 2015).

Preventing and managing conflicts that can hinder the decision-making process. Johnson (1993) contends that with the increased stakeholders' influence in public decision-making, conflicts are inevitable. In his view, promoters of public involvement will achieve better decisions, gaining a competitive advantage that allows them to get on with a specific policy intervention. Bai et al. (2010) underline that participatory governance can reduce conflicts as, through dialogue, citizens start to appreciate different positions amongst themselves. Furthermore, engaging the public is expected to lead to reduced misunderstandings, an increase in trust and, thus, to fewer conflicts and delays (Roniotes et al., 2015). By preventing/overcoming conflicts in policymaking, participatory governance aims to increase the social legitimacy of the decision-making process.

Empowering and educating the community. Participatory governance supports reinforcement of the communities' capacities through the acquisition of knowledge on a specific issue and an improved understanding of bureaucratic jargon, laws and regulations. Opening the black box of decision-making and implementation processes allows citizens to understand the complexity of these processes. Furthermore, it allows citizens to gain insights into the functioning of public institutions and the timing of both decision-making and implementation of decisions. Increasing citizens' knowledge on public policy processes not only increases their trust in public institutions, but also contributes to increasing their trust in their own capacity to meaningfully take part in public policy processes. In turn, this supports citizens' participation in the policymaking process (Lee, 2020).

4.2   Goals of Restorative Justice

Three main goals of restorative justice, including in the environmental field, are: (a) repair harm done to individuals, communities and environment; (b) empower communities and promote their well-being; (c) reaffirm community values in the justice process. These are well-documented by the literature as detailed below.

Repair harm done to individuals, communities, environment. Harm reparation and restoration of relations is the primary goal of restorative justice, through addressing the damage caused, holding offenders responsible for the damage, addressing victims' needs, making the community aware of its responsibilities in preventing further harm and increasing the community's participation, on the one hand, in offenders' social reintegration and, on the other hand, in victims' empowerment. This focus makes restorative justice particularly relevant for the environmental field, an area where 'reporting, prosecution and sentencing is extremely low' (Pali & Aertsen, 2021). Furthermore, in the environmental field, restorative justice aims to protect the environment, heal current damage and prevent future harm, by giving the environment a voice, validity and respect and, through these mechanisms, transform the relationship between humans and the environment (Gal, 2016).

Empower communities and promote their well-being. Restorative justice aims to contribute to communities' well-being and empowerment through mending social fractures and healing wounds, rebuilding trust in each other and society, enhancing individual and collective responsibilities towards preventing and managing harm, enhancing social support and strengthening relations, increasing communities' collective efficacy and tackling the causes behind social tensions or conflicts (Gerkin, 2012; Patrizi, 2019). Further, restorative practices aim to improve knowledge and build new shared practices in achieving justice, which includes development of conflict resolution skills and individual and community empowerment (Patrizi, 2019).

Reaffirm community values through increasing citizens' ownership of the justice process. Several authors (e.g. Zehr, 1990; Christie, 1977) have noted that the community's full delegation of conflict management and justice process to public administrations has resulted in its weakening and in the distancing of victims, families and the society at large from the justice process. According to Christie (1977), restitution of conflict resolution back to the people directly involved and their respectful and safe management of disputes aims to make communities stronger, to empower victims, to support offenders' reintegration and to clarify social norms regulating community life.

Restorative justice represents an approach through which communities return to the centre of the justice process, as it offers people the opportunity to participate in the justice process, to collectively define its needs, to reinforce community values and to reaffirm its collective consciousness. Furthermore, through active participation in the justice process, community members can discover each other, understand social injustice and learn not to fear 'the other' different from 'us' and to deal peacefully with conflicts/crimes and social injustices. In turn, this is expected to contribute to strengthening the community, restoring social relations, healing current harm and preventing future harms (Gerkin, 2012; McCold, 2004; Zehr & Mika, 1998; Zehr, 1990). Through recognising and enhancing the community's responsibility in the justice process, restorative justice aims to democratise it (Preston, 2011).

4.3   Summary

Participatory governance and restorative justice have in common the aim of empowering communities and democratising the decision-making process. Nevertheless, both follow different pathways to achieving these aims: ensuring social legitimacy, efficiency and effectiveness of a specific project/decision through citizens' participation in participatory governance versus citizens' participation as stakeholders responsible for the well-being of their communities and being in possession of relevant resources needed to fully participate in a dialogical construction of a shared vision for a community impacted by harm. Furthermore, while both approaches aim to prevent conflicts, their vision of conflict differs: in participatory governance, conflict is an obstacle to policymaking, needing to be reduced/overcome to increase the efficiency of policymaking; whereas in restorative justice, conflict is an opportunity to enhance integration and social inclusion (Chapman, 2019), and to create social cohesion to heal communities and prevent future harm. Acknowledging conflict as a learning and relational opportunity rather than as an obstacle allows for enhancing the dialogue between multiple social values and needs. This is particularly relevant for the environmental field characterised by actors' competing interests (Paavola, 2007). Thus, while restorative justice aims to achieve social justice, participatory governance pursues an instrumental goal of legitimising, improving and increasing the efficiency of policymaking (Castro, 2007; Lee et al., 2015).

5   Values of Participatory Governance and Restorative Justice

5.1   Values of Participatory Governance

Participatory governance relies on the following values:

* *Transparency of the process*, which implies the diffusion of clear, reliable and accessible information on the process, communication of all participants' interests and participants' involvement in all stages and decisions regarding the process, including in the definition of the rules of participation.
* *Openness of the process*, implying that the process is open to all those interested in a specific decision.
* *Participation on equal basis*, consisting of ensuring equal access to participation of all parties interested in the process, with a particular focus on people affected by the decisions but who are generally distant from the public participation process (e.g. marginalised groups, women).
* *Fairness*, providing citizens with the opportunity to assert and challenge claims and participate in decision-making.
* *Dialogue* between multiple needs and social values based on positive communication and active listening.
* *Affecting decisions*, public participation bases its logic of intervention on the promise that citizens' contribution will influence the contents of the final decision regarding a specific intervention.

5.2   Values of Restorative Justice

Restorative justice draws on two grounding principles: (i) (human) beings are worthy, irrespective of their features and actions; (ii) (human) beings are interconnected with each other and with the world (Evans & Vaandering, 2016). Core values of restorative justice stem from and have regard to these two principles:

* *Respect*, meaning 'looking again from the point of view of the other, to putting one's self in the other's shoes and then respond[ing]' (Evans & Vaandering, 2016, p. 32). Respecting others directly relates to the 'all people are worthy principle'. As all (human) beings are worthy, all (human) beings should be respected.

* *Dignity*, referring to 'worth … that cannot be substituted. People have dignity because of the essence of who they are and cannot be replaced' (Evans & Vaandering, 2016, p. 32). Dignity and respect are interconnected, and both reflect the principle of the worth of all (human) beings.

* *Mutual concern*, consisting of '(…) the reciprocal, interconnected caring' (Evans & Vaandering, 2016, p. 33), which concerns the interconnected nature of our societies. It assumes that a society ensuring the well-being of all its components is a society in which each one cares for others and actively strives to ensure their well-being.

* *Solidarity*, consisting of 'the interdependence and diversity of people and the critical importance of the quality of relationships to individual's wellbeing and social cohesion. It provides an opportunity to reconnect and to learn how to fulfil one's obligations to each other's wellbeing' (EFRJ, 2021).

* *Truth*, referring to recognising everyone's truth and everyone's need for his/her truth to be listened to and recognised.

* *Justice*, an intergenerational relational ethic engaging people in just relations and making them more accountable for their actions to prevent injustices and to repair existing harm. To achieve restorative justice, the process needs to be fair, open, transparent and not dominated by any party. Furthermore, it needs to involve all those affected, that is, wrongdoers, victims and community members (EFRJ, 2021).

5.3 Summary

While restorative justice embeds most of the values of the participatory process, it goes beyond them, proposing a new 'pair of lenses' with which to see the world (Zehr, 1990). The perspective proposed by restorative justice does not restrict the view to reaching a specific decision, as in the case of the participatory process, but expands it to reveal our way of being human, of connecting with each other and with the surrounding environment, and of belonging to the society in which we live. Thus, wearing the lenses of restorative justice allows people to acknowledge the potentialities and vulnerabilities of others; engage with each other in order to overcome and prevent wrongdoings and injustices; and see life as 'an opportunity for social engagement rather than for social control' (Evans & Vaandering, 2016, p. 36). Wearing either the participatory governance 'lenses' or the restorative ones can significantly impact how people participate in the social, political and civic life, for example, social engagement for nurturing the well-being of society in the case of restorative justice versus activism for shaping a specific decision in the case of participatory governance. Furthermore, wearing the restorative or the participatory 'lenses' impacts on the type of participation: that is, bottom-up participation, often outside of the institutional arena, with communities having a relevant say in shaping their participation in the case of restorative justice versus mostly top-down participation, more often limited to the institutional arena in the case of public participation in policymaking. In fact, one of the criticisms raised in the literature against participatory policymaking consists of the fact that their activation relies too much on the initiative of public institutions and on the level of power they are willing to share with the citizens as owners of the community's engagement process (Carpentier & Jenkins, 2013).

6   Outcomes of Participatory Governance and of Restorative Justice

6.1   Outcomes of Participatory Governance

The literature on participatory policymaking reveals several advantages and shortfalls of participatory policymaking. With regards to advantages, the literature acknowledges that participatory policymaking can help develop decision-makers' and citizens' capacity to deal with complex social problems, through learning about each other's views about a specific issue, opening the 'black box' of policymaking and providing understanding on the logic behind public decisions. In turn, this favours an increased mutual trust between public institutions and citizens and better framed and more innovative public policies. Higher levels of mutual trust between and amongst public institutions and citizens involved in participatory policymaking corresponds an increase in the existing level of social capital.

Furthermore, citizens' engagement in policymaking increases their ownership of the respective process, contributing to reducing opposition to the final decision reached, even for stakeholders who might be less satisfied with the final decision. This triggers diminishment of the economic and social costs associated with protesting against public decisions, making policymaking more efficient. Participatory policymaking also contributes to citizens' increased awareness of their role and responsibilities and of their potentiality as a civic group.

When it comes to gaps in participatory policymaking, the literature (Lee et al., 2015; Groves, 2017; Carpentier & Jenkins, 2013) reveals that the participatory process can be undemocratic and reproduce the same inequalities it aims to challenge, when it is decoupled from the local context and culture, ignores power distribution and structures and is used as a top-down management tool in which the citizens' role is reduced to advising public institutions on a specific issue instead of challenging social inequalities. For instance, investigation of environmental remediation processes, which include participatory processes, has revealed that when local history, cultural identity and heritage of local inhabitants failed to be acknowledged in the environmental remediation process, social injustices were reinforced (Lee, 2020). Moreover, critics of participatory policymaking contend that while the participatory process tries to avoid conflicts, conflict should be acknowledged as an 'essential prerequisite to consensus because it forces recognition of groups, issues and interests that might otherwise be marginalized' (Lee et al., 2015, p. 15). Wamsler et al. (2020) shows that in several cases contestation resulted in positive sustainable outcomes (i.e. adoption of nature based solutions). According to Teles (2010) and Walker (2009), often the participatory process fails to empower the broader society, as participation is instrumentally used as a counter-pressure to citizens' power and future movements. Furthermore, this also puts at risk the sustainability outcomes of participatory processes, as often municipalities lack the necessary capacities to favour a citizens' involvement that goes beyond interactions as a 'technocratic compromise' and encourages both democratic approaches and sustainability results (Wamsler et al., 2020).

The capacity of participatory policymaking to empower and democratise society is also limited by the 'induced' (Wampler & McNulty, 2011) or 'invited' (Cornwall, 2000) nature of most of the participatory policymaking processes, fostered by the wide institutionalisation of participatory policymaking. 'Induced/invited' participation refers to participatory spaces opened up and driven by public authorities, where citizens are invited to contribute to the decision-making process but where the rules of the process and the extent to which citizens' contributions are taken up in the final decision remain in the hands of public institutions that drive the participatory process. To truly empower societies, participatory governance should favour development of 'created participation', defined as spaces for engagement created by citizens themselves and rooted in their shared identity and common interests (Stockholm Environment Institute, 2019). Such processes have started to emerge, especially in environmental policymaking, even though they still constitute a limited minority.

Another critique raised in the literature regards the increased professionalisation of the participatory policymaking, which focuses on short-term citizens' mobilisation and individualised action, failing thus to truly empower communities and democratise the decision-making process. In other words, it is argued that 'deliberation consultants build public legitimacy for the retrenchment of programs, they enhance the reputational capital of the consultants' clients, and they encourage citizen mobilisation focused on short-term, individualised action' (Lee et al., 2015, p. 17).

6.2   Outcomes of Restorative Justice

In contrast to participatory policymaking, which is generally 'owned' by public institutions and shaped by professionals, in most restorative justice interventions it is participants (i.e. community members, offenders, or victims) themselves who have a decisive role in shaping the engagement process and the final decision regarding the agreement, including restorative actions to be implemented by the offender (Stark, 2016) In restorative justice interventions, community members are often involved as facilitators of the processes instead of professionals (Gal, 2016). This favours community trust building and participation and empowers the community to deal with such processes on its own in the future. Through active engagement in restorative interventions, the community shares knowledge about community values, principles and social injustices, constructs a shared vision of the truth and of restoration actions, develops social relations and social support and acquires resources and competencies to peacefully deal with social injustices by itself in the future (Gal, 2016). Ultimately, the community's active engagement and shared dialogue in restorative interventions triggers a sense of community, including for wrongdoers, which favours citizens' assumption of responsibility for community care (Gal, 2016; Preston, 2011). This assumption of responsibility favours the democratisation of the decision-making process and criminal justice, as citizens act as equal partners at each level of decision-making (Stark, 2016).

Again, in contrast to participatory policymaking, which often reinforces social injustice, restorative interventions aim to create spaces where issues of social injustices can be questioned through activating a healing process of the harm suffered by both direct victims and the wider community and the recalibrating of power relations between those involved in such injustices. In this healing process, ideally victims become more empowered, as they acquire a sense of worth and social esteem and wrongdoers become accountable and responsible for their actions. As underlined by Al-Alosi and Hamilton (2019), acknowledgement of offenders' responsibilities in restorative justice interventions is superior to any sentence imposed by a court of law, as it considers the individual circumstances and is part of a larger process of restoration of community relations and understanding of the causes behind the wrongdoing, rather than focusing on a single intervention (as is the case in participatory policymaking). This holds true in the environmental field, where direct victims are the landscape and animals, which do not have the ability to explain the harm suffered directly by themselves and where the harm caused impacts extensively on the life of the entire community. Thus, through giving the environment a voice and recognising it as a victim for the healing and the repair of which the wrongdoer is held accountable, restorative justice allows for a transformative relationship between wrongdoers and the environment, and between the wrongdoer and the harmed community (Gal, 2016).

6.3   Summary

The outcomes of participatory governance are characterised by light and shadow. While participatory governance favours improvement in the capacity of the community as a whole to deal with complex issues in a more legitimised and efficient way, and to find more innovative solutions to these issues, it also triggers professionalisation of the decision-making and implementation of participatory processes, induced/invited participation, reproduction of social inequalities, when power imbalances are not embedded into the participatory process, and conflict avoidance.

Through its focus on community well-being grounded in peaceful and positive social relations and community responsibility over social injustices/wrongdoings, restorative justice moves beyond the single issue and manages to address social injustice and democratisation of decision-making processes holistically, filling the gaps not covered by participatory governance. By engaging the community as an equal partner in deciding on the shape of the engagement process and on the agreement to be reached, restorative justice not only empowers the community, but also makes it responsible for its collective well-being, including from an environmental point of view. Furthermore, through investigating causes underlying the harm done to the community, restorative justice questions social justices and activates transformative processes. This transformative element is even more powerful in the case of the environment, as restorative justice gives a voice to the environment and acknowledges it as a victim with which relationships have to be restored.

7 Conclusions: Towards a Just Ecological Transition on the Ground

As noted above, the goals and values of participatory governance and restorative justice crisscross. When it comes to goals, both participatory governance and restorative justice aim to empower communities, democratise the decision-making process and prevent conflict. Furthermore, restorative justice embeds most of the values of the participatory process. However, both approaches pursue a different outcome on the ground: community well-being and social injustice/crime reduction (restorative justice) versus an increase in the efficiency, effectiveness and quality of decision-making processes (participatory governance). Nevertheless, both approaches can—and should—complement each other on the ground to achieve a just ecological transition.

Restorative justice can support the goals and delivery of participatory governance in various ways. Restorative justice, particularly in the environmental field, bases its rationale for intervention on procedural inequalities inconsistent with the objectives of participatory democracy (Stark, 2016). In restorative justice, 'power is embodied in the support and expectations that people have for one another's humanity' (Evans & Vaandering, 2016, p. 63) and people are viewed as actors to be honoured. When providing support in a balanced and mutual way, as in restorative interventions, power is used in a constructive manner and is available to all, triggering people's engagement with each other (Evans & Vaandering, 2016) and with the authorities. Thus, applying restorative justice allows for the fulfilment of the right of communities to participate as equal partners at all stages of the policymaking process (Stark, 2016).

Even when not purposely pursued, restorative justice can favour participatory governance, particularly in disadvantaged communities where participatory governance may be significantly undermined by the lack of trust between local institutions and the community (Lee, 2020). Restorative justice does this through (re)building trust in communities by investigating the harm a community feels and promoting an empathic, peaceful and constructive dialogue between those considered the harmers, those suffering and the community at large, to jointly identify solutions for dealing with the harm that account for the needs of all actors engaged in the dialogue. It also implements a process of accountability for harm done and for solutions. In restorative justice, accountability is not passive, but rather active, meaning that all community members are accountable to a 'shared set of expectations and to the others within that community' (Evans & Vaandering, 2016, p. 92). This is particularly relevant for participatory governance, as trust is one of the main ingredients for its effectiveness (Depoe et al., 2004).

Furthermore, restorative justice can support the delivery of participatory governance through paying attention to understanding the justice perspectives of all stakeholders involved in the process. For instance, in analysing community redevelopment projects, based on collaborative partnership between citizens and authorities in Spartanburg (US), Lee (2020) points out that it was largely stakeholders' different understandings of justice that impeded delivery of redevelopment practices. According to Lee (2020), this happened because stakeholders' justice perspectives impacted how they formulated and delivered community redevelopment practices, as well as their motivation to participate in the community redevelopment process. Lee (2020) concludes that the concept of justice should not be defined by a single actor, but by all actors involved in community redevelopment projects. Through provision of a peaceful forum of discussion and commitment, targeted at all community members, where all parties are equal, valuable, and whereby everyone's values, principles and needs are acknowledged, listened to, respected and elaborated jointly to find a way to meet all of them individually and collectively, restorative justice can favour the creation of a shared vision of justice for all stakeholders. This favours delivery of specific interventions (e.g. community sustainable development) that meet community members' needs. In restorative justice, achieving justice means ensuring that 'people get what they need' (Evans & Vaandering, 2016, p. 51). This supports the delivery of participatory interventions and enhances the well-being of the respective community. As argued by Sullivan and Tifft (2001, p. 167): 'We develop our potentialities as human beings and enhance our collective wellbeing when our needs are respected, expressed, listened to, defined with care and ultimately met'.

Through paying specific attention to active listening, empathic communication and connecting people, restorative justice can contribute to the achievement of some of the critical success factors of the participatory process identified in the literature (Depoe et al., 2004, p. 32), such as 'active listening: courtesy or an absence of discounting verbal and non-verbal behaviour; early and ongoing voice; [...] reflection of genuine empathy for the concerns of other perspectives, dialogue, debate and feedback'.

Moreover, restorative justice can support participatory governance through the acknowledgement of conflict as a learning and relational opportunity. Through involvement in restorative interventions, people learn to manage conflict in a positive and constructive way, paying particular attention to understanding all parties' positions and meeting all parties' needs. Learning to deal with conflicts in a positive manner is also beneficial to participatory governance, as minimal and controlled levels of conflict when associated with debate can favour a higher effectiveness of the participatory process (Maiello et al., 2013).

When it comes to how participatory governance can support restorative justice, some insightful warnings can be derived from their implementation, for careful consideration in any large-scale promotion of restorative justice:

Mandated top-down institutionalisation and over-professionalism. Particular attention should be paid to the institutionalisation of restorative justice. As learnt from participatory governance, while institutionalisation can offer opportunities for expanding restorative processes and for ensuring their sustainability over time, it also risks turning restorative justice into a top-down process governed by public administrations, thereby weakening community's ownership of the restorative process. Restorative justice should, therefore, search for a balance between bottom-up and top-down processes that can reinforce its value within the community and avoid being left exclusively in the hands of public administrations. Awareness-raising interventions, which allow stakeholders, including citizens, to understand the individual and collective benefits of participating in restorative justice, should be strengthened.

Analysis of participatory governance also shows that institutionalisation often comes with over-professionalisation of the participatory process. Over-professionalisation may act as a self-censor for citizens, particularly for those at risk of social exclusion, as they perceive science and expertise as having a higher authority compared to them (Stockholm Environment Institute, 2019). While the development of specific skills and competences of restorative justice is particularly relevant for the delivery and diffusion of restorative justice, over-professionalisation may put at risk its objectives of bringing conflicts back to be dealt with at the community level. One way to approach this issue is to train citizens who can act as community facilitators in restorative justice interventions on a volunteer basis.

Participatory governance analysis reveals that institutionalisation is not free of conflicts. Fear that the participatory process may be appropriated by special interests, or may not be representative, or that citizens lack adequate technical knowledge to make decisions, and/or potential loss of the policy agenda, can give rise to opposition from civil servants. In the quest for a balanced bottom-up and top-down approach, restorative justice should foresee spaces of dialogue, learning and knowledge-sharing between bureaucratic staff and stakeholders, including citizens, to enhance mutual trust between them. Furthermore, restorative justice should encourage the participation of bureaucratic staff as 'ordinary' community members to bottom-up restorative processes, contributing to such processes with their 'technical' knowledge.

Lessons from participatory governance reveal that participatory policymaking may face resistance from politicians. Political opposition is particularly relevant in the case of restorative justice, which, against a background of increased penal populism, remains a sensitive topic on the political agenda, despite being embedded in international juridical and policy frameworks. Restorative justice should pay particular attention to creating spaces of dialogue with opponents, not only at micro-level (i.e. community), but also at macro-level (e.g. regional, national, European) and aim to widely disseminate its benefits for the whole community.

Under-resourced and short-term ad-hoc process. To be effective and sustainable over time, participatory policymaking requires a cultural shift towards shared policymaking, adequate (human, financial, time, legal, political) resources and embedment in a long-term strategic framework. These elements continue to be a challenge in participatory governance.

Paying attention to these aspects is equally paramount for restorative justice. Restorative justice implies a community change in how justice is perceived and delivered in order to fully deploy its benefits. Such changes need continuous maintenance of the restorative process, and this takes time and requires significant resources (Marsh, 2017). Their embedment in a long-term strategic framework can contribute to 'protect' restorative justice from changes in the political agenda, and ensure the necessary resources for its continuous implementation over time. However, in framing a long-term strategic framework for restorative justice, attention should be paid to fostering bottom-up restorative processes, while ensuring top-down support (i.e. resources) for their delivery.

Quick solving process. Another warning derived from the analysis of participatory governance involves the expectations of the promoters that the participatory process solves all issues (particularly conflicts) in a short period of time. Restorative justice also faces this challenge. To avoid participants' disengagement, the timeline of restorative interventions should be clarified from the outset and participants provided with continuous feedback on the results achieved throughout the entire process.

Context blind process. The analysis of participatory policymaking emphasises that failure to consider the contextual features in shaping the participatory process may put at risk its effectiveness and reinforce social inequalities. Such risk is particularly relevant for restorative justice. Thus, promoters of restorative interventions should think about contextual elements: for example, political and legal framework of restorative justice; social and economic development and potential imbalances; level of social capital and civic culture; institutional trust and corruption; power structures and inequalities; type of community (e.g. hostile or apathetic; divided or united); types of conflicts; actors' priorities, resources, interest in the process and influence over it; supporting/opposing actors and those that might be missing from the process. The reputation of the promoter of restorative interventions is also significant for their effective delivery.

Disconnection with essential concepts such as democracy, empowerment and equality. As discussed earlier, participatory governance risks being undemocratic and destructive when it is power-blind and decoupled from the democratic values and principles at its basis. While this risk seems to be much more limited in the case of restorative justice, its scaling up and potential use as a mere tool for managing single conflicts may push its application away from original restorative values and principles. Particular attention should be paid not only to the effectiveness of the intervention itself but also to respecting the full array of restorative values and principles in the implementation process.

References

Al-Alosi, H. & Hamilton, M. (2019). The ingredients of success for effective restorative justice conferencing in an environmental offending context. *UNSW Law Journal,* 142(4), 1460–1488.

Bai, X., McAllister, R. R.J., Matthew, R., & Taylor, B. (2010). Urban policy and governance in a global environment: complex systems, scale mismatches and public participation. *Current Opinion in Environmental Sustainability,* 2, 129–135.

Bobbio, L., & Zeppetella, A. (1999). *Perché proprio qui? Grandi opere e opposizioni locali.* Milano: Franco Angelli Press.

Boehmer-Christiansen, S. (1995). Reflections on the politics linking science, environment and innovation. *Innovation: The European Journal of Social Science Research,* 8 (3), 275–287.

Carpentier, N., & Jenkins, H. (2013). Theorising participatory intensities: A conversation about participation and politics. *Convergence: The International Journal of Research into New Media Technologies,* 19(3), 265–286.

Castro, J. E. (2007). Water governance in the twentieth-first century. *Ambiente & Sociedade,* 10 (2), 97–118.

Chapman, T. (2019). La giustizia riparativa in Europa: Sfide e opportunità. In P. Patrizi (Ed.), *La giustizia riparativa* (pp. 41–61). Roma: Carocci Editore.

Christie, N. (1977). Conflicts as property. *British journal of Criminology,* 17, 1–15.

Cornwall, A. (2000). Beneficiary, consumer, citizen: Perspectives on participation for poverty reduction. *Sidastudies no. 2.* Retrieved from: https://www.researchgate.net/publication/44827244_Beneficiary_Consumer_Citizen_Perspectives_on_Participation_for_Poverty_Reduction (last accessed 26 October 2021).

Creighton, L. J. (2005). *The public participation handbook. Making better decisions through citizen involvement.* San Francisco: Jossey-Bass.

Depoe, S. P., Delicath, J. W., & Aepli Elsenbeer, M. F. (2004). *Communication and public participation in environmental decision making*. New York: Sunny series.

Dobrynina, M. (2017). The roots of "penal populism": The role of media and politics. Retrieved from: https://www.researchgate.net/publication/317960480_The_Roots_of_Penal_Populism_the_Role_of_Media_and_Politics (last accessed 26th October 2021).

Dunn, W. N. (1981). *Public policy analysis: An introduction*. New Jersey: Pearson Education.

European Commission (2019a). The European Green Deal sets out how to make Europe the first climate-neutral continent by 2050, boosting the economy, improving people's health and quality of life, caring for nature, and leaving no one behind. European Commission Press Release, Brussels, 11 December 2019. Retrieved from: https://ec.europa.eu/commission/press-corner/detail/e%20n/ip_19_6691 (last accessed 26 October 2021).

European Commission (2019b). The European Green Deal COM/2019/640 final. Retrieved from: https://eur-lex.europa.eu/legal-content/EN/TXT/?qid=1588580774040&uri=CELEX:52019DC0640 (last accessed 26 October 2021).

European Forum for Restorative Justice (2018). Practice guide on values and standards for restorative justice practices. Retrieved from: https://www.euforumrj.org/sites/default/files/2019-11/efrj-values-and-standards-manual-to-print-24pp.pdf (last accessed 26 October 2021).

European Forum for Restorative Justice (2021). Manual on restorative justice values and standards for practice. Retrieved from: https://www.euforumrj.org/sites/default/files/2021-11/EFRJ_Manual_on_Restorative_Justice_Values_and_Standards_for_Practice.pdf (last accessed 26 October 2021)

Evans, K., & Vaandering, D. (2016). *The little book of restorative justice in education*. New York: Good books.

Gal, T. (2016). 'The conflict is ours': Community involvement in restorative justice. *Contemporary Justice Review*, 19 (3), 289–306.

Gerkin, P. M. (2012). Who owns this conflict? The challenge of community involvement in restorative justice. *Contemporary Justice Review*, 15 (3), 277–296.

Gibbs M., (2009). Using restorative justice to resolve historical injustices of Indigenous peoples, *Contemporary Justice Review*, 12:1, 45–57.

Graham, S., & Healey, P. (1999). Relational concepts of space and place: Issues for planning theory and practice. *European Planning Studies*, 7 (5), 623–646.

Groves, C. (2017). Remaking participation: Science, Environment and Emergent Publics, *Science as Culture,* 26 (3), 408–412.

Involve (2005). People and participation: How to put citizens at the heart of decision-making. London: Involve.

Johnson, P. T. (1993). How I turned a critical public into useful consultants. *Harvard Business Review, Reprint No. 93103.* Retrieved from: https://hbr.org/1993/01/how-i-turned-a-critical-public-into-useful-consultants (last accessed 26 October 2021).

Lee, T.-P. (2020). Pursuing justice in a community experiencing environmental injustice: The practice of community revitalisation. *Contemporary Justice Review,* 23 (4), 337–353.

Lee C. W., McQuarrie, M., & Walker, E. T. (2015). Democratising inequalities: Dilemmas of the new public participation. New York: NYW Press.

Maiello, A., Christovão, A. C., Nogueira de Paiva, A. L., Britto, A., & Frey M. (2013). Public participation for urban sustainability: investigating relations among citizens, the environment and institutions—an ethnographic study. *Local Environment,* 18 (2), 167–183.

Marsh, V.L. (2017). Restorative practice: History, successes, challenges & recommendations. Retrieved from: https://www.rochester.edu/warner/cues/wp-content/uploads/2020/12/Restorative-Practices-Brief-1_marsh_final.pdf (last accessed 26 October 2021).

McCold, P. (2004). What is the role of community in restorative justice theory and practice? In H. Zehr & B. Toews (Eds.), *Critical issues in restorative justice* (pp. 155–172). Monsey, NY: Criminal Justice Press.

McCauley, D., & Heffron, R. (2018). Just transition: Integrating climate, energy and environmental justice. *Energy Policy,* 119, 1–7.

Nabatchi, T., & Leighninger, M. (2015). Public participation for 21st century democracy. Retrieved from: https://onlinelibrary.wiley.com/doi/book/10.1002/9781119154815 (last accessed 26 October 2021).

Pali, B. & Aertsen, I. (2021). Inhabiting a vulnerable and wounded earth: Restoring response-ability. *The International Journal of Restorative Justice,* 4(1), 3–16.

Paavola, J. (2007). Institutions and environmental governance: a reconceptualization. *Ecological Economics,* 63 (1), 93–103.

Patrizi, P. (2019). *La giustizia riparativa.* Roma: Carocci Editore.

Pimbert, M., & Wakeford, T. (2001). Overview—Deliberative democracy and citizen empowerment. Retrieved from: https://www.researchgate.net/publication/256840297_Overview_-_Deliberative_Democracy_and_Citizen_Empowerment (last accessed 26 October 2021).

Pratt, J., & Miao, M. (2017). Penal populism: The end of reason. Retrieved from: https://www.researchgate.net/publication/312594772 (last accessed 26 October 2021).

Preston, B. (2011). The use of restorative justice for environmental crime. *Criminal Law Journal*, 136, 2–25.

Roniotes, A., Malotidi, V., Virtanen, H., & Vlachogianni, T. (2015). A handbook on the public participation process in the Mediterranean. A tool for achieving sustainable development. Retrieved from: https://www.researchgate.net/publication/312054744_A_handbook_on_the_Public_Participation_Process_in_the_Mediterranean (last accessed 26 October 2021).

Sartori, G. (1987). *Elementi di teoria politica*. Bologna: Il Mulino.

Scottish Parliament (2004). Participation handbook. Edinburgh: Scottish Parliament.

Skinnider E. (2011). Victims of environmental crime: Mapping the Issues. Retrieved from: https://globalinitiative.net/wp-content/uploads/2017/12/ICCLR-Victims-of-Environmental-Crime.pdf (last accessed 26 October 2021).

Stark, A. (2016). Environmental restorative justice. *Pepperdine Dispute Resolution Law Journal*, 16 (3), 435–462.

Stockholm Environment Institute (SEI) (2019). Making space: How public participation shapes environmental decision-making. Retrieved from: https://cdn.sei.org/wp-content/uploads/2019/01/making-space-how-public-participation-shapes-environmental-decision-making.pdf (last accessed 26 October 2021).

Sullivan, D., & Tifft, L. (2001). *Restorative justice: Healing the foundations of our everyday lives*. Monsey, NY: Willow Tree Press.

Suna, L., Zhua D., & Chanb, E. H.W. (2016). Public participation impact on environment NIMBY conflict and environmental conflict management: Comparative analysis in Shanghai and Hong Kong. *Land Use Policy* 58, 208–217.

Teles, S. M. (2010). *The rise of the conservative legal movement*. Princeton: Princeton University Press.

United Nations (2021). Press release—Climate change 'Biggest threat modern humans have ever faced', World-renowned naturalist tells Security Council, calls for greater global cooperation. Retrieved from: https://www.un.org/press/en/2021/sc14445.doc.htm (last accessed 26 October 2021).

UNDOC (2006). *Handbook on restorative justice programmes*. New York: United Nations. Retrieved from: https://www.unodc.org/pdf/criminal_justice/ Handbook_on_Restorative_Justice_Programmes.pdf (last accessed 26 October 2021).

Walker, E. T. (2009) Privatising participation: Civic change and the organisational dynamics of grassroots lobbying firms. *American Sociological Review, 74*, 83–105.

Wampler, B., & McNulty, S. (2011). Does participatory governance matter? Exploring the nature and impact of participatory reforms. Retrieved from: https://www.alnap.org/system/files/content/resource/files/main/ cusp-110108-participatory-gov.pdf. (last accessed 26 October 2021).

Wamsler, C., Alkan-Olsson, J., Björn, H., Falck, H., Hanson, H., Oskarsson, T., Simonsson, E., & Zelmerlow, F. (2020). Beyond participation: When citizen engagement leads to undesirable outcomes for nature-based solutions and climate change adaptation. *Climatic Change, 158*, 235–254. Retrieved from: https://link.springer.com/article/10.1007/s10584-019-02557-9 (last accessed 26 October 2021).

Webler, T. (1995). 'Right' discourse in citizen participation: An evaluative yardstick. In O. Renn, T. Webler, & P. Wiedemann (Eds.), *Fairness and competence in citizen participation: Evaluating models for environmental discourse* (pp. 35–77). Boston: Kluwer Academic Press.

Wright Mills, C. (2000). *The power elite*. New York: Oxford University Press.

Zehr, H., & Mika, H. (1998). Fundamental concepts of restorative justice. *Contemporary Justice Review*, 1, 47–55.

Zehr, H. (1990). Changing lenses: A new focus for crime and justice. Scottsdale: Herald Press.

9

Climate Reparations, Compensation, and Intergenerational Restorative Justice

Ben Almassi

1   Introduction

Restorative justice might initially seem an odd fit for reckoning with the implications of climate change. For one thing, the constitutive practices of restorative justice were first built for relational repair in local communities, while anthropogenic climate change is truly a global phenomenon. From desertification to island nations and other low-lying regions threatened by rising sea levels, the people and places most burdened by climate change are often distant, physically if not ecologically speaking, from those most responsible for it. Climate change occurs not just internationally but also intergenerationally; its causes and effects are both spatially and temporally diffuse. Amongst other things, this means that the perpetrators of climate injustices are often spared from having to see the consequences of what they have done, and the victims of climate injustices are often unable to confront those who have wronged them directly.

B. Almassi (✉)
Governors State University, University Park, IL, USA
e-mail: balmassi@govst.edu

Despite these complications, I am optimistic about the prospects for a thoroughly international and intergenerational restorative response to climate injustice. I begin by contrasting Weiss's (1984, 1992) account of *planetary rights* and Gardiner's (2011) *pure intergenerational problem*, then build on Baier's (1981) discussion of a *cross-generational community* including past, present, and future people. Together these ideas enable us to make sense of climate justice as an asynchronous yet still relational process. This then raises the non-ideal moral, social, and political question of what we as members of a cross-generational community owe each other if, or more accurately, *given that* climate injustices have been, are being, and will be perpetrated, and the relationships that hold this community together are strained and torn by the ecological, economic, social and political consequences of global climate change. For her part, Baier takes a restitutive or compensatory approach to corrective justice when the rights of past and future persons are violated. While I am otherwise sympathetic and indeed indebted to her work on relational feminist ethics, this is one place where we diverge. My own account of *reparative environmental justice* recommends a restorative rather than restitutive response to climate injustice and other environmental wrongdoing.

On an intergenerational approach to restorative justice, I argue that climate reparations offer a better way to acknowledge and ameliorate the injustices of anthropogenic climate change than compensation for 'loss and damage', which the United Nations Framework Convention on Climate Change defines as both economic and non-economic harms to human or natural systems resulting from sudden onset events, like hurricanes, or slow onset processes, like rising sea levels (UNFCCC, 2020; cf. McShane, 2017). This is so for several reasons. First, framing climate justice in terms of compensation invites a risky, self-serving sense of moral relief. Seeing compensation as erasing injustice can be hard to resist when the effects of carbon emissions might not be felt for decades. Restorative justice by contrast does not hide our climate injustices from future generations' recognition nor encourage present or future generations to forget histories of climate change. Intergenerational restorative justice also avoids some of thornier metaphysical problems plaguing narrowly harm-based conceptions of climate injustice and amelioration. Compensation is made additionally problematic in climate justice because of how entangled greenhouse gas emissions are in their effects. It can be difficult to discern the comparable compensation owed by individual or group emitters from earlier generations to future generations (not to mention what is owed past generations) when assessed in isolation. On intergenerational restorative justice, an ameliorative response to injustice is not compensation but rather accountability and making amends to repair and renew the conditions upon which cross-generational relationships can be rebuilt. Restorative justice also enables meaningful responses to climate-related wrongs for which restitution and compensation are incongruous, for example, in seeking amelioration for epistemic injustices against Indigenous knowers and traditional ecological knowledge. Finally, a restorative approach to climate justice can give us direction as to how to achieve relational repair for a wider array of relationships that have been damaged by perpetrations of actions that contribute to climate change. My own contributions to climate change undermine healthy relationality with those experiencing the harmful effects of climate change, even when I did not cause those specific effects. Even if I have not harmed you and do not owe you compensation, I am still responsible for the impact that you are experiencing. This calls for relational repair in a way that restorative justice accommodates better than compensation for loss and damage.

2   Justice in a Cross-generational Community

In earlier work, I argued for reparative environmental justice as a non-ideal approach to environmental ethics, one which takes the aftermath of environmental injustices, destruction, degradation, and wrongdoings generally as an important, often neglected context for ethical analysis and action (Almassi, 2020). I began with the idea that environmental injustice and other environmental wrongdoing do moral damage to our relationships—intercultural, intergenerational, interspecies, and otherwise—which thus calls for a process of moral repair if these ecologically significant relationships are to be made healthier than they've become in the wake of wrongdoing. This approach is situated alongside and indebted to other relational environmental work by Hourdequin and Wong (2005), Kimmerer (2011, 2013), Emmerman (2014, 2019), and Fredericks (2019, 2021) and draws upon general accounts of moral repair, reparative justice, and restorative justice by Zehr (1990, 2013), Radzik (2009), Thompson (2009, 2015), Woolford (2009), and especially Walker (2006a, b, 2010, 2015).

Responses to wrongdoing may be distinguished by their objects of emphasis. If *retributive* justice focuses on perpetrators' offences, and *restitutive* justice focuses on victims' losses, *reparative* and *restorative* justice focus more on the relationships amongst perpetrators, victims, and other community members, relationships damaged by wrongdoing and in need of an ameliorative response—as Walker (2006a, p. 28) puts it, 'restoring or creating trust and hope in a shared sense of value and responsibility'. Constitutive practices of relational repair include acknowledgements of wrongdoing, apologies, making amends, and eventually rebuilding trust and extending forgiveness. They include the individuals directly involved but also other community members who help to hold wrongdoers accountable for their actions and recognise victims' standing to call for accountability. Despite popular assumptions, reparative and restorative justice is not about monetary payment. Reparations are not compensation for losses incurred but communicative acts: expressions of apology and amends towards renewed trust and cooperation. As Ross (2006, p. xvii) puts it, 'the harm is only peripherally about "stuff". Instead, the harm is understood in the relational realm'.

When we make amends through a reparative process, we do not aim to offer proportionate compensation for wrongdoing or an apologetic partial compensation. Reparations are given direction not by making victims whole based on what they have lost but by expressing what victims need to see occur in order to repair the damaged relationship. As Walker (2015, p. 217) puts it, reparations are 'a medium for the contentious yet hopeful negotiation in the present of proper recognition of the past and proper terms of relation in the future'. Understood in this way, reparative justice does not ask us to look backward but to do what we can to repair our relationships and communities that have been hurt by injustice. Admitting and apologising for our part in wrongdoing and making amends are meaningful because of the expressive burden (Walker, 2013) they carry: their ability to convey regret, to acknowledge wrongdoing, and most of all, to recognise those who have been hurt as equal members of our shared community, and thus equally deserving of respect and consideration.

It bears emphasis that theories, practices, and processes of reparative and restorative justice need not treat all of our varied relationships the same way. And as we consider environmental relationality in particular, that reminder remains important. We stand in different kinds of relationships with the members of what Leopold (1966) calls our *biotic community*. There is no one way to understand relational damage or enact relational repair that applies uniformly across all of our environmentally significant relationships. This is no less true for climate injustice than other environmental wrongdoing, where the relationships between perpetrators and victims cut across not only international, intercultural, and interspecies boundaries, but intergenerational ones as well.

Reparative environmental justice asks us to take seriously the relevance of relationality in the aftermath of wrongdoing. Yet many climate ethicists frame climate justice in terms of ideal theory (Page, 2006, Garvey, 2008, Hiskes, 2009), and amongst those who do discuss what to do *given* climate injustice, many otherwise worthwhile analyses neglect relationality (Shue, 1999; Jamieson, 2014; Meyer, 2015). I am reminded of Walker's (2001, p.112) observation that philosophers often see ethics as 'answering the question, "What ought I to do?," which implies a set of choices on a fresh page. Yet one of our recurrent ethical tasks is better suggested by the question, 'What ought I to do *now*?' after the page is blotted or torn by our own or others' wrongdoing'.

A recent challenge to ideal moral theorising is Heyward and Roser's 2016 edited collection *Climate Justice in a Non-Ideal World*, the contributions to which loosen the presumptive ideal-theoretic constraints on climate ethics debates in various ways. As the editors note (Heyward & Roser, 2016, p. 6), Rawls's distinction between ideal and non-ideal theory in political philosophy is built on two main idealising assumptions: full compliance (such that all individuals and institutions in a society act in accordance with the principles of justice) and favourable circumstances (such as moderate rather than extreme scarcity). One example of ideal theory in climate ethics is the assumption of full compliance with greenhouse-gas egalitarianism. In theorising about the extent and scope of our moral obligations concerning global climate change, it's typical to begin by assuming that everyone limits their emissions to their fair share. This isn't remotely true, but the idea is to start there and consider what changes, ethically speaking, if the idealising assumption is removed. When and if others fail to meet their climate responsibilities, Caney (2016) asks, how does this change the ethical landscape? Does noncompliance mean complying parties should carry heavier ethical burdens? Does it mean we should lower our standards for compliance? Should we develop theoretical and practical resources to induce higher rates of compliance?

These questions rarely arise for ideal theories on climate justice. But what I find notable is something that Caney and many other worthwhile contributions to *Climate Justice in a Non-Ideal World* do not directly address. If and when we fail to meet our climate duties, what sorts of ameliorative responsibilities follow from this moral failure? The question here is not only what should be done to make up the difference between full and partial compliance, but beyond that, how to reckon with our (longstanding, persisting, international, intergenerational) moral failings themselves. We need practices of corrective justice given climate change; specifically, I would suggest, we need practices of intergenerational repair.

Weiss (1992) builds her theory of planetary rights on the notion that 'all generations are linked by their on-going relationship with the earth'. Building from this starting point, Weiss argues that violations of intergenerational obligations by one generation can create further intergenerational obligations for others. Contrast this with the pure intergenerational problem (PIP) that Gardiner (2011) identifies as a central problem for intergenerational climate ethics. For the PIP, generations are ideally modelled as completely temporally non-overlapping, and the causal and temporal asymmetries between earlier and later generations preclude any possibility of intergenerational reciprocity (Gardiner, 2011, p. 166). The pure problem is of course an idealisation, yet Gardiner sees it as useful in making sense of the ethical challenge of intergenerational buck-passing in more realistic conditions as well (pp. 170–174). Weiss and Gardiner theorise climate justice differently due in part to their respective emphasis on and abstraction from intergenerational relationality as explanatorily significant.

For her part, Baier's (1981, p. 177) intergenerational relational ethic is built on the notion that past, present, and future persons are all 'members of a cross-generational community, a community of beings who look before and after, who interpret the past in light of the present, who see the future as growing out of the past, who see themselves as members of enduring families, nations, cultures, traditions'. Here she joins a long tradition in political philosophy spanning from Burke's cross-generational community to Rawls's social union across generations and continuing with De Shalit (1995, 2011), Thompson (2009), and Heath (2013), amongst others. One feature of Baier's cross-generational community worth emphasising is that it includes past as well as present and future persons. We might contrast this with Barry's (1999) account of intergenerational justice as strictly between present and future generations or Hiskes's (2009) similarly future-facing account. For example, our duty 'to leave as much and as good of the public goods previous generations have bequeathed', Baier (1981, p. 176) argues, 'rises as much from a right of past persons to have their good intentions respected as it does from any right of future persons'. Notice that the rights of past persons are not limited to the past but forward-looking in the sense that they create what Meyer (2006, p. 413) calls *surviving duties*: 'The rights imply duties that are (also) binding after the death of the bearer of the right if the appropriate bearer of the duty is identified'. Initially this might seem mysterious: how can we have duties to people who *no longer exist*? Even if they don't yet exist, at least future people's future lives might be affected (and in the case of global climate change, significantly so) by our present action or equally disastrous inaction. Those who once existed but no longer do would seem to be beyond our reach.

It's not that past persons know whether present or future persons respect their wishes, keep our promises, or continue on with intergenerational projects that mattered to them. The point is that even if they cannot know and they cannot be made happier or sadder, disappointed or satisfied, past persons can nonetheless still be indirectly affected by what present and future persons do (or do not do). Acknowledging previous generations as victims of wrongdoing cannot change their well-being, yet it is still possible that fulfilling our surviving duties to them *changes the relationship* between us. Meyer (2006, p. 415) explains:

> It will be true of the past victims of these injustices that they have the posthumous property that we fulfilled our surviving duty toward them [...] To be sure, a change of the relation between a currently living person and a dead person does not bring about or rely upon a real change of the latter person. Rather the relational change is based upon the real change of the person who carried out the act.

Past persons may be victims of historical injustices and targets of intergenerational restorative justice depending on present and future persons' restorative practices. And to the extent that present and future persons fail to respect the rights of past persons as members of our cross-generational community, they may be victims of additional injustices and targets of intergenerational restorative justice in that respect as well.

Baier argues that because interdependency is transitive, our cross-generational community is cross-cultural also. (And while our focus here is on intergenerational human relationality and human victims of climate injustice, there is no reason this cross-generational community could not be expanded to include animals and ecosystems too.) The full relational picture here is truly global and quite different from most theories of climate justice.

> We, like most of our forbearers, are the unconsulted beneficiaries of the sacrifice of past generations, sometimes seen by them as obligatory, often in fact nonobligatory. If we owe something in return, what is it, and what can we do for those who benefited us? The most obvious response is to continue the cooperative scheme they thought worth contributing to, adapting our contributions to our distinctive circumstances' (Baier, 1981, p. 180).

Like Weiss, Baier recognises the possibility of intergenerational ethical failure and finds that our failure to meet our obligations to past and future generations creates more obligations to them. Specifically, she argues, 'We incur obligations to compensate our victims in a future overcrowded world for the harm we have thereby done them' (Baier, 1981,

p. 177). In framing this as a matter of compensation, Baier differs from the restorative response to intergenerational injustice I am advocating here, though our approaches both emphasise relationality and interdependency. With that difference in mind, let's consider the benefits of reparations and restorative justice compared to restitutive models of compensation for loss and damage due to anthropogenic climate change.

3   Reparations and Compensation for Climate Change

In my earlier work, I argued that framing intergenerational climate justice in terms of compensation invites a risky, self-serving sense of moral relief (Almassi, 2020, p. 75; cf. Emmerman, 2014). Perhaps, when all has been accounted for and adequate compensation has been made, we will try to convince ourselves that we really avoided doing anything wrong. Our actions *would* have been wrong, absent the compensation, we might allow, but in paying a fair price for our climatological actions, we have met our intergenerational and international obligations and so have nothing to apologise for.

It is not difficult to imagine some segment of the present generation advancing this sort of rationalisation. It can be especially tempting for perpetrators to regard compensation as balancing and thus cancelling out some prior injustice. Unlike apology, reparations, and amends, the act of compensation in itself need not imply wrongdoing: indeed, we routinely compensate each other within fair and equal exchanges each day. Reading compensation as successfully erasing injustice may be especially hard to resist for anthropogenic climate change, where the full extent of climatological effects of today's greenhouse gas emissions might not be felt for decades. Earlier generations trying to account for how their wrongful actions affect later generations will generally do so in a years-long interim between the actions themselves and their felt effects. Rather than victims experiencing an injustice *and then* experiencing some reparations or compensation for it, climate reparations or compensation might precede or run concurrent to—and so seemingly obviate the effects of—climate injustice. Because of that, perpetrators and their apologists might encourage us to see this chain of events as catching what *would have been* a climate injustice just in time, so to speak.

Intergenerational restorative justice, by contrast, does not mask the reality of climate injustice from future victims or present-day perpetrators. Nor does it encourage either future or present generations to forget our histories of climatological wrongdoing. The restorative aim is forgiveness not forgetting, renewed relationships not reimbursement. The work of intergenerational restorative justice is dynamic and ongoing, just as intergenerational relationships are themselves dynamic and ongoing. And if we mean to repair the moral conditions of a cross-generational global community that climate change has strained and torn, apologies and acknowledgements for wrongdoing are no less essential than are material responses on their own.

Consider how contemporary discussions position mitigation, adaptation, and compensation as pillars of climate policy, three kinds of responsibility for addressing climate change. The first includes strategies to reduce greenhouse gas sources and increase greenhouse gas sinks. Two decades now into the twenty-first century, climate scientists agree that mitigation alone is an inadequate response. 'Given the volume of greenhouse gases that have been emitted, we are already committed to some climate change,' Caney (2012, p. 257) explains, and further says: 'In virtue of this, there is an ethical imperative that we—humanity—take steps necessary to ensure that any climate changes that occur do not undermine what people are entitled to do as a matter of justice. In the terminology employed by climate scientists, "Adaptation" is required'. Caney identifies inoculations against infectious disease, drainage systems, and sea walls as examples of adaptation. Such actions will not reduce greenhouse-gas levels or stop their atmospheric effects, but if they are successful, they can stop (some of) the harmful implications of now inevitable climate change. Finally, compensation comes into play when mitigating and adaptive measures are insufficient, and climate change prevents people from, as Caney (2012, p. 258) puts it, 'doing that which they are entitled to do'.

My concern is how we understand this taxonomy of climate responsibilities on a restorative rather than restitutive model of justice. If corrective justice is about restoring economic, ecological, or other material conditions to some prior state, or alternatively, providing compensation for material changes that cannot be reversed, each of these pillars of climate policy may indeed by instrumentally important. But what has been side-lined is recognition and acknowledgement of our responsibility for perpetrating climate change as itself amongst our responsibilities for addressing climate change. To the extent that strategies for climate amelioration neglect the reparative significance of acknowledgement, apology, and amends, they neglect the reality of climate change as not only a social-ecological *challenge* but also an (existing and ongoing) *injustice* to be rectified.

In her paper 'Reparative Justice for Climate Refugees', Buxton (2019, p. 199) argues that one thing that makes climate reparations different from and preferable to compensation is the question of who can do it. Where compensation may be provided by a third party, reparations by their nature must be made by those responsible for the injustice in question. I would add that, in cases in which those responsible are as yet unable or unwilling to apologise and make amends for their wrongdoing, third parties may play a much-needed role in recognising a wrongdoing, providing aid, and affirming victims' moral standing to call for accountability. As important as such third-party measures could be, Buxton is right that third party responses alone are not enough for repair. They do not relieve those responsible for climate injustices from their ameliorative duties, do not require perpetrators' realisation or acknowledgement of what they have done, and do not repair the moral conditions of relationships between perpetrators and victims that climate injustices have damaged or destroyed.

Compensation is additionally problematic for global climate change because of how entangled our partial contributions are in their collective ecological effects. What the debate between Sinnott-Armstrong and his critics on the ethics of individual emissions shows is that neither individuals' nor groups' emissions function in isolation (see Sinnott-Armstrong, 2005; Nolt, 2011; Hiller, 2011; Almassi, 2012; Kingston & Sinnott-Armstrong, 2018; Broome, 2019). Other parties' contributions significantly affect how harmful or benign our own contributions will prove to be. It is difficult if not impossible to determine the comparable compensation owed by individual or group emitters from an earlier generation to future generations (not to mention what is owed to past persons) when considered in isolation. For climate reparations, by contrast, the response to climate injustice is not compensation but rather accountability and amends needed to renew the conditions on which our relationships may be rebuilt. To the extent that intergenerational amelioration is appropriate, it is not to make some sort of comparable compensation. Indeed, the aspiration to determine an appropriate compensation amount for intergenerational repayment would itself contradict the humility and accountability needed for successful reconciliation. What we are doing in restorative climate justice is making amends towards cross-generational trustworthiness and forgiveness.

Amongst philosophers, the question of whether anthropogenic climate change is properly understood as an intergenerational injustice is often thought to turn on whether or not future generations are harmed by past and present greenhouse gas emissions. The main point of debate here is not about whether climate change is human-caused, or about how extensive the resulting ecological, economic, social, and political changes will be, but rather about whether any future individuals will be *harmed* by these changes. We might think that these future individuals clearly would have been better off if not for anthropogenic climate change, but this then raises what philosophers following Parfit (1987) call the *non-identity problem*. (On the non-identity problem, future generations, and climate ethics specifically, see Broome, 1992, and Parfit, 2017.) In what sense can a person truly be said to be harmed by an action if and when she *would not have existed* had the action not been done? The global reach of anthropogenic climate change makes this an especially relevant question. The future world will be different in countless ways because of climate change and the people who will live in this world will be different from the people who would have come into existence if not for climate change. Consider, for example, climate-related migration, which will bring individuals into contact with other individuals they otherwise never would have met, and likewise derail other encounters that would have happened, which means (amongst many other differences) that different people couple up and different children are born. And if people are born owing to climate change who would not otherwise have been born, this line of reasoning goes, how can we say that climate change harms them?

And if these future people have been not harmed, why do we owe them anything for the changes we have wrought?

Many philosophers have taken on the non-identity problem (see e.g., Harman, 2004, and Boonin, 2008), and I do not mean to suggest that it is unavoidable for every compensatory view of climate justice. But restorative intergenerational justice offers us a way around the metaphysical and ethical problems that this problem raises. For one, intergenerational relational damage is not limited to ameliorating harms that individuals experience: harm is only one of the ways that we wrong each other, and thus not the only wrongdoing that calls for a restorative response. Nor are intergenerational relationships limited to individual relationality. We stand in relation not only to other individuals but also to other groups, some of which span generations, others of which came before us, others of which will come later. Even as individual membership of future generations might be changed by our actions (or inactions), we can still make sense of our relations to this future group, as a teacher might consider her obligations to next year's class even before the individual members of that class have been assigned. Furthermore, following Baier's (1981) characterisation of our cross-generational community, we have relational obligations to not just future but also past and present people. Though our reparative efforts to make amends for climate change will of course be future-oriented in their effects, they need not be limited to repairing our relationships with future people.

What I find particularly promising about a restorative approach to climate justice is that it accommodates and gives direction to relational repair for a wider array of relationships damaged indirectly by the parties' respective perpetrations and experiences of climate change. As Gardiner and Hartzell-Nichols (2012) remind us, the temporal diffusion of climate change presents distinct ethical challenges. As we slowly start to recognise the anthropogenic nature of recent climate changes (frequency and severity of hurricanes, rising sea levels and temperatures, etc.), it can be tempting to blame contemporary emitters for attendant harms that humans, other animals, and the rest of the world are experiencing today. And yet those most responsible for the climatological changes felt now might be long dead and those most affected by current greenhouse-gas emissions may not yet be alive. This is not exclusively so, to be sure; the asynchronicity of emissions and their effects is not so protracted as to preclude *some* temporal overlap between those who cause and those who experience particular climatological harms. This was part of the power of Thunberg's (2019) UN address: our world leaders are old enough and she herself is young enough that, temporal diffusion of climate change notwithstanding, in this case a future victim is able to confront her assailants.

While they may be most causally responsible, restorative justice for anthropogenic climate change need not be limited to members of the eldest living generation. Others of us must work to repair the damaged conditions of our relationships too—with past and future persons, yes, but also with those currently feeling the harmful effects of climate change, even if our emissions did not cause the changes they will soon experience or already are experiencing. Contributing to climate change undermines healthy relationality with those experiencing the effects of climate change, and this is so even when our respective experiences are not directly causally related. I have not harmed you and so I do not owe you compensation, but I am nevertheless responsible for the very sort of thing that you are experiencing, and this fact calls for acknowledgement and relational repair. If this seems opaque, consider an analogy involving restorative justice in a criminal context. When a police officer commits racist violence against one person, they owe reparations to not only that specific person and their loved ones but other victims of racist police violence in the community as well. No doubt, the reparations owed (acknowledgements of wrongdoing, apologies, amends, and so on) may be different, because the relationships and the ways this police officer's wrongful actions have damaged or destroyed these relationships may be different. But to limit the moral and social scope of the injustice as only affecting the relationship between a perpetrator and their direct victim is to fail to appreciate—or at least fail to admit and acknowledge—the interconnections amongst individual cases of racist police violence.

Lastly, intergenerational restorative justice enables constructive ameliorative responses to climate-related injustices and other wrongdoing for which restitution or compensation would be incongruous. Consider for example *epistemic injustices*, which following Fricker (2007) we can define as occurring when people are wronged in their capacities as knowers. Epistemic injustices include a wide variety of phenomena in myriad contexts of individual or collective knowledge, expertise, and experience, including but not limited to credibility deficits and excesses, hermeneutical marginalisation, speech-act misattributions, testimonial silencing, gaslighting, exclusion, and epistemic exploitation (see Hookway, 2010; Dotson, 2011; Medina, 2011; Pohlhaus, 2012; Kukla, 2014; Berenstain, 2016; Davis, 2016; McKinnon, 2016; Almassi, 2018). In contexts involving ecological knowledge generally and climatological knowledge specifically, historically marginalised and oppressed people have routinely had their lived experiences and perspectives dismissed, devalued, misunderstood, misattributed, and appropriated by dominant knowers and knowledge systems. Such polyvalent perpetrations of epistemic injustice against traditional ecological knowledge and knowers can be found in the histories of imperialist and settler colonialist impositions on Indigenous communities, and more specifically in the fraught relationship between Indigenous and scientific ecological knowledge (see Kimmerer, 2002; Whyte, 2013, 2018; Almassi, 2020, pp. 87–107).

Processes of social-epistemic repair in the aftermath of historical and ongoing climate-related epistemic injustices should build on Davis's (2016, p. 494) reminder that improving one's own epistemic character is not sufficient for epistemic justice. We must repair not only ourselves but also the social-epistemic environmental conditions in which perpetrators and victims are situated, the communities in which they are members. Whatever else it is, epistemic repair (the restoration of conditions for trustful, collaborative knowledge-making) must be a multi-agential process, not just something that perpetrators, victims, and third-party community members realise, but something we actively do. By acknowledging our responsibility for perpetrating epistemic injustices we do not therein complete the full process of epistemic repair, nor does such acknowledgement license renewed trust on its own. But it is an important step nonetheless, without which victims of unacknowledged epistemic injustice are otherwise deprived of the considerable practical and epistemic benefits of healthy trust relationships (cf. Grasswick, 2017, p. 319).

A retributive response to epistemic injustice might demand punishment; a restitutive response might call for perpetrators to compensate victims. Yet, neither punishment nor compensation has clear applications in epistemic contexts. One might try to make a case for epistemic compensation in terms of giving credit previously withheld, filling hermeneutical gaps, reassessing misattributed speech acts, and so on. And yet while each of these measures might play a meaningful role in processes of epistemic repair, as epistemic *compensation* they fall short. Victims cannot really be repaid or made whole in the aftermath of epistemic wrongdoing. Traditional ecological knowledge regarding climate change that was silenced by testimonial quieting or smothering might be communicated again and better appreciated on some future occasion; but to see this as compensation making up for the earlier testimonial injustice would be to ignore the dynamic and iterative nature of epistemic agency and knowledge production. Filling a hermeneutical gap owing to the social-epistemic marginalisation of Indigenous ecological knowledge may indeed be beneficial going forward, but it cannot 'repay' Indigenous knowers for the practical and epistemic harms they experienced in having their knowledge obscured and misunderstood in the interim.

To argue against compensation for epistemic injustice towards traditional ecological knowledge on climate change is not to insist that an apology alone is enough for epistemic repair either. We would do better to understand corrective actions taken in the aftermath of epistemic injustice as *amends,* communicative acts, rather than attempted compensation. Victims of epistemic injustice are being asked to trust in the wake of degrading or demoralising experiences that would seem to constitute powerful evidence to the contrary—that their previous unjust treatment notwithstanding, they will be believed, heard, understood, and valued. Apologies and amends are what allow us to bridge that gap. This is why a restorative process centred on victim subjectivities is so important for determining which actions and outcomes are required for making amends in the aftermath of epistemic injustice. Those who commit epistemic injustices might think they know best how to show their contrition and renewed epistemic trustworthiness, and yet to substitute their own ideas for victims' perspectives on the precipitating epistemic injustice and its amelioration would not only miss the point of epistemic repair but also risk committing further epistemic injustices in the process.

4   Clarifications, Complications and Conclusions

To conclude, let me acknowledge and address some concerns for a restorative-justice response to climate change. The first such concern is that intergenerational relational repair is a distraction, a luxury that we cannot afford given the existential threat posed by global climate change. I appreciate the urgency behind this concern, but I would contest the underlying assumption that relational repair is at best extraneous and at worst contrary to stopping climate change. The drive to find climate solutions unhampered by the work of restorative justice steers us towards shortcuts with pitfalls and impediments of their own. Whyte (2020) reminds us that global climate justice without Indigenous environmental justice thereby undermines both: prioritising ecological problems without regard for relational ones perpetuates and exacerbates unresolved injustices and, in doing so, undermines the conditions for trustful intercultural collaboration necessary for a global problem like climate change. Trust is what enables us to do together what we cannot alone, after all, and when trust is damaged or eroded through betrayal, corruption, and wrongdoing, our relationships cannot support collective action. Global climate change is one such case of collective action, not only in its causes but also in its solutions: the presence (or absence) of trust and trustworthiness is what enables (or impedes) critical coordinated international-intergenerational policies of emission reduction, mitigation, and social adaptation. To the extent that climate justice requires sustained collaboration, and collaboration requires resilient relationships, intercultural, international, and intergenerational restorative justice is not a distraction but rather an essential part of the process of addressing climate change.

Conceiving of climate change in terms of intergenerational restorative justice means taking victims and their relationships seriously. Some might wonder, however, whether the challenge of identifying victims across generations is simply too great. Prioritising victims' subjectivity provides normative force and direction to restorative practices, yet to the extent that amends are made by earlier generations to later generations, the former may be unable to access the latter's subjectivities. Earlier generations may be tempted to appeal to their own subjectivities through misplaced empathy, thereby giving the impression that perpetrators have deferred to future people even as their actual subjectivities remain inaccessible to them. But recall Baier's cross-generational community: if the victims of climate injustice include past persons *and* future persons, then the work of restoring the conditions of our damaged cross-generational community may be directed also in part by the subjectivities of past and present persons, whose perspectives, preferences, and values are comparatively better known to us. The relative inaccessibility of the future victims' subjectivities matters, to be sure, and it must be acknowledged as a serious ethical and epistemic challenge. Here I would second Alcoff's (1991) warning about the ethical and epistemic risks involved in speaking for others: the risks are real, and yet categorically refusing to ever speak for others (or to allow others to speak for you) is not a workable solution either. What future victims of climate injustice need to see from us may not be easy to ascertain, yet refusing to consider their subjectivities (alongside those of past and present victims) would leave our restorative practices biased and incomplete.

The spatially and temporally diffuse nature of global climate change certainly complicates relationships between perpetrators, victims, and their communities. One might wonder whether the process of repairing one relationship might directly or indirectly damage others, or alternatively whether restorative practices might impede restitutive or retributive responses to injustice. It does indeed seem likely that practices of intergenerational restorative justice will sometimes diverge from and conflict with intergenerational justice understood on different terms. That said, there is surely room for convergence. While compensation is not itself the goal, it can play a significant role in a process of making amends. And while climate reparations in making amends are not primarily about repayment, if we as perpetrators truly acknowledge our wrongdoing and recognise ourselves as being accountable to past, present, and future victims, part of being accountable means fixing what we've broken when we can and when this is what victims need. Further still, refraining from repairing an intergenerational relationship can also damage others, perhaps by calling into question partially penitent perpetrators' trustworthiness or sowing division amongst those whose standing in our intergenerational community has been reaffirmed and those whose standing has not. The wider lesson is not that intergenerational restorative justice in the aftermath of climate injustice should be avoided, but that processes of intergenerational restorative justice themselves should be subject to critical scrutiny and reflection.

References

Alcoff, L. (1991). The problem of speaking for others. *Cultural Critique, 20*, 5–32.

Almassi, B. (2012). Climate change and the ethics of individual emissions. *Perspectives, 4*(1), 4–21.

Almassi, B. (2018). Epistemic injustice and its amelioration. *Social Philosophy Today, 34*, 95–113.

Almassi, B. (2020). *Reparative environmental justice in a world of wounds.* Lanham: Lexington Books.

Baier, A. (1981). The rights of past and future persons. In E. Partridge (Ed.), *Responsibilities to future generations* (pp.171–83). New York: Prometheus Books.

Barry, B. (1999). Sustainability and intergenerational justice. In A. Dobson (Ed.), *Fairness and futurity* (pp. 93–117). Oxford: Oxford University Press.

Berenstain, N. (2016). Epistemic exploitation. *Ergo, 33*(2), 569–590.

Boonin, D. (2008). How to solve the non-identity problem. *Public Affairs Quarterly, 22*(2), 129–159.

Broome, J. (1992). *Counting the cost of global warming.* Cambridge: The White Horse Publishing.

Broome, J. (2019). Against denialism. *Monist, 102*(1), 110–129.

Buxton, R. (2019). Reparative justice for climate refugees. *Philosophy, 94*(2), 193–219.

Caney, S. (2012). Just emissions. *Philosophy and Public Affairs, 40*(4), 255–300.

Caney, S. (2016). Climate change and non-ideal theory. In C. Heyward & D. Roser (Eds.), *Climate justice in a non-ideal world* (pp. 21–42). Oxford: Oxford University Press.

Davis, E. (2016). Typecasts, tokens, and spokespersons. *Hypatia, 31*(2), 485–501.

De Shalit, A. (1995). *Why posterity matters.* New York: Routledge.

De Shalit, A. (2011). Climate change refugees, compensation, and rectification. *Monist, 94*(3), 310–28.

Dotson, K. (2011). Tracking epistemic violence, tracking practices of silencing. *Hypatia*, 26, 236–257.

Emmerman, K. (2014). Sanctuary, not remedy. In L. Gruen (Ed.), *Ethics of captivity* (pp. 213–230). Oxford: Oxford University Press.

Emmerman, K. (2019). What's love got to do with it? An ecofeminist approach to inter-animal and inter-cultural conflicts of interest. *Ethical Theory and Moral Practice*, 22, 77–91.

Fredericks, S. (2019). Climate apology and forgiveness. *Journal of the Society of Christian Ethics*, 39(1), 143–59.

Fredericks, S. (2021). *Environmental guilt and shame*. Oxford: Oxford University Press.

Fricker, M. (2007). *Epistemic injustice*. Oxford: Oxford University Press.

Gardiner, S. (2011). *A perfect moral storm*. Oxford: Oxford University Press.

Gardiner, S. M., & L. Hartzell-Nichols. (2012). Ethics and global climate change. *Nature Education Knowledge*, 3(10), 5.

Garvey, J. (2008). *The ethics of climate change*. London: Continuum.

Grasswick, H. (2017). Epistemic injustice in science. In I. Kidd, G. Pohlhaus Jr, & J. Medina (Eds.), *Routledge handbook of epistemic injustice* (pp. 313–323). London: Routledge.

Harman, E. (2004). Can we benefit and harm in creating? *Philosophical Perspectives*, 18, 89–113.

Heath, J. (2013). The structure of intergenerational cooperation. *Philosophy & Public Affairs*, 41, 31–66.

Heyward, C., & D. Roser. (2016). Introduction. In C. Heyward & D. Roser (Eds.), *Climate justice in a non-ideal world* (pp. 1–17). Oxford: Oxford University Press.

Hiller, A. (2011). Climate change and individual responsibility. *Monist*, 94, 349–68.

Hiskes, R. (2009). *The human right to a green future*. Cambridge: Cambridge University Press.

Hookway, C. (2010). Some varieties of epistemic injustice. *Episteme*, 7(2), 151–163.

Hourdequin, M., & D. Wong. (2005). A relational approach to environmental ethics. *Journal of Chinese Philosophy*, 32, 19–33.

Jamieson, D. (2014). *Reason in a dark time*. Oxford: Oxford University Press.

Kimmerer, R. (2002). Weaving traditional ecological knowledge into biological education. *BioScience*, 52, 432–438.

Kimmerer, R. (2011). Restoration and reciprocity. In D. Egan, E. Hjerpe, & J. Abrams (Eds.), *Human dimensions of ecological restoration* (pp. 257–276). Washington, DC: Island Press.

Kimmerer, R. (2013). *Braiding sweetgrass.* Minneapolis, MN: Milkweed.

Kingston, E., & Sinnott-Armstrong, W. (2018). What's wrong with joyguzzling? *Ethical Theory and Moral Practice, 21,* 169–186.

Kukla, Q. (2014). Performative force, convention, and discursive justice. *Hypatia, 29*(2), 440–457.

Leopold, A. (1966). *A sand county almanac.* New York: Ballantine Books.

McKinnon, R. (2016). Epistemic injustice. *Philosophy Compass, 11*(8), 437–446.

McShane, K. (2017). Values and harms in loss and damage. *Ethics, Policy and Environment, 20*(2), 129–142.

Medina, J. (2011). The relevance of credibility excess in a proportional view of epistemic injustice. *Social Epistemology, 25*(1), 15–35.

Meyer, L. (2006). Reparation and symbolic restitution. *Journal of Social Philosophy, 13*(3), 406–22.

Meyer, L. (2015). Intergenerational justice. In E. Zalta (Ed.), *Stanford Encyclopedia of Philosophy.* Retrieved May 15, 2021, from https://plato.stanford.edu/entries/justice-intergenerational/

Nolt, J. (2011). How harmful are the average American's greenhouse gases? *Ethics, Policy, and Environment, 14*(1), 3–10.

Page, E. (2006). *Climate change, justice, and future generations.* Cheltenham, UK: Edward Elgar.

Parfit, D. (1987). *Reasons and persons.* Oxford: Clarendon Press.

Parfit, D. (2017). Future people, the non-identity problem, and person-affecting principles. *Philosophy and Public Affairs, 45*(2), 118–157.

Pohlhaus, G. (2012). Relational knowing and epistemic injustice. *Hypatia, 27*(4), 715–735.

Radzik, L. (2009). *Making amends.* Oxford: Oxford University Press.

Ross, R. (2006). *Returning to the teachings.* Toronto: Penguin.

Shue, H. (1999). Global environment and international inequality. *International Affairs, 75*(3), 531–545.

Sinnott-Armstrong, W. (2005). It's not my fault: Global warming and individual moral obligations. In W. Sinnott-Armstrong & R. Howarth (Eds.), *Perspectives on climate change* (pp. 4–21). Amsterdam: Elsevier.

Thompson, J. (2009). *Intergenerational justice.* New York: Routledge.

Thompson, J. (2015). Reparative claims and theories of justice. In K. Newman & J. Thompson (Eds.), *Historical justice and memory* (pp. 45–62). Madison, WI: University of Wisconsin Press.

Thunberg, G. (2019). Speech at the UN Climate Action Summit. September 23. Retrieved May 15, 2021, from https://www.npr.org/2019/09/23/763452863/transcript-greta-thunbergs-speech-at-the-u-n-climate-action-summit

United Nations Framework Convention on Climate Change (UNFCCC). (2020). Introduction to Loss and Damage. Retrieved September 8, 2021, from https://unfccc.int/topics/adaptation-and-resilience/the-big-picture/introduction-to-loss-and-damage

Walker, M. (2001). Moral repair and its limits. In T. Davis & K. Womack (Eds.), *Mapping the ethical turn* (pp. 110–129). Charlottesville, VA: University of Virginia Press.

Walker, M. (2006a). *Moral repair: Reconstructing moral relations after wrongdoing.* Cambridge: Cambridge University Press.

Walker, M. (2006b). Restorative justice and reparations. *Journal of Social Philosophy, 37*(3), 377–95.

Walker, M. (2010). *What is reparative justice?* Milwaukee, WI: Marquette University Press.

Walker, M. (2013). The expressive burden of reparations. In A. MacLachlan & A. Speight (Eds.), *Justice, responsibility, and reconciliation in the wake of conflict* (205–225). Dordecht: Springer.

Walker, M. (2015). Making reparations possible: Theorizing reparative justice. In C. Corradetti, N. Eisikovits, & J. Rotondi (Eds.), *Theorizing transitional justice* (pp. 211–223). London: Ashgate.

Weiss, E. (1984). The planetary trust: Conservation and intergenerational equity. *Ecology Law Quarterly, 11*(4), 495–582.

Weiss, E. (1992). In fairness to future generations and sustainable development. *American University International Law Review, 8*(1), 19–26.

Whyte, K. (2013). On the role of traditional ecological knowledge as a collaborative concept. *Ecological Processes, 2*(7), 1–12.

Whyte, K. (2018). What do indigenous knowledges do for indigenous peoples? In M. Nelson & D. Shilling (Eds.), *Traditional ecological knowledge* (pp. 57–83). Cambridge: Cambridge University Press.

Whyte, K. (2020). Too late for indigenous climate justice. *WIRES Climate Change, 11*(1), e603.

Woolford, A. (2009). *The politics of restorative justice.* Halifax, NS: Fernwood Publishing.

Zehr, H. (1990). *Changing lenses: A new focus for crime and justice.* Scottdale, PA: Herald.

Zehr, H. (2013). Retributive justice, restorative justice. In G. Johnstone (Ed.), *A restorative justice reader* (pp. 23–35). New York: Routledge.

10

Meeting on Thin Ice: The Potential for Restorative Climate Justice in Deglaciating Environments

Tanya Jones

1   Introduction

Climate injustice is a contemporary reality for millions of people, entwined with other forms of global and local inequality and harm. The scientific realities of climate change are creating and intensifying global relationships of responsibility, which are insufficiently acknowledged or addressed by conventional legal and political processes. This chapter argues for a potential role for restorative justice in expressing and exploring these relationships and thereby facilitating transformational practical and attitudinal change. It builds on the analysis of Margaret Urban Walker (2006) to suggest a framework for restorative climate justice praxis applicable to the specific climate injustices experienced in the Peruvian Andes. The chapter uses the terms 'majority' and 'minority' worlds, in place of the more familiar Global South or North, or 'developing' and 'developed' countries. As Samantha Punch (2003) and Timothy

T. Jones (✉)
University of Dundee, Dundee, UK
e-mail: t.w.jones@dundee.ac.uk

Doyle (2005) note, these terms, though they still oversimplify the complexities of global relations, are less geographically inaccurate or culturally loaded than their alternatives. They also, crucially, help to clarify the reality of global imbalances: that those regions which hold the preponderance of resources and power generally represent the fewest people.

2   The Crisis of Climate Justice

It is now widely acknowledged, in institutional as well as activist circles, that the world is in a climate emergency (Maslin, 2020). But this is not only a crisis of greenhouse gas emissions and temperature rises; it is a crisis of global justice. Across the world, those least responsible for climate change are suffering its severest impacts with, overwhelmingly, the fewest financial resources and the least access to decision-making power (Tokar, 2019). It is this imbalance, and the collective failure to rectify it, that constitutes climate injustice. Lack of action on climate mitigation (the reduction of greenhouse gas emissions), on adaptation (adjustment to a changing climate) and on compensation for loss and damage (those impacts which cannot adequately be adjusted for) represent particular manifestations of climate injustice. These intersect with underlying inequalities, legacies of colonialism, exploitation and oppression (Okereke, 2014). Ironically, a widening acknowledgement of the climate crisis itself brings new injustices, such as where net zero pledges are used to justify new land and resource grabs in the name of offsetting 'nature-based solutions' (Stabinsky, 2021). As the United Nations Environmental Programme (UNEP) recognises, deep and broad changes are urgently needed, involving radical action and fundamental shifts in attitudes and policies, especially by the minority world (UNEP, 2019). Like climate change itself, climate injustice is a wicked problem, requiring polycentric and multi-layered responses (Hulme, 2000; Jordan et al., 2018). Countering it requires a range of concepts that are salient and compelling and mechanisms that are fair and effective, together with committed and tenacious actors.

The climate justice movement, which includes Indigenous and peasant farmers' organisations, environmental NGOs, trade unions and anti-racist and debt-cancellation campaign groups, provides both dedicated activists and compelling concepts (Tokar, 2019; Warlenius, 2018). Central to its analysis is the existence of a climate debt owed by the industrialised minority to majority states, communities and individual victims. Rikard Warlenius (2018, p. 32) defines this concept as 'the idea that climate change is caused by rich people while mainly harming people that are poor, and, therefore, the former should take the burden of mitigation and adaptation costs'. The movement calls upon the minority world to repay this debt by means of financial transfers but also by lifting barriers to justice created by intellectual property rights, capacity and technology shortfalls, immigration controls and human rights violations (Warlenius, 2018). The demands are based on values of equality and sustainability, on legal concepts including the polluter pays principle, historical responsibility, the global commons and the right to sustainable development, but also on practical expediency. Global transformation cannot happen without the majority world, which in turn cannot play its part without the resources it needs (Thanki, 2019).

The concept of climate debt, and the actions which it requires, are morally, intellectually and practically compelling from a global perspective. And yet, after decades of activism, the movement's voices are still silenced on the stage of international decision-making. For powerful states and the corporate lobbies at their shoulders, the implications of climate debt are simply too wide, too challenging, too dangerous (Warlenius, 2018). One part of the problem is the intransigence of minority elites, but another involves issues of proximity and distance. Legal and political systems are based on national sovereignty, and as long as their self-defined versions of the rule of law are met within their borders, it is near impossible to enquire further. Most societies have identified moral duties as being owed, with diminishing force, first to family, then to neighbours and compatriots, but only rarely further (Singer, 1993). Human rights obligations are primarily binding upon governments in relation to their own citizens, while legal liability generally requires proximity of standing, causation and foreseeability. Even environmental degradation was until recently largely a matter of geographical neighbourhood, with ascertainable loci of pollution and impact.

Yet, the necessary connection between proximity and responsibility is now being dismantled, partly by processes of globalisation but overwhelmingly by the scientific realities of climate change. With no directly ascertainable link between particular greenhouse gas emissions and particular impacts and harms, we are all flung into a global network of responsibility and interdependence, forming new connections that are inexorable but too numerous and complex for our structures to contain. And yet, each of those new connections is a new human relationship. As Jeremy Waldron (2003) notes, although we are natural Good Samaritans, willing to help those with whom we have no previous relationship, beyond the borders of nation and community, we respond best to appeals from those we encounter 'on the road'. We need not only new ways of thinking about climate justice, new mechanisms for achieving it, and new narratives, discourses and coalitions (Barry, 2012: Ghosh, 2017) but also new and renewed ways of recognising, across distances of space and culture, those with whom we share our widened boundaries.

3   The Potential for Restorative Climate Justice

Restorative justice, which focuses precisely upon acknowledging, building and restoring relationships, offers the potential for recognising and exploring these connections, thriving in the fertile spaces between political negotiation, legal liability and voluntary aid. Restorative climate justice itself is a concept not quite over the horizon, urged and commended by a range of perspectives and disciplines (e.g. Huggel et al., 2016; Lunn, 2011; McCauley & Heffron, 2018; Motupalli, 2018), but with, as yet, little systemic analysis. Its strongest academic presence derives from the philosophy of Margaret Urban Walker (2006) as developed by Ben Almassi (2020), especially in relation to intergenerational climate injustice. The term 'restorative climate justice' has also been proposed as a concept in the assessment of aid and development projects, combining elements of restorative and climate justice with agroecology and food sovereignty criteria (Earth in Common, 2021). It can draw upon other emergent accounts of environmental restorative justice, upon green criminology and many suppressed and marginalised forms of Indigenous knowledge and experience (e.g. Biffi & Pali, 2019; Cook & Powell, 2003; Gibbs, 2009: Parola, 2017; Preston, 2011; White, 2014). Restorative climate justice, as it evolves, can also learn from post-conflict transitional justice and mediation in a range of contexts (e.g. Bakiner, 2016; Roberts & Palmer, 2005).

In *Moral Understanding*, Walker (1998) argues for an 'expressive-collaborative' approach to ethics, focusing on responsibilities, including those where factors such as colonialism and exploitation create relationships between distant actors (Walker, 1998, pp. 85–86). Walker later builds upon this insight in her application of restorative and reconciliatory justice to historic injustices, specifically slavery and its continued legacy of racism and violence (Walker, 2006). She identifies a 'profound distortion of relationship', which prevents white communities from recognising their own complicity and from acknowledging the need for apology (Walker, 2006, p. 388). The central aim of restorative justice is to correct this distortion, restoring relationships, not in the sense of returning them to a benevolent status quo, which may never have existed, but of achieving what Walker calls 'moral adequacy' (Walker, 2006, p. 384). It is not primarily the creation of personal emotional connections, though these may ensue, but the resetting of a moral compass based upon respect, responsibility and attention to the specificity of others' identity, experience and story.

The impacts and injustices of the climate crisis create new relationships of responsibility as well as exposing and intensifying those which already exist. These include the intergenerational and interspecies relationships discussed by Almassi (2020), and also the response of the safe and comfortable to the plight of climate refugees whose displacement by sea-level rise has deprived them of state protection. Exploring this issue, Rebecca Buxton (2019) advocates a restorative form of reparations, the meeting of practical and cultural needs, including the establishment of a sense of place for future generations. The rationale for such assistance is the polluter pays principle but also an acknowledgement of 'affinity' between those responsible for climate harm and those suffering its impacts (Buxton, 2019, p. 199). Similar forms of affinity can be identified between the minority world states, institutions and corporations which have significantly contributed to the climate crisis and the majority world communities which bear its most destructive consequences. These actors may, but need not, have conscious existing relationships arising out of geographical proximity, colonial history, trade and debt, armed conflict, migration or shared culture. Whether or not such histories are shared, wherever a significant imbalance of emissions, resources and power exists, a central climate relationship can be discerned. And in every case, it will be the human representatives of those entities, their citizens, employees, members and residents, whose shared participation can build or rebuild relationships of shared experience, solidarity and active transformational justice.

4   A Restorative Framework and its Application to the Deglaciating Peruvian Andes

Walker (2006, p. 383) identifies six principles of restorative justice which are applicable to cases of group injustice, and these will be used here as a guiding framework for restorative climate justice praxis. The first is that restorative justice aims to 'repair the harm caused by wrong, crime, and violence' (Walker, 2006, p. 383), assuming a consensus that wrong has occurred, and seeking to address its aftermath (Almassi, 2020; Pavlich, 2002). The second principle places the needs of victims (or, as this chapter will call them, 'the harmed') at the centre of restorative practice. The third concerns those responsible for the harm, obliging them to accept their responsibility and to act appropriately towards the harmed. The fourth concerns the participants in restorative processes, asserting that 'ownership of the resolution' should be shared between the harmed, harmers and appropriate communities (Walker, 2006, p. 383). The fifth principle is that restorative repair and accountability should enable harmers to recover self-respect and to be reintegrated into their communities, and the sixth is that restorative justice builds individual and communal capacity 'to do justice actively' (Walker, 2006, p. 383). Together these principles represent key elements in restorative practice: harm, needs, responsibility, process, repair and transformation. While, as will be seen, Walker's analysis cannot be mapped uncritically onto the terrain of climate justice, it provides an appropriate initial structure for future development.

Climate injustice is a matter of specifics, particular intersections of environmental harm with structural inequalities. In constructing a framework for restorative climate justice from Walker's principles, it will be helpful, therefore, to focus on a specific site. The Peruvian Andes combines high certainty of climate-induced impacts with deep socio-economic inequalities. It is located within a national economy dominated by foreign capital, in which imbalances between regions and groups have been exacerbated by the grave impacts of the coronavirus pandemic (Crabtree, 2020; Gianella et al., 2020). It illustrates the extent to which existing tensions and dilemmas relating to water insecurity are exacerbated by climate change, and the manner in which they represent some of its most contested arenas. Finally, it exemplifies the ways in which Indigenous people are often both the most acutely affected by climate change and the best informed to lead adaptive responses.

4.1   Climate Wrongs and Climate Harms

The restorative process begins with the recognition of wrong, also characterised as injustice (Woolford & Nelund, 2019). From a climate justice perspective, disproportionate emissions, especially when combined with exploitation of people and natural resources, suppression of, or misinformation about climate science and failure to attend to majority world voices or needs, constitute 'wrong' in this sense. The focus then moves to specific harms, the direct and indirect effects of this 'triggering' injustice (Woolford & Nelund, 2019, p. 37). In the Andean context, these harms comprise both immediate climate impacts and their environmental and social consequences. The global phenomenon of glacier recession is one of the most dramatic and clearest effects of climate change, with continued losses likely even if the climate were to stabilise, due to the lag times in glacier responses (Hugonnet et al., 2021; Zemp et al., 2015). Tropical

Andean glaciers such as those in Peru are especially sensitive and are amongst the fastest shrinking in the world (Buytaert et al., 2017). There have been considerable ice volume losses recorded since the late 1980s, and many Peruvian glaciers are expected to disappear within the next few decades (Vuille et al., 2018).

Deglaciation itself produces cascading impacts upon sea level, regional water cycles and local hazards (Zemp et al., 2015) especially glacial lake outburst floods (GLOFs) which have become prevalent since the Little Ice Age ended in the late nineteenth century (Carey, 2010). Glacial lakes are formed as glaciers retreat, leaving basins filled with meltwater and dammed by bedrock, moraines, ice, or a combination of these (Carrivick & Tweed, 2013; Cook & Quincey, 2015). These lakes are susceptible to a variety of impacts, including waves from higher lakes, ice and rock falls, landslides and earthquakes, which can cause the dam to collapse or degrade, or the lake to overtop its impounding dam (Cook et al., 2016; Emmer & Vilímek, 2013; Kougkoulos et al., 2018; Westoby et al., 2014). The resulting floods typically carry with them large quantities of rock and debris, travelling at high speed, with catastrophic effects for people, animals and structures in their path. While it is very difficult to attribute any individual GLOF to the effects of climate change,[1] increased temperatures over the past few decades have certainly led to the generation of more and larger glacial lakes, and diminished the stability of the slopes that surround them (Carey et al., 2012).

One of the principal tensions in relation to glacial lakes relates to their dual status as both flood risk and valuable resource in an area of increasing water stress. (Drenkhan et al., 2019; Walker-Crawford, 2017). This stress has multiple causes, many of which are political rather than hydrological, and predate the immediate crisis of deglaciation (Buytaert et al., 2017). They include the expansion of hydropower (Drenkhan et al., 2019) of agribusiness, growing thirsty export crops in dry coastal areas, fed by Andean water (Buytaert et al., 2017), and of mining, which may pollute water supplies with arsenic, lead and other contaminants (Vuille et al., 2018). Demographic changes, urbanisation, deforestation and tourism all add to the strains upon a system in which the water needs of

---

[1] Though this has now been done in relation to Laguna Palcacocha (Stuart-Smith et al., 2021).

local people, for domestic use and subsistence farming, rank very low (Bolin, 2009). Deglaciation has itself masked some of these stresses, as in the early stages of glacier shrinkage the amount of meltwater increases, until 'peak water' is achieved, after which glaciers become too small to sustain the same level of runoff. However, the peak has already passed for many of Peru's glaciers, and will do so soon for the others, with rapidly diminishing glaciers producing less and less water (Huss & Hock, 2018). In Peru, meltwater plays a critical role as a buffer in the dry season and during droughts. With more extreme dry seasons expected in the future, groundwater depletion and saline intrusions, the importance of that buffer role will be critical, just as it becomes much more difficult to fulfil (Drenkhan et al., 2015).

4.2   The Climate Harmed and Their Needs

Just as restorative justice requires initial consensus on the reality of an injustice, so it needs the assignment, at least broadly, of categories of harmer and harmed. The climate crisis causes many human beings, however privileged, to suffer, even if not physical loss and ill-health, then forms of sadness, guilt, lost experience and awareness of their descendants' burden. These experiences represent important common ground upon which relationships can be built. But while we are all hurt by climate change, we are not all wounded by climate injustice. While the category of those harmed includes non-human nature and future generations (Almassi, 2020; Motupalli, 2018; Preston, 2011), this chapter will limit its exploration to the needs of contemporary communities, mainly in the majority world, who are facing the most acute immediate impacts with the least access to countervailing capacity, and who experience practical, recognition and participatory needs. Amongst such communities are those of the largely Quechua-speaking inhabitants of the Peruvian high Andes. They experience ongoing racism and structural disadvantage which is both a legacy of colonialism and a result of more recent political and economic decisions and conflicts (Crabtree, 2002). The 2021 election of President Pedro Castillo, himself from the Andean region of Cajamarca, might be expected to benefit local and Indigenous people, but substantial barriers to reform continue to exist (Burt, 2021).

Against this backdrop, and in addition to the exacerbation of water stresses, hazards and risks, Andean communities are experiencing other climate impacts, including frost- and heat-induced crop failures, excess rain during the wet season (which prevents the drying of the potato harvest), decline and disappearance of medicinal plants, changes to lake ecosystems and an increase in invasive species and diseases (Bolin, 2009: Vuille et al., 2018). Assailed by these impacts, traditional subsistence farming becomes increasingly unsustainable, and consequent migration of the young to the cities is already the norm in many places (Wrathnall et al., 2014). Their practical needs may therefore include assistance with a range of appropriate adaptive measures. They are also likely to require additional scientific resources to address the complexities involved in identifying and measuring often inaccessible glacial lakes, predicting their degree of hazard and risk and likely impacts in the event of rupture (Cook et al., 2016; Drenkhan et al., 2018, 2019; Emmer & Vilímek, 2013). These needs will include compensation for the loss and damage that cannot be avoided by adaptation, and genuine mitigation on the part of the minority world in order to limit future harms.

Recognition needs include those relating to identities and shared narrative, both of which can be undermined by climate-induced changes. Deglaciation represents a profound spiritual, cultural and emotional loss for Andean people whose communal identity is closely bound up with relationships of interdependence with their neighbouring lakes and glaciated mountains, and who have been forced to alter rituals and religious understandings as a result of this loss (Bolin, 2009). Recognition also encompasses expertise, especially Indigenous ecological knowledge, such as the risk-spreading terracing agricultural techniques and sustainable water allocations preserved by Andean communities (Almassi, 2020; Drenkhan et al., 2019; Paerregaard, 2016; Vuille et al., 2018). Respect for such knowledge, and opportunities to utilise it unhindered may be at least as beneficial as the transfer of new technology.

Participatory, or procedural, needs include the necessity for sufficient and accurate observation, measurement and reporting of impacts for these to be adequately considered in international scientific contexts (Huggel et al., 2016). As has been highlighted by the coronavirus pandemic, without additional financial, political and cultural resources, majority world and marginalised, especially Indigenous, groups are excluded from effective participation in global, including climate, conferencing (Tollefson, 2021). Local communities are frequently denied real consultation in environmental and energy decision-making (Dalglish et al., 2018; Heffron & McCauley, 2017). For example, even where it is possible to identify those Andean glacial lakes with a high risk of outburst flooding, there are often conflicting views as to their management, with local people and authorities contesting options that include drainage, infrastructure, warning systems—and even relocation of communities in the floodpath (Drenkhan et al., 2018; Walker-Crawford, 2017).

4.3   Climate Harmers and Their Responsibilities

The rhetoric of the Anthropocene has sometimes been used to suggest that the climate crisis is the fault of humanity as a whole, arising from an original sin of greed or curiosity (see Malm, 2016). Such a blanket conviction, minimising particular responsibilities, especially of minority world states and corporations, is neither accurate nor helpful. To see climate injustice as created by the intersecting forces of colonialism and capitalism is a more useful starting point, though lacking the full specificity and capacity for relationship required by restorative justice. Individual decision-makers can sometimes be identified, key actors whose choices, actions or inactions have significantly contributed to the current state of climate injustice. Yet, even if their culpability could accurately be distinguished from their context, few would be available, willing and permitted to enter into a contemporary restorative process.

A more fruitful basis for restorative dialogue may be to invite minority world institutions, such as states, corporations, municipalities, religious bodies and educational establishments, to reflect upon their own complicity in climate injustice. This approach would obviously encompass a spectrum of moral responsibility, both amongst the institutions themselves and the individuals who comprise their human members, passive beneficiaries as well as active polluters. This breadth of inclusion would provide the opportunity for restorative processes to take place even where the primary harmers will not participate. As both Walker (2006) and

Almassi (2020) argue, such restorative initiatives by representative or proxy can serve to model the restoration of healthy relationships, to express experiences of collective responsibility and to facilitate progress towards wider justice outcomes.

In the context of the Peruvian Andes, there are existing, and often contested, relationships between local communities and minority world corporations, especially in the mining sector. While most communities are not opposed to mining projects per se, growing threats to both the quantity and quality of water, overlapping and conflicting land rights, and failures of consultation and promise-keeping have led to numerous and intensifying conflicts (Crabtree & Durand, 2017). These local conflicts offer at least a theoretical space for restorative approaches in the use of *mesas di diálogo*, roundtable mediation processes between corporations and communities. However, climate change is not perceived as central to these conflicts, although it is, of course, exacerbating their water-related conflicts. Such relationships are not, therefore, the primary focus of this chapter.

A more relevant connection for climate restorative justice is that recognised by the legal case of *Luciano Lliuya, S.A. v. RWE AG.* (2015), which at the time of writing is proceeding through the German courts. Saúl Luciano Lliuya lives in the small Peruvian city of Huaraz, at the foothills of the Andes. In 1941, a glacial lake outburst flood from Laguna Palcacocha hit Huaraz and killed at least 1800 people (Mark Carey (2010) estimates 5000). Despite various preventative measures, the water level in the lake has again reached dangerous levels. Both the 1941 flood and the current hazardous condition of Lake Palcacocha have been shown to result from the retreat of Palcaraju glacier, attributable to human-induced climate change (Stuart-Smith et al., 2021). The case has been brought against the German energy corporation RWE on the basis that it is responsible for 0.47% of global greenhouse gas emissions and ought to contribute proportionately to the costs of preventative action against a further flood. Whatever the eventual outcome of the case, which despite its relatively tiny financial value is being adamantly resisted by RWE, its claim exemplifies the relationship of climate responsibility between minority world actors and the people of the Peruvian Andes.

The concept of responsibility within restorative justice is rich and deep, relevant to accountability for past acts or omissions, to present participation in the restorative process, and to future commitments (Wallis, 2014). In relation to climate justice for Indigenous communities, its past-focused elements might include responsibility for historic emissions, for concealing the truth about climate science, subsidising and investing in the fossil fuel industry, excluding majority world voices and concerns from climate negotiations and failing to assist with adaptation, as well as broader global strategies of colonialism and exploitation. Institutions and individuals may have played an active part in these acts of injustice, or may have been bystanders, constrained by mechanisms of denial from acknowledging the full implications of their involvement (Cohen, 2001). An example of a climate harmer in this sense might be a minority world university. Such an institution might have invested in coal, oil and gas corporations and legitimised their activities, benefited from unjust financial transfers from the majority world, employed academics who denied or minimised the severity and effects of climate change, and been responsible for its own excessive emissions. Similar responsibilities could be identified on the part of financial, religious and cultural institutions, extractive corporations, vehicle and other manufacturers, and minority world cities, towns and regions, especially those which have benefited from the fossil fuel industry.

4.4   The Restorative Climate Process

In envisioning a genuinely participatory process, care must be taken in translating restorative approaches and insights from a criminal justice to a climate context, in which balances of power may be substantially different. Facilitation must ensure that powerful harmers do not dominate, while recognising that the harmed are not a single homogenous unit with identical perspectives and needs. The concept of wider community involvement, transferred to a global stage, could be a means whereby other harmed communities, and allies in solidarity, share experiences, perceptions and visions to enhance the restorative process. But it could also open the door to powerful international interests opposed to a robust and equitable outcome. Restorative climate practice would aim for an inclusion that is horizontal, welcoming all who can contribute positively, and also vertical, paying particular attention to unheard voices within each community. As in any restorative process, inclusion also requires active consent by the harmed community, both to participation at all, and to its timing and depth (Christie, 1977). This does not mean that every individual would have to agree to the existence of the process, or that it necessarily requires official sanction, but that there is real consent by those most acutely affected.

The form of any particular restorative climate process will depend upon the nature of the parties, the resources available (time and commitment as well as finances), the specificities of their relationship and the kind of outcome sought. At the most ambitious end of the scale are Maxine Burkett's (2009) proposal for inter-state climate reparations or Fadhel Kaboub's (2020) suggestion of a restorative Global Truth and Reconciliation Commission. At the other are informal conversations between majority and minority world activist groups, such as those already facilitated within the climate justice movement.[2] A restorative process between an Andean community and a minority world institution might, for example, draw upon the work of the University of Glasgow in investigating its slave-owning donation history and committing to restorative reparation (Mullen & Newman 2018). Restorative circles or assemblies could take place in person but also, bearing in mind both the climate impacts of long-distance travel and vulnerabilities to the continuing pandemic, could be largely or entirely virtual. Whatever the precise form of the process, it would seek to provide space for respectful and honest conversations between members of climate harmed communities and representatives of climate harming institutions.

4.5   Accountability and Repair

Accountability can be seen as an outward-facing aspect of responsibility, an opportunity for the harmer, having examined their own behaviour

---

[2] For example, https://cop26coalition.org/global-gathering/ https://climatefringe.org/events/from-the-ground-up-global-gathering-for-climate-justice/

and its impacts on the harmed, to express direct regret and repentance. An authentic apology requires what Walker calls the 'leveraging' of responsibility, transforming a superficial awareness of harm into a deeper understanding of its nature and consequences (Walker, 2006, p. 385). In the context of climate injustice, restorative practice may be able to combine the formal structure of official apology with the empathetic content of those individual members who have participated in the process. And even where an official apology may not be forthcoming, a representative group of such members may choose to formulate their own.

It must, of course, be open to the harmed to respond to the apology in whichever way they see fit, including ignoring it (Minow, 1998). In some contexts, it may be appropriate for leaders to answer on behalf of the harmed community, while in other situations, individuals or smaller groups may make their own responses. As in other parts of the process, care must be taken to acknowledge the full spectrum of voices, especially those which differ from the statements of communal gatekeepers. Acceptance of apology is not necessarily synonymous with forgiveness, nor does it preclude other means of seeking justice (Minow, 1998). There will be cases in which continued and combative resistance by the harmed is appropriate, not only for themselves but in order to bring about wider change (Cunneen, 2002), where court judgments can do more than mediated settlements to bring about systemic change (Fiss, 1984), and where the harmer's corporate conduct has been so egregious that the only morally decent option is its dissolution (Almassi, 2020).

Restorative justice requires repair as well as accountability, radical action that achieves tangible outcomes to address practical, recognition and participatory needs. In the case of complex wrongs such as climate injustice, it will not be possible simply to put the parties back in a position as though the wrong had never occurred. Neither can this gap be fully made up by the payment of damages, although compensation and restitution will often be a necessary part of repair (Walker, 2006). Many types of loss cannot be compensated, and call for symbolic as much as material reparations, involving what Martha Minow (1998) describes as different lexicons of value. If then, restorative repair is neither a return to a former status quo nor a simple compensatory award, what does it look like? The Japanese tradition of *kintsugi*, the art of repairing broken ceramics with gold or silver decorated resin, offers a useful metaphor (Domingo, 2019). Rather than trying to hide cracks and imperfections, the intention of the technique is to accentuate them, commemorating not only the original beauty of the piece but also its history of damage and restoration. Its transformative repair simultaneously evokes memories of pre-trauma integrity and of damage, acknowledgement of care and awareness of continued fragility (Keulemans, 2016). Just as the *kintsugi* master must begin with the challenge of each unique crack, so also, restorative justice, at its best, rather than imposing a pre-existing template for repair, focuses on the particularity of the actual existing damage. Duncan McLaren (2018), discussing the ethics of climate repair, uses *kintsugi* as one of his models, along with other forms of restoration, including the healing of bodies, ecosystems and relationships. He contrasts the 'business as usual' domination of geoengineering with the potential for restorative and reconciliatory justice to build solidarity, recognition and care between majority and minority worlds. As he suggests, the restoring of relationships and the repair of harms are not entirely separate imperatives for restorative climate justice. Transformational outcomes will be as much about changing attitudes and dynamics of power as about technology and resources.

Walker's analysis suggests that this repair will benefit not only the harmed, but also the harmers, enabling them to recover self-respect and to be reintegrated 'without stigma' into their community (Walker, 2006, p. 383). In the climate context, self-respect might appropriately be expressed as 'integrity' of institutions, but the aim of reintegration is more problematic. A climate-harming institution which has fully acknowledged its responsibility will not easily be able to settle back into its privileged minority world position, but must seek a new, perhaps less comfortable, global community in which to act.

4.6   Transformative Climate Justice

Walker's final principle of 'active justice' speaks to visions of restorative justice as insurgent and subversive (Sullivan & Tifft, 2006), and as a social justice movement, 'interrupting' injustice and building foundations for future healing (Stauffer & Turner, 2018, p. 447). Here restorative justice becomes transformative justice, conscious of its responsibility to seek fundamental change to societal structures, processes, values and relationships (Harris, 1987; Sullivan & Tifft, 2005; Zehr, 2018). Climate injustice is not a discrete problem which can be managed without disturbing the status quo. Rather, it intertwines with other global and local injustices, giving restorative justice both the opportunity and the obligation to amplify common voices and shared struggles (Woolford & Nelund, 2019). In the context of the Peruvian Andes, a restorative process would not only explore patterns of climate-induced harm but would also illuminate the underlying and overlapping injustices which lie behind simplistic narratives of glaciers as disappearing water towers (Carey, 2010). Growing awareness of the climate crisis within the minority world provides a potentially accessible and not overtly political space in which pre-existing inequalities and oppressions can be understood and challenged.

5 Conclusions

Climate justice requires recognition, not just of the unequal effects of climate change and the oppressions which underlie them, but of global relationships of responsibility, involving people, communities and institutions of the majority and minority worlds. Restorative justice offers a potential means of expressing, exploring and acting upon those responsibilities. Its sensitive praxis could enact healthy relationships of respect, solidarity and care, while simultaneously exerting moral pressure for just policies at national and international levels. This chapter has sought to suggest, based on the principles identified by Margaret Urban Walker, a framework in which global restorative climate justice could consciously, carefully and conscientiously act. The example of the Peruvian Andes suggests both the potential opportunities for such initiatives in majority world deglaciating environments, and the challenges which they face. Even where climate impacts are clearest, as in glacier recession, they do not stand alone, but intersect with other forms of damage and injustice. Climate change itself is a complex process creating various, cascading and interacting effects, some of which may temporarily mask the operation of others. The needs and priorities of those harmed by climate injustice are therefore not necessarily those anticipated, or those expressed by popular narratives. Restorative justice should be well placed to respond to the articulated needs of the harmed, including where the harmed call for redress of other related forms of environmental and social injustice. This process, however, may involve some painful enlightenment on the part of the harmers, all the more so as they acknowledge the moral relationships which already connect majority and minority worlds in countries such as Peru. Some of these connections are positive and healthy, but many display recurring patterns of oppression and exploitation, rooted in unresolved colonial histories and enabled by complicit domestic decision-making.

For restorative climate practice to flourish in these spaces, it will need to draw upon its deepest ethical and experiential resources. It will need to be radically relational, context-specific, focused upon the actual needs of the harmed, and prepared to interrogate the full responsibility of the harmers. It will need to enact genuinely participatory processes, making silence for the quietest voices, while resisting domination by the powerful. It will need to be patient and meticulous, while at the same time aware of the desperate urgency of the climate crisis. It will need to be insurgent, inspiring participants, informed by their restored relationships, to speak experienced truth to complacent power. These are challenging requirements, but we live in challenging times. Restorative climate justice, drawing upon the wisdom and commitment of both the restorative and climate justice movements, has the potential to be part of a truly transformational change.

References

Almassi, B. (2020). *Reparative environmental justice in a world of wounds.* Lexington: Lexington Books.

Bakiner, O. (2016). *Truth commissions: Memory, power and legitimacy.* Philadelphia: University of Pennsylvania Press.

Barry, J. (2012). *The Politics of actually existing unsustainability: Human flourishing in a climate-changed, carbon constrained world.* Oxford: Oxford University Press.

Biffi, E., & Pali, B. (Eds.) (2019). *Environmental justice: Restoring the future—Towards a restorative environmental justice praxis*. Leuven: European Forum for Restorative Justice.

Bolin, I. (2009). The glaciers of the Andes are melting: Indigenous and anthropological knowledge merge in restoring water resources. In S. A. Crate & M. Nuttall, M. (Eds.), *Anthropology and climate change: From encounters to actions* (pp. 228–239). New York: Routledge.

Burkett, M. (2009). Climate reparations. *Melbourne Journal of International Law*, 10, 509–542.

Burt, J-M. (2021). Peru's turbulent elections lay bare systemic crisis. *NACLA Report on the Americas*, 53 (3), 215–219.

Buxton, R. (2019). Reparative justice for climate refugees. *Philosophy*, 94, 193–219.

Buytaert, W., Moulds, S., Acosta, L., De Bièvre, B., Olmos, C., Villacís, M., Tovar, C., & Verbist, K.M.J. (2017). Glacier melt content of water use in the tropical Andes. *Environmental Research Letters*, 12 (114014). Retrieved from: https://doi.org/10.1088/1748-9326/aa926c (last accessed 21 January 2022).

Carey, M. (2010). *In the shadow of melting glaciers: Climate change and Andean society*. New York: Oxford University Press.

Carey, M., Huggel, C., Bury, J., Portocarrero, C., & Haeberli, W. (2012). An integrated socio-economic framework for glacier hazard management and climate change adaptation: Lessons from Lake 513, Cordillera Blanca, Peru. *Climatic Change*, 112, 733–767.

Carrivick, J. L., & Tweed, F. S. (2013). Proglacial lakes: Character, behaviour and geological importance. *Quaternary Science Reviews*, 78, 34–52.

Christie, N. (1977). Conflicts as property. *The British Journal of Criminology*, 17(1), 1–15.

Cohen, S. (2001). *States of denial: Knowing about atrocities and suffering*. Cambridge: Polity.

Cook, K., & Powell, C. (2003). Unfinished business: Aboriginal reconciliation and restorative justice in Australia. *Contemporary Justice Review*, 6(3), 279–291.

Cook, S.J., & Quincey, D. J. (2015). Estimating the volume of Alpine glacial lakes. *Earth Surface Dynamics*, 3(4), 559–575.

Cook, S.J., Kougkoulos, I., Edwards, L.A., Dortsch, J., & Hoffmann, D. (2016). Glacier change and glacial lake outburst flood risk in the Bolivian Andes. *The Cryosphere*, 10, 2399–2413.

Crabtree, J. (2002). *Peru*. Oxford: Oxfam GB.

Crabtree, J. (2020). Democracy, elite power and civil society: Bolivia and Peru compared. *Revista CIDOB d'Afers Internacionals*, 126, 139–161.

Crabtree, J. & Durand, F. (2017). *Peru: Elite power and political capture*. London: Zed Books.

Cunneen, C. (2002). Restorative justice and the politics of decolonisation. In E.G.M. Weitekamp & H.-J. Kerner (Eds.), *Restorative justice: Theoretical foundations* (pp. 32–49). Cullompton: Willan.

Dalglish, C., Leslie, A., Brophy, K., & Macgregor, G. (2018). Justice, development and the land: the social context of Scotland's energy transition. *Landscape Research*, 43(4), 517–528.

Domingo, V. (2019). La justicia restaurativa transforma, *La Justicia Restaurativa*. Retrieved from: https://www.lajusticiarestaurativa.com/la-justicia-restaurativa-transforma. (last accessed 27 September 2021).

Doyle, T. (2005). *Environmental movements in minority and majority worlds: A global perspective*. Rutgers University Press.

Drenkhan, F., Carey, M., Huggel, C., Seidel, J., & Oré, M.T. (2015). The changing water cycle: Climatic and socioeconomic drivers of water-related changes in the Andes of Peru. *WIREs Water*, 2, 715–733.

Drenkhan, F., Guardamino, L., Huggel, C., & Frey, H. (2018). Current and future glacier and lake assessment in the deglaciating Vilcanota-Urubamba basin, Peruvian Andes. *Global and Planetary Change*, 169, 105–118.

Drenkhan, F., Huggel, C., Guardamino, L., & Haeberli, W. (2019). Managing risks and future options from new lakes in the deglaciating Andes of Peru: The example of the Vilcanota-Urubamba basin. *Science of the Total Environment*, 665, 465–483.

Earth in Common (2021). Restorative climate justice: A concept to place at the heart of foreign aid and international development? Retrieved from: https://tinyurl.com/rcjust (last accessed 21 September 2021).

Emmer, A., & Vilímek, V. (2013). Review article: Lake and breach hazard assessment for moraine-dammed lakes: an example from the Cordillera Blanca (Peru). *Natural Hazards and Earth System Sciences*, 13, 1551–1565.

Fiss, O.M. (1984). Against settlement. *The Yale Law Journal*, 93 (6), 1073–1090.

Ghosh A. (2017). *The great derangement: Climate change and the unthinkable*. Chicago: University of Chicago.

Gianella, C., Iguiñiz-Romero, R., Romero, M., & Gideon, J. (2020). Good health indicators are not enough: Lessons from COVID-19 in Peru. *Health and Human Rights*, 22(2), 317–319.

Gibbs, M. (2009). Using restorative justice to resolve historical injustices of Indigenous peoples. *Contemporary Justice Review*, 12(1), 45–57.

Harris, M. K. (1987). Moving into the new millennium: Toward a feminist vision of justice. *The Prison Journal,* 67(2), 27–38.

Heffron, R., & McCauley, D. (2017). The concept of energy justice across the disciplines. *Energy Policy,* 105, 658–667.

Hugonnet, R., McNabb, R., Berthier, E., Menounos, B., Nuth, C., Girod, L., Farinotti, D., Huss, M., Dussaillant, I., Brun, F., & Kääb, A. (2021). Accelerated global glacier mass loss in the early twenty-first century. *Nature,* 592, 726–747.

Huggel, C., Wallimann-Helmer, I., Stone, D., & Cramer, W. (2016). Reconciling justice and attribution research to advance climate policy. *Nature Climate Change,* 6, 901–908.

Hulme, M. (2000). *Why we disagree about climate change: Understanding controversy, inaction and opportunity.* Cambridge: Cambridge University Press.

Huss, M., & Hock, R. (2018). Global-scale hydrological response to future glacier mass loss. *Nature Climate Change,* 8(2), 135–140.

Jordan, A., Huitema, D., van Asselt, H., & Forster, J. (2018). *Governing climate change: Polycentricity in action?* Cambridge: Cambridge University Press.

Kaboub, F. (2020). Green and just? *New Internationalist* 527, 24–25.

Keulemans, G. (2016). The geo-cultural conditions of kintsugi. *The Journal of Modern Craft,* 9(1),15–34.

Kougkoulos, I., Cook, S. J., Jomelli, V., Clarke, L., Symeonakis, E., Dortch, J. M., Edwards, L.A., & Merad, M. (2018). Use of multi-criteria decision analysis to identify potentially dangerous glacial lakes. *Science of the Total Environment,* 621, 1453–1466.

Luciano Lliuya, S.A. v RWE AG. (2015). *Statement of claim.* Retrieved from: https://germanwatch.org/sites/default/files/announcement/20822.pdf. (last accessed 27 September 2021)

Lunn P. (2011). *Costing not less than everything: Sustainability and spirituality in challenging times.* London: Quaker Books.

Malm, A. (2016). *Fossil capital: The rise of steam power and the roots of global warming.* London: Verso.

Maslin, M.A. (2020). The road from Rio to Glasgow: a short history of the climate change negotiations. *Scottish Geographical Journal,* 136, 5–12.

McCauley, D., & Heffron, R. (2018). Just transition: Integrating climate, energy and environmental justice. *Energy Policy,* 119, 1–7.

McLaren, D. (2018). In a broken world: Towards an ethics of repair in the Anthropocene. *Anthropocene Review,* 5(2), 136–154.

Minow, M. (1998). *Between vengeance and forgiveness: Facing history after genocide and mass violence.* Boston: Beacon.

Motupalli, C. (2018). Intergenerational justice, environmental law and restorative justice. *Washington Journal of Environmental Law and Policy*, 8(2), 333–361.

Mullen, S., & Newman, S. (2018). *Slavery, abolition and the University of Glasgow: Report and recommendations of the University of Glasgow History of Slavery Steering Committee*. Retrieved from: https://www.gla.ac.uk/media/Media_607547_smxx.pdf (last accessed 27 September 2021).

Okereke, C. (2014). Cough up: Rich nations can't dodge bill for historic pollution University of Reading Geography and Environmental Science blog 15 Dec 2014. Retrieved from: https://blogs.reading.ac.uk/geography-and-environmental-science/?p=319 (last accessed 27 September 2021)

Pavlich, G. (2002). Deconstructing restoration: The promise of restorative justice. In G.M. E. Weitekamp & H.-J. Kerner (Eds.), *Restorative justice: Theoretical foundations* (pp. 90–109). Cullompton: Willan.

Paerregaard, K. (2016). Making sense of climate change: Global impacts, local responses, and anthropogenic dilemmas in the Peruvian Andes. In S. A. Crate & M. Nuttall, M. (Eds.), *Anthropology and climate change: from actions to transformations* (pp. 250–260). New York: Routledge.

Parola, G. (2017). Restorative justice: A legally binding instrument to implement our ecological duties. *Environmental Liability, Law Practice and Policy*, 25(2), 80–92.

Preston, B. (2011). The use of restorative justice for environmental crime. *Criminal Law Journal* 35, 136–153.

Punch, S. (2003). Childhoods in the majority world: Miniature adults or tribal children? *Sociology, 37*(2), 277–295.

Roberts, S., & Palmer, M. (2005). *Dispute processes: ADR and the primary forms of decision-making*. Cambridge: Cambridge University Press.

Singer, P. (1993). *How are we to live? Ethics in an age of self-interest*. Amherst, N.Y: Prometheus Books.

Stabinsky, D. (2021). *Chasing carbon unicorns: The deception of carbon markets and "net zero"*. Amsterdam: Friends of the Earth International.

Stauffer, S., & Turner, J. (2018). The new generation of restorative justice. In T. Gavrielides, (Ed.), *Routledge international handbook of restorative justice* (pp. 442–461). Abington: Routledge.

Stuart-Smith, R. F., Roe, G. H., Li, S., & Allen, M. R. (2021). Increased outburst flood hazard from Lake Palcacocha due to human-induced glacier retreat. *Nature Geoscience*, 14(2), 85–90.

Sullivan, D. & Tifft L. (2005). *Restorative justice: Healing the foundations of our everyday lives*. Boulder: Lynn Rienner.

Sullivan, D. & Tifft, L. (2006). Introduction: The healing dimension of restorative justice: a one-world body. In D. Sullivan & L. Tifft (Eds.), *Handbook of restorative justice: A global perspective* (pp. 1–16). Abington: Routledge.

Thanki, N. (2019). A new chance for climate justice? *Open Democracy* Retrieved from: https://www.opendemocracy.net/en/opendemocracyuk/new-chance-climate-justice/ (last accessed 21 January 2022).

Tokar, B., (2019). On the evolution and continuing development of the climate justice movement. In T. Jafry (Ed.), *Routledge Handbook of Climate Justice* (pp. 13–25). Abington: Routledge.

Tollefson, J. (2021). Pivotal climate summit dogged by concerns about Covid and equity. *Nature*, 597, 315.

UN Environment Programme (UNEP) (2019). *Emissions Gap Report.* Retrieved from: https://www.unep.org/resources/emissions-gap-report-2019 (last accessed 27 September 2021)

Vuille, M., Carey, M., Huggel, C., Buytaert, W., Rabatel, A., Jacobsen, D., ... & Sicart, J-M. (2018). Rapid decline of snow and ice in the tropical Andes—Impacts, uncertainties and challenges ahead. *Earth-Science Reviews*, 176, 195–213.

Waldron J. (2003). Who is my neighbor? Humanity and proximity. *Monist*, 86, 333–354.

Walker, M. U. (1998). *Moral understandings: A feminist study in ethics.* London: Routledge.

Walker, M. U. (2006) Restorative justice and reparations. *Journal of Social Philosophy*, 37(3), 377–395.

Walker-Crawford, N. (2017). Shifting climates of responsibility: Facing environmental disaster in the High Andes. In J. Taks & S. Atzugaray (Eds.), *Anthropological contributions for sustainable futures: Research and interventions in the fields of environmental needs, gender equity, human rights and knowledge in South America and the United Kingdom.* Retrieved from: https://www.fhuce.edu.uy/images/comunicacion/publicaciones/Taks-Alzugaray-2019-06-23-todo.pdf. (last accessed 27 September 2021).

Wallis, P. (2014). *Understanding restorative justice: How empathy can close the gap created by crime.* Bristol: Policy Press.

Warlenius, R. (2018). Decolonizing the atmosphere: The climate justice movement on climate debt. *Journal of Environment & Development*, 27, 131–155.

Westoby, M. J., Glasser, N. F., Brasington, J., Hambrey, M. J., Quincey, D. J., & Reynolds, J. M. (2014). Modelling outburst floods from moraine-dammed glacial lakes. *Earth-Science Reviews*, 134, 137–159.

White, R. (2014). Indigenous communities, environmental protection and restorative justice. *Australian Indigenous Law Review*, 18, 43–54.

Woolford, A., & Nelund, A. (2019). *The politics of restorative justice*. Black Point: Fernwood.

Wrathnall, D.J., Bury, J., Carey, M., Mark, B., McKenzie, J., Young, K., Baraer, M., French, A., & Rampini, C. (2014). Migration amidst climate rigidity traps: Resource politics and social–ecological possibilism in Honduras and Peru, *Annals of the Association of American Geographers*, 104(2), 292–304.

Zehr, H. (2018). Foreword. In T. Gavrielides (Ed.), *Routledge international handbook of restorative justice* (pp. xxvii–xxviii). Abingdon: Routledge.

Zemp, M., Frey, H., Gärtner-Roer, I., Nussbaumer, S.U., Hoelze, M., Paul, F., …& Vincent, C. (2015). Historically unprecedented global glacier decline in the early 21st century. *Journal of Glaciology*, 228, 745–762.

11

Environmental Restorative Justice in Transitional Settings

Rachel Killean

1   Introduction[1]

Since 2019, the Colombian *Jurisdicción Especial para la Paz* (JEP),[2] a specialist court established through Colombia's 2016 peace accord, has passed five resolutions recognising the territories of Indigenous peoples and Black communities as victims of the conflict (Huneeus & Rueda Sáiz, 2021). In doing so, the JEP offers judicial recognition of the fact that the territories, defined as 'living wholes' and the 'sustenance of

---

[1] This chapter builds on work I first developed with my colleague Dr Lauren Dempster (see Killean & Dempster, 2022). My thanks go to her for her input and for feedback on this chapter. Thanks also to the organisers and participants at the Oñati Environmental Restorative Justice workshop in June 2021 for their inspiring insights.

[2] See https://www.jep.gov.co/Paginas (last accessed 25 January 2022).

R. Killean (✉)
The University of Sydney Law School, Sydney, NSW, Australia
e-mail: Rachel.killean@qub.ac.uk

© The Author(s), under exclusive license to Springer Nature Switzerland AG 2022   **247**
B. Pali et al. (eds.), *The Palgrave Handbook of Environmental Restorative Justice*,
https://doi.org/10.1007/978-3-031-04223-2_11

identity and harmony' (Decree 4633, 2011, Articles 3, 8, 29 and 45)[3] are subjects of Colombia's 2016 peace treaty regime (Ruiz Serna, 2017). Such acknowledgment transforms the territories from objects into 'legal subjects with rights to truth, justice, reparation, and guarantees of non-repetition,' as well as the right to participate in legal processes (Huneeus & Rueda Sáiz, 2021, p.1). The JEP's resolutions have been praised for 'breaking new ground' (Huneeus, 2020), giving legal expression to the idea that conflicts not only harm individuals and communities, but can victimise ecosystems and other-than-human entities, as well as the spiritual relationships that can exist between humans and their natural environments.

Furthermore, the JEP promotes practices which resonate with restorative justice concepts, such as seeking 'truth through consciousness, reconciliation, healing, and harmonization between victims and accused that allows for the strengthening of the community fabric, as well as the harmonization of the territory' (General Regulation 2018, Article 44).[4] While it is unclear at the time of writing how this recognition of territory-as-victim will influence measures of restoration (Huneeus & Rueda Sáiz, 2021, p. 15), Colombia is arguably developing one of the first examples of an environmental restorative justice approach within the field of 'transitional justice'.

Transitional justice is a field characterised by the 'set of practices, mechanisms and concerns that arise following a period of conflict, civil strife or repression' (Roht-Arriaza, 2006). Dominant 'practices and mechanisms' include 'prosecutions, reparations, truth-seeking, institutional reform, vetting and dismissals' (UNSG, 2004). Since emerging in the context of regime change in South America in the 1980s, it has developed beyond its original conception of addressing the 'wrongdoings of repressive predecessor regimes' (Teitel, 2003, p. 69) to 'advancing nation-building as an alternative source of rule of law and the goal of preserving

---

[3] Minister del Interior, Decreto Ley No. 4633, DE 2011, Por medio del cual se dictan medidas de asistencia, atención, reparación integral y de restitución de derechos territoriales a las víctimas pertenecientes a los Pueblos y Comunidades indígenas.

[4] Jurisdicción Especial Para la Paz, Reglamento General, Acuerdo No. 001 DE 2018, Por el cual se adopta el Reglamento General de la Jurisdicción Especial para la Paz; For a brief description of a harmonisation ritual see Cabrera (2019).

peace' (Hulme, 2017, p. 121). Indeed, transitional justice has shown itself to be a field capable of significant expansion, in terms of the actors engaged in its work, the scenarios to which it applies, the formulations of justice it espouses, and the mechanisms it employs (Zunino, 2019).

While it has often prioritised retributive justice (Lambourne, 2009), restorative justice mechanisms have also emerged (Clamp, 2016). High-profile examples include attempts to address system-wide offences through Truth and Reconciliation Commissions, perhaps most famously in South Africa (Tutu, 1999). At a community level, examples are diverse, including Gacaca courts in Rwanda (Clark, 2010) and alternatives to paramilitary violence in Northern Ireland (McEvoy & Mika, 2001). There is a well-developed literature which explores the challenges and possibilities of pursuing restorative justice in transitional contexts, drawing from a range of case studies and perspectives (e.g., McEvoy & Newburn, 2003; Natalya Clark, 2008; Lambourne, 2010; Clamp, 2014, 2016; Zernova, 2017). However, an *environmental* restorative justice response to conflict is arguably long overdue.

Growing international recognition of the need to protect the environment during conflict have been reflected in, for example, the International Committee for the Red Cross's recently revised 'Guidelines on the protection of the natural environment in armed conflict' (ICRC, 1994, 2020), and the UN International Law Commission's draft principles on 'Protection of the environment in relation to armed conflict', which are due to be adopted in 2022 (Bothe, 2017). The growth of environmental peacebuilding has also played an important role in highlighting the environmental consequences of war, the use of natural resources to finance armed conflict, the potential for cooperation around natural resources, and the links amongst post-conflict peacebuilding, climate resilience, and natural resource management (see Ide et al., 2021 for an overview). In the legal sphere, there is increased recognition of the need to protect the environment in the context of *jus post bellum* (Stahn, Iverson & Easterday, 2017), calls for a crime of ecocide have gained momentum (Stop Ecocide Foundation, 2021; Higgins, 2010; Higgins, Short, & South, 2013), and the Prosecutor at the International Criminal Court has indicated a willingness to prioritise cases that involve environmental destruction (ICC, 2016).

In this context of growing environmental consciousness, it is notable that the field of transitional justice has historically failed to meaningfully engage with environmental harm (Natalya Clark, 2016). This is beginning to change. There is a growing literature which explores the advantages of including environmental considerations in the transitional justice agenda (e.g., Harwell & Le Billon, 2009; Nichols, 2014; Pelizzon, 2015; Harwell, 2016; Ong, 2017; Killean & Dempster, 2022). Some have considered the potential for 'greener' forms of transitional justice mechanisms such as memorials, truth commissions and reparations (Natalya Clark, 2016; Cusato, 2017; Farrar, 2020; Varona, 2020; Killean, 2021). Others have drawn attention to the specific harms of colonialism, land expropriation and large-scale environmental degradation (Brinton Lykes & Van Der Merwe, 2017; Murdock, 2018), the intertwined relationships between victims and their environments (Mitchell, 2014; Natalya Clark, 2020), and the severe impacts of environmental harm for Indigenous peoples (Izquierdo & Viaene, 2018). Following the Colombian JEP's resolutions detailed above, others have begun to explore what it means to recognise territories as victims of conflict (Ruiz Serna, 2017; Huneeus & Rueda Sáiz, 2021).

Against this backdrop, this chapter considers the value of adopting an environmental restorative justice approach to transitional justice. To do so, it engages with the emerging critical transitional justice scholarship which seeks to challenge dominant perspectives for 'the lives they ignore, the injustices they fail to see and the patriarchal and racialized power structures that remain intact and unexamined' (Rooney & Ní Aoláin, 2018). Key critiques have included the 'dominance of legalism' in the field (McEvoy, 2007), the prioritisation of civil and political rights to the exclusion of other forms of harm (Nagy, 2008), and transitional justice's exclusion of Indigenous and local traditions of justice (An-Na'im, 2013). The chapter brings an environmental restorative justice lens to these critiques, drawing from the growing body of 'critical knowledge and practice' preoccupied with 'how to conceive of violence and harms against ecosystems and non-human beings' (Varona, 2021, p. 42), how to use restorative justice principles to respond to those harms (Pali, 2019, p. 13), and how to 'infuse' restorative justice principles with an environmental sensibility (Forsyth et al., 2021, p. 18). Throughout the chapter, I understand environmental restorative justice principles to include: (1) an orientation towards healing, inclusive of nature; (2) participation of those with power to take responsibility and those who have suffered harm, inclusive of other-than-humans; (3) the use of storytelling and dialogue; (4) the identification of harm, victims, and those responsible; and (5) accountability in the pursuit of relational justice (Forsyth et al., 2021, p. 20).

I am sensitive to the critique that transitional justice 'clumsily applies the same thinking and tools across a range of contexts and transitional types as if they were the same thing' (Sharp, 2019, p. 571). What will be appropriate in any transitional context will be shaped by the specificities of the harm experienced, the actors involved, the stage of the transition, and a multitude of other factors. Therefore, I do not advocate for specific models of environmental restorative justice in the aftermath of conflict. Rather, I explore how an environmental restorative justice approach might benefit transitional justice by challenging some of the field's anthropocentric and neo-colonial tendencies. I argue that an environmental restorative approach invites recognition and representation of other-than-human victims of conflict and enables greater responsiveness to Indigenous people's experiences of conflict. Going further, I consider the possibilities of environmental restorative justice as a tool through which to better engage with Indigenous understandings of justice.

2 The Environment as a Victim of Conflict

The environmental costs of conflict have been documented by the United Nations Environment Programme since 1999. Their Post-Conflict Environmental Assessments have highlighted, for example, widespread loss of vegetation and wildlife in Afghanistan (UNEP, 2003), unsustainable resource exploitation in Sudan (UNEP, 2007), land and water contamination in the Gaza Strip (UNEP, 2009a) and severe damage to national parks in Côte d'Ivoire (UNEP, 2015). Conflict destroys human lives and habitations, but also plants, animals, landscapes, cultural practices, and the range of 'complex linkages' or 'webs of relationships' that exist between these phenomena (Mitchell, 2014, p. 5; Ruiz Serna, 2017).

These harms are particularly acute when conflicts involve Indigenous communities, who may have specific historical, cultural, social, and emotional ties to land, and who may consider the destruction of the environment and its inhabitants an act that 'literally eliminates, disfigures, and maims' other-than-human members of Indigenous communities (Eichler, 2020, p. 104).

These victimisations are often exacerbated by pre-existing structural inequalities and socio-economic injustices, which can act as a source of conflict and a barrier to lasting peace (Hulme, 2017; Cusato, 2020). Empirical studies have demonstrated that states emerging from natural resource-related armed conflicts are more than twice as likely to relapse into conflict as those prompted by other motivations (UNEP, 2009b). As climate change and mass extinctions continue to place additional pressures onto vulnerable communities, the likelihood of environmental conflicts increases, while the environmental impacts of conflict become more difficult to withstand (Guess, 2020). Transitions in turn bring their own environmental harms, including unsustainable resource extraction, deforestation, and subsequent land conflicts resulting from displacement and loss of territory (Suarez et al., 2018; Huneeus & Rueda Sáiz, 2021) On the other hand, well-managed natural resources and environmental cooperation can consolidate peace, enable social reconstruction, and improve the livelihood of the population following conflict (Milburn, 2014; Cusato, 2017).

2.1   Transitional Justice and Anthropocentrism

Transitional justice has rarely recognised other-than-human victimisation as worthy of recognition and redress. This is in part a reflection of the dominance of legalism within transitional justice (McEvoy, 2007) which has given anthropocentric legal frameworks a central role in shaping how victimhood is understood (Killean & Dempster, 2022). Indeed, laws designed to address past atrocities often specifically define who can be recognised as a victim (see e.g., García-Godos & Lid, 2010; Jankowitz, 2018). The field is particularly grounded in 'human rights law, criminal law and international law of war' (Teitel, 2003), meaning that transitional justice often centres victims of 'gross violations of civil and political right' or 'criminal acts (property destruction, abuse of children, etc)' (Nagy, 2008, p. 284). Environmental law is rarely listed 'within the wide range of legal disciplines deemed to be both applicable and already encompassed by the growing field' (Ong, 2017, p. 218).

The development of international criminal law as a means of addressing atrocity has further entrenched the dominance of a 'crime-driven lens' (Borda, 2020), leading to an emphasis on actions which harm 'humans from a very anthropocentric and economic perspective' (Walters, Westerhuis, & Wyatt, 2013, p. 7). Such understandings of victimhood arguably emphasise the 'biological, mental and moral superiority of humans over other living and non-living entities' (White, 2014b, p. 12) creating exclusionary hierarchies of victimhood which overlook human victims of environmental harm and other-than-human entities (White, 2018, pp. 240–247). Far from being perceived as a victim, the natural environment and its other-than-human inhabitants are generally framed as a 'disputed resource' (Varona, 2020, p. 670), to be managed appropriately rather than granted redress (Motupalli, 2018, p. 345).

In practice, limited references to environmental harm can be found in transitional justice mechanisms. For instance, in addition to the Colombian example highlighted above, Peru has acknowledged the pain inflicted upon Mother Earth in a memorial sculpture of the ancestral goddess Pachamama (Falcón, 2018). However, environmental harm is more often overlooked, or framed in terms of harms against people and/ or property. The international criminal law framework established to prosecute mass atrocities has been critiqued for its failure to address environmental crime (Drumbl, 1998; Weinstein, 2005). There is only one explicit eco-centric war crime (Lawrence & Heller, 2007; Gillett, 2017), and in practice its high evidentiary thresholds render it almost unusable (Killean, 2021). In the context of truth commissions, few have made recommendations relating to environmental harm, and those that have done so have usually framed the harm as an economic loss or attack against the person (Cusato, 2017; Crosby & Brinton Lykes, 2011; Izquierdo & Viaene, 2018).

2.2   Recognising Other-than-Human Victims

Environmental restorative justice offers a framework for looking beyond those harms traditionally considered 'criminal' (Wijdekop, 2019). As Kershen argues, 'Because restorative justice starts from a different set of questions [than criminal law], it presents an opportunity to bring into the picture a wider set of concerns and to ensure environmental harm is more effectively addressed' (Kershen, 2019, pp. 41–42). While in practice restorative justice often remains 'strongly anthropocentric…it has the potential, more than any other justice approach, to incorporate ecocentric perspectives' (Pali & Aersten, 2021, p. 5) which accept victims as including humans, specific environments and animals and plant life (White, 2018, p. 241). For example, Preston's (2011, pp. 10–11) influential categorisation of environmental victimhood includes not only individuals, communities, and classes of people (including future generations), but the environment other than humans, such as the biosphere and nonhuman biota. As he explains, the 'biosphere and nonhuman biota have intrinsic value independent of their utilitarian or instrumental value for humans' and can therefore also be victims (Preston, 2011, p. 12). Thus, environmental restorative justice offers the possibility of 'extending notions of victim and offender' (Pavlich, 2005, p. 85) and 'imagining a community that includes non-human beings' (Komatsubara, 2021).

Through such a perspective, previously 'invisible and unconsidered harms' of conflict become visible (Varona, 2021, p. 46). Consider, for example, the poisoning of vital water wells and drinking water as a tactic of violence, a harm which has previously been perpetrated by the Arab Janjaweed militia in Darfur, Sudan (Freeland, 2005). In such a case, an anthropocentric approach would recognise that the humans who relied upon the water source have been victimised, as have any communities for whom the water held cultural or sacred significance, and individuals or corporations who may have relied on the water source for commercial activities (Hamilton, 2021, p. 127). However, an environmental restorative justice approach would go further, acknowledging that the water source itself can be considered a victim, as can the flora and fauna for which it was a home.

In recognising other-than-human victims of conflict, environmental restorative justice might facilitate a more complete understanding of the harms of conflict, in turn enabling fuller forms of redress. As Ruiz Serna (2017, p. 103) argues, recognising territories as victims with legal rights 'makes possible a field of interlocution and transaction…in which different communities can describe the effects that go beyond human beings'. This might in turn facilitate responses in which the human is no longer 'the exclusive subject of reparation' (Ruiz Serna, 2017, p. 104). As I have explored elsewhere, the design of environmentally conscious reparations might be of particular benefit in the aftermath of conflict (Killean, 2021). For example, the design of peace parks might provide a space in which both the environment and communities could recover and recuperate, while environmentally conscious income-generating initiatives could potentially assist a wide range of human and other-than-human victims (Milburn, 2014). As Forsyth et al. note, 'relational and emotional healing is central to all restorative justice (2021, p. 30). Through an environmental restorative justice process, outcomes could be agreed which acknowledge the symbiotic relationship between restoring natural environments and repairing the harm experienced by individuals and communities during conflict (Lawry-White, 2017).

2.3   Representing Other-than-Human Victims

Such an approach raises issues of victim representation. As Verona observes, 'in restorative programmes we have to give voice to non-human forms of life' who 'can neither talk nor exercise rights' (Varona, 2021, p. 47). Citing Stone's influential work *Should Trees Have Standing* (1972), Preston (2011, p. 14) argues that the 'fact that the environment and non-human biota are not able to vocalise their claims and concerns' fails to present an 'insuperable problem', as long as a surrogate can 'bring to the restorative process an eco-centric and not an anthropocentric perspective'. There is precedent for this; natural objects such as rivers and trees have been represented successfully in domestic environmental restorative justice conferences in numerous jurisdictions (Hamilton, 2021).

It is notable that in the Colombian context, the recognition of Indigenous people's territory as a victim has been framed as a 'political victory for the Indigenous peoples' organizations' (Izquierdo & Viaene, 2018). The organisations intervened to express dissatisfaction with the Law on Victims of the Internal Armed Conflict (Ruiz Serna, 2017), and were successful in obtaining a separate Decree which frames the territory as 'a living whole and sustenance of identity and harmony' that suffers damage requiring 'spiritual healing' (Decree 4633, 2011, Articles 3, 8, 29 and 45), forming the basis of subsequent JEP resolutions. As this example demonstrates, Indigenous peoples are often at the forefront of advocating on behalf of land and other-than-human victims. Thus, the claim that Indigenous peoples' 'strong connection with the land and their skills in listening to the non-verbal communication from nature… uniquely positions them to be spokespersons for the harmed environment is a strong one' (Wijdekop, 2019, p. 45). However, it would be problematic to make such assumptions without meaningful and genuine collaborative engagement with the relevant peoples. The risks of other-than-human victimisation becoming 'immersed in idealism' (Komatsubara, 2021, p. 125), with Indigenous peoples expected to act as 'convenience representatives' for all victims (Hamilton, 2021, p. 217), have also been flagged in the literature and would require careful attention. As White notes, 'Indigenous people do not "think the same way" just because they are Indigenous and/or because they have a connection to country' (White, 2017). Indeed, overly simplified narratives also risk overlooking the tensions that might emerge between Indigenous peoples' and conservationist perspectives, for example. I return to this point below.

In cases without strong cultural links between human and other-than-human victims, an environmental restorative justice approach might nonetheless involve representation for that environment. Kershen (2019, p. 50) highlights that a range of inter-species communicators might communicate with and articulate on behalf of other-than-human victims through 'electronic, non-verbal or intuitive means'. It may be that relevant organisations who have worked with the land prior to, during, or after the conflict might act in this capacity, or ecologists or other scientific professionals specifically appointed to act as representatives (Forsyth et al., 2021, pp. 33–34). However, this would mark a shift towards a more 'facts and figures' voice (Hamilton, 2021, pp. 226–228), potentially reconceptualising ideas of harm and repair. More fundamentally, the introduction of 'experts' may run counter to what restorative justice aims to achieve, that is, the emergence of empathy for the other and the rebuilding of relationships (White, 2014a/2015; Forsyth et al., 2021). It is therefore probable that environmental restorative justice would be most appropriate and effective in transitional contexts involving victimised communities with a 'special relationship' to land and other-than-human victims (Izquierdo & Viaene, 2018). This theme is explored further in the following section, which considers in greater depth the interconnected links between destruction of environments and the frequent victimisation of Indigenous peoples during conflict, and their marginalisation within the field of transitional justice.

3   Indigenous Victims of Armed Conflict

Indigenous peoples are often victims of periods of armed conflict, facing attacks on their individual and collective rights, territories, and cultural and natural heritage (Tauli-Corpuz, 2017). They can become caught up in violence in several ways: they can be active participants in conflict, particularly following encroachment onto their territories (Kipuri, 2017). They can also be forcibly recruited into conflict by insurgent groups and they can find themselves caught between insurgencies and governmental forces as a 'third party' (Barume, 2017). While the 2007 United Nations Declaration on the Rights of Indigenous Peoples prohibits military activities taking place on Indigenous territories without justification or free agreement, in practice Indigenous communities often experience the forced militarisation of their territories (Tauli-Corpuz, 2017). As explored above, the experiences of conflict in these cases are not limited to individual rights violations but are 'inscribed in the myriad of beings that inhabit their territories' (Ruiz Serna, 2017). Indeed, Indigenous peoples are those most acutely victimised by environmental conflict (Wijdekop, 2019, p. 32), due to the complex links that exist between 'the loss of intimate relationships with non-human beings; violence toward Indigenous faiths; and loss of habitat, culture, and community' (Komatsubara, 2019, p. 68).

3.1   Transitional Justice's Neo-colonial Tendencies

Indigenous peoples often find themselves overlooked in transitional justice processes, which can fail to address their issues and grievances (Shakya, 2014; Kipuri, 2017) and leave their specific harms undocumented (Barume, 2017; Antkowiak, 2014). For example, Cambodia's official transitional justice mechanisms failed to account for the experiences of Cambodia's Indigenous minorities during the Khmer Rouge regime, precluding them from receiving reparations and acknowledgement of harm (Killean & Moffett, 2020). This overlooking can occur even when Indigenous peoples are some of the most significantly victimised by the conflict. For example, in the case of Guatemala, most transitional justice initiatives have taken place in the capital and they are shaped by 'Western views of human rights' (Izquierdo & Viaene, 2018). This overlooks the reality that 'the vast majority of victims are indigenous people living in rural areas' (Izquierdo & Viaene, 2018). In practice, the end of conflict can simply return Indigenous peoples to pre-existing positions of subjugation and discrimination (Tauli-Corpuz, 2017; Lawry-White, 2017), while transitions can signal the commencement of new practices of resource extraction and environmental damage to their territories (Izquierdo & Viaene, 2018; Tauli-Corpuz, 2017; Kipuri, 2017).

In light of these intersecting challenges, there is an emerging realisation within transitional justice scholarship of the need for transitional justice efforts to be 'informed by Indigenous worldviews' (Balint et al., 2014). An-Na'im has critiqued the tendency to grant preference to those standards of justice mandated by the international community, while deeming Indigenous or 'traditional' practices inconsistent with 'universal' human rights norms (An-Na'im, 2013). As a result of this 'neo-colonial tendency' (An-Na'im, 2013) transitional justice programmes implemented in the global south are often conceived in the global north (Madlingozi, 2010; Fletcher & Weinstein, 2018; Benyera, 2019). Reflecting the discussion above, this often leads to the prioritisation of legal (and colonial) forms of individual accountability and retribution (Okello, 2012; Garzón, 2019). One effect of this tendency is that 'even the possibility of an Indigenous alternative conception of justice is not taken seriously at a theoretical or empirical level' (An-Na'im, 2013, p.197), and as such, Indigenous perspectives are largely absent 'within the human rights and transitional justice fields' (Izquierdo & Viaene, 2018, p. 15; see also Viaene, 2020).

3.2   Indigenous Harm and Indigenous Justice

Environmental restorative justice presents one possible tool for challenging Indigenous marginalisation within transitional justice. While by no means synonymous, there are many similarities between restorative justice and Indigenous justice practices (Braithwaite, 1999; Lambourne, 2016; Wijdekop, 2019, p. 84), including their grounding in relational understandings of justice. Environmental restorative justice in particular has been framed as sprouting 'from soil tended and fertilised by generations of Indigenous communities' (Braithwaite et al., 2019, p. 11) and 'is culturally familiar to—and borrows from—Indigenous and aboriginal traditions of conflict resolution' (Wijdekop, 2019, pp. 79–80). Rather than being grounded in legalism and 'universal norms' as interpreted by the global north, environmental restorative justice practices can move 'beyond legal formats' (Varona, 2021, p. 53), enabling diverse 'cultural, emotional and spiritual values to be expressed' (Wijdekop & van Hoek, 2019, p. 22). This can be done in a range of diverse ways, including through performance, song, storytelling, or visits to places to invoke experiences (Hamilton, 2021, pp. 226–228; Forsyth et al., 2021, p. 32). This adaptability is more inclusive of diverse Indigenous worldviews and modes of expression, which may be grounded in oral rather than written forms of knowledge production and sharing (Matsunaga, 2016).

This inclusivity had been highlighted as a means of creating space for 'eco-centric' as well as Indigenous approaches to harm, participation, and restoration (Wijdekop & van Hoek, 2019, p. 22). For example, the exclusion of other-than-human victims from dominant transitional justice practices is reflective of particular understandings of agency; for some Indigenous peoples, the other-than-human has agency (Nash, 2005; Watts, 2013), and expresses themselves in diverse ways. As described by Izquierdo & Viaene (2018, p. 15) 'of course the territory speaks and expresses its feelings. A mountain gets angry, it gets sad, and it expresses this through signs in the dreams of the elders, fire ceremonies or because accidents occur with people.' Thus, by attending to 'difference', environmental restorative justice might invite not only diverse cultural frameworks, but also diverse ways of seeing the world itself (Ruiz Serna, 217, p. 101).

One further benefit of an Indigenous-informed environmental restorative approach could be its ability to draw out 'pre-existing social-political tensions' and facilitate an 'in-depth analysis of each conflict' (Randazzo, 2019, p. 33). As noted by 'structural-economic transitional justice scholars' (Huneeus & Rueda Sáiz, 2021), transitional justice can often fail to account for those structurally embedded injustices that restrict Indigenous peoples' ability to access, use and care for their territories. Restorative justice has been highlighted as a potential tool for pursuing environmental justice (Pali, 2019, p. 13; Minguet, 2021), due to its ability to highlight historic structural inequalities and environmental harm's disproportionate impacts on marginalised communities (Stark, 2016, pp. 455–456). As Forsyth et al. (2021, p. 33) note, a restorative justice approach which centralises 'acknowledgement of, and responsiveness to, previous injustice', is arguably crucial for achieving just futures on Indigenous lands.

3.3   'Decolonising' Transitional Justice?

It may be that the design of environmental restorative justice mechanisms could both contribute towards and be the result of increasing calls for a 'decolonisation' of transitional justice (Balint et al., 2014; Park, 2020; Van der Merwe & Brinton Lykes, 2021). As Huneeus & Rueda Sáiz (2021, p. 5) note, such a process involves not only incorporating Indigenous worldviews, but understanding them as legitimate and repairing 'broken relations not only between humans, but also between humans and non-humans.' It is telling that those involved in designing the Colombian Decree 4633 (which first recognised Indigenous territories as victims) described its formulation as requiring a 'dialogue of knowledge' inclusive of 'different variants in Indigenous thought' (Ruiz Serna, 2017,

p. 101). As this case illustrates, meaningful dialogue with Indigenous 'conceptions of the environment' and 'understandings of harm and healing' (McMillan & Rigney, 2018) can lead to the design of environmentally restorative transitional justice mechanisms. If transitional justice scholars and practitioners heed the call to direct their 'energies towards developing justice alternatives' (Matsunaga, 2016, p. 39) centring 'Indigenous visions, aspirations and aims' (Smith, 2007, p. 75), it may be that more environmentally restorative practices might be adopted in the aftermath of conflict.

In exploring these possibilities, it would be crucial to avoid the imposition of 'orientalist' or 'neo-folklore' approaches which seek to 'pacify' Indigenous peoples and consume 'other' cultures (Blagg, 1997; Cumes, 2019). Izquierdo & Viaene (2018, p. 16) caution against 'ancestral practices and norms' simply becoming 'another tool of the transitional justice toolbox' which promote simplistic, romantic and disconnected notions of Indigenous practices, while denying reparation or reconciliation with other-than-human beings. While environmental restorative justice offers greater cultural inclusivity, this should not lead to the trivialisation of 'Indigenous culture and law in the name of universalising claims about restorative justice' (Cunneen, 2002). As Tauri (2018) has observed, settler colonial states have been guilty of marketising restorative justice practices in ways that seek to regulate and impose upon Indigenous peoples rather than addressing their justice needs. The incorporation of Indigenous worldviews must be understood as 'a ground up process rather than dictated from above' (Blagg, 1997). Attention would need to be paid to historical contexts and the power dynamics which might exist between communities and transitional justice 'elites' (Doak, 2016, pp. 82–83) as well as within the communities themselves. Steps to mitigate against the risk of 'top-down' tokenism could be informed by the substantial literature exploring transitional justice 'from below' which goes beyond merely listening to diverse perspectives and requires the 'hard work' of building connections between transitional justice practitioners, Indigenous communities, and other marginalised victim groups (McEvoy & McGregor, 2008; Sium, Desai & Ritskes, 2012).

4   Concluding Thoughts

Proponents of environmental restorative justice have expressed a belief that it is capable of 'being employed for all environmental crimes that cause harm' (Preston, 2011, p. 25). This chapter has explored some potential benefits of environmental restorative justice as a response to conflict-related environmental harm. It has argued that such an approach might challenge transitional justice's anthropocentrism by recognising other-than-human victims of periods of conflict. It has further argued that it might provide one tool for challenging transitional justice's neo-colonial tendencies, by better reflecting Indigenous peoples' experiences of conflict, acknowledging the links between environmental harms and the harms experienced by Indigenous communities, and creating space for Indigenous justice.

Undoubtedly, a range of practical challenges would arise in designing and implementing environmental restorative responses to conflict. For example, as Wijdekop (2019, p. 32) notes, environmental harm can be 'victim-full', a reality exacerbated by the widespread and diverse harms of armed violence. Questions would arise as to how to enable participation by large groups of diverse victim groups (Forsyth et al., 2021, p. 34). Social and ecological contexts and power dynamics might cause conflict between victim groups, and there may be differing views as to how victims are represented (Hamilton, 2021, p. 218). Challenges of victim hierarchy, conflict, and voice are not new to transitional justice (e.g., Madlingozi, 2010; McEvoy & McConacchie, 2013; Killean & Moffett, 2017). However, additional hierarchies could result from the inclusion of 'new' categories of victim such as other-than-human beings, while engagement with the historic injustices experienced by Indigenous peoples might create additional tensions when determining appropriate restoration. While I have drawn out the possible complementarities of incorporating other-than-human victims' perspectives and Indigenous worldviews, it would be a mistake to over-simplify these possibilities or the relationships Indigenous peoples have with their territories. Indeed, there is scope for significant tensions between such perspectives. For example, the restoration of access to land for traditional hunting practices may be culturally significant for an Indigenous population, but clash with conservationist perspectives (Hamilton, 2021, p. 217; White, 2018).

Yet, restorative justice principles suggest that conflicts are not insurmountable, or even unwelcome. In keeping with Christie's (1977) notions of 'conflicts as property', proponents of restorative justice suggest that navigating disagreements can present opportunities for participatory decision making and the development of deeper understandings of justice (Clamp, 2016, p. 3). This may be particularly beneficial in transitional contexts, where 'creating safe spaces for... difficult conversations around what has happened and what must happen in the future' (Pali & Aersten, 2021, p. 5) may reduce feelings of alienation and social distance, combat dominance and increase community cohesion. Environmental restorative justice may even present opportunities to engage with postconflict tensions between pursuing environmental exploitation for economic recovery, respecting individual and collective human rights, and protecting nature (Varona, 2020, p. 668). This is particularly important in conflicts involving Indigenous communities, given the 'deep and fundamental disconnects' that can arise between Indigenous and non-Indigenous worldviews (Pali & Aersten, 2021, p. 5). It may be that when Indigenous communities represent their own harm in appropriately designed processes, 'the social fabric is enriched by providing opportunities to share, understand and celebrate Indigenous values... as the knowledge holders and keepers of natural and cultural heritage' (Wijdekop, 2019).

Transitions often accompany or follow moments of rupture, creating space for rethinking previously entrenched societal structures (Bell & O'Rourke, 2007; Gheciu & Welsh, 2009). It is worth considering whether these moments of rupture may present a nexus of opportunity for rethinking relationships between Indigenous and non-Indigenous peoples, and between humans and the natural world. As Komatsubara (2021, p. 138) concludes, 'when the concept of community is conceived on the basis of victims' voices, a new form of community may emerge. By entrenching an environmentally restorative ethos, transitional justice may present one effective vehicle for recognising the intrinsic value of the environment, contributing to 'healing earth systems, healing relationships of humans with nature, humbling humans' domination of nature, and in doing so healing ourselves as humans' (Braithwaite, Forsyth & Cleland, 2019, pp. 8–9).

References

An-Na'im, A.A. (2013). Editorial note: From the neocolonial 'transitional' to Indigenous formations of justice. *International Journal of Transitional Justice,* 7(2), 197–204.

Antkowiak, T. M. (2014). A dark side of virtue: The Inter-American Court and reparations for Indigenous peoples. *Duke Journal of Comparative and International Law,* 25(1), 1–80.

Balint, J., Evans, J., & McMillan, N. (2014). Rethinking transitional justice, redressing Indigenous harm: A new conceptual approach. *International Journal of Transitional Justice,* 8(2), 194–216.

Barume, A. (2017). Unaccounted for: Indigenous peoples as victims of conflict in Africa. In E. Stamatopoulou (Ed.), *Indigenous peoples' rights and unreported struggles* (pp. 55–67). New York: Colombia University.

Bell, C. & O'Rourke, C. (2007). The people's peace? Peace agreements, civil society and participatory democracy. *International Political Science Review,* 28(3), 293–324.

Benyera, E. (2019). *Indigenous, traditional, and non-state transitional justice in Southern Africa.* Washington D.C.: Lexington Books.

Blagg, H. (1997). A just measure of shame? Aboriginal youth and conferencing in Australia. *British Journal of Criminology,* 37 (4), 481–501.

Borda, A.Z. (2020). History in international criminal trials: The 'crime-driven lens' and its blind spots. *Journal of International Criminal Justice,* 18(3), 533–546.

Bothe, M. (2017). The ILC's Special Rapporteur's preliminary report on the protection of the environment in relation to armed conflict: An important step in the right direction. In P. Acconci et al. (Eds.), *International law and the protection of humanity. Essays in honor of Flavia Latanzi* (pp. 211–224). Leiden: Brill.

Braithwaite, J. (1999). Restorative justice: Assessing optimistic and pessimistic accounts. *Crime and Justice,* 25, 1–127.

Braithwaite, J., Forsyth, M., & Cleland, D. (2019). Restorative environmental justice: An introduction. In E. Biffi & B. Pali (Eds.), *Environmental justice restoring the future: Towards a restorative environmental justice praxis* (pp. 8–12). Leuven: European Forum for Restorative Justice.

Brinton Lykes, M. & Van Der Merwe, H. (2017). Exploring/expanding the reach of transitional justice. *International Journal of Transitional Justice,* 11(3), 371–377.

Cabrera, A. (2019). En ritual de armonización la JEP recibe acreditación de las víctimas de los pueblos indígenas. Radio Santa Fe. Retrieved from https://www.radiosantafe.com/2019/12/07/en-ritual-de-armonizacion-la-jep-recibe-acreditacion-de-las-victimas-de-los-pueblos-indigenas/ (last accessed 25 January 2022).

Christie, N. (1977). Conflicts as property. *British Journal of Criminology,* 17(1), 1–15.

Clamp, K. (Ed). (2016). *Restorative justice in transitional settings.* London: Routledge.

Clamp, K. (2014) *Restorative justice in transition.* London: Routledge.

Clark, P. (2010). *The Gacaca courts, post-genocide justice and reconciliation in Rwanda: Justice without lawyers.* New York, NY: Cambridge University Press.

Colombian Legislative Decree No. 4633 of 3 December 2011.

Crosby, A. & Brinton Lykes, M. (2011). Mayan women survivors speak: The gendered relations of truth telling in postwar Guatemala. *International Journal of Transitional Justice,* 5(3), 456–476.

Cumes, A. (2019). "Lo indígena" como circo en el regreso de la interculturalidad. Retrieved from: https://tujaal.org/lo-indigena-como-circo-en-el-regreso-de-la-interculturalidad/ (last accessed 16 August 2021).

Cunneen, C. (2002). Restorative justice and the politics of decolonisation. In E. G. M. Weitekamp & H. Kerner (Eds.), *Restorative justice: Theoretical foundations* (pp. 32–49). Devon: Willan Publishing.

Cusato, E. (2017). Back to the future? Confronting the role(s) of natural resources in armed conflict through the lenses of Truth and Reconciliation Commissions. *International Community Law Review,* 19(4–5), 373–400.

Cusato, E. (2020). International law, the paradox of plenty and the making of resource-driven conflict. *Leiden Journal of International Law,* 33(3), 649–666.

Doak, J. (2016). Stalking the state: The state as a stakeholder in post-conflict restorative justice. In K. Clamp (Ed.), *Restorative justice in transitional settings* (pp. 74–94). London: Routledge.

Drumbl, M. (1998). Waging war against the world: The need to move from war crimes to environmental crimes. *Fordham International Law Journal,* 22(1), 122–153.

Eichler, L.J. (2020). Ecocide is genocide: Decolonising the definition of genocide. *Genocide Studies and Prevention: An International Journal,* 14(2), 104–121.

Falcón, S. (2018). Intersectionality and the arts: Counterpublic memory-making in post-conflict Peru. *International Journal of Transitional Justice,* 12(1), 26–44.

Farrar, S. (2020). Islamic ethics and Truth Commissions in the Muslim world: Towards a just and ecologically sustainable peace? In J. Camilleri & D. Guess (Eds.), *Towards a justice and ecologically sustainable peace* (pp.135–165). Singapore: Palgrave Macmillan.

Fletcher, L. & Weinstein, H. (2018). How power dynamics influence the "North-South" gap in transitional justice. *Berkeley Journal of International Law,* 36 (2), 190–217.

Forsyth, M., Cleland, D., Tepper, F., Hollingworth, D., Soares, M., Nairn, A., & Wilkinson, C. (2021). A future agenda for environmental restorative justice? *The International Journal of Restorative Justice,* 4(1), 17–40.

Freeland, S. (2005). Human rights, the environment and conflict: Addressing crimes against the environment. *International Journal on Human Rights,* 2(2), 113–141.

García-Godos, J. & Lid, K. (2010). Transitional justice and victims' rights before the end of a conflict: The unusual case of Colombia. *Journal of Latin American Studies,* 42(3), 487–516.

Garzón, P. (2019). Legal pluralism, Indigenous law and legal coloniality: Rethinking law from the coloniality of power. *Ius Inkarri. Journal of the Faculty of Law and Political Science,* 8, 215–226.

Gheciu, A., & Welsh, J. (2009). The imperative to rebuild: Assessing the normative case for post-conflict reconstruction. *Ethics and International Affairs,* 23, 121–146.

Gillett, M. (2017). Eco-struggles. In C. Stahn, J. Iverson & J. Easterday (Eds.), *Environmental protection and transitions from conflict to peace* (pp. 220–253). Oxford: Oxford University Press.

Guess, D. (2020). Introduction. In J. Camilleri, & D. Guess (Eds.), *Towards a justice and ecologically sustainable peace* (pp.1–16). Singapore: Palgrave Macmillan.

Hamilton, M. (2021). *Environmental crime and restorative justice.* Palgrave Studies in Green Criminology, Basingstoke: Palgrave Macmillan.

Harwell, E. (2016). Building momentum and constituencies for peace: The role of natural resources in transitional justice and peacebuilding. In C. Bruch, C. Muffett, & S. Nichols (Eds.), *Governance, natural resources and post-conflict peacebuilding* (pp.633–664). London: Routledge.

Harwell, E. & Le Billon, P. (2009). Natural connections: Linking transitional justice and development through a focus on natural resources. In P. De Grieff & R. Duthie (Eds.), *Transitional justice and development: Making connections* (pp. 283–330). New York: Social Science Research Council.

Higgins, P. (2010). *Eradicating ecocide*. London: Shepheard-Walwyn.

Higgins, P., Short, D., & South, N. (2013). Protecting the planet: A proposal for a law of ecocide. *Crime, Law and Social Change*, 59(3), 251–266.

Hulme, K. (2017). Using a framework of human rights and transitional justice for post-conflict environmental protection and remediation. In C. Stahn, J. Iverson & J. Easterday (Eds.), *Environmental protection and transitions from conflict to peace* (pp.119–142). Oxford: Oxford University Press.

Huneeus, A. (2020). Territory as a victim of Colombia's war. *EJIL: Talk!* 7 May 2020. Retrieved from: www.ejiltalk.org/territory-as-a-victim-of-colombias-war/#:~:text=In%20two%20recent%20resolutions%2C%20Colombia's, Colombia's%2050%2Dyear%20civil%20war. (last accessed 16 August 2021).

Huneeus, A., & Rueda Sáiz, P. (2021). Territory as a victim of armed conflict. *International Journal of Transitional Justice*. https://doi.org/10.1093/ ijtj/ijab002

ICC (2016). Office of the Prosecutor Policy Paper on Case Selection and Prioritisation, 15 September 2016. Retrieved from: www.icc-cpi.int/items-Documents/20160915_OTP-Policy_Case-Selection_Eng.pdf (last accessed 16 August 2021).

International Committee of the Red Cross. (1994). Guidelines for Military Manuals and Instructions on the Protection of the Environment in Times of Armed Conflict, endorsed by UN GA Res. 49/50 (Geneva: ICRC, 1994, updated 2021).

Ide, T., Bruch, C., Carius, A., Conca, K., Dabelko, G. D., Matthew, R., & Weinthal, E. (2021). The past and future(s) of environmental peacebuilding. *International Affairs*, 97(1), 1–16.

ICRC (2020). *Guidelines on the Protection of the Natural Environment in Armed Conflict: Rules and Recommendations Relating to the Protection of the Natural Environment under International Humanitarian Law, with Commentary*. Geneva: ICRC. Retrieved from: https://reliefweb.int/report/world/ guidelines-protection-natural-environment-armed-conflict-rules-and-recommendations (last accessed 25 January 2022).

Izquierdo, B. & Viaene, L. (2018). Decolonising transitional justice from Indigenous territories. *Peace in Progress*, 34, 11–19.

Jankowitz, S. (2018). The 'hierarchy of victims' in Northern Ireland: A framework for critical analysis. *International Journal of Transitional Justice,* 12(2), 216–236.

Kershen, L. (2019) Implementing restorative justice to environmental harm. In E. Biffi & B. Pali (Eds.), *Environmental justice restoring the future* (pp. 40–53). Leuven: European Forum for Restorative Justice.

Killean, R., & Dempster, L. (2022, forthcoming). Greening transitional justice? In M. Evans (Ed.), *Beyond transitional justice? Transformative justice and the state of the field (or non-field).* London: Routledge. Accepted version available at: https://papers.ssrn.com/sol3/papers.cfm?abstract_id=3769180.

Killean, R. (2021). From ecocide to eco-sensitivity: 'Greening' reparations at the International Criminal Court. *International Journal of Human Rights,* 25(2), 323–347.

Killean, R., & Moffett, L. (2017). Victim legal representation before the ICC and the ECCC. *Journal of International Criminal Justice,* 15(4), 713–740.

Killean, R., & Moffett, L. (2020). "What's in a Name? Reparations at the Extraordinary Chambers in the Courts of Cambodia". *Melbourne Journal of International Law,* 21(1), 115.

Kipuri, N. (2017). Indigenous peoples' rights, conflict and peace building: Experiences from East Africa. In E. Stamatopoulou (Ed.), *Indigenous peoples' rights and unreported struggles* (pp. 68–79). New York: Colombia University.

Komatsubara, O. (2021). Imagining a community that includes non-human beings: The 1990s Moyainaoshi Movement in Minamata, Japan. *The International Journal of Restorative Justice,* 4(1), 123–140.

Komatsubara, O. (2019). Restorative justice and environmental crime: The case of Minamata disease in Japan in 1970. In E. Biffi and B. Pali (Eds.), *Environmental Justice Restoring the Future* (pp. 67–71). Leuven: European Forum for Restorative Justice.

Lambourne, W. (2009). Transitional justice and peacebuilding after mass violence. *International Journal of Transitional Justice,* 3(1), 28–48.

Lambourne, W. (2010). Transitional justice after mass violence: Reconciling retributive and restorative justice. In H Irving, J Mowbray & K Walton (Eds.), *Julius Stone: A study in influence* (214–237). Sydney: Generation Press.

Lambourne, W. (2016). Restorative justice and reconciliation: The missing link in transitional justice. In K. Clamp (Ed.), *Restorative justice in transitional settings* (pp. 56–73). London: Routledge.

Lawrence, J. C., & Heller, K. J. (2007). The first ecocentric environmental war crime: The limits of Article 8(2)(b)(iv) of the Rome Statute. *Georgetown International Environmental Law Review, 20*(1), 61–95.

Lawry-White, M. (2017). Victims of environmental harm during conflict: The potential for justice. In C. Stahn, J. Iverson & J.S. Easterday (Eds.), *Environmental protection and transitions from conflict to peace* (pp. 367–395). Oxford: Oxford University Press.

Madlingozi, T. (2010). On transitional justice entrepreneurs and the production of victims. *Journal of Human Rights Practice, 2*(2), 208–228.

Matsunaga, J. (2016). Two faces of transitional justice: Theorising the incommensurability of transitional justice and decolonisation in Canada. *Decolonisation: Indigeneity, Education and Society, 5*(1), 24–44.

McEvoy, K. (2007). Beyond legalism: Towards a thicker understanding of transitional justice. *Journal of Law and Society, 34*(4), 411–440.

McEvoy K. & McConacchie, K. (2013). Victims and transitional justice: Voice, agency and blame. *Social and Legal Studies, 22*(4), 489–513.

McEvoy, K. & Mika, H. (2001). Punishment, policing and praxis: Restorative justice and non-violence alternatives to paramilitary punishments in Northern Ireland. *Policing and Society, 11*(3–4), 359–382.

McEvoy, K. & McGregor, L. (2008). Transitional justice from below. In K. McEvoy & L. McGregor (Eds.), *Transitional justice from below: Grassroots activism and the struggle for change* (pp. 1–14). Oxford: Hart Publishing.

McEvoy, K. & Newburn, T., (2003). *Criminology, conflict resolution and restorative justice*. Basingstoke: Palgrave Macmillan.

McMillan, M. & Rigney, S. (2018). Race, reconciliation and justice in Australia. *Ethnic and Racial Studies, 41*(4), 759–777.

Milburn, R. (2014). The roots to peace in the DRC: Conservation as a platform for green development. *International Affairs, 90*, 871–887.

Minguet, A. (2021). Environmental justice movements and restorative justice. *The International Journal of Restorative Justice, 4*(1), 60–80.

Mitchell, A. (2014). Only human? A worldly approach to security. *Security Dialogue, 45*(1), 5–21.

Motupalli, C. (2018). Intergenerational justice, environmental law, and restorative justice. *Washington Journal of Environmental Law and Politics, 8*(2), 333–361.

Murdock, E. (2018). Storied with land: 'transitional justice' on Indigenous lands. *Journal of Global Ethics, 14*(2), 232–239.

Nagy, R. (2008). Transitional justice as global project: Critical reflections. *Third World Quarterly,* 29(2), 275–289.

Nash, L. (2005). The agency of nature or the nature of agency? *Environmental History,* 10(1), 67–69.

Natalya Clark, K. (2008). The three Rs: Retributive justice, restorative justice and reconciliation. *Contemporary Justice Review,* 11(4), 331–150.

Natalya Clark, J. (2016). Are there "greener" ways of doing transitional justice? *International Journal of Transitional Justice,* 20(8), 1199–1218.

Natalya Clark, J. (2020). Re-thinking memory and transitional justice. *Memory Studies,* 1–18.

Nichols, S. (2014). Reimagining transitional justice for an enduring peace. Accounting for natural resources in conflict. In D. Sharp (Ed.), *Justice and Economic Violence in Transition* (pp. 203–231). New York: Springer.

Okello, M.C. (2012). Where law meets reality: Forging African transitional justice. Nairobi: Pambazuka Press.

Ong, D. M. (2017). Prospects for transitional environmental justice in the socio-economic reconstruction of Kosovo. *Tulane Environmental Law Journal,* 30(2), 217–272.

Pali, B. (2019). Restorative responses to environmental harm? Yes, we must! In E. Biffi & B. Pali (Eds.), *Environmental Justice Restoring the Future* (pp. 13–17). Leuven: European Forum for Restorative Justice.

Pali, B. & Aersten, I. (2021). Inhabiting a vulnerable and wounded earth: Restoring response-ability. *The International Journal of Restorative Justice,* 4(1), 3–16.

Park, A.S.J. (2020). Settler colonialism, decolonisation and radicalising transitional justice. *International Journal of Transitional Justice,* 14(2), 260–279.

Pavlich, G. (2005). *Governing paradoxes of restorative justice.* London: Routledge.

Pelizzon, A. (2015). Transitional justice and ecological jurisprudence in the midst of an ever-changing climate. In N. Szablewska & S.D Bachmann (Eds.), *Current issues in transitional justice* (pp. 317–338). Springer Series in Transitional Justice. New York: Springer.

Preston, B. J., (2011). The use of restorative justice for environmental crime. *Criminal Law Journal,* 35, 136–161.

Randazzo, S. (2019). Interview to Angèle Minguet. In E. Biffi & B. Pali (Eds.), *Environmental Justice Restoring the Future* (pp. 29–34). Leuven: European Forum for Restorative Justice.

Roht-Arriaza, N. (2006). The new landscape of transitional justice. In N. Roht-Arriaza & J. Mariecurrena (Eds.), *Transitional justice in the twenty-first century* (pp.1–16). New York: Cambridge University Press.

Rooney, E., & Ní Aoláin. F. (2018). Transitional justice from the margins: Intersections of identities, power and human rights. *International Journal of Transitional Justice, 12*(1), 1–8.

Ruiz Serna, D. (2017). Territory as victim. Political ontology and the laws for victims of Indigenous and Afrodescendant communities in Colombia. *Colombian Journal of Anthropology, 53*(2), 85–113.

Shakya, P. (2014). Indigenous peoples: Expanding the realm of justice. Peace Insight, 26 March 2014, https://www.peaceinsight.org/en/articles/indigenous-peoples-expanding-realm-justice/?location=&theme=human-rights. Accessed 16 August 2021.

Sharp, D. N. (2019). What would satisfy us? Taking stock of critical approaches to transitional justice. *International Journal of Transitional Justice, 13*, 570–589.

Sium, A., Desai, C., & Ritskes, E. (2012). Towards the 'tangible unknown': Decolonisation and the Indigenous future. *Decolonization: Indigeneity, Education & Society, 1*(1), I-XIII.

Smith, L.T. (2007). Getting the story right—Telling the story well, Indigenous activism—Indigenous research. In A. Te Pareake Mead & S. Ratuva (Eds.), *Pacific genes and life patents: Pacific Indigenous experiences and analysis of the commodification and ownership of life* (pp.74–81). Call of the Earth LLamado de la Tierra and The United Nations University Institute of Advanced Studies.

Stahn, C., Iverson, J., & Easterday, J. (2017). *Environmental protection and transitions from conflict to peace.* Oxford: Oxford University Press.

Stark, A., (2016). Environmental restorative justice. *Pepperdine Dispute Resolution Law Journal, 16*(3), 435–462.

Stone, C. D. (1972). *Should trees have standing?* Oxford: Oxford University Press.

Stop Ecocide Foundation. (2021). Independent Expert Panel for the Legal Definition of Ecocide: Commentary and Core Text, June 2021. Retrieved from: https://www.stopecocide.earth/legal-definition (last accessed 25 January 2022).

Suarez, A., Arias-Arevalo, P., & Martinez-Mera, E. (2018). Environmental sustainability in post-conflict countries: Insights for rural Colombia. *Environmental Development Sustainability, 20*, 997–1015.

Tauli-Corpuz, V. (2017). Conflict, peace and the human rights of Indigenous peoples. In E. Stamatopoulou (Ed.), *Indigenous peoples' rights and unreported struggles* (pp.1–19). New York: Colombia University.

Tauri, J. M. (2018). Restorative justice as a colonial project in the disempowerment of Indigenous peoples. In T. Gavrielides (Ed.) *Routledge International Handbook of Restorative Justice* (pp. 342–358). London: Routledge.

Teitel, R. G., (2003). Transitional justice genealogy. *Harvard Human Rights Journal,* 69–94.

Tutu, D. (1999). *No future without forgiveness.* New York: Doubleday.

UNSG (2004). *Report on the rule of law and transitional justice in conflict and post-conflict societies.* Report of the United Nations Secretary General, 23 August 2004, Un Doc. S/2004/616.

UNEP (2003). Afghanistan: Post-conflict environmental assessment. Nairobi: United Nations Environment Programme.

UNEP (2007). Sudan: Post-conflict environmental assessment. Nairobi: United Nations Environment Programme.

UNEP (2009a). Environmental assessment of the Gaza Strip following the escalation of hostilities in December 2008–January 2009. Nairobi: United Nations Environment Programme.

UNEP (2009b). From conflict to peacebuilding: The role of natural resources and the environment. Nairobi: United Nations Environment Programme.

UNEP (2015). Côte d'Ivoire: Post-conflict environmental assessment. Nairobi: United Nations Environment Programme.

Van der Merwe, H., & Brinton Lykes, M. (2021). Racism and transitional justice. *International Journal of Transitional Justice,* 14(3), 415–422.

Varona, G. (2021). Why an atmosphere of transhumanism undermines green restorative justice concepts and tenets. *The International Journal of Restorative Justice,* 4(1), 41–59.

Varona, G. (2020). Restorative pathways after mass environmental victimisation: Walking in the landscapes of past ecocides. Oñati Socio-Legal Series, 10(3), 664–685.

Viaene, L. (2020). Transitional justice, dialogue and critical reflexivity: weaving thoughts. *Eunomics. Journal on Culture of Legality,* 18, 416–423.

Walters, R., Westerhuis, D.S., & Wyatt, T. (Eds.) (2013). *Emerging issues in green criminology.* Basingstoke: Palgrave Macmillan.

Watts, V. (2013). Indigenous place-thought and agency amongst humans and non-humans (first woman and sky woman go on a European world tour!). *Decolonisation: Indigeneity, Education and Society,* 2(1), 20–34.

Weinstein, T. (2005). Prosecuting attacks that destroy the environment. *Georgetown International Environmental Law Review,* 17(4), 697–722.

White, R. (2018). Green victimology and non-human victims. *International Review of Victimology,* 24 (2), 239–255.

White, R. (2017). The four ways of eco-global criminology. *International Journal for Crime, Justice and Social Democracy,* 6 (1), 8–22.

White, R. (2014a/2015). Indigenous communities, environmental protection and restorative justice. *Australian Indigenous Law Review*, 18(2), 43–54.

White, R. (2014b). *Environmental harm: An eco-justice perspective*. Bristol: Policy Press.

Wijdekop, F. (2019). Restorative justice responses to environmental harm. Amsterdam: IUCN.

Wijdekop, F., & van Hoek, A. (2019). Green criminology and restorative justice: Natural allies? In E. Biffi & B. Pali (Eds.), *Environmental Justice Restoring the Future* (pp.18–25). Leuven: European Forum for Restorative Justice.

Zernova, M. (2017). Restorative justice in the Basque peace process: Some experiments and their lessons. *Contemporary Justice Review*, 20(3), 363–391.

Zunino, M. (2019). *Justice framed: A genealogy of transitional justice*. New York: Cambridge University Press.

12

The Importance of Environmental Restorative Justice for The United Nations Decade on Ecosystem Restoration (2021–2030)

Felicity Tepper

1   Introduction

The United Nations Decade on Ecosystem Restoration 2021–2030 (the UN Decade) is a global initiative aimed at placing ecosystem restoration firmly at the heart of environmental protection and care activities worldwide. Its overarching goal is to 'prevent, halt and reverse the degradation of ecosystems' (UNEP/FAO, 2022; UNEP, 2021) and its vision is to 'restor[e] the relationship between humans and nature' (United Nations, 2020, p. 6). This initiative coincides with many scientists, policymakers, and world leaders believing that the coming decade is crucial for restoring 'the route back to stability and balance' (Buttfield & Hughes, 2021,

F. Tepper (✉)
RegNet School of Regulation and Global Governance,
Australian National University, Canberra, ACT, Australia
e-mail: felicity.tepper@anu.edu.au

p. 12), to ensure we stay within Earth's planetary boundaries.[1] Concerns about our disconnectedness from nature (Attenborough, 2020; Fischer et al., 2021) have brought about a cascade of demands for humans to consciously resume a relationship with nature ('living with nature') and cease destroying it (Attenborough, 2020). Reconnection to our environment through ecosystem restoration is one important means for us to see and stop environmental degradation.

We *all* have a role to play in ecosystem restoration and in restoring our relationship with nature. Ecosystem restoration is an inherently inter- and transdisciplinary pursuit (Fischer et al., 2021) and requires every citizen, practitioner, academic and workplace to be doing their part, whether as citizen scientists, civic ecologists (Büscher & Fletcher, 2020; Krasny & Tidball, 2015), or as bureaucrats-by-day/ecologically inclined activists by night. In this chapter, I suggest that environmental restorative justice (ERJ) will help both to promote this breadth of involvement and connect the global initiative to local action and knowledge through giving us practices and values acutely suited to the care-work of restoration and reconnection. Specifically, restorative justice's focus to 'heal and put things as right as possible' (Zehr, 2002, p. 37) is reflective of the purpose of ecosystem restoration. More broadly, in seeking to re-engage with nature through restoration work, ERJ can help us co-create an energised 'restorative culture' (Blignaut & Aronson, 2020; Cross et al., 2019), based upon restorative justice's principles, values and philosophy, and practices. A restorative culture embodies values that support healthy, multidimensional relationships with each other and nature. It 'revitalises the nature and culture union that is then embedded into the social, political and educational fabric' (Cross et al., 2019, p. 927), helping turn ecosystem restoration into an intentional, conscious, and everyday practice inclusive of everyone and all ecosystems.

I propose three principal ways in which ERJ can contribute to strengthening implementation of the global initiative and foreground local knowledge and restorative action. I see these three contributions as: First,

[1] The 'planetary boundaries' framework developed by Johan Rockström, Will Steffen and others (Rockström et al., 2009) describes the nine processes regulating the Earth system, keeping it stable and resilient. Within these boundaries, humans have a 'safe operating space' but pushing past them would destabilise Earth's system into effects beyond human capabilities to manage.

by helping to sustain a social-ecological, relational ethos that can infuse and guide ecosystem restoration activities across society (General Assembly Resolution 73/284, clauses 3 (a), (b) & (c)) and foster 'a holistic view' that encourages a restorative culture (General Assembly Resolution 73/284, clause 3(e)). By approaching ecosystem restoration from an ethos of care and accountability towards each other *and* our environment, this can help us to better understand how to live *with* nature (as opposed to exploiting, abandoning, or ignoring it). Framing ecosystem restoration as a social-ecological practice draws on relationality and the interwoven connectedness between humans, more-than-humans and nature,[2] to achieve both social *and* environmental benefits, human *and* environmental healing/health. Second, through education and dialogue to 'raise awareness of the importance of successful ecosystem restoration' (General Assembly Resolution 73/284, clause 1); this includes facilitating local participants to share their experiences to 'scale up good practices' (General Assembly Resolution 73/284, clause 3(d)) and 'promote sharing of experiences and good practices' (General Assembly Resolution 73/284, clause 3(f)). In this way, the local informs the global. Third, by helping to support the 'full involvement of all' … 'to actively support the implementation of the Decade' (General Assembly Resolution 73/284, clauses 5, 6), through providing practices to collaboratively work through misunderstandings or conflict that may arise through the course of restoration activities. With ERJ, this is not just a one-off, but is an ongoing, relational practice. In these three ways, ERJ can contribute towards creating and strengthening relational, knowledge-sharing, and collaborative approaches to repair and care for the ecosystems upon which we all depend.

---

[2] Living beings (animals, plants) and geological, water, atmospheric and chemical processes all form a part of 'nature', following Lynn Margulis' and James Lovelock's concept of 'Gaia' (Grinspoon, 2016). As authors Neale and Kelly (2020, p. 46) explain in relation to Australian Aboriginal beliefs, 'humans are equal with all things animate and inanimate', an understanding held by many Indigenous peoples that we must not lose sight of.

2   Orienting Environmental Restorative Justice in the UN Decade

Prior to proclaiming the UN Decade on 1 March 2019 at the United Nations General Assembly, several global ecological assessments had revealed a pressing urgency for focusing on ecosystem restoration as key to repairing environmental harm (Aronson et al., 2020). This prompted an initial proposal from El Salvador at a 2018 high-level biodiversity meeting in Brazil, after which General Assembly Resolution 73/284 (the Resolution) was adopted, with the UN Decade launched in 2021. The principal goal to 'prevent, halt and reverse the degradation of ecosystems on every continent and in every ocean' (UNEP/FAO, 2022) in the coming decade is further linked with relevant Sustainable Development Goals (General Assembly Resolution 73/284, clause 2).[3] Additionally, the UN Decade aims to stop biodiversity loss, mitigate climate change, and improve food security (UNEP, 2021), and serves as a 'public health intervention' (Reid, 2020), since human health is intimately entwined with ecosystem health (FAO et al., 2021; Robinson, 2020; UNEP, 2021). Many see the UN Decade as a source of hope in transforming our relationship with ecosystems, moving 'ecosystem restoration from a fringe movement to an essential response' (WSP USA, 2020) and offering 'an unprecedented opportunity—perhaps the last—for humanity to address multiple environmental problems at once' (Watson et al., 2020).

A draft strategy to support the Resolution and the UN Decade with goals, pathways, and guidance was launched in October 2019, seeking comments. A colleague[4] and I, noticing the absence of restorative justice in this draft, wrote a submission requesting its inclusion be considered, submitting that 'a commitment to restorative values—such as dialogue, participation and accountability—can form the platform for building and strengthening the relationships between stakeholders that will be necessary to make the UN Decade a success'.[5] We were not alone in this

---

[3] Cristina Oliveira, a member of the European Forum for Restorative Justice ERJ Working Group, is currently working on entry points for restorative justice in the Sustainable Development Goals.

[4] Deborah Cleland, co-author of Chap. 18 in this handbook.

[5] Cited from the Australian National University's submission.

request, with the European Forum for Restorative Justice also making a submission, amongst others, requesting inclusion of restorative justice (United Nations, 2020, Annex 3).

When the final version of *The United Nations Decade on Ecosystem Restoration Strategy* (the Strategy) was released in August 2020, we were pleased to see inclusion of restorative justice as a principle (United Nations, 2020, p. 1), alongside acknowledgement of its collaborative potential for development of context-specific restoration plans for landscapes 'achieved through dialogue, participation and accountability' (United Nations, 2020, p. 10) and importantly for local participation, 'applying restorative justice approaches to engage civil society actors, especially youth organisations, women and Indigenous peoples, in national policy planning and implementation of ecosystem restoration' (United Nations, 2020, p. 35).

This formal acknowledgement of the importance of restorative justice alongside other ecosystem restoration principles and practices is a significant and desirous inclusion that opens up discussion about what ERJ means for those participating in, facilitating, and running activities related to the UN Decade. This chapter is an initial attempt to think through some of the implications, options, and potential significance of ERJ's inclusion. The restorative values, principles, and practices I have selected for discussing are neither exhaustive nor hierarchical. Those chosen are presented as robust examples of what I consider will support inclusive, care-oriented, and co-designed approaches to ecosystem restoration projects that actively involve citizens and community, serving as an entryway for future discussion about their applicability.

To orient terms such as 'activities' or 'work' relating to ecosystem restoration, I envisage these as projects of environmental repair and care directly involving community and local stakeholders, covering a 'continuum of restorative activities' (FAO et al., 2021) that are locally contextualised for healing degraded ecosystems (UNEP, 2021), and that make relational values a primary consideration.[6] This interpretation fits with

---

[6] It is acknowledged that privately run and government-run ecosystem restoration activities must also pursue restorative values. This chapter is concerned mostly to highlight *wide* involvement in the UN Decade.

the Strategy's broad envisioning of participation, and more specifically to 'applying restorative justice approaches to engage civil society actors, especially youth organisations, women and Indigenous peoples, in national policy planning and implementation of ecosystem restoration' (United Nations, 2020, pp. 10, 35; also see the Resolution, clauses 5 & 6).

3   Ecosystem Restoration as a Social-ecological, Relational Ethos

Whilst various scientific definitions of ecosystem restoration are properly about scientific techniques and measures for doing and assessing restoration, for the purposes of this chapter, ecosystem restoration is viewed as a social-ecological practice. A social-ecological approach recognises the nexus connecting the social and the environmental, and 'makes clear that although the biologic may set the basis for the existence of humans and hence our social life, it is this social life that sets the path along which the biologic may flourish—or wilt' (Krieger, 1994, cited in Friel, 2019, p. xvi). A social-ecological restoration approach appreciates the interdependence and interactivity of social and ecological processes (Fernández-Manjarrés et al., 2018); acknowledges the complex adaptive systems nature of our environment (Fischer et al., 2021); and encourages interdisciplinary approaches to ecosystem restoration, especially to unearth the potential interlinked environmental and social/human health benefits (Fischer et al., 2021; Reid, 2020). Recent framings of the social-ecological approach acknowledge that 'restoration can be more effective if it engages with the relational values of diverse actors' because this can reveal factors that either encourage restoration (e.g. love of being outdoors) or discourage it (e.g. financial benefits from leaving things as they are) (Fischer, et al., 2021, p. 23). Understanding these motivations is invaluable for effective collaborative problem-solving and conflict resolution.

Ecosystem restoration carries within it a potential for restoring cultural and traditional approaches to living with nature. Akhtar-Khavari and Richardson (2017, p. 48) note that '[r]estoring nature … is also about restoring culture, to refashion the connections between human communities and their natural environs in the way that many Indigenous peoples already demonstrate through their embedded spiritual connections to the land'. This brings forth possibilities for healing people and communities, and for emphasising the human stewardship role, 'as an ethic of caring about all living beings while recognising their interconnectedness' (Fischer et al., 2021, p. 22). Taking a social-ecological perspective enhances the relational connection of humans to nature, encouraging conscious, caring, and ongoing involvement in Earth repair.

A key aspect of ecosystem restoration as a social-ecological practice is ensuring that *everyone* is involved. It needs policymakers, ecology specialists, scientists, and other experts, for sure, but it won't be effective or meaningful unless it also involves local communities, Indigenous people, committed environmentalists, citizen scientists, neighbours—and you. Whilst the Resolution acknowledges 'that restoration needs to be carried out [...] with the engagement of *relevant stakeholders*, including Indigenous peoples and local communities' (General Resolution, clause 6, my emphasis), it is important that every person is considered relevant, for we all rely upon ecosystems for our health and well-being, we all have a responsibility to reconnect with our environment, we are *all* planetary stakeholders. So, despite this seemingly narrow phrasing, I suggest that the UN Decade's Restoration Principles (FAO et al., 2021, p. 6) and the Resolution's acknowledgment elsewhere of 'the emergence of voluntary restoration initiatives and commitments at all levels, created to spur ambition and action to restore ecosystems across the world' (General Assembly Resolution 73/284, preambular clause), indicate a wide reading of participation. This is supported by the United Nations Environment Programme's (UNEP) key message 6 that 'everyone has a role to play in ecosystem restoration' at all scales (UNEP, 2021, p. 4 and pp. 18, 40, 44). This broader interpretation supports my insistence that ecosystem restoration is everybody's business. To not be naïve, however, since some people and entities may interpret their own involvement as being about preventing restoration or controlling everything, broad engagement requires that experts, businesses, and laypeople must acknowledge and respect each other whilst working together on ecosystem restoration in a 'kind of civic virtue that keeps [... everyone] involved' in a hands-on, learning practice informed by each other's knowledge and experiences of Earth's needs (Nichols, 2017, p. 238).

Together, these elements of broad engagement, interconnected human-nature relationship, plurality of views and motivations, systems under-standing, interdisciplinarity, and potential for restoring culture form a socio-ecological, relational ethos. At the international level, this ethos can draw not only from the UN Decade Resolution and associated guidance but also from UN human rights instruments that reflect restorative jus-tice values. Braithwaite (2002, p. 569) notes that 'restoration of the envi-ronment' is clearly a restorative value worthy of maximising, along with 'restoration of communities … restoration of compassion or caring … restoration of a sense of duty as a citizen'. At the local level, this ethos is supported by ERJ values and practices, such as respectful listening/dia-logue, accountability, citizen empowerment, and healing (Braithwaite, 2002; Braithwaite, 2016). Equally, this ethos can inform ERJ by keeping it focused on the interconnectedness of humans and nature. How ERJ can action the social-ecological, relational ethos is discussed next.

3.1   Relational Connectedness: In Partnership with Nature and Each Other

Relationality and relational justice are important attributes of ERJ (Forsyth et al., 2021), just as they are for the social-ecological approach above. Reference to, and potential for, relational approaches to ecosystem restoration are found in both the Resolution and the Strategy. In the Resolution, relationality potential is revealed through its call for awareness-raising, inclusion of all levels of society (including civil society and local levels), a call to share experiences and practices, and broad involvement of 'relevant' stakeholders (in particular, clauses 1, 3, 5 and 6). As discussed, references to 'relevant' should not be viewed as a limita-tion on the imperative to encourage every person's involvement. Furthermore, throughout the Strategy, there is a clearly stated intention that the UN Decade will 'set the stage for a new trajectory for the rela-tionship between humans and nature through the 21st century' (United Nations, 2020, para. 8, p. 4) and for 'a rediscovered relationship with nature that results in improved livelihoods and health for current and future generations' (United Nations, 2020, para. 28, p. 11). The United Nations Environment Programme echoes this, stating 'we need to recreate a balanced relationship with the ecosystems that sustain us' (UNEP, 2021, p. 3).

Restoration of the environment is a relational process, 'a reciprocal and generous relationship' (Pearce, 2018, p. 177). The starting point is to acknowledge that human beings are *a part of* nature as well as being connected socially with one another. The connected-to-nature dimension of relationality matters because ecosystem restoration occurs within the places where people live, work/create, play/recreate, worship,[7] build, produce and extract resources, and form communities, and it is the natural environment that underlies our sense of place. Yet, we often don't acknowledge this or see environmental harm. Shifting baseline syndrome—in which each generation accepts ever-growing depletion of nature as normal because we have nothing else to compare it with—leads us to assume our current environment is healthy (Soga & Gaston, 2018). As natural historian David Attenborough (2020, p. 100) says, '[w]e have become accustomed to an impoverished planet', increasing the urgency to repair our relationship with the environment. Rebuilding our connectedness-to-nature though the process of ecosystem restoration can help us to better comprehend environmental degradation and awaken 'an impetus for protection' (France, 2008, p. 5). Direct involvement with ecosystem restoration activities can form a virtuous circle, in which gaining a deeper relationship with nature tends to promote even greater involvement in caring for it (Robinson, 2020; Whitburn et al., 2019).

The UN Decade 'aims to create a platform for societies globally to put their relationships with nature on a new trajectory for centuries to come' (United Nations, 2020, p. ii). This 'new trajectory', that by necessity must be enduring and form the new outlook of all humanity, should include strengthening community bonds and regaining an understanding of and having respect for Indigenous thinking and approaches to

---

[7] This term is used broadly, to encompass 'the moment when we discover outside, beyond and above ourselves, values that are more important to us than we ourselves are' (Anthony of Sourozh, 1970, p.120), including a realisation of our duties towards nature and Gaia.

environmental care. For many Indigenous cultures, the land is not a separate entity from people (Neale & Kelly, 2020; Robinson, 2020) and their interwoven relationship with land and waters 'as alive and therefore worthy of thoughtful and respectful care' (Long et al., 2020, p. 74) is a relationality perspective everyone can learn from. In our relationships with each other, relational justice for citizens, communities, governance bodies, and business is cooperative, and 'doesn't divide into winners and losers; every voice still has a right to be heard' (Zournazi & Williams, 2021, p. 96). This requires 'integrating multiple viewpoints and knowledge types, including science, lay, managerial and Indigenous knowledge used by stakeholders and decision-makers' (Kenchington et al., 2012). In our relationship with our environment, we must gain a better understanding of humans as 'self-aware world changers with the good sense to work with the planet and not against it' (Grinspoon, 2016, p. 197).

In this next decade, we have 'an opportunity to reimagine our relationships with our environment and with each other' (PWP, 2021) and the work to be done in 'ecological recovery should facilitate the development of stronger positive human relationships with ecosystems, and increasingly address social justice within the restoration framework' (Cross et al., 2019, p. 925). Such reimagining would benefit from creating what Carolyn Merchant (1996, p. 217) calls 'partnership ethics [...] grounded in the concept of relation'. A partnership ethic focused on living with nature celebrates and reifies connectedness with each other, with our environment and more-than-human communities, is inclusive and ethical, respects diversity in culture and biota, and requires us as humans to reflect on the morality of humanity's ability to degrade ecosystems. It requires us to seek to co-work with nature, to effect 'a new balance in which both humans and nonhuman nature are equal partners, neither having the upper hand, yet cooperating with each other' (Merchant, 1996, p. 218). This partnership ethic will help us acknowledge our relational responsibility for repairing environmental harm, replacing the Holocene domination-over-nature narrative with Anthropocene human-as-partner, taking responsibility for the complex task of restoring what humanity has broken. The partnership ethic requires ongoing effort, because we still need to attend to unequal power dynamics, address differences between North and South lifestyles, respect Traditional

Knowledge, push back against business as usual, and make sure unheard voices are amplified (Merchant, 1996, p. 222; Stott et al., 2019). Bearing these needs in mind, an applied ethical partnership, relational approach can help to foster a restorative culture supported by ERJ values.

3.2   Relationality Realised: Accountability as Social-ecological Care

David Attenborough (2020, p. 5) laments that 'a damaging lack of care and understanding […] affects everything we do' in relation to our environment. Yet, he also acknowledges that 'we have a choice to make' and that 'we could change' (Attenborough, 2020, p. 6). In ERJ, accountability must be underpinned by a restorative ethos of attentive, conscious care, the 'restoration of compassion or caring' (Braithwaite, 2002). This restorative value gains expression, *inter alia*, in the concept of environmental stewardship, also promoted in the UN Decade (FAO et al., 2021). Krasny and Tidball (2015, p. xv), emphasise 'ongoing stewardship of land, life, and community' as 'caring actions' that lead to 'an ongoing relationship with the rest of nature that contributes to their own and their community's well-being, and even survival'. This ethos of care through stewarding is restorative both spiritually and in practice, encouraging a sense of repair and care as a way of life, and is often carried out by volunteers.

Whilst the Resolution envisages local, voluntary restoration projects, voluntary initiatives can be plagued by disengagement, poor commitment, and fadeout unless good, strong relationships are in place with a committed forum for re-engaging whenever there is a need to update everyone, discuss challenges and reroute initial solutions. ERJ could help to provide this accountability and some of its practices could be used to keep relationships between industry, government and community working together on restoration activities intact and responsive by providing forums that give everyone the opportunity to stay informed, feel safe airing concerns and empowered to suggest renewed or innovative approaches to repair (Forsyth et al., 2021, p. 35). Being accountable for environmental degradation includes ensuring that all who are affected by this harm have the opportunity to participate in restoration work, which includes the ability to veto, adapt or change restoration suggestions that do not align with community health, safety or connectivity to land, waters and nature (Forsyth, et al., 2021).

A particular aspect of the restorative value of accountability related to volunteer restoration work is the need for those with resources to assist those who have the will, but not the funds. Expectations that people will continue to give their emotional labour or pay for equipment or resources out-of-pocket, for the love or intrinsic value of restoring the environment can become exhausting and unjust. This is not a call for paying volunteers to undertake ecosystem restoration activities, which risks 'crowding out' passionate altruistic motivations (Telesetsky, 2013, p. 520). Rather, it is about suggesting that balance is needed between harnessing the innate desire to patiently restore the environment, and the reality that there is a cost to involvement, whether it be transportation, childcare, or community hall hire for holding restorative circles (Stott et al., 2019). Voluntary commitments to ecosystem repair must be funded and resourced in-kind by those able to do so, otherwise the risk is that those involved are skewed towards those who can afford it financially or time-wise, potentially leaving out some of the crucial people the UN Decade wants to be most involved, such as Indigenous people, the young and women (Pearce, 2018).[8] To be accountable, to ensure relational justice, those with resources must ensure they are appropriately redistributed in aid of healing communities and environments through ecosystem restoration.

The above discussion has aimed to show how ERJ supports and it is guided by a social-ecological and relational ethos in the context of ecosystem restoration, especially through connectedness-with-nature,

---

[8] In an environmental restorative justice research project lead by Professor Miranda Forsyth in which I am also participating (discussed in Chap. 6), one community volunteer interviewee from a river care group explained to us the difficulty of recruiting new volunteers, especially from amongst the young, ' the under 45s. During these years, they have children to raise, sport, ballet lessons, during the week they're working whilst on the weekends, they're catching up with stuff'. In another interview with an Indigenous organisation, we were told: 'Traditional Owners give their time a lot, they don't get paid for their time, they might get their petrol costs and accommodation covered if they're lucky. The government goes away, ticks the box of "cultural inclusion" and nothing changes'. Both of these comments struck me as clear indications that whilst people are happy to volunteer, they are aware of the limitations of their time and finances, and how this is detrimental to future involvement.

relationship building, caring, and a partnership ethic. The next part will examine how some ERJ values can support ecosystem restoration.

4 Using ERJ Values to Promote Knowledge, Sharing and Trust

I now turn to the ERJ values of dialogue, participation, and empowerment to highlight how they might be used in an educational way. Here, I interpret education as being about participants sharing local social-ecological knowledge between each other; deliberative discussion that increases understanding both of nature's role in community lives and other's viewpoints about Earth repair; and empowerment of citizens through ecosystem restoration activities.

4.1 Dialogue for Learning, Mending and Agreement

Dialogue as a restorative practice and value includes narratives, stories, and explanations. In the context of ecosystem restoration, dialogue could occur as a means of resolving conflicts (conferencing, circles) (Umbreit & Peterson Armour, 2011), top-down organised participatory events (open houses, town hall meetings, consultation groups) and informal, bottom-up talk (meetings, morning teas and chats on-the-fly during restoration activities). Dialogue in ecosystem restoration includes its use to ensure full participation of local communities; provide transparency by explaining what is happening; work through conflict; and listen to the stories of those harmed and their requests for repair. Of importance is that the dialogue is constructive, inclusive, and respectful, that it involves deep listening, and that it takes place during *all* stages of reparative work. Another important aspect of dialogue is its ability to build mutual trust (Blignaut & Aronson, 2020), both in terms of relations with experts and to ensure open access to information and resources, to support the unpaid and freely given time and knowledge from volunteers committed to environmental care (Umbreit & Peterson Armour, 2011). Trust is also built between participants through 'learning about diverse perspectives' (Krasny & Tidball, 2015, p. 125).

Ecological restoration is itself a form of conversation (Pearce, 2018). Talking about our ecosystems raises our awareness of their importance and makes evident that nature is one's home, not an exotic destination. Dialogue about human connectedness to local ecosystems can help us to 'give natural systems the degree of respect accorded to other members of [… our own] species' (Cairns, 2001, p. 186) and 'remind humans that we are part of ecosystems rather than separate from them' (Summers & Vivian, 2018, p. 2). Sharing local explanations about ecosystem processes and social-ecological memories can help participants feel more connected to nature and each other (Krasny & Tidball, 2015). Through the narratives we create around ecosystem restoration, including its potential to create a coherent, collective restorative narrative that reflects a broad range of goals, values, and knowledge (Blignaut & Aronson, 2020), we nurture the much-needed restorative culture.

ERJ provides both the impetus and the forum for dialogue, and respectful listening and storytelling, through which people can share their stories of environmental harm, and environmental hope. Dialogue can elicit what people want to see happen in ecosystem restoration activities, and how they see their own role, allowing for a 'plurality of values' and 'deliberative and inclusive practices' (Macdonald & King, 2018, p. 56) on what ecosystem restoration means for a particular community. This latter awareness is essential to legitimising restoration work and ensuring that local and Indigenous understandings are accounted for; it cannot be assumed that a top-down restoration approach will meet with community acceptance without dialogue about its intent, adjustment to adapt to local histories and knowledge, and clarity as to citizen involvement. Dialogue can help us to uncover thinking about existing paradigms that encourage or normalise practices that degrade our ecosystems and 'affective experiences of restoration within place can open dialogue to plural ontological and epistemological perspectives and shift the path of dominant histories' (Pearce, 2018, p. 176). Upfront and ongoing dialogue quells 'rumour and speculation that may otherwise influence [community] when forming a view about what is or should be happening' (Donovan, 2013, p. 243).

Of deep importance is the potential for dialogue to enable what scholar Jake Robinson (2020) terms 'hyper-localised' knowledge of an ecosystem to be heard, respected, and supported. This is especially relevant for Indigenous communities and peoples (Robinson, 2020), local communities, women and youth, and others whose voices are often side-lined or even captured to rubber-stamp projects via ritualised 'check-a-box' consultation. Through opening dialogue and 'talking-with' rather than talking down to or about Indigenous and local communities and people, through learning to be open to new ways of seeing and understanding, it becomes possible to hear and heed the intricacies, nuances and truths being spoken, revealing 'the dynamics that are often hidden to less experienced observers, such as how the architecture of a tree reveals a history of human engagement with the land' (Long et al., 2020, p. 79). Margo Neale and Lynne Kelly (2020, p. 47), discussing Australia's Martu people of the Pilbara, note 'they know their Country so intimately that they are part of its story'. Listening to this story, and other localised ones, is an important means through which worldviews 'rooted in "kincentricity"' and '[i]ntimate interactions with plants and animals over countless generations … can guide restoration' (Long et al., 2020, p. 74). Through deep listening and non-tokenistic learning from Indigenous stories of care for land and waters, the 'connectivity between people and place, past and present' (Neale & Kelly, 2020, p. 31) becomes clear and can help us also to hear and respect past and future generations.

One final possibility for ERJ in the context of dialogue during restoration projects is its potential to provide a forum for everyone to come together and reach a shared consensus or common ground on what 'recovery' looks like.[9] Following environmental harm, the question of what is involved in restoration is likely to be contentious, given the varying theories and preferences for whether or not it's possible/desirable to restore an ecosystem to an historical benchmark or to accept what it has become (albeit healthy) (Macdonald & King, 2018; Watson et al., 2020). Where experts in the room are drawn from a wide range of inter- and

---

[9] Donovan (2013, p. 247) cautions that use of the word 'recovery' may itself be at issue if it is taken to mean a return what was rather than inclusive of a new spirit of renewal. In the ecosystem restoration context, renewal should be inclusive of the social-ecological reconnectedness and conscious awareness of living with nature.

transdisciplinary backgrounds, where all relevant regulatory agencies are interacting, where Indigenous people, community groups and members, NGOs, and others with an interest in the ecosystem in question, are all present, a much broader pool of knowledge and ideas exists to draw upon to decide when the restorative work is sufficient or not, when it's time to leave the ecosystem to its own devices or continue the repair work. Stewardship of land (discussed above) requires us to be 'open to new avenues for more holistic, pluralistic and innovative visions' (Macdonald & King, 2018, p. 157) and it is through restorative dialogue that we can come together to reach some agreement.

4.2 Participation and Empowerment in the UN Decade

Participation of those responsible for harm and those affected by it is a core restorative principle, to ensure a thorough exchange of apologies, reparative promises/ actions and collective discussion of ways forward. Participation as a restorative value also involves acceptance that different forms of knowledge have something to contribute to a holistic, more complete understanding. By bringing together a broad spectrum of participants who hold varied knowledge and understandings, there is an increased chance of doing the right thing by a harmed ecosystem and to respond to what Lilian Pearce (2018, p. 186) refers to as 'contested landscapes layered with histories'. Participation fulfils the partnership ethic and is 'critical to build[ing] genuine collaborative space' that helps to 'bridg[e] top-down:bottom-up disconnects' (Stott et al., 2019, p. 112). Local knowledge, both Indigenous and grassroots, can bring forth localised understandings that might otherwise be overlaid with assumptions based solely on scientific assessments and policy pressures (Pearce, 2018) and may result in more effective environmental and social outcomes (Baker et al., 2014). Moreover, participation can increase tolerance towards differing values, thereby reducing conflict, and can legitimise restoration efforts and promote social acceptance of the desirability of restoring ecosystems (Baker et al., 2014). Participation enhances relational connectivity, especially where it contributes to people's sense of place, belonging, and attachment (Telesetsky, 2013). Another important aspect of participation is the ability to harness community and citizen support for long-term monitoring and assessment of restoration (Young & Schwartz, 2019). This links directly with accountability, another restorative value explored above, whereby citizen involvement in monitoring restoration outcomes can improve knowledge, assess restoration effectiveness, and aid collaborative integration of community and practitioners (Young & Schwartz, 2019).

Empowerment of citizens and communities participating in ecosystem restoration is essential to fully realised participation, community connectedness, and motivating ownership of ongoing environmental care. Wijdekop (2019, p. 6) explains that '[i]nvolvement in restorative processes strengthens community identity and resilience and empowers change from the bottom up, because it is a way for communities to develop social capital, social networks and civic interconnectedness'. Empowerment includes ensuring that citizens gain access to and be supported by the experts, information, and the institutions where they need to carry out restoration work (Donovan, 2013). Citizen empowerment is best achieved through involvement and doing, 'to take control of their shared surroundings by incrementally building up experience of designing and managing interventions' (Donovan, 2013, p. 248). This acting and doing could be through citizen science (data production, monitoring, awareness-raising, etc.), co-design of restorative activities, pulling weeds or planting trees, rewilding, and similar activities.

In what could also be viewed as 'restoration of a sense of duty as a citizen' (Braithwaite, 2002, p. 569), democratic engagement with ecosystem restoration can empower individuals and communities to re-engage with both civic duty and environment. Citizen empowerment occurs through ensuring that 'local residents [are] integrated into planning and decision-making as central voices from start to finish' (Büscher & Fletcher, 2020, p. 196) and inclusion of 'a broad spectrum of actors whose interests encompass multiple (and often conflicting) values ranging from biodiversity to cultural identity and aesthetic delight' (Macdonald & King, 2018, p. 156). ERJ values and practices can highlight and support expectations that local people and communities help determine and co-design restorative needs and are not simply told what a top-down authority considers appropriate. Through a restorative process, people could come together to hear the 'authoritative' science, policy, and financial issues then, in turn, present their preferences based on what Büscher and Fletcher (2020) call 'embedded' values, which encapsulate their own authority. This embedded authority can reference restorative values like citizen empowerment, participation, and healing. In this way, democratic engagement would legitimise ecosystem restoration work, encourage long-lived engagement from locals and would also, in time as people become more attuned to living with nature, be inclusive of multidimensional relationships with both humans and more-than-humans.

Participatory empowerment entails the responsibility of working constructively with each other too, 'appreciating but also politically confronting and agonistically struggling with each other' (Büscher & Fletcher, 2020, p. 173). Thus, through learning to 'honor diverse perspectives' (Krasny & Tidball, 2015, p. 126), empowerment means ensuring that nobody is side-lined or alienated (Donovan, 2013; Larson Sawin & Zehr, 2007). Lastly, participation and empowerment as restorative values must further acknowledge the interplay between international and local, to help develop a coherency of direction and togetherness arising from being part of the UN Decade, relaying local experiences to help inform and shape the global.

5   Conflict and Harm Prevention through ERJ

In this last part, I consider the third way ERJ can contribute to ecosystem restoration, by looking at its potential for finding common ground, assisting healing and preventing ecosystem harm. The Strategy envisages restorative justice as being able, in part, to 'assist [...] countries in resolving conflicts over natural resources and reduce the need for communities to migrate as a result of such conflicts' (UN Strategy, 2020, para. 1.a). Whilst this imagines a role of ERJ in the context of environmental disputes at a global scale, ERJ can also play a valuable role in assisting in addressing conflict at the local, grassroots level. In an ecosystem restoration project context, such conflict is to be expected (indeed, as noted above, democratic engagement invites conflicting viewpoints). A variety of reasons can trigger conflict, including people feeling left out or unheard; feeling pressured to accept lesser standards of environmental care than they had envisaged; intra-community disputes; inconsistencies in goals and methods of institutions/experts (Baker et al., 2014; Stott et al., 2019); differences in valuing what to restore and what is considered 'alien' or unworthy of environmental care (Baker et al., 2014), amongst other reasons.

ERJ can provide 'a space for exploring why such a situation has surfaced and the different perspectives' (Stott et al., 2019, p. 121). It can help to resolve conflict through providing a forum for deep listening and heart-felt apology from those responsible for environmental harm, promoting discussion that is reflective, inclusive, healing, and in which people's stories of their relationship with the environment are centralised, heard, and heeded. ERJ can provide the respectful forum wherein conflicting viewpoints can be harnessed to 'foster a culture of restoration' (Blignaut & Aronson, 2020, p. 8) reflective of everyone's desired contributions to ecosystem restoration.

Restorative justice's future-focus on relationship maintenance further requires keeping participatory dialogue open and ongoing in a conscious effort to decrease conflict through regular reviews 'to explore how partners are working together' (Stott et al., 2019, p. 121). This may include ensuring that companies or government bodies set up ongoing facilitated open houses, community reference groups,[10] or other forms of community gathering at which information, updates and sharing of stories takes place on a regular basis. At government level, conflict may even be addressed through a restorative enquiry process, through which long-term structural and prior injustices that impede ecosystem health can be discussed, with intent to remove legislative and administrative roadblocks.

More creatively, using ERJ to resolve conflict can enable unheard voices to be included, such as future generations, animal and plant species, and rivers, landscapes, and other environmental beings, which is not

---

[10] See further Forsyth, Chap. 6, for more details about such bodies in the context of the State of Victoria and its Environment Protection Agency.

the case for most judicial and administrative resolution processes. In such ways, ERJ may help establish multidimensional 'democratic legitimacy of restoration' (Baker et al., 2014, p. 517) through respecting a breadth of voices from affected social and ecological communities.

5.1   Healing People and Environment

Healing is an important attribute of ERJ, 'of communities, of relationships, of institutions and of ecosystems damaged by human action or inaction' (Forsyth et al., 2021, p. 30). Broadly, healing is 'relational, reparative, procedurally fair, nurturant of apology and forgiveness' (Braithwaite, 2016). More specifically, 'ecological restoration is […] a way to heal humans' relationship with nature', whereby 'persons can become, in important ways, restored to land' (Van Wieren, 2008, p. 238). The UN Decade represents 'a worldwide attempt to heal the environment, biodiversity and human health' (Mikolič-Berrios, 2019, p. 2).

Healing in the context of ecosystem restoration must encompass the healing of our damaged relationship with the environment and more-than-humans (Braithwaite et al., 2019). Since ecosystem restoration is a 'process of healing' (Maathai, 2010, p. 46), ERJ's emphasis on healing (Forsyth, et al., 2021) can play a valuable role in providing 'genuine healing of relationships and the natural environment' (Forsyth et al., 2021, p. 22). It can provide a forum where people can come together to talk about the impacts of environmental harm on their lives and well-being and reflect upon ways to recreate connectivity with nature and each other, reintegrate the human steward in place of the human destroyer, and rethink how to cooperate and live with nature.

Imagine a restorative circle wherein a community (including Indigenous persons and representatives of voices of the environment) impacted by ecosystem degradation sits together with those taking responsibility for the harm and asks to be a part of designing what that community views as being needed to heal. Jenny Donovan's (2013, p. 239) 'designing to heal' works here; she says that everyone is responsible for this activity (reflecting that everyone needs to be involved). Healing will only occur if the wounds are acknowledged and tended—through direct involvement with designing, discussing, and doing ecosystem restoration work, it becomes possible for community and ecosystem wounds to be healed as a shared journey, as part of 'collective recovery' (Donovan, 2013, p. 247). Gerry Johnstone and Daniel Van Ness (2007, p. xxi) state 'a core idea of restorative justice is that the people most affected by a problem decide among themselves how it should be dealt with'; thus, when people provide their own solutions, the autonomy and problem-solving skills they gain aid in the healing. ERJ forums can help ensure this happens, through dialogue and openness, placing healing at the forefront of everyone's focus during repair of environmental harm, and 'cultivating the changes needed in hearts and minds and on the ground to help people recover and communities to renew themselves' (Donovan, 2013, p. 239).

As noted earlier, involvement in caring for nature results in an increased desire to care for nature, emphasising the importance of giving everyone impacted by environmental degradation the choice to be involved mending it. Time spent in nature promotes healing (Summers & Vivian, 2018) and increases ecological literacy, an essential skill for overcoming our embeddedness in shifting baseline syndrome, making us more aware of and willing to heal ecological wounds (Cairns, 2001). Healing as a restorative value is underpinned by the earlier discussed active principles of dialogue, participation, and accountability.

5.2   Harm Prevention: The Flipside to Restoration

Philosopher Elizabeth Spelman (2008, p. 127) terms ecosystem restoration, amongst other reparative efforts, 'repair work' and notes that we are 'repairing animals'. After listing off the good that this entails, she notes that this can lead to making errant and presumptuous judgments about what to repair or how to repair it, as we try to turn back the clock or endure brokenness (Spelman, 2008). In many instances, we cannot return an ecosystem to what it was once but instead must accept that change has occurred beyond its original state and the adapted ecosystem (cleaned up and healthy) that it has become (Pearce, 2018; Watson et al., 2020), is what we must learn to live with. Choices about which ecosystem functions to prioritise will require 'deliberative and inclusive practices' (Macdonald & King, 2018, p.156) to avoid unintentional additional damage and unrealistic hopes (Spelman, 2008).

However, there is one other choice—it is our responsibility to protect what is still flourishing, the remaining biodiversity hotspots, unpolluted rivers, buffer zones, and healthy lands. The UN Decade's preventive aspect must be promoted too, as part of our reconnection to nature. ERJ's attentiveness to harm prevention, its forward-looking orientation, encouragement for reflection and focus on building strong relationships can support knowledge-sharing, advocacy, and dialogue aimed at preventing future ecosystem degradation (Forsyth et al., 2021).

6   Conclusion

The UN Decade provides an important opportunity for local community-led and community-involved ecosystem care and repair to be influenced and guided by environmental restorative justice. This opportunity is both practical—through ensuring broad, reflective, and ongoing democratic engagement in ecosystem restoration—and motivational, in that it creates a restorative culture whereby 'the principles, ethics, and standards of holistic ecological restoration are embedded in all aspects of human existence and endeavor' (Cross et al., 2019, p. 926). Viewing ecosystem restoration as relational, partnership work based on ERJ values will benefit both human and ecological communities, foregrounding the merits of cooperating, living with nature, and being astutely aware of our role as 'conscious shapers of our world' (Grinspoon, 2016, p. xv) with the ability to heal or harm, depending on the choices we make. Now that 'humankind has the knowledge and the conscience to ensure that the relationship with natural systems enriches rather than despoils' (Cairns, 2001, p.188), the restorative values of ongoing learning and collaboration must underpin our future efforts to restore environments and prevent future environmental harm.

Local ERJ is ready to inform the global too—as Bolivar, Guerra and Martínez suggest in Chap. 22, the time is apt for restorative justice to set its sights towards infusing the international level with restorative values and practices. After all, environmental 'crises don't read maps, they don't stop at national boundaries' (Zournazi & Williams, 2021, p. 102). Like reaching up through the layers of the soils, forests, and oceans that support the ecosystems we are responsible for restoring, ERJ can be scaled up to share local learnings and experiences of its use in ecosystem restoration, between communities, nationally, and with other regions and countries, alongside sharing the 'good practices in ecosystem restoration and conservation' (General Assembly Resolution 73/284, clause 3(f)).[11]

In this chapter, I have explored how ERJ can contribute significantly to helping us reconnect to and restore ecosystems, through enthusiastic embracement of broad community participation and leadership. By instilling a social-ecological, relational ethos, by providing a space for dialogue to share stories and improve comprehension of nature's value in our lives, and by offering practices to collaboratively work through conflict and prevent future harm, ERJ furnishes us with a way of thinking and acting restoratively throughout the continuum of Earth repair activities. It brings in the potential for careful listening, shared vision and increased sense of responsibility towards the environment and all species, reflecting a key UN Decade aim of 'nature being respected across society' (United Nations, 2020, p. ii). In the UN Decade's hope to restore (and maintain) the relationship between humans and nature, ERJ's values, practices and emphasis on care for people, nature, and Gaia will be vital for lighting the way.

References

Akhtar-Khavari, A., & Richardson, R.J. (2017). Ecological restoration and the law: recovering nature's past for the future. *Griffith Law Review*, 26(2), 47–53.

Anthony of Sourozh, M. (1970). Worship in a secular society. *Studia Liturgica*, VII(2–3), 120–130.

Aronson, J., Goodwin, N., Orlando, L., Eisenberg, C., & Cross, A.T. (2020). A world of possibilities: Six restoration strategies to support the United Nation's Decade on Ecosystem Restoration. *Restoration Ecology*, 28(4), 730–736.

---

[11] Both scaling up and sharing of learnings aligns with Goals 1, 2 and 3 of the UN Decade and Pathways I-III (UNEP, 2021) and the Resolution's call to 'share experiences and practices'.

Attenborough, D. (2020). *A life on our planet. My witness statement and a vision for the future*. London: Penguin Random House.

Baker, S., Eckerberg, K., & Zachrisson, A. (2014). Political science and ecological restoration. *Environmental Politics*, 23(3), 509–524.

Blignaut, J., & Aronson, J. (2020). Developing a restoration narrative: A pathway towards system-wide healing and a restorative culture. *Ecological Economics*, 68, 1–9.

Braithwaite, J. (2002). Setting standards for restorative justice. *Brit. J. Criminology*, 42, 563–577.

Braithwaite, J. (2016). *Restorative justice and healing the environment*. Retrieved from: http://johnbraithwaite.com/2016/09/14/restorative-justice-and-healing-the-environment/ (last accessed 22 January 2022).

Braithwaite, J., Forsyth, M. & Cleland, D. (2019). Restorative environmental justice: An introduction. In E. Biffi & B. Pali (Eds.), *Environmental justice: Restoring the future*, EFRJ Newsletter (September 2019). Retrieved from: https://earthrestorativejustice.org/article/36455/booklet-environmental-justice-restoring-the-future (last accessed 22 January 2022).

Büscher, B. & Fletcher, R. (2020). *The conservation revolution: Radical ideas for saving nature beyond the Anthropocene*. London: Verso.

Buttfield, C., & Hughes, J. (2021). *Earthshot: How to save our planet*. Falkirk: John Murray (Publishers).

Cairns, Jr., J. (2001). Healing the world's ecological wounds. *Int. J. Sustain. Dev. World Ecol.*, 8, 85–89.

Cross, A.T., Nevill, P.G., Dixon, K.W., & Aronson, J. (2019). Time for a paradigm shift toward a restorative culture. *Restoration Ecology*, 27(5), 924–928.

Donovan, J. (2013). *Designing to heal*. Collingwood: CSIRO Publishing.

FAO., IUCN., CEM., & SER. (2021). *Principles for ecosystem restoration to guide the United Nations Decade 2021–2030*. Rome: FAO.

Fernández-Manjarrés, J.F., Roturier, S., & Bilhaut, A-G. (2018). The emergence of the social-ecological concept. *Restoration Ecology*, 26(3), 404–410.

Fischer, J. Riechers, M., Loos, J., Martin-Lopez, B., & Temperton, V.M. (2021). Making the UN Decade on Ecosystem Restoration a social-ecological endeavour. *Trends in Ecology & Evolution*, 36(1), 20–28.

Forsyth, M., Cleland, D., Tepper, F., Hollingworth, D., Soares, M., Nairn, A., & Wilkinson, C. (2021). A future agenda for environmental restorative justice? *The International Journal of Restorative Justice*, 4(1), 17–40.

France, R.L. (2008). Swamped! A tale of two restorations: part I: The view from home. In R.L. France (Ed.), *Healing natures, repairing relationships: New perspectives on restoring ecological spaces and consciousness* (pp. 1–6). Sheffield: Green Frigate Books.

Friel, S. (2019). *Climate change and the people's health.* New York: Oxford University Press.

General Assembly Resolution 73/284, United Nations Decade on Ecosystem Restoration (2021–2030), A/RES/73/284 (6 March 2019). Retrieved from: https://digitallibrary.un.org/record/3794317?ln=en   (last   accessed   22 January 2022).

Grinspoon, D. (2016). *Earth in human hands: Shaping our planet's future.* New York: Grand Central Publishing.

Johnstone, G. & Van Ness, D.W. (Eds.) (2007). *Handbook of restorative justice.* Cullompton: Willan Publishing.

Kenchington, R., Stocker, L., & Wood, D. (2012). Lessons from regional approaches to coastal management in Australia: a synthesis. In R. Kenchington, L. Stocker, & D. Wood (Eds.), *Sustainable coastal management and climate adaptation* (pp. 193–209). Collingwood: CSIRO Publishing.

Krasny, M.E., & Tidball, K.G. (2015). *Civic ecology: Adaptation and transformation from the ground up.* Cambridge: MIT Press.

Larson Sawin, J., & Zehr, H. (2007). The ideas of engagement and empowerment. In G. Johnstone & D.W. Van Ness, *Handbook of restorative justice* (pp. 41–58). Cullompton: Willan Publishing.

Long, J.W., Lake, F.K., Goode, R.W., & Burnette, B.M. (2020). How traditional tribal perspectives influence ecosystem restoration. *Ecopsychology,* 12(2), 71–82.

Maathai, W. (2010). *The world we once lived in.* Australia: Penguin Random House.

MacDonald, E., & King, E.G. (2018). Novel ecosystems: A bridging concept for the consilience of cultural landscape conservation and ecological restoration. *Landscape and Urban Planning,* 177, 48–59.

Merchant, C. (1996). *Earthcare: Women and the environment.* New York: Routledge.

Mikolič-Berrios, A. (2019). *When ecosystems suffer, so do humans: To heal people we need to heal the planet.* In These Times. Retrieved from: https://inthese-times.com/article/ecosystems-gondwana-waterton-ecohealth-network-environmental-health/ (last accessed 22 January 2022).

Neale, M., & Kelly, M (2020). *Songlines: The power and the promise.* Melbourne: Thames & Hudson.

Nichols, T. (2017). *The death of expertise.* New York: Oxford University Press.

Pearce, L.M. (2018). Affective ecological restoration, bodies of emotional practice. *International Review of Environmental History*, 4(1), 167–189.

PWP (2021). *The UN Decade of Ecosystem Restoration launches on World Environment Day*. Retrieved from: https://plantwithpurpose.org/decade-of-ecosystem-restoration/ (last accessed 22 January 2022).

Reid, L. (2020). *Healthy societies built from healthy ecosystems: How Australia and Aotearoa New Zealand are working at the intersection of human health and ecological restoration for a healthier world*. Retrieved from: https://mbgecologicalrestoration.wordpress.com/2020/07/23/healthy-societies-built-from-healthy-ecosystems-how-australia-and-aotearoa-new-zealand-are-working-at-the-intersection-of-human-health-and-ecological-restoration-for-a-healthier-world/ (last accessed 22 January 2022).

Robinson, J.M. (2020). *Four reasons why restoring nature is the most important endeavour of our time*. Retrieved from: https://theconversation.com/four-reasons-why-restoring-nature-is-the-most-important-endeavour-of-our-time-147365 (last accessed 22 January 2022).

Rockström, J., Steffen, W., Noone, K., Persson, Å., Chapin, III, F. S. Lambin, E., Lenton, T. M., Scheffer, M., Folke, C., Schellnhuber, H., Nykvist, B., De Wit, C. A., Hughes, T., van der Leeuw, S., Rodhe, H., Sörlin, S., Snyder, P.K., Costanza, R., Svedin, U., Falkenmark, M., Karlberg, L., Corell, R. W., Fabry, V.J., Hansen, J., Walker, B., Liverman, D., Richardson, K., Crutzen, P., & Foley, J. (2009). Planetary boundaries: Exploring the safe operating space for humanity. *Ecology and Society*, 14(2): 32. Retrieved from: http://www.ecologyandsociety.org/vol14/iss2/art32/ (last accessed 26 January 2022).

Soga, M., & Gaston, K.J. (2018). Shifting baseline syndrome: Causes, consequences, and implications. *Frontiers in Ecology and the Environment*, 16(4), 222–230.

Spelman, E. (2008). Embracing and resisting the restorative [sic] impulse'. In R. L. France (Ed.), *Healing natures, repairing relationships: New perspectives on restoring ecological spaces and consciousness* (pp. 127–138). Sheffield: Green Frigate Books.

Stott, L., Dwonczyk, M., & Pyres, J. (2019). Going local: Partnering with citizens and communities. In L. Stott (Ed.), *Shaping sustainable change: The role of partnership brokering in optimising collaborative action* (pp. 110–126). Oxford: Routledge.

Summers, J.K., & Vivian, D.N. (2018). Ecotherapy—a forgotten ecosystem service. *Frontiers in Psychology*, 9, 1–13.

Telesetsky, A. (2013). Ecoscapes: The future of place-based ecological restoration laws. *Vermont Journal of Environmental Law*, 4(4), 492–548.

Umbreit, M., & Peterson Armour, M. (2011). *Restorative justice dialogue. An essential guide for research and practice.* New York: Springer Publishing Company.

UNEP/FAO (2022). *Preventing, halting and reversing the degradation of ecosystems worldwide.* Retrieved from: https://www.decadeonrestoration.org/ (last accessed 22 January 2022).

United Nations (2020). *The United Nations Decade on Ecosystem Restoration: Strategy.* Retrieved from: https://www.decadeonrestoration.org/strategy (last accessed 22 January 2022).

United Nations Environment Programme (UNEP) (2021). *Becoming #GenerationRestoration: Ecosystem restoration for people, nature and climate.* Retrieved from: https://www.unep.org/resources/ecosystem-restoration-people-nature-climate (last accessed 22 January 2022).

Van Wieren, G. (2008). Ecological restoration as public spiritual practice. *Worldviews*, 2, 237–254.

Watson, J.E.M., Keith, D.A., Strassburg, B.B.N., Venter, O., Williams, B., & Nicholson, E. (2020). Set a global target for ecosystems. *Nature*, 578, 360–362.

Whitburn, J., Linklater, W., & Abrahamse, W. (2019). Meta-analysis of human connection to nature and proenvironmental behavior. *Conservation Biology*, 180–193.

Wijdekop, F. (2019). *Restorative justice response to environmental harm.* Amsterdam: IUCN.

WSP USA (2020). *Helping the UN guide ecosystem restoration into the mainstream.* Retrieved from: https://www.wsp.com/en-AU/insights/2020-un-guide-ecosystem-restoration-into-mainstream (last accessed 22 January 2022).

Young, T.P., & Schwartz, M.W. (2019). The Decade on Ecosystem Restoration is an impetus to get it right. *Conservation Science and Practice*, 1–3.

Zehr, H. (2002). *Little book of restorative justice.* New York: Good Books, Inc.

Zournazi, M., & Williams, R. (2021). *Justice and love: A philosophical dialogue.* London: Bloomsbury Academic.

Part II

Applications of Environmental Restorative Justice

13

Restorative Justice for Illegal Harms Against Animals: A Potential Answer Full of Interrogations

Gema Varona

1   Introduction

Despite some inclusive uses in art, literature and other disciplines (Simons, 2002; Lönngren, 2021), and despite the lesser legal protection of some humans in comparison to some animals in different times and contexts (Bourke, 2011), in both criminology and victimology the term 'animal' has usually carried exclusionary connotations. Within cultural criminology, popular culture and media studies, the use of the word 'animal' can be traced to the purpose of dehumanisation, the Holocaust being the paramount example of a state crime where animal names were deployed to refer to innocent people who were massively criminalised and eliminated (Andrighetto et al., 2016). Today, many persons still call certain offenders 'animals' and by doing so, in a contradictory way, try to justify the need for dehumanising criminal sanctions (Stevenson et al., 2015). In the field of victimology, in Romance and Germanic languages,

G. Varona (✉)
Basque Institute of Criminology, Donostia/San Sebastian, Spain
e-mail: gemmamaria.varona@ehu.eus

a victim is etymologically an animal that can be sacrificed (Van Dijk, 2009), denoting how victims have traditionally been instrumentalised in the criminal justice system.

Mainstream uses or evocations of the term 'animal' contain a pejorative and anthropocentric view of animals, depicting them as brutal, irrational, inferior and instrumental to human interests and forgetting that, amongst living creatures, we are human animals. In current socio-legal studies, taking the question of non-human animals seriously is sometimes met with contempt and suspicion, arguing that human rights—or ecology in general—are much more important concerns. Aware of this controversy, in this chapter, it is contended that, if interdependent human rights are to be taken seriously, it is not coherent to forget about the human responses to animal harm. In these pages, the term 'animal' refers to non-human animals. From a socio-legal perspective (Cao & Wyatt, 2016), animal harm is understood as any formal rule-breaking of norms that protect companion, domestic and wild animals, including criminal, administrative and civil laws and regulations in relation to animal abuse and mistreatment. This rule-breaking is not victimless (Spapens, 2016)— the animals being the direct victims—which raises complex political and justice issues around the interests at stake. Higher legal education can 'work towards a better representation of animal interests in the political process' (Peters, 2020, p. 11) and, it could be added that victimological research and training might do the same in the restorative justice processes as well. However, introducing animals into the restorative justice debate (United Nations, 2020) is a complex matter for two main reasons.

Firstly, it requires envisioning animals as vulnerable victims (Sunstein & Nussbaum, 2004), notwithstanding the fact that this vision is not legally accepted today by the European Union nor, to cite the specific country which I use in this chapter as an example, by the Spanish legislation on victims' rights (Varona, 2020a). Nevertheless, the European Union has acknowledged that animals are sentient beings that can be harmed, which brings positive obligations to protect their welfare accordingly both where and how animals live or in the ways they are used by humans. Moreover, there is no single common position on this issue held by different animal welfare and rights movements and certainly not by society at large.

Secondly, restorative justice aims at connecting victims, offenders and communities for reparation in a meaningful, egalitarian and empathetic conversation. Knowing that a historical and devastating characteristic of humans is our will to discriminate and exclude other humans from our own species (Antelme, 2001), in the case of animal harm, restorative justice is even more challenging. Here reparation might entail inter-species communication in a scenario where animals are usually considered morally inferior to human beings and use different communication systems that we still can scientifically decipher, in part due to outdated notions still dominating the public imagination.

The objective of this chapter is to tackle these two complex matters from the standpoint of green victimology (Hall, 2014) and animal studies (Giménez-Candela, 2019) within the Spanish context. It will start by describing to what extent the demands of most animal activists claiming for animal welfare and rights might be contradictory to restorative justice as a set of principles critical of punitiveness. The chapter will later concentrate on how restorative justice in this field can be developed in theory and in practice.

2   Representing Animal Interests in Victim Activism, Victimological Studies and Law

Approaching the so-called 'animal question' (namely, how much do animals matter morally?) implies a myriad of neuroscientific, bioethical, political, cultural, religious and economic elements. Law, as a normative system expressing 'both interests and ideals' (Peters, 2020, p. 9, 2021) should be part of the disciplines and conceptual bases that consider the animal question. Broadly speaking, only since the end of the sixties, victims' interests have been included specifically in criminal law, after the pressure of victims' and human rights activists' movements. That inclusion has received much criticism by some criminal legal scholars, either because it is viewed as being merely symbolic (Van Dijk, 2009) or because it has brought about more punitiveness (Snacken and Dumortier 2012). However, there is some evidence that the entry of victims into restorative justice programmes, especially where these programmes respect international standards (United Nations 2020), might make the penal system less punitive (Johnstone & Klaasen, 2015) because many victims who go through the restorative justice process seem to be more interested in solidarity and transformation than in pain reciprocity (Van Camp, 2014).

In the case of animal victimisation, animals' interests and welfare are being promoted through legal reforms by victim activism (animal welfare and rights parties and NGOs), academic green victimology (Vegh Weis & White, 2020) and general animal studies (Giménez-Candela, 2019). Many animal welfare and rights activists in Spain tend to ask for more criminalisation and punishment. This is partially true in the case of animal studies too, but not so much in the emerging Spanish literature on green criminology and victimology.

3   The Spanish Animal Welfare and Rights Movement as a Victim Movement

Animal activism in Spain revolves around the concepts of animal welfare and anti-speciesism (Sollund, 2021). One result of this activism can be seen in the 2017 inclusion in the Spanish Royal Academy dictionary of the terms 'speciesism' and 'anthropocentrism'. The term 'speciesism' means discrimination against animals by considering them to be an inferior species and the second term, 'anthropocentrism' refers to the understanding of human beings as the most significant entity of the universe and by whose values the rest of reality or experience can be interpreted (Steiner, 2005).

Obviously, if animals themselves are to be considered as victims, there are no victim associations as such in this field, but rather victims' interests support groups. With the precedent of the refuges for abandoned animals at the beginning of the twentieth century, the activism of several Spanish groups gained momentum after the end of the Franco dictatorship (Lorente, n.d.). This can be perceived as an evolution from the idea of protection to the idea of justice; animal rights (Francione & Garner, 2010) and anti-speciesm as an abolitionist movement against discrimination (Méndez, 2020). The first *Animal Protection Society in Defence of Animal Rights* (ADDA) was created in Barcelona in 1975; it participated in the drafting of the non-legally binding Universal Declaration of Animal Rights in 1978 and fostered the first Catalan legislation on animal rights in 1988. In 1985, the group *Alternative for Animal Liberation* was founded. In 2002, *Rights for Animals* was created, followed by the *Animal Welfare and Rights Party against Maltreatment* in 2003. Today, several Spanish organisations belong to the EU-wide *Eurogroup for Animals* that tries to influence international and European policies. Most organisations seek more effective prosecution of and increased criminalisation for acts of animal harm.

Spanish animal welfare or rights groups have benefitted from globalised movements that conceptualise animals as victims. Examples of such movements have included the *Save* movement that aspires to have humans bearing witness to the transportation of farm animals to slaughterhouses (Freeman & Tulloch, 2013; Fernández, 2020), *Meat the Victims, Anonymous For The Voiceless, Action For Liberation* and *Animal Liberation Front*. Some of these groups promote legal disobedience or other protest actions (Giménez-Candela & Cersosimo, 2020; Ellefsen & Busher, 2020). In any case, animal liberation movements differ within the movements themselves on animals as rights holders (Regan, 1983; Luke, 1992; Munro, 1999; Kim, 2011; Johnston & Johnston, 2017) and subjects of protection (Singer, 1975, 2009) in utilitarian or egalitarian terms, with some holding even more radical abolitionist positions beyond current legality (Vázquez & Valencia, 2016).

Today's groups and parties advocating animal rights and interests are not only single-issue entities, but they are also developing the concepts of compassion, equality, intrinsic value and interdependence (Lucardie, 2020), all of which are relevant concepts for restorative justice. Moreover, many crimes against animals might also be related to human victimisation, and vice versa, resulting in multiple and related victims (Sollund, 2013), whether or not labelled as crimes against animals.

4   Animals in Animal Law, Animal Studies and Green Victimology

Since the end of the twentieth century, the political and legal achievements of the animal welfare and rights movements in Spain can be seen in legal reforms in civil, administrative and criminal areas of law, as well as in the creation of the Congress Society on Animal Rights in 2007, the General Unit for Animal Protection in the central government in 2020, and, particularly since 2015, specific units in several bar associations.

Even though there is no mention of animal welfare in the Spanish Constitution, in line with the EU legislation and the sectorial and dispersed legislation by the Autonomous Communities and local authorities, the Judicial Decision 81/2020 of July 15, 2020 of the Spanish Constitutional Court on the constitutionality of the Animal Protection Act of the Autonomous Community of La Rioja, provides for an interpretation of the constitutional framework for the legal regulation of animal welfare public policies (Medina, 2020). Inherited from the Roman legal tradition, like the majority of European civil codes, Spain conceptualises animals as property (*res*), but, following other countries like Portugal, the Czech Republic and France, as well as the EU legislation (on production, experimentation, transport, and shows), Spain has decided to change this conceptualisation by approving bills in 2017 and 2021 to transcend its civil code understanding of animals as objects. Giménez-Candela (2019) refers to an 'animal transition', like the political transition, where legal and non-legal professionals defend animals, and Spanish society has experienced a change towards the need for a basic understanding of the need for a better treatment of animals.

In the country of bullfighting and other cruel *fiestas* with animals, together with activist movements, after the end of dictatorship some Spanish philosophers and jurists advanced the thinking on animals from protection to rights and justice (Riechmann, 2005; Tafalla, 2007; Horta, 2010; Mosterín, 1999; Baltasar, 2015). In the animal rights debate, mainly since the 1990s, Spanish ethologists, psychologists and jurists have also participated internationally in *The Great Ape Project* and the *Nonhuman Rights Project*.

Beyond veterinary specialities, in academic studies, Spain is the location of the first European University to create an 'Animal Law and Society' Master's programme in 2011 at the Autonomous University of Barcelona (Giménez-Candela & Cersosimo, 2020). Spanish academia has also produced significant publications in this field. Even if critical of its real impact, the academic position in this realm is calling for more effective criminal law in dealing with harm to animals.

In 2021, a working group on green criminology was created within the Spanish Society for Criminological Research, a significant event because no such group existed previously in the European Society of Criminology, despite keynote presentations and panels in several international conferences (Vander Beken et al., 2021). Although still marginal, green victimology has been developed by some Spanish authors (Ríos, 2021) and annual interdisciplinary conferences have been held, since 2015, at the Basque Institute of Criminology (Varona, 2020b). Within Spanish green victimology, more critical viewpoints can be considered towards the use of criminal law to defend animals' welfare or interests.

Across disciplines, the scientific and legal core notion upon which the advancement of academic knowledge is based is the notion of animals as sentient beings. Even if it has to be clearly stated for companion and wild animals, the notion of sentient beings has given rise to:

a standard of animal treatment, which entails the recognition of their capacity not only to feel physical pain, but also suffering, pleasure, and enjoyment ... animal sentience as a regulatory parameter of a dignified life (and death) of an animal, understood to be a public responsibility. (Giménez-Candela, 2017, pp. 6–7)

Although in the quoted text, the author does not explicitly mention the need for more punishment, de-objectification of animals is linked to effectively protecting and enforcing the obligations of individuals, 'a task that appeals to the responsibility of the State, to the responsibility of citizens, to the responsibility of Public Administration, and to the responsibility of the Security Forces' (Giménez-Candela, 2019, p. 21).

5   Crimes Against Animals in the Criminal Justice System: Can Restorative Justice Offer Better Responses?

In the Spanish criminal code, the term 'animal' is mentioned 30 times and three Articles refer to crimes against 'species of wild fauna' (Articles 333–336). Restorative justice is not mentioned in that code for any kind of crime. A general reference can only be found within the Statute of the Victim of 2015, transposing the Directive 2012/29/EU, establishing minimum standards on victims' rights. For that Directive, only natural persons can be victims.

5.1   Punishing Crimes in Relation to Animal Harm in the Spanish Criminal Code

In general, animal harm is punished with penalties of up to two years' imprisonment, fines and being barred from certain professions or activities.[1] Since animals do not have the right to physical or psychological integrity, the legal interest being protected in these crimes has been interpreted as private property, public domain, nature for its own sake, or general interests, including the bond between human beings and animals (Hava, 2021; Adams, 1994, 2018). In recent judgements, the aspect of animal welfare concerned with the avoiding of unnecessary suffering has been recognised too (Toribio, 2020).

Articles 325, 326 and 326bis establish several crimes against the environment where animals are or can be harmed. Here penalties can reach

---

[1] According to the Ministry of Interior (2020, p. 278), in 2019, 3, 629 persons were detained or investigated for environmental crimes by the specialised police force called the Civil Guard. Of those, 601 were for animal abuse, 134 for illegal trafficking of protected species and 403 for forest fires. That same year, 104,090 administrative offences against the environment were investigated by police, mainly about spills (20,084), breaches against fishery rules (12,491) and alleged mistreatment of companion animals (10,332). According to the General Prosecutor's Office (2020, p. 856), crimes against the environment and against animals have increased in the last two? decades. In 2019, 1180 sentences were imposed (57 for crimes against the environment, 290 against plants and wild animals, 136 for forest fires, and 293 for animal abuse, the kind of crime in this group that has experienced the greatest increase).

five years of imprisonment. Moreover, some Spanish jurisprudence has interpreted these provisions beyond anthropocentric considerations to take into account animal welfare (Varona, 2020a).

Articles 343 and 345 establish punishments of up to 12 years of imprisonment for crimes related to nuclear energy and ionising radiation that might endanger animals. Forest fires that significantly alter the conditions of animal life can by punished up to six years of imprisonment, according to Article 353. Within offences against public health, administering prohibited substances to animals employed for human consumption can lead to sentences of up to 10 years of imprisonment (Article 364). Inter-species reproduction techniques are also criminalised.

For animals under human control or dependence, Article 337 on animal abuse criminalises unjustified serious attacks against the health of the animal with penalties up to one and a half years of imprisonment (Ramos & Fuentes, 2021). Before 2003, that crime was considered a minor one and only punished with a fine (Olmedo, 2021). The 2015 reform included the prohibition of sexual exploitation of animals for economic benefit and sexual practices causing animal suffering. A very influential criminal law professor, Gimbernat (2015), criticised the 'animal welfare and rights lobby' for the inclusion of moral questions if no harm could be demonstrated. For him, this law brought to mind the prohibition and cruel punishment of bestiality in the mediaeval ages, particularly against those accused of being witches (Beirne, 1997). However, other criminal law professors have argued that it is not a question of morality, but of animal suffering and that, according to scientific studies, many animals can experience emotional suffering and not just pain (Toribio, 2020).

With regard to animal abuse in general, other abuses can also be sanctioned by administrative Autonomous Community and local legislation, sometimes resulting in harsher fines (Requena, 2021). For this reason, some offenders might prefer the criminal justice system where the prison penalty is usually substituted by community service or suspended (Toribio, 2020), which, together with the lack of prosecution and low penalties, has been interpreted as impunity by some activists and experts.

5.2   Restorative Justice

Harm against animals is often culturally justified for utilitarian reasons, particularly in Spain: harm might be tacitly acknowledged, but not its injustice, which prevents animals from being viewed as victims (Maglione, 2017) by most members of society. Due to the existing power imbalance, because animals themselves cannot report and exercise their interest to be free of illegal harm, defined as abuse and mistreatment, and the fact that the animal question has been traditionally a culturally minor one, many activists find that restorative justice might lead to more impunity in this field. However, in this chapter, the view is defended that if animal welfare is an emancipatory or liberation movement, it would be contradictory to aim for more punishment.

Humane treatment, care and responsibility principles (Pelluchon, 2018), which are supposed to improve the lives of animals according to the animal welfare and rights movement, understood in the first section of this chapter as a victim movement, and the academic thoughts on animal rights and welfare, might be more consistent. This could be achieved if claims for criminalisation (including harmful behaviours in the criminal code—as long as the *ultima ratio* is respected—(Cuerda, 2021; Cervello, 2021)), are provided along with alternatives to classical punishment. A substantive amount of criminological evidence questions the idea that more punishment brings reduced victimisation (Caruso, 2021). In particular, when it concerns the protection of animals, harsher punishment might bring about more human *and* non-human harm, mainly because the attention is focused on the perpetrator, and not on the victim (in the case of an animal, barely understood as such), and the resources are put at the service of the process of punishment instead of prevention of future harm or reparation of the animals' welfare, when possible.

Restorative justice, understood as a dialogue to make the harm and its consequences visible and to engage all those affected in its reparative prevention, can provide the forum in which we can question distant suffering and foster the morality of accountable proximity (Bauman, 2004), where we are responsible for those whom we cannot hear or listen to (Jacobsen, 2021).

Restorative justice might offer a way to help foster that difficult conversation with the animal 'Other', amplifying the circle of empathy and the moral community (Cantor, 2014). It could provide a platform for understanding the harms produced and the way they can be repaired, entailing the questioning of classical ideological frameworks and the need for a transformation of structural conditions that engender or normalise abuse, harm and neglect of animals. However, even if it recognises the inherent value of animals, their preference for autonomy and their right to respectful treatment, how can animals, as victims, be one of the stakeholders in restorative justice? The endeavour to respond to this question should provide for thinking through the ways in which we might be able to have inter-species communication on the issues of who participates, what the harm is that has to be repaired and how. The offender would always be human (sometimes part of a powerful corporation and even the owner of the animal) and the victims, together with humans who take responsibility for acting in the interests of the animals and even if not legally recognised as such, the animals affected by the wrongdoing. Of course, animals by themselves cannot articulate a voice meaningful for humans (García Cano, 2012) and cannot denounce that reparations are not being duly fulfilled. For this reason, there would be an irreducible anthropocentric view on what reparation might mean, at least as an attempt to stop causing harm within a given context and provide the conditions for future living that respects animal welfare.

Hence, because the question of justice is relevant, it is necessary to think about animals as agents or recipients of that justice. In terms of pre-reflective self-awareness, Rowlands underlines that 'personhood extends into the animal realm as far as experiencing *us*' (2019, p. 15). This might help us develop imaginative thinking on the inclusion of experiences, instead of voices or narratives, coming from companion, farm or wild animals (Herman, 2021; de Froideville, 2021). The point of departure could be animal welfare on the basis of their being sentient beings. In committing harm against animals, the human-animal relationship has been broken because a human has caused unnecessary suffering to an animal. Following Bentham ([1789] 1907), the fundamentals for animals' role as victims does not rest on whether they can reason or talk, but on whether they can suffer and because science confirms that they can, the human obligation to protect them has been breached and some-one should be held accountable for reparation.

Martha Fineman (2008) defines vulnerability as the possibility of becoming dependent, a quality that, even if unequally distributed in time and space, is inherent to all living beings. Our bodies, and their associated emotions, are fragile and have a beginning and an end. Awareness of fragility, precariousness and vulnerability (embodied finitude) amongst human and non-human animals might create connections and entail obligations. According to Susi (2017), the vulnerability theory can be combined with other capabilities theories (Sunstein & Nussbaum, 2004) applied to animals as vulnerable, yet capable, subjects with a life that is worth living beyond basic freedoms from harm (Mellor, 2016), and in relation to other lives.

The element of interdependence at the core of restorative justice might be found within the One Health transdisciplinary approach to animal harm (Tarazona et al., 2020; Destoumieux-Garzón et al., 2018; Sinclair, 2019). This approach is defined as designing and implementing pro-grammes, policies, legislation and research in which multiple sectors communicate and work together to achieve better public health out-comes, also considering animal welfare. This might be done by using a bioecological systems model (Jegatheesan et al., 2020), where the health interconnectedness between very diverse people, animals, plants, and their shared environment is recognised. This does not mean it is so in an egalitarian way or that tensions might not arise even within a broad notion of bioecological justice (Morelle, 2021; O'Neil, 2000), where individual and species' collective aspects might enter into the discussion from a bioecological standpoint (Walters, 2019). However, if focused on prevention and reparation, this public health model could reduce the punitive standpoint to be taken towards animal harm. At the same time, it might also fail to account for the relevance of justice to respond to animal harm and environmental harm in general.

With this permanent critical perspective, restorative justice should apply its inherent flexibility and responsiveness, as a creative process, to involve common undervalued facets of humans and animals. For exam-ple, according to the psychologist Burghardt (2020, 2019), who coined the concept of critical anthropomorphism (Burghardt 1991), playing is an integral part of animal life (including humans) and can be defined as a repeated behaviour, voluntarily initiated, in a context without stress. It exists in many animals and is key for their health, learning and group relationships. In this regard, following the work of Graeber, (Pemberton, 2015, p. 43) reminds us of the playful drive to counter injustice. Thus, the flexible and creative practices of restorative justice, adapted to each context and individual, might entail some playing with companion animals, for example in relation to therapy as reparation, as will be mentioned in the following section. In any case, further research is needed on the restorative justice capacity to be truly inclusive with animals and to minimise anthropocentrism or other biases that focus the imaginary of the restorative justice participant mainly on the homo *economicus, sociologicus* or *sentiens* (Archer, 2000). Restorative justice would in some way remain a speciesist practice (Singer, 2009) that is mainly focused on the welfare of humans, despite the moral consideration of animals as sentient beings. However, this inevitability can be restoratively tempered by awareness and acknowledgement of human responsibility.

6   Envisioning Restorative Justice Programmes Within the Spanish Criminal Framework

Even though there are no statistics on how many cases involving harm to animals have been developed through restorative justice programmes, the Basque Institute of Criminology has been debating this topic with different stakeholders and proposing specific programmes since 2015 (Varona, 2020b). Considering the complexities of animal harm and the risks of banalising it, as well as the increasing numbers of cases that reach the criminal justice system (despite the high hidden level of victimisation in this field), in this section, an adaptation of the Restorative Circles developed by Dominic Barter and Duke Duchscherer is proposed to address such crimes. Restorative Circles are based on nonviolent response to conflict and the construction of community spaces for communication where, usually, many participants are to be involved (Dzur, 2017). Three phases could be distinguished within this framework.

6.1   Circle Phases

Three phases could be distinguished within this framework: a pre-circle to identify the participants and their needs; the restorative circle where a more or less direct encounter takes place; and final post-circle(s) where some follow up of agreements is possible.

Pre-circle: Identifying the Participants and Their Needs

Because this kind of circle involves more diversity and a greater number of participants than in mediation, co-facilitation should be encouraged. In any case, facilitators should have some basic information on animal studies and green victimology.

In the pre-circle phase, with individual or subgroup encounters, the facilitator would identify the protagonists and their supporters and clarify the basics of what happened and its meaning to human and non-human life in terms of needs (Gibney & Wyatt, 2020). It would include informing participants about the dialogue process and its principles or rules, according to international standards. Representatives of the affected community, human and non-human (Safina, 2020), should be selected considering different forms of proximity and levels of harm. This might include bringing humans and animals into the circle, depending on the contexts of the harm and the conditions of participants, always under the principle of not causing secondary victimisation or further harm; that is, under a perspective of safety and care for both humans and the animals involved. Due to the limitations of this chapter, these specificities should be further developed in concrete projects and related action research papers.

Well-being is usually a term deployed to human victims. In the case of animals, it is a debatable topic (Lerner, 2008; Broom, 2017) and the term animal welfare is usually preferred. Animal welfare has been defined as the physical and mental state of an animal in relation to the conditions in which it lives and dies. The World Organisation for Animal Health (OIE, n.d.) refers to the welfare of *terrestrial animals*, as society's expectations for the conditions animals should experience when under human control. These are: freedom from hunger, malnutrition, and thirst; freedom from fear and distress; freedom from heat stress or physical discomfort; freedom from pain, injury and disease; and freedom to express normal patterns of behaviour. In relation to aquatic animals, the OIE refers to the use of handling methods appropriate to the biological characteristics of the fish and a suitable environment to fulfil their needs. In general, if we consider that animals are sentient beings with both biological and emotional needs, we can understand animal health and animal welfare as synonyms, that is, we can talk about health with prevention in focus, just like human health.

The Circle: Including Animals in an Itinerant Dialogue Beyond Words

In a second phase, the circle, the protagonists (including the harmed animal itself or representations of the animal through images or symbolic elements) would meet for a dialogue where all senses are relevant beyond verbal communication. The point of departure could be trying to explain how every protagonist feels at that very moment and, particularly, when thinking about the harm produced. The relevant issue is what matters to them at the moment of the encounter: meeting in the present to travel to a past that has consequences, also for the future. The facilitators should make sure that a basic connection is being produced, in some cases asking one participant to select another, as a principal listener in the circle, and later asking protagonists in conversation about what has been understood at different levels of communication. This process requests one-on-one respectful listening, being done within the extended group so that all can listen to the exchange. Humans supporting the animal (or who care for or work with similar animals) can try to explain the harm and its impact from their perspective and experience of being with those animals frequently. The classical human victim's question of 'why me?' would be absent as such, but it might be substituted with images and sounds of some animals and restorative walks through spaces of the animal's habitat, where the harm was produced, where it can be repaired or where some link exists (Natali & de Nardin, 2019). The offender can reflect on this and the whole circle would develop around awareness of interconnection for reparation, on the side of the offender, with the help of some supporters (family members, friends, other professionals, etc.).

Establishing a connection, understanding, and self-responsibility between human and non-human animals with the help of supporters might avoid further harm and allow reparation with a focus on agreed actions that facilitate future collective well-being. Here the idea of mutuality can be replaced by interconnectedness, with a more active role being taken on by the offender with the support of the rest of the participants. Restorative agreements might entail, perhaps cumulatively, forms of compensation, memorialisation aimed at messaging non-repetition, apologies to the potential humans affected, following some kind of suitable rehabilitation programme, community service, and so on.

Post-circle: Because Restorative Justice Is Not Just Therapy Or Symbolism

The post-circle would seek to keep track of those agreed actions, once initiated or completed. This third phase is important considering the suspicion by some animal welfare and rights movements that restorative justice might be a lesser form of or soft justice. Animals as victims are not aware of the labelling of the harm as crime, nor of the notion of human time (past-present-future), but they have experienced the harm where a human is responsible for repairing it. If restorative justice aspires to be transformative, it should involve the community and require that those involved take the time to evaluate that the conditions conducive to ensuring non-repetition have been changed. Besides considering the link between interpersonal and collective violence and violence against animals (Bernuz, 2015), preventing animal harm is particularly important.

6.2   Referrals and Penal Impact of Restorative Programmes

Following the suggestions made by the United Nations Handbook on Restorative Justice into ways to assure the participation of victims, four actions should be promoted (United Nations, 2020, pp. 100–101):

1. Fostering action research and publishing real case studies where restorative justice has been used to repair harms done to animals.
2. Increasing public awareness through general and targeted education and by expanding the ways open to stop, prevent and repair harms, to include restorative justice in regulations and legislation and informing the public through social media.
3. Working with criminal justice professionals and organisations for the protection of animals and for animal rights in order to seek referrals to restorative justice services.
4. Cooperation between stakeholders working at the frontiers of criminal and administrative law in a field of blurred frontiers on the notion of harm.

As we have seen, in Spain, most penalties for crimes against animals usually amount to up to two years of imprisonment, meaning that a sentence for imprisonment can be substituted by community service or suspended. Even though many animal rights' activists consider that alternative sentencing might entail impunity and revictimisation, it might also be the entrance point for restorative justice programmes that ensure accountability, reparation and prevention, mainly developed at the trial, sentencing stage and post-sentencing phases. At the trial and sentencing stage, the mitigating reparation circumstance of Article 21 of the Spanish Criminal Code can be considered. At the post-sentencing stage, Article 83. 1. 6 allows for the suspension of the execution of the sentence under the obligation of 'participation in environmental or animal protection programs', and Article 84 permits conditioning the suspension of the execution of the sentence to the fulfilment of the 'agreement reached through mediation' or the development of community service as a way of 'symbolic reparation'. Article 49 provides community service to be done through 'reparation work of the harm caused or support or assistance to victims'.

With the necessary support, again avoiding secondary victimisation or revictimisation, offenders could also participate in restorative therapies with animals, including those who have been themselves victims of animal abuse (Bernuz, 2020). Spanish prisons already have more than 20 programmes of assisted therapy with animals, conceived as complementary to treatment for inmates with affectivity and self-esteem problems, including those with psychiatric pathologies. What is proposed here is a different thing; developing a restorative engagement as a core element of programmes, based on the fact that these programmes help develop empathy and prosocial behaviour for animals and human beings (Komorosky and O'Neal, 2015).

Finally, in the Spanish post-sentencing stage, a new treatment programme called PROBECO, for economic crimes (including those against the environment and animals), might facilitate an early release from prison for more serious crimes. PROBECO envisages restorative panels or workshops to promote prosocial behaviour and, specifically, the prevention of animal maltreatment. Carried out in groups, this programme has a length of nine months and it could be complemented with other activities.

7   Conclusion

The animal question has entered into the political and justice debate (Kymlicka & Donaldson, 2011). Green victimology can study any non-human animal victimisation, beyond a strict legal conceptualisation of environmental crime, allowing for new ways of thinking about victimhood, vulnerability, responsibility and rights, all of them key for restorative justice. Within criminal policy and justice, the human-animal relationship allows us to conceive of two moments in criminal policy. First, if according to the democratic principles of minimum intervention and *ultima ratio*, criminalisation, understood as describing a behaviour as a crime, can only be used to protect the most important public goods when other less harmful mechanisms of social control have failed. We need to reconsider the use of criminal law in cases of severe animal abuse and cruelty, acknowledging that the cultural tolerance towards animal harm is changing positively. Second, in accordance with the humanity principle, the ethics of care, non-violence and harm minimisation in responding to crime and victimisation, sanctioning in criminal law should be more open to restorative justice where a more reparative and preventive response could be achieved.

The animal welfare and rights movement questions the tenets of our society and that should include the criminal justice system at large too. Without questioning the current penal system, more human and non-human animal harm can be provoked. Criminal punishment as a chosen response to animal harm might prove ineffective, inefficient and harmful (for societies and animals). At the same time, the restorative justice movement and academic literature would benefit from considering the ethical, theoretical and practical challenges of including animals as agents in restorative justice and criminology (Cole, 2020; Hillyard et al., 2004; Canning & Tombs, 2021). Moreover, we need future research on the potential of restorative justice programmes for restorative governance within Earth Jurisprudence (Lampkin, 2020) which stresses the interconnection between human and non-human life that must be acknowledged as basic in our legal system. Restorative justice's imaginative frontiers can be questioned as to their coherence with ethics and scientifically and legally recognised evidence regarding the suffering of animals and their status as sentient beings. That status is already recognised when animals are being used in criminal rehabilitation programmes to provide for justice beyond punitivism.

A restorative conversation implies the exchange of information, the unearthing of uncertainties and the need to respect affection in relation to minimising dominion over animals that fails to address harm to animals meaningfully or expansively. Envisaging emancipatory rules for cohabitation where the animal question can be included matters because it is political in nature and linked to justice outcomes. Animals are dependent agents in the world of humans, with differing margins of autonomy dependent on their perceived role. Restorative justice must rise to the challenge of promoting better ways to approach radical Otherness and to provide a forum in which we can question the cultural conditions that allow the production of harm. Understanding the Other is not a matter of identification—quite the contrary—it is a matter of discovery within and outside of our own skin. Restorative justice could be one vital means through which humans start to socially support and monitor the individual, group and collective duty of care towards animals, before and after the illegal harm. Finally, restorative justice can highlight not only the impact of the harm but the need for involvement in the response by means of restoration. To do this effectively, it will have to help ensure that the restoration can be achieved through real and tangible programmes.

References

Adams, C. J. (1994). Bringing peace home: A feminist philosophical perspective on the abuse of women, children, and pet animals. *Hypatia*, 9(2), 63–84.
Adams, C. J. (2018). *Neither man nor beast: Feminism and the defense of animals*. London: Bloomsbury Publishing.
Andrighetto, L., Riva, P., Gabbiadini, A., & Volpato, C. (2016). Excluded from all humanity: Animal metaphors exacerbate the consequences of social exclusion. *Journal of Language and Social Psychology*, 35(6), 628–644.
Antelme, R. (2001). *La especie humana*. Madrid: Arena.
Archer, M. S. (2000). *Rational choice theory: Resisting colonisation*. New York: Routledge.
Baltasar, B. (Ed.) (2015). *El Derecho de los animales*. Madrid: Marcial Pons.
Bauman, Z. (2004). *Wasted lives: Modernity and its outcasts*. Cambridge: Polity Press.
Beirne, P. (1997). Rethinking bestiality: Towards a concept of interspecies sexual assault. *Theoretical Criminology*, 1(3), 317–340.
Bentham, J. (1907) [1789]. *Introduction to the principles of morals and legislation*. Oxford: Clarendon.
Bernuz, M. J. (2015). El maltrato animal como violencia doméstica y de género. Un análisis sobre las víctimas. *Revista de Victimología/Journal of Victimology*, 2, 97–123.
Bernuz, M. J. (2020). Castigos (eficaces) para delitos contra los animales? Repensando la respuesta al maltrato animal. *Indret: Revista para el Análisis del Derecho*, 1, 1–14.
Bourke, J. (2011). *What it means to be human: Historical reflections from the 1800s to the Present*. London: Virago.

Broom, D. M. (2017). *Animal welfare in the European Union*. Brussels: Directorate General for Internal Policies.

Burghardt, G.M. (1991). Cognitive ethology and critical anthropomorphism: A snake with two heads and hognose snakes that play dead. In C. A. Ristau (Ed.), *Cognitive ethology: Essays in honor of Donald R. Griffin* (pp. 53–90). London: Psychology Press.

Burghardt, G. M. (2019). Play: A neglected factor in ritual, religion, and human evolution. In T. B. Henley, M. J. Rossano & H. P. Kardas (Eds.), *Handbook of cognitive archaeology* (pp. 120–134). New York: Routledge.

Burghardt, G. M. (2020). Insights found in century-old writings on animal behaviour and some cautions for today. *Animal Behaviour*, 164, 241–249.

Canning, V., & Tombs, S. (2021). *From social harm to zemiology: A Critical introduction*. New York: Routledge.

Cantor, D. (2014). Beyond humanism, toward a new animalism. In W. Tuttle (Ed.), *Circles of compassion: Essays connecting issues of justice* (pp. 22–36). Danvers: Vegan Publishers.

Cao, A. N., & Wyatt, T. (2016). The conceptual compatibility between green criminology and human security: a proposed interdisciplinary framework for examinations into green victimisation. *Critical Criminology*, 24(3), 413–430.

Caruso, G. D. (2021). *Rejecting retributivism. Free will, punishment, and criminal justice*. Cambridge: Cambridge University Press.

Cervello, V. (2021). La penalidad en los delitos de maltrato y abandono de animales. In Cuerda, M. L. (Ed.), *De animales y normas. Protección animal y derecho sancionador* (pp. 80–112). Valencia: Tirant lo Blanch.

Cole, M. (2020). Criminology, harm and non-human animals. In L. Copson, E. Dimou, & S. Tombs (Eds.), *Crime, harm and the state* (pp. 11–143). Milton Keynes: Open University Press.

Cuerda, M. L. (Ed.) (2021). *De animales y normas. Protección animal y derecho sancionador*. Valencia: Tirant lo Blanch.

de Froideville, S. M. (2021). Storied experiences of the Havelock North drinking water crisis: A case for a 'narrative green victimology'. *International Review of Victimology*. https://doi.org/10.1177/02697580211005013.

Destoumieux-Garzón, D., Mavingui, P., Boetsch, G., Boissier, J., Darriet, F., Duboz, P., Fritsch, C., Giraudoux, P., Le Roux, F., Morand, S., Paillard, C., Pontier, D., Sueur, C. & Voituron, Y. (2018). The one health concept: 10 years old and a long road ahead. *Frontiers in Veterinary Science*, 5, 1–13. https://doi.org/10.3389/fvets.2018.00014.

Dzur, A. (2017). Conversations on restorative justice: A talk with Dominic Barter. *Restorative Justice*, 5(1), 116–132.

Ellefsen, R., & Busher, J. (2020). The dynamics of restraint in the Stop Huntingdon animal cruelty campaign. *Perspectives on Terrorism, 14*(6), 165–179.

Fernández, L. (2020). Images that liberate: Moral shock and strategic visual communication in animal liberation activism. *Journal of Communication Inquiry,* https://doi.org/0196859920932881.

Fineman, M. A. (2008). The vulnerable subject: Anchoring equality in the human condition. *Yale Journal of Law & Feminism, 20*, 1–24.

Francione, G. L., & Garner, R. (2010). *The animal rights debate: Abolition or regulation?* New York: Columbia University Press.

Freeman, C. P., & Tulloch, S. (2013). Was blind but now I see: Animal liberation documentaries' deconstruction of barriers to witnessing injustice. In A. Pick & G. Narraway (Eds.), *Screening nature: Cinema beyond the human* (pp. 110–126). New York: Berghahn Books.

García Cano, D. M. (2012). La voz de los que no tienen voz: Movimiento animalista y acción colectiva. *Revista Kogoró, 4*, 24–32.

General Prosecutor's Office (2020). *Memoria elevada al gobierno de S. M. presentada al inicio del año judicial por la Fiscal General del Estado.* Madrid: Fiscalía General del Estado.

Gibney, E., & Wyatt, T. (2020). Rebuilding the harm principle: Using an evolutionary perspective to provide a new foundation for justice. *International Journal for Crime, Justice and Social Democracy, 9*(3), 100–115.

Gimbernat, E. (2015, April 24). La reforma del Código Penal. *Diario del Derecho.* Retrieved from: https://www.iustel.com/diario_del_derecho/noticia.asp?ref_iustel=1139913 (last accessed 23 January 2022).

Giménez-Candela, M. (2017). The de-objectification of animals (I). *derecho ANIMAL, 8*(2), https://doi.org/10.5565/rev/da.318.

Giménez-Candela, M. (2019). *Transición animal en España.* Valencia: Tirant lo Blanch.

Giménez-Candela, M., & Cersosimo, R. (2020). *La enseñanza del derecho animal.* Valencia: Tirant lo Blanch.

Hall, M. (2014). Environmental harm and environmental victims: Scoping out a 'green victimology'. *International Review of Victimology, 20*(1), 129–143.

Hava, E. (2021). La tutela penal del bienestar animal. In M. L. Cuerda (Ed.), *De animales y normas: Protección animal y derecho sancionador* (pp. 190–225). Valencia: Tirant lo Blanch.

Herman, D. (2021). Narratology beyond the human: Self-narratives and interspecies identities. In S. McHugh, R. McKay, & J. Miller (Eds.), *The Palgrave handbook of animals and literature* (pp. 51–64). New York: Palgrave Macmillan.

Hillyard, P., Pantazis, C., Tombs, S., & Gordon, D. (Eds.). (2004). *Beyond criminology: Taking crime seriously.* London: Pluto Press.

Horta, O. (2010). Igualitarismo, igualación a la baja, antropocentrismo y valor de la vida. *Revista de Filosofía, 35,* 133–15.

Jacobsen, M. H. (2021). Suffering in the sociology of Zygmunt Bauman. *Qualitative Studies, 6*(1), 68–90.

Jegatheesan, B., Enders-Slegers, M. J., Ormerod, E., & Boyden, P. (2020). Understanding the link between animal cruelty and family violence: The bio-ecological systems model. *International Journal of Environmental Research and Public Health, 17*(9), 3116, https://doi.org/10.3390/ijerph17093116.

Johnston, G., & Johnston, M. S. (2017). 'We fight for all living things': Countering misconceptions about the radical animal liberation movement. *Social Movement Studies, 16*(6), 735–751.

Johnstone, G., & Klaasen, E. (Eds.). (2015). *Building bridges: Restorative dialogues between victims and offenders.* Retrieved from: http://restorative-justice. eu/bb/wp-content/uploads/sites/3/2016/02/WS-2-D2.4-Building-Bridges-Guidebook-.pdf (last accessed 23 January 2022).

Kim, C. J. (2011). Moral extensionism or racist exploitation? The use of Holocaust and slavery analogies in the animal liberation movement. *New Political Science, 33*(3), 311–333.

Komorosky, D., & O'Neal, K.K. (2015). The development of empathy and prosocial behavior through humane education, restorative justice, and animal-assisted programs. *Contemporary Justice Review, 18,* 395–406.

Kymlicka, W., & Donaldson, S. (2011) *Zoopolis: A political theory of animal rights.* Oxford: Oxford University Press.

Lampkin, J. (2020). *Uniting green criminology and Earth Jurisprudence.* New York: Routledge.

Lerner, H. (2008). *The concepts of health, well-being and welfare as applied to animals: A philosophical analysis of the concepts with regard to the differences between animals.* Linköping: LiU-tryck.

Lönngren A.-S. (2021). Metaphor, metonymy, more-than-anthropocentric. The animal that therefore I read (and follow). In S. McHugh, R. McKay & J. Miller J. (Eds.), *The Palgrave handbook of animals and literature* (pp. 37–50). New York: Palgrave Macmillan.

Lorente, C.-J. (n.d.). Movimientos y activismo por los derechos animales. Retrieved from: https://www.abogacia.es/publicaciones/blogs/blog-de-derecho-de-los-animales/movimientos-y-activismo-por-los-derechos-animales/ (last accessed 23 January 2022).

Lucardie, P. (2020). Animalism: A nascent ideology? Exploring the ideas of animal advocacy parties. *Journal of Political Ideologies*, 25(2), 212–227.

Luke, B. (1992). Justice, caring, and animal liberation. *Between the Species*, 8(2), https://doi.org/10.15368/bts.1992v8n2.11.

Maglione, G. (2017). Embodied victims: An archaeology of the 'ideal victim'of restorative justice. *Criminology & Criminal Justice*, 17(4), 401–417.

Medina, E. M. (2020). Regulación jurídica de las políticas públicas de protección, bienestar o "derechos" de los animales. *Revista General de Derecho Animal y Estudios Interdisciplinares de Bienestar Animal*, 6, 1–101.

Mellor, D.J. (2016). Updating animal welfare thinking: Moving beyond the "five freedom" towards "a life worth living". *Animals*, 6, https://doi.org/10.3390/ani6100059.

Méndez, A. (2020). América Latina: Movimiento animalista y luchas contra el especismo. *Nueva Sociedad*, 288, 45–57.

Ministry of Interior (2020). *Anuario estadístico del Ministerio del Interior*. Madrid. Retrieved from: http://www.interior.gob.es/web/archivos-y-documentacion/anuario-estadistico-de-2020 (last accessed 24 January 2022).

Morelle, E. (2021). El bienestar animal frente al equilibrio ecológico desde el derecho penal: el caso de la eliminación de cabras en Es Vedrà (Eivissa). In M. L. Cuerda, (Ed.), *De animales y normas. Protección animal y derecho sancionador* (pp. 334–371). Valencia: Tirant lo Blanch.

Mosterín, J. (1999). Resumen de mis principales tesis en ¡Vivan los animales! *Teorema: Revista Internacional de Filosofía*, 18(3), 1–8.

Munro, L. (1999). Contesting moral capital in campaigns against animal liberation. *Society & Animals*, 7(1), 35–53.

Natali, L., & de Nardin Budó, M. (2019). A sensory and visual approach for comprehending environmental victimization by the asbestos industry in Casale Monferrato. *European Journal of Criminology*, 16(6), 708–727.

O'Neil, R. (2000). Animal liberation versus environmentalism. *Environmental Ethics*, 22(2), 183–190.

OIE (World Organisation for Animal Health). (n.d.). Retrieved from: https://www.oie.int/en/what-we-do/animal-health-and-welfare/animal-welfare/ (last accessed 23 January 2022).

Olmedo, E. (2021). Pasado, presente y futuro de los delitos de maltrato animal en España. In M. L. Cuerda (Ed.), *De animales y normas: Protección animal y derecho sancionador* (pp. 372–395). Valencia: Tirant lo Blanch.

Pelluchon, C. (2018). *Manifesto animalista: Politizar la causa animal*. Barcelona: Reservoir Books.

Pemberton, A. (2015). *Victimology with a hammer. The challenge of victimology: Inaugural lecture.* Tilburg: Tilburg University.

Peters, A. (2020). Preface. In M. Giménez-Candela, & R. Cersosimo (Eds.), *La enseñanza del derecho animal* (pp. 9–12). Valencia: Tirant lo Blanch.

Peters, A. (2021). *Animals in international law.* City: Brill Nijhoff.

Ramos, J. A., & Fuentes, M. A. (2021). El maltrato ¿justificado? de animales. In M. L. Cuerda (Ed.), *De animales y normas. Protección animal y derecho sancionador* (pp. 396–417). Valencia: Tirant lo Blanch.

Regan, T. (1983). *The case for animal rights.* Oakland: University of California Press.

Requena, A. (2021). Las entidades de protección de animales ante el maltrato: posibilidades y límites de actuación. In M. L. Cuerda (Ed.), *De animales y normas. Protección animal y derecho sancionador* (pp. 418–441). Valencia: Tirant lo Blanch.

Riechmann, J. (2005). *Todos los animales somos hermanos.* Madrid: Catarata.

Ríos, J. M. (2021). La consolidación de la Victimología verde a propósito del abandono y del maltrato animal. In M. L. Cuerda (Ed.), *De animales y normas. Protección animal y derecho sancionador* (pp. 442–481). Valencia: Tirant lo Blanch.

Rowlands, M. (2019). *Can animals be persons?* Oxford: Oxford University Press.

Safina, C. (2020). *Becoming wild: How animal cultures raise families, create beauty, and achieve peace.* New York: Henry Holt and Company.

Simons, J. (2002). *Animal rights and the politics of literary representation.* New York: Palgrave Macmillan.

Sinclair, J. R. (2019). Importance of a One Health approach in advancing global health security and the Sustainable Development Goals. *Rev. Sci. Tech.,* 38, 145–154.

Singer, P. (1975). *Animal liberation.* New York: HarperCollins.

Singer, P. (2009). Speciesism and moral status. *Metaphilosophy,* 40, 567–81.

Snacken, S., & Dumortier, E. (Eds.) (2012). *Resisting punitiveness in Europe? Welfare, human rights and democracy.* New York: Routledge.

Sollund, R. (2013). The victimisation of women, children and non-human species through trafficking and trade: Crimes understood through an ecofeminist perspective. In N. South, & A. Brisman (Eds.), *Routledge international handbook of green criminology* (pp. 333–346). New York: Routledge.

Sollund, R. (2021). Nonspeciesist criminology, wildlife trade, and animal victimisation. *Oxford research encyclopedia of criminology and criminal justice,* https://doi.org/10.1093/acrefore/9780190264079.013.608.

Spapens, T. (2016). Invisible victims: The problem of policing environmental crime. In T. Spapens, R. White, & M. Kluin (Eds.), *Environmental crime and its victims* (pp. 239–254). New York: Routledge.

Steiner, G. (2005). *Anthropocentrism and its discontents: The moral status of animals in the history of western philosophy*. Pittsburgh: University of Pittsburgh Press.

Stevenson, M. C., Malik, S. E., Totton, R. R., & Reeves, R. D. (2015). Disgust sensitivity predicts punitive treatment of juvenile sex offenders: The role of empathy, dehumanisation, and fear. *Analyses of Social Issues and Public Policy*, 15(1), 177–197.

Sunstein, C.R., & Nussbaum, M.C. (2004). *Animal rights, current debates and new directions*. Oxford: Oxford University Press.

Susi, M. A. (2017). Empowering animals with fundamental rights – the vulnerability question. *East-West Studies*, 8, 76–90.

Tafalla, M. (2007). La defensa de los animales: Razones para un movimiento moral. *Revista Crítica*, 941, 58–61.

Tarazona, A. M., Ceballos, M. C., & Broom, D. M. (2020). Human relationships with domestic and other animals: One health, one welfare, one biology. *Animals*, 10(1), 43, https://doi.org/10.3390/ani10010043.

Toribio, A. (2020). La explotación sexual de animales y la zoofilia en el código penal español. *Crítica penal y poder: una publicación del Observatorio del Sistema Penal y los Derechos Humanos*, 20. Retrieved from: https://www.raco.cat/index.php/CPyP/article/view/374897 (last accessed 23 January 2022).

United Nations (2020). *Handbook on restorative justice programmes*. Vienna: United Nations Office on Drugs and Crime.

Vander Beken, T., Vandeviver, C., & Daenekindt, S. (2021). Two decades of European criminology: Exploring the conferences of the European Society of Criminology through topic modelling. *European Journal of Criminology*, https://doi.org/10.1177/14773708211007384.

Van Camp, T. (2014). *Victims of violence and restorative practices: Finding a voice*. New York: Routledge.

Van Dijk, J. (2009). Free the victim: A critique of the western conception of victimhood. *International Review of Victimology*, 16(1), 1–33.

Varona, G. (2020a). Restorative pathways after mass environmental victimisation: Walking in the landscapes of past ecocides. *Oñati Socio-Legal Series*, https://doi.org/10.35295/osls.iisl/0000-0000-0000-1044.

Varona, G. (2020b). *Victimidad y violencia medioambiental y contra los animales: Retos de la Victimología verde*. Granada: Comares.

Vázquez, R., & Valencia, Á. (2016). La creciente importancia de los debates antiespecistas en la teoría política contemporánea: del bienestarismo al abolicionismo. *Revista Española de Ciencia Política*, 42, http://orcid.org/0000-000 2-7888-5692.

Vegh Weis, V. V., & White, R. (2020). Environmental victims and climate change activists. In J. Tapley & P. Davies (Eds.), *Victimology: Research, policy and activism* (pp. 301–319). New York: Palgrave Macmillan.

Walters, R. (2019). Green justice. In P. Carlen, & L. A. França (Eds.), *Justice Alternatives* (pp. 42–59). New York: Routledge.

14

Towards Environmental Restorative Justice in South Africa: How to Understand and Address Wildlife Offences

Ashleigh Dore, Annette Hübschle, and Mike Batley

1   Introduction

South Africa is one of the most biodiverse countries in the world. Beyond a multitude of endemic species of fauna and flora, it is home to biodiversity hotspots such as the Cape Floristic Region, the Succulent Karoo and parts of the Maputaland-Pondoland-Albany landscape. Human activities

The research for Ashleigh Dore's contribution is funded through the project implemented by Endangered Wildlife Trust, under the World Wide Fund for Nature South Africa Khetha Programme and supported by the United States Agency for International Development.
The research for Annette Hübschle's contribution to the chapter was funded by the European Research Council (ERC) under the European Union's Horizon 2020 research and innovation program (Grant Agreement No. 804851).
The authors would like to thank Harriet Davies-Mostert for her comments on an earlier draft of the chapter and Diana Berzina for editorial assistance. We are also grateful to participants in roundtable discussions and consultations who shared their experiences and insights on wildlife and conservation 'harmscapes' and harms.

A. Dore (✉)
Endangered Wildlife Trust, Johannesburg, South Africa
e-mail: ashleighd@ewt.org.za

333

such as deforestation, illegal hunting, unsustainable harvesting and over-fishing, pollution and climate change are affecting these species and bio-diversity hotspots as well as the broader socio-ecological landscape (IPBES, 2019). Conserving these natural wonders is of critical importance but does bring a barrage of historical baggage going to the core of conservation paradigms, mentalities and practices.

Like elsewhere in the Global South, South Africa has a painful conservation history that is deeply intertwined with the colonial and apartheid regimes. Indigenous Peoples and Local Communities (IPLCs) lost land, natural resource user rights (including hunting rights), as well as access to cultural and ancestral sites during colonisation. Local people were evicted from their land to make space for the new settler economies, which included the reservation of land and forests that were exclusively earmarked for wildlife conservation or hunting by the colonial elite. Dowie (2009) estimates that more than 14 million Africans were evicted during the colonial period. Protected areas—one of South Africa's key conservation tools—were designed to provide a sanctuary in which certain species of wildlife could prosper, free from all human interference (Carruthers, 1993). Conservation benefits schemes and income were inequitable, privileging economic and political elites. While the powerful and connected strata of colonial society benefited from the conservation economy, IPLCs bore most of the costs and only a few individuals found sustainable employment in or near protected areas. Conservation and nature management became tools for economic and social exclusion of IPLCs who, being grouped in communities, had no land ownership and hence no land rights. Conservation also became a tool for colonial governments to exert administrative control over remote areas (Dlamini, 2020), where livelihood strategies of IPLCs such as hunting or harvesting

A. Hübschle
Global Risk Governance Programme, University of Cape Town,
Cape Town, South Africa
e-mail: Annette.Hubschle@uct.ac.za

M. Batley
Restorative Justice Centre, Pretoria, South Africa
e-mail: mike@rjc.co.za

wildlife were criminalised under the guise of conservation and wildlife protection.

Against this background, this chapter explores the history and legacy of conservation in South Africa first in part 2, before considering wildlife offences and structural and other drivers of wildlife offences in parts 3 and 4. By using the analytical lens of 'harm landscapes' or 'harmscapes' (Berg & Shearing, 2018), we consider the harmscape associated with wildlife offences in part 5 and in part 6 explore restorative justice as an appropriate approach to addressing wildlife offences and the underlying harmscapes that continue to affect South Africa. It is the overall conclusion of this chapter that restorative justice is an appropriate response not only to achieve justice for the victims of individual wildlife offences but also to address broader structural injustices in South Africa.

2   The History and Current Context of Conservation in South Africa

The following section delves into the history of conservation in South Africa. Prior to European colonisation in 1652, the use of wildlife was regulated in several ways: areas were demarcated for specific purposes and the use of wildlife[1] was determined by cultural norms and religious and spiritual beliefs. People were banned from hunting or using totem animals, and valuable products (like elephant ivory and leopard skin) were reserved for rulers in the precolonial era (Dlamini, 2020).

The scales tipped towards overexploitation of the still abundant wildlife shortly after the European colonisers arrived in South Africa. Colonial settlers introduced firearms, agricultural technologies and access to overseas markets (Carruthers, 1988). The early settlers survived through hunting, which served the purposes of land clearance, income generation and provision of meat to avoid having to slaughter their own livestock. MacKenzie (1988, p. 7) argues that the colonial frontier 'was also a hunting frontier and the animal resource contributed to the expansionist urge'

---

[1] Wildlife is used in this chapter in its widest sense and includes wild animals, birds, reptiles and plants.

where hunting was 'ritualised and occasionally a spectacular display of white dominance'. Although local people and colonial settlers both contributed to the decimation of wildlife numbers, colonial regulators stamped local people and their cultural heritage as intrusive and destructive and chose to preserve what was left of the 'wilderness' without local influences (Meskell, 2012, p. 117).

Colonial-era laws and regulations were employed to explicitly criminalise subsistence hunting, an act that was associated with IPLCs who used hunting and fishing as a livelihood strategy. Meanwhile, colonial settlers were free to carry on with hunting as they pleased. The first colonial administrator, Jan van Riebeeck, decreed the first conservation laws delineating hunting restrictions in the area of the Cape colony a mere five years after landing at the Cape of Good Hope in South Africa. According to Roman Dutch law, wild animals had the status of *res nullius*, a legal principle that whoever captured or killed a wild animal, owned it if they were in possession of the right permits. The hunting or harvesting of wild animals would therefore not amount to theft and once captured or killed, the wild animal remained the property of the hunter or captor (Couzens, 2003). However, the objective of van Riebeeck's *Placaat* of 1657[2] was to rein in hunting by Indigenous peoples and slaves. Subsequent laws further delineated who was allowed to hunt and who was proscribed from doing so; annual 'close seasons' were introduced, during which the hunting of certain species was proscribed and the hippopotamus, elephant and bontebok were declared royal game, for which a special hunting permit was required.

Protected areas have been used as a conservation method since the late 1800s (Paterson, 2007, p. 2) and are considered cornerstones of biodiversity conservation and vital to the culture and livelihoods of IPLCs. They also deliver clean air, ecosystem services and benefits through tourism and affiliated industries to millions of people as well as protection from climate change and natural disasters (Dudley & Stolton, 2008). However, seen from a historical perspective, the establishment of protected areas often involved the forced removal and exclusion of Indigenous and local

---

[2] A *placaat* is a proclamation issued at the Cape of Good Hope during the rule of the Dutch East India Company.

communities in South Africa. The enactment of the *Native Land Act* of 1913 left IPLCs with only 10 per cent of the total land area in South Africa (Thwala, 2006). The apartheid era began in 1948, ushering in a period of racial discrimination at a level South Africa had not experienced in its past, lasting for the next 46 years. Apartheid reinforced the division of land approach that had started during the colonial period with a continued policy of exclusion (both in terms of access and of benefit sharing). So-called fortress conservation (Brockington, 2002) was based on the premise that effective conservation within protected areas required the exclusion of IPLCs from that area.

Commencing in the 1960s, the development of wildlife ranching contributed to the commodification and privatisation of wildlife in general; further entrenching property rights of the white elite while depriving black communities of the same. Through legislative changes in 1991, game ranchers were granted ownership over wildlife and the right to derive income from consumptive and non-consumptive utilisation, such as the killing of wild animals for profit (Lindsey et al., 2007, p. 463).

The apartheid regime came to an end in 1994. While apartheid institutions have largely been dismantled, the old approach of fortress conservation continues to permeate conservation practices and protected area management.

The main objective behind the post-apartheid environmental framework legislation was to develop a human-centred approach to conservation. In the aftermath of the first free and fair elections in 1994, a new Constitution cleared the way for the transformation of institutional arrangements, policy frameworks and the apartheid bureaucracy. Environmental rights, with the promotion of conservation and ecological sustainable development and use of natural resources as tools for environmental protection, became enshrined in the new Constitution (Republic of South Africa, 1996, p. 6). In the immediate period following the end of apartheid, several significant events impacted the Department of Nature Conservation, subsequently known in turn as the Department of Environmental Affairs, the Department of Environment, Forestry and Fisheries (DEFF) in 2019 and then the Department of Forestry, Fisheries and the Environment (DFFE) in 2021. The new Constitution opened the floor for the clearing of a store of draconian apartheid laws and institutions relating to all sectors of public and private life. Concurrently, the wildlife ranching, safari and game industries experienced massive growth as the end of apartheid enabled access to previously untapped international markets of hunters and tourists, who had boycotted the country in the past. The new environmental bureaucracy was transformed, with many former public servants from the old regime opting out by accepting retrenchment packages, early retirement or job opportunities in the private sector (Hübschle & Jojo, 2021, p. 18). While the apartheid regime endorsed the notion of 'sustainable use'[3] by creating incentives for white landowners, the new democratic regime developed a legislative framework, the *National Environmental Management Act 107 of 1998* (NEMA), puts greater emphasis on sustainable use linked to community empowerment and social development, as envisaged by the Constitution.

The South African Constitution is exemplary in the world, as it provides constitutional protection of the environment and promotes conservation practices. Section 24 of the Constitution (Republic of South Africa, 1996) requires that the environment be protected for present and future generations through legislation, policies and practices that, amongst other things, promote conservation. NEMA and the specific environmental management Acts specifically, the *National Environmental Biodiversity Act 10* of 2004 (NEMBA) and the *National Environmental Management Protected Areas Act 57* of 2003 (NEMPAA) and their regulations, create the framework legislation within which environmental protection, regulation and management operate. NEMBA, together with the Threatened or Protected Species regulations, specifies a number of restricted activities in relation to listed species, which includes consumptive use. NEMA establishes the Environmental Management Inspectorate (EMI) to enforce the environmental legislation. It contains a list of principles, which apply to all organs of the state, and must be considered in the implementation and enforcement of environmental law.

---

[3] Sustainable use refers to the use of natural resources in a way and at a rate that does not lead to the long-term degradation of the environment, thereby assuring that present and future generations continue to benefit from such resources. The sustainable use paradigm in conservation is the counterpoint to preservationism which allows no or limited use of natural resources.

The provisions comply with international best practices and they are widely regarded as progressive and socially just. However, public enforcement and oversight bodies are chronically underfunded and are sometimes mismanaged (Rademeyer, 2016). This can have negative effects on oversight and accountability, which is particularly concerning in the current environment of increasing employment of paramilitary and military strategies, tactics and military-trained staff in the broader conservation sector.

NEMPAA provides the legal framework for protected areas[4] and determines who has the mandate to protect and enforce compliance within these areas. It also specifically provides for the continued existence of South African National Parks (SANParks) as the management and enforcement authority for national parks. Even though NEMPAA provides for co-management agreements with local communities and landowners, these provisions have not been implemented to the extent it was hoped. Of further importance is that the NEMPAA makes provision for the sustainable utilisation of protected areas for the benefit of people. Thus, natural resources should be accessible to local communities in protected areas, as long as the ecological character of the area is preserved, and the conservation status of area permits such use.

South African laws, especially the national framework legislation, specify communities as key constituencies; however, stewardship programmes have not been fully realised. Essentially the legislation is progressive on paper, but the enforcement mechanisms are contradictory to the NEMA principles and they over-emphasise the command and control approach (Kidd, 2002, p. 24).[5] The enforcement bodies and the implementation plans are also not adhering to the more inclusive conservation approaches mandated by the framework legislation and the

---

[4] At 2021, South Africa has a protected area coverage of 8.69% for terrestrial areas and 15.5% for marine protected areas (UNEP-WCMC, 2021) and with motion 101 passed by majority vote in the 2021 IUCN World Conservation Congress, there is a global effort to effectively and equitably protect and conserve at least 30% of terrestrial areas, inland waters and coastal and marine areas, respectively, by 2030.

[5] The 'command and control' mechanism prescribes the legal requirement and then ensures the compliance through an array of enforcement mechanisms.

Constitution. Meanwhile the government bodies are not capacitated to oversee and implement the framework legislation.

The new South Africa needed to constructively address a 400-year-old legacy of land dispossession and exclusion and, to do so, embarked on a massive land reform programme to redress past inequalities. Two key components of this land reform programme were, and continue to be, land tenure reform and land restitution (Paterson, 2011, p. 15). Land restitution claims settled in protected areas have failed to achieve an equitable balance between conservation and land reform imperatives (Paterson, 2011, p. 15). The South African government has largely relied on a narrow and inflexible approach to co-management with conservation authorities controlling and managing protected areas. IPLCs have received financial compensation for relinquishing access and user rights in some instances. With a few notable exceptions (e.g. the Makuleke land claim in the Kruger National Park), IPLCs remain largely excluded from the management and benefit schemes of protected areas (Paterson, 2011, pp. 15, 17).

In showing the long and often difficult history of conservation in South Africa, it is essential to acknowledge 'the danger of a single story' (Adichie, 2009). Although the establishment of protected areas often meant loss, exclusion and marginalisation for IPLCs, they were not victims of their destiny but learnt to live with different forms of deprivation and exclusion (Dlamini, 2020). As much as the apartheid regime tried to hide IPLCs' heritage and right of belonging in the landscape, IPLC's presence should not be reduced to fulfilling the role of poachers or labourers in conservation narratives. There were many different individual and community experiences of conservation in the landscape. Given the limited literature on indigenous conservation systems, mentalities and practices, several research projects are in the making to capture IPLC history in the landscape (Hübschle & Jojo, 2021, p. 24). Conservation is however not a colonial import but was practised in precolonial times and continues to be promoted today, as envisaged by the South African Constitution.

3   Wildlife Offences

While the South African government has made progress, implementation of socially just conservation programming has been slow due to both internal and external constraints. South Africa has a solid legislative framework, but more work needs to be done to affect implementation. As an example, the South African government statistics on the number of conservation crime prosecutions is used to indicate heightened conservation agency guardianship. A proactive approach would gauge high levels of voluntary compliance, which, according to Herbig (2008), would be a better indicator of success. The command-and-control approach also provides little incentive for local communities to protect the environment. Criminal measures and criminal sanctions are forms of punishment, yet they do not encourage positive action. Civil and administrative measures focus on compelling persons to cease the harmful activity and to take measures to stop, prevent, remediate or mitigate the harm. Traditionally, environmental authorities have relied almost exclusively on criminal measures to compel compliance with wildlife and marine law contexts. The command-and- control approach requires well-resourced and capacitated enforcement authorities to be effective because the control functions are time-consuming and expensive. These mechanisms are also inflexible in that they do not allow discretion to tailor compliance to suit specific situations (Craigie et al., 2009). Recognising the weakness of the current command-and-control approach, this chapter argues that restorative justice, discussed in part 6, enables a more inclusive, flexible and people- and harm-centred response to wildlife offences.

Wildlife offences are considered one of the leading threats to South Africa's wildlife. A wildlife offence can be defined as '*the taking, trading (supplying, selling or trafficking), importing, exporting, processing, possessing, obtaining and consumption of wild fauna and flora, including timber and other forest products, in contravention of national or international law*'.[6] Wildlife offences include illegal wildlife hunting, which refers to the harvesting, catching, extraction or killing of wildlife that is not authorised by the state or private owners of wildlife (commonly referred to as poaching).

---

[6] CITES, 'Wildlife crime', https://cites.org/eng/prog/iccwc/crime.php.

The illegal killing of rhinoceros is a well-documented wildlife offence which does not only extend to the killing and dehorning of a rhino/rhinos but may include trespassing on private or public land, possession and use of a hunting rifle on protected land and the transport without valid permits of rhino body parts.

It is not only South Africa's megafauna that are threated. In fact, due to the diversity of wildlife offences taking place in South Africa, one can make the distinction between syndicated wildlife offences (those offences motivated or facilitated by organised criminal syndicates, operating internationally or nationally within South Africa) and non-syndicated wildlife offences (offences not perpetrated by organised criminal networks, including, for example, snaring for personal consumption or retaliatory killing of wildlife due to human wildlife conflict) (Hübschle et al., 2021, p. 147). What must also be noted is that wildlife offences (both syndicated and non-syndicated) seldom take place in isolation and are often accompanied by offences relating to interpersonal violence and animal welfare.[7]

Responses to disrupt rhino poaching have been likened to fighting a 'war on poaching' (Duffy, 2014) as military and paramilitary personnel, training, technologies and partnerships are used in the pursuit of conservation efforts. Several hundred poaching suspects have been killed on

---

[7] This legislative framework includes:

* For wildlife offences: National Environmental Management Biodiversity Act 10 of 2004, read with the Convention on International Trade in Endangered Species Of Wild Fauna and Flora Regulations published in Government Notice R173 in Government Gazette 33002 and the Threatened or Protected Species Regulations published in Government Notice R152 in Government Gazette 29657 (read with Lists of Critically Endangered, Endangered, Vulnerable and Protected Species, published in Government Notice R151 in Government Gazette No 29657), Marine Living Resources Act 18 of 1998 (read with its regulations) and provincial legislation that includes but is not limited to: Cape Nature and Environmental Conservation Ordinance 19 of 1974, Nature Conservation Ordinance 8 of 1969 (Free State), Nature Conservation Ordinance 12 of 1983 (Gauteng), Nature Conservation Ordinance 15 of 1974 (KwaZulu-Natal), Kwa-Zulu Nature Conservation Act 29 of 1992, Limpopo Environmental Management Act 7 of 2003, Mpumalanga Nature Conservation Act 10 of 1998, Northern Cape Nature Conservation Act 9 of 2009, North West Transvaal Nature Conservation Ordinance 12 of 1983.
* For offences relating to interpersonal violence and organised crime: Criminal Procedure Act 51 of 1977, Firearms Control Act 60 of 2000 and Prevention of Organised Crime Act 121 of 1998.
* For animal cruelty offences: Animals Protection Act 71 of 1962.

private and public conservation land since 2010 (Hübschle, 2016). Why then, at the risk of being killed, do people still participate in wildlife offences?

4 Pathways to Poaching and the Notion of Contested Illegality

Having introduced wildlife offences above, Duffy et al. (2016) point to the importance of context when considering illegal hunting: the shooting of wildlife may well be illegal in a protected area but once a wild animal breaks loose and crosses into communal or private land, its killing might not only be legal but life-saving. This ties in with the notion of 'contested illegality' developed by Hübschle (2016, 2017b) during the course of her fieldwork on the illegal rhino horn economy. While interviewing people who were engaging in illegal economic activities linked to the illegal wildlife trade, it became clear that many did not accept the label of illegality imposed upon their activities. Rural hunters legitimised bushmeat poaching in protected areas by putting forth cultural and economic justice reasons for their offences. In their eyes, hunting was as much a rite of passage as it was an expression of cultural conventions, practices and traditions, which are however often not recognised in statute books and Western societies. Others rejected outright the label of illegality, arguing that hunting was not an illegal activity, as they were hunting on land that used to belong to them but that had been taken away by colonial settlers, the state or private investors. Game farmers and wildlife professionals legitimised breaking the law or exploiting regulatory loopholes by criticising the rule-makers who were seen as illegitimate rulers overreaching their remit. They also critiqued the rules per se in terms of fairness and impact on conservation and their own economic objectives. Meanwhile traders and consumers of illegally harvested wildlife were often aware of the illegality attaching to their activities but did not accept prohibition due to cultural and social legitimacy of wildlife consumption and use. It was thus socially acceptable to break the rules. Scholars have long looked at the interface between legality and illegality and the blurred boundaries between what is considered legal or illegal, licit or illicit, and legitimate or illegitimate (Heyman, 2013; Heyman & Smart, 1999; Van Schendel & Abraham, 2005; Hall, 2013). Moreover, these boundaries while situated in time and space are also geographically, socio-politically, economically and culturally mediated and the processes of delimiting these boundaries differentially impact social groups along axes of race, gender, caste and class. In fact, the production and regulation of these boundaries affect and transform the (natural) environments (and non-human actors) that mediate their possibilities as licit or illicit activities.

Contested illegality of traditional hunting activities aside, there is a variety of reasons why local people deliberately engage in what is undisputed illegal wildlife hunting, trade and trafficking. These reasons include: The lack of sanctions and penalties for illegal activities and hence limited disincentives not to engage in poaching, trafficking or trade; limited benefits from wildlife stewardship and hence limited incentive to conserve wildlife rather than poach it; human-wildlife conflict and the associated high cost of living with wildlife and/or near protected areas resulting in resentment towards conservation and motivation to undermine it; and limited alternative ways to make a living and meet basic needs (Roe & Booker, 2019).

Protected areas and environmental policies under the fortress conservation paradigm define exclusionary rights which grant tourists and scientists (powerful and influential actors) access whereas the original inhabitants of the area are stamped as intruders. Once livelihood strategies that are reliant on protected natural resources are diminished, local people must seek out alternatives. These alternatives often do not align with the existing skillsets of IPLCs who have been historically and traditionally engaging in certain forms of livelihoods. Such alternatives may not always be viable—for example, farmers might have to deal with predators attacking livestock or wildlife crop-raiding fields. Fishers may have to expand their fishing operations to more dangerous maritime zones or move elsewhere. The loss of livelihoods is often cited as a motivation to join illegal poaching, logging or fishing operations. However, we must be careful of monocausal fallacies, such as the hypothesis that poverty leads to poaching—the so-called conservation poverty hypothesis. Especially in South Africa, members of wealthy farming communities, wildlife veterinarians and professional hunters have been involved in rhino poaching networks (Hübschle, 2017a, 2019), contradicting the widespread notion that poor rural residents are driving rhino extinction.

Some poor rural residents support or participate in wildlife poaching because of feelings of frustration and anger triggered by historic and current exclusion from protected areas and their benefit schemes. However, the incentives for local people to participate in illegal wildlife economies often outweigh benefits from legal equivalents and alternative livelihood strategies. In such cases, the profits linked to illegal activities may be greater than legal alternatives or the profits exceed what is earned from existing conservation benefit and rural development schemes (Roe & Booker, 2019). As an example, a single poaching event involving rhino can earn a rhino poacher more than the average annual income of rural citizens in southern Africa (Hübschle, 2016).

Scholars have started to acknowledge the historical context of land expropriation, loss of natural resource use rights, contested illegality as well as the forced removals during colonial times to explain why IPLCs may support or engage in poaching economies in southern Africa (Hübschle, 2016, 2017b; Hübschle & Shearing, 2018; Moneron et al., 2020). Beyond poaching for the 'cooking pot and pocket book' (Kahler & Gore, 2012), convicted poachers in South Africa cited feelings of stress, disempowerment, anger, peer pressure and emasculation leading to poaching decisions. While younger poachers (late teens to late twenties) espoused anomic and individualistic desires, older offenders wanted to take care of their families and the community (Hübschle, 2017b). Some convicted wildlife offenders wanted to achieve social upward mobility and used illegal hunting to attain political influence or provide social welfare to community members. Structural violence, the generational pain of dispossession and marginalisation played a facilitating milieu (Hübschle & Shearing, 2018, p. 32), while unhappiness with rule-makers and the perceived illegitimacy of the rules (contested illegality) highlighted the distrust of past and present state authority. A study undertaken amongst convicted wildlife criminals in Namibia found that some were driven by curiosity as they were previously unaware of the species and wanted to find out more about it (Prinsloo et al., 2021). Moneron et al. (2020, p. 26) categorised influencing factors that led to the commission of wildlife offences among convicted wildlife criminals in South Africa neatly into individual, community and societal factors. In addition to the earlier factors, the study identified a skewed perception of risk and the provision of employment to others as key individual drivers while opportunism and peer pressure were added to the list of community drivers.

5 The Harm Landscape Associated with Wildlife Offences in South Africa

Having contextualised wildlife offences as experienced in South Africa and why these offences are committed, this part considers the harm experienced as a result. Berg and Shearing (2018) coined the concept of 'harmscapes' (harm landscapes) to capture the notion that contemporary risks and associated harms require a departure from traditional crime and justice models. Contemporary harmscapes are characterised by both radical uncertainty and unpredictability (Mutongwizo et al., 2021, p. 2). We loosely apply the concept of harmscape in the context of wildlife offences to demonstrate how harms resulting from the exclusionary legacy of conservation, the commission of wildlife offences and responses thereto are multifaceted, inter- and intragenerational, even circular.

The graphic violence that goes hand in hand with the commission of some wildlife offences (most notably rhino poaching) is often cited as a leading cause for the escalation in enforcement responses. In South Africa's Kruger National Park, 90 per cent of a ranger's duties used to relate to conservation-related tasks and 10 per cent of their time would be dedicated to enforcement activities. With the onset of the latest rhino poaching crisis in the late 2000s (as discussed in more detail below), rangers started dedicating more of their time to anti-poaching related duties while conservation duties took up less of their time (Hübschle & Jooste, 2017). This change in priorities was in response to the real problem of rhino poaching but could also be attributed to a more militarised approach, as more military and private security professionals were placed at the helm in the Greater Kruger landscape, an area which also includes private reserves and commercial farms (Annecke & Masubelele, 2016). The Endangered Wildlife Trust, under the World Wide Fund for Nature (WWF) South Africa Khetha Programme and supported by the United States Agency for International Development (USAID), launched a pilot project in August 2019, seeking to apply restorative justice approaches to wildlife offences in South Africa (The Restorative Justice Pilot Project). In recent engagements with stakeholders in and around the Kruger National Park, conducted under the Restorative Justice Pilot Project, two of the authors were able to qualify the wide degree of harm resulting from wildlife offences. A key take-home message is that wildlife offences are not 'victimless'. Victims can be divided into three board categories of people, wildlife and society as a whole.

The people victimised by wildlife offences are diverse and include private wildlife owners, rangers, law enforcement officials, community members and broader South African society. The harm suffered by people includes physical harm and intimidation (threats of personal violence, intimidation and death threats made against community members, rangers and private wildlife owners), harm due to trauma (including the mental health impact of living in a state of perpetual conflict), and financial and material losses to owners of wildlife (this includes the loss of or injury to the animal itself due to poaching but also the costs of fence repair, protection services and veterinary care). These harms occur as a direct result of the wildlife offence itself but there is also harm that occurs because of the offence being committed, as illustrated by the following example.

When poachers trespass into a protected area, they usually have to cut fence lines (South Africa utilises a fence system to establish ownership over wildlife and to prevent disease spreading from wildlife to domestic stock and vice versa). Once cut, these downed fences allow wildlife to migrate between protected areas, commercial farms, villages and towns, posing danger and harm to lives and livelihoods. Linked to this harm is the potential for greater conflict between communities and protected areas due to increased (and often unresolved) human-wildlife conflict. Other forms of associated harm include the spread of zoonoses and infectious wildlife diseases. Finally, the harm of living near individuals associated with organised criminal syndicates, increased exposure to criminal behaviour and social harm (such as gambling and illicit drug use) needs to be considered.

The second category relates to harms to wildlife and the environment. To illustrate this harm, we use the current rhinoceros poaching crisis as an example. Over the period 2011 to 2020 there were a confirmed 8349 rhinos poached in South Africa (DFFE, 2021, p. 63). Kruger National Park used to be home to the greatest number of rhinos in the world; towards the end of 2021, less than 3000 rhinos survive there (Interview with KNP conservationist by one of the authors, 2021). First responders provide accounts of traumatic crime scenes with rhinos on rare occasions surviving poaching events. The harm to the animal also extends to the calves who may not be poached themselves but are left orphaned and unborn embryos that are not carried to term—the gestation period for rhinos is 16 months.

Beyond syndicated wildlife offences, non-syndicated offences such as snaring also result in widespread harm to wildlife and the broader ecosystem. This hunting method, prohibited in South Africa, is non-discriminatory, which means any animal who runs through the snare line is impacted. Snaring has resulted in the illegal killing of critically endangered, endangered and threatened wildlife species. This form of hunting is considered cruel and unethical as the trapped animal often takes hours to strangle to death (if the neck is caught), gets attacked by predators and/or scavengers or bites off its own legs to free itself.

In addition to harm to the individual animal, there is also wider harm to the environment that must be considered when wildlife offences occur. The poaching of keystone species (such as rhino and elephant) removes the ability of these individuals to support 'habitat architecture' (Van de Water et al., 2021, in preparation). Further, the environment plays a key role in informing climate change adaptation and mitigation by providing ecosystem services. When wildlife numbers are severely impacted (or species go extinct), the ability of the environment to fulfil these functions is jeopardised (ibid.).

The third victim category is society as a whole. A recent paper has qualified 16 categories of benefits from elephant conservation specifically that illustrate how society suffers harm as a result of wildlife offences (ibid.). These benefits include: employment opportunities, spiritual and cultural benefits, social benefits, the facilitation of learning and inspiration, provision of an intergenerational legacy and the facilitation of human wellbeing (both physical and psychological) (ibid.). The inverse of these benefits is the harm that society faces in their loss. Elephants across their range are described as decreasing and have gone locally extinct in two countries already; when they go extinct in an area, these benefits go with them (Gobush et al., 2021). While the example focuses on elephants, it is submitted that this is equally applicable to other keystone species.

The harm landscape associated with wildlife offences is multifaceted, personal and intergenerational, creating great volumes of victims. It is important to note the complex motivations driving individuals and communities to support wildlife poaching economies and the associated harmscape. Enforcement responses, including green militarisation, have added another layer of harm to the harmscape. To escape the cycle of violence, harms and trauma, the following section will consider how a restorative justice process can provide a future-oriented alternative to retributive justice.

6 The Usefulness of Restorative Justice Approaches to Wildlife Offences

'The inefficiency of our criminal justice system, and disturbing overcrowding and lack of rehabilitation in our prisons can truly be termed a crisis' (Cameron, 2020, p. 52). These are the words of Justice Cameron, a retired justice of the Constitutional Court of South Africa following his investigation into the crisis of criminal justice in South Africa. As with many jurisdictions globally, South Africa relies on a predominately retributive justice system. Cameron's research shows the progression of retribution in the justice system during the apartheid regime and into the new democratic era of South Africa. Under the apartheid system prisons were referred to as 'universities of crime' and the death penalty was used regularly (at its height more than three times a week) (Cameron, 2020, p. 33). In the late 1980s, South Africa began reconsidering its approach to crime and punishment, starting with a shift in mandate to corrections and rehabilitation, to a flurry of policy documents in the mid-to late nineties advocating a restorative approach (Cameron, 2020, p. 33; Batley, 2005, pp. 120–126). Unfortunately, this was not to become an entrenched approach in the new South Africa. In the late 1990s, due to an escalation in crime across South Africa, a 'tough on crime' approach was adopted (Cameron, 2020, p. 39). War terminology became common parlance (similar to the green militarisation discussed above), with the former Minister of Safety and Security stating, 'criminals have obviously declared war against the South African public' (ibid.). The consequences of the tough on crime approach include prison overcrowding: When Cameron (2020, pp. 41–42) was writing his report in 2020, there were over 160,000 prisoners in South Africa with more than 18,000 prisoners serving life sentences. South Africa not only has one of the highest rates of incarceration but also one of the highest recidivism rates in the world (estimated range from between 60–90 per cent). Cameron (2020, p. 45) concluded that:

> [a]fter formal apartheid, the alleged 'crime wave' led to our dumping a restorative justice approach and embracing a punitive one. The 'tough on crime' approach led to an increase in the prison population—but without a matching increase in infrastructural and institutional capacity, and with utterly no benefit regarding crime. The result has been overcrowding. We are essentially shovelling prisoners into a funnel and they are not being rehabilitated.

The solutions proposed by Cameron include adopting a restorative justice approach that gives victims an increased role in criminal proceedings (although Cameron limited this to sentencing) (Cameron, 2020, p. 68). Restorative justice, although not the dominant approach, has found application in South Africa through legislation, court judgments and sentencing frameworks, specifically holding that sentences must 'consider the interests of victims' (including improved provision for victim involvement) and with the same framework proposing a sentence of reparation (including elements of restitution and compensation)

(Hübschle et al., 2021, p. 142; Cameron, 2020, p. 69).[8] Globally and in South Africa, there are calls to address the harm that the environment suffers in environmental crime. Compensation orders have been available for environmental offences since 1998 through the inclusion of section 34(1) of the NEMA, which provides:

> Whenever any person is convicted of an offence under any provision listed in Schedule 3 and it appears that such person has by that offence caused loss or damage to any organ of state or other person, including the cost incurred or likely to be incurred by an organ of state in rehabilitating the environment or preventing damage to the environment, the court may in the same proceedings at the written request of the Minister or other organ of state or other person concerned, and in the presence of the convicted person, inquire summarily and without pleadings into the amount of the loss or damage so caused. (Kidd, 2003, p. 58)

There are other similar provisions found in South African legislation relating to water, forests and heritage and these are important because they are aimed at remediation of environmental damage (Kidd, 2003, p. 58). Certain legislation in South Africa, such as the *National Heritage Resources Act 25* of 1999, also provides for reparation orders, requiring the offenders to carry out the reparation him or herself (Kidd, 2003, p. 59). Therefore, the legislative framework providing for reparation exists within the environmental legal framework and by applying this, with the broader view of restorative justice and together with the principles of engagement, empowerment of all participants through dialogue and problem-solving, a more appropriate, effective and enduring resolution to wildlife offences can be found (Kershen, 2019, pp. 47–50).

As has been well documented in the restorative justice literature, the conventional Western paradigms for approaching crime—whether of retribution for its own sake or in the belief that it will deter the individual or other potential offenders from committing crime, and of

---

[8] The South African Law Reform Commission in their discussion paper 'A New Sentencing Framework' (2000) provided that sentences must 'consider the interests of victims, victim impact statements, informing victims of the release or sentence of their accused and other mechanisms' (Cameron, 2020, p. 69).

rehabilitation, have clearly become inadequate for responding to crime generally, but all the more so in cases of environmental (and specifically wildlife) offences. It is submitted that this is the case because these approaches are too narrow, and that by focusing only on the offence they fail to properly include the voice of the victims and the broader community, they fail to properly consider harm and (in the case of the retributive paradigm) they rely excessively on imprisonment and fines as the predominate sanctions. Seeking to address this problem, the Restorative Justice Pilot Project was launched in South Africa to explore the application of restorative justice approaches to wildlife offences (both syndicated and non-syndicated as discussed above). By applying restorative justice in this context, the Restorative Justice Pilot Project seeks to confirm who the victims of environmental harm are, what harms have been suffered, and who should have a voice in restorative processes. This is illustrated by the following case example:

> Three people (under the age of 18) are apprehended in a protected area with dogs. Dog hunting is a prohibited form of hunting in South Africa but has a long history as a hunting practice by local communities. The dogs were confiscated and euthanised and the offenders charged with trespassing.

The inadequacies of the retributive approach become clear with this case study: the harm (direct and associated) that occurs due to dog hunting is not addressed (specifically as the charges laid do not speak to this offence at all yet the dogs were confiscated), neither the victim of the offence nor community leadership are afforded an opportunity to participate in the proceedings, there is no opportunity to address the underlying reasons for the offence and there is no opportunity to engage on dog hunting specifically (thus perpetuating sentiments of contested illegality). A restorative justice approach would have addressed all of these elements.

Additional benefits of the application of restorative justice to respond to wildlife offences include identifying who and on what basis can speak on behalf of future or past generations and other non-human victims (animals, plants, rivers, land, places, etc.) and facilitating their involvement in the restorative justice process. Finally, and central to the aims of environmental law, as restorative justice facilitates reparation of harm, particular focus will be on the degree of environmental harm that may be repaired and restoration of the environment that may be facilitated through restorative justice processes.

Given the historical context relating to conservation and with due consideration to socio-economic realities in South Africa, at the launch of the Restorative Justice Pilot Project, the project team considered the most appropriate conceptual framing for the project. Drawing on the three conceptions of restorative justice of encounter, reparation and transformation articulated by Johnstone and Van Ness (2007, pp. 9–16), three frameworks were identified for the project. The first is based solely on the transformative conception, the second seeks to make the criminal justice system more effective and responsive and the third is a combination of the first two, addressing both dimensions by seeing restorative justice as a social movement (Hübschle et al., 2021, pp. 145–146). Within this third framing, consideration was given to the works of Stauffer (2015) and Henkeman (2012) who held that by concentrating on restorative justice as interpersonal interactions, there has been a tendency to overlook structural injustice. Structural injustice is defined as, 'Disparities, disabilities and deaths result when systems, institutions, policies or cultural beliefs meet some people's needs and human rights at the expense of others' (Schirch, 2004, p. 8). Writers within the peacebuilding tradition have articulated the connection between structural injustice and personal, community and national destruction (Schirch, 2004, pp. 9–24). This connection can be responded to with the understanding of conflict transformation, which creates 'constructive change processes that reduce violence, increase justice in direct interaction and social structures and responds to real-life problems in human relationships' (Lederach, 2003, p. 14). Our view is that a response to wildlife crime involves an acknowledgement of the immediate and historical context in which the offence has occurred. The constructive change processes referred to in the definition would typically involve creating opportunities for acknowledgement and validation of historical harm and for problem-solving that addresses the deep-rooted issues local communities adjacent to protected areas grapple with. Ultimately, under the Restorative Justice Pilot Project in its initial phases, restorative justice approaches will be applied to individual harms, while at the same time the project team will address the three core tasks identified by Stauffer (2015) to enable restorative justice to develop as a social movement. These are the formation of institutional alliances, the development of strong localised practice and finally, creating mechanisms for collaboration that drive transformation of the system as a whole (Stauffer, 2015). It is our considered view that restorative justice approaches provide the depth and breadth necessary for an appropriate response to harms at the macro and micro levels of a landscape.

7   Conclusion

This chapter has provided context to the history and legacy of conservation in South Africa, with specific focus given to the drivers of wildlife offences. Consideration was then given to the harm resulting from these offences, through the analytical lens of harmscapes. The chapter has also highlighted the impact of the 'war on X' campaigns with specific reference to the war on poaching and the war on crime, calling for a more holistic, inclusive, harm- and people- centred approach to justice. For us an approach anchored in restorative justice presents a viable alternative. The chapter confirmed that in addition to creating a more enabling and responsive justice system, restorative justice has the potential to address persistent structural injustices. Ultimately, we have shown that restorative justice is an appropriate approach to addressing wildlife offences and the underlying harmscapes that continue to affect South Africa. We conclude with the following quote '[t]hus, when any environmental issue is pursued to its origins, it reveals an inescapable truth—that the root cause of the crisis is not to be found in how [people] interact with nature, but in how they interact with each other—that to solve the environmental crisis we must solve the problems of poverty, racial injustice and war' (Paterson, 2011, p. ii).

References

Adichie, C. N. (2009). The danger of a single story. *TEDGlobal*. Retrieved from: https://www.ted.com/talks/chimamanda_ngozi_adichie_the_danger_of_a_single_story (last accessed 8 July 2021).

Annecke, W., & Masubelele, M. (2016). A review of the impact of militarisation: The case of rhino poaching in Kruger National Park, South Africa. *Conservation and Society*, 14(3), 195–204.

Batley, M. (2005). Outline of relevant policies. In T. Maepa (Ed.), *Beyond retribution: Prospects for restorative justice in South Africa* (pp. 120–126). Pretoria: Institute for Security Studies.

Berg, J., & Shearing, C. (2018). Governing- through-harm and public goods policing. *The Annals of the American Academy of Political and Social Science*, 679, 72–85.

Brockington, D. (2002). *Fortress conservation: The preservation of the Mkomazi Game Reserve, Tanzania*. Bloomington: Indiana University Press.

Cameron, E. (2020). The crisis of criminal justice in South Africa. *South African Law Journal*, 137 (Part 1), 1–183.

Carruthers, J. (1988) *Game protection in the Transvaal 1846 to 1926*. Cape Town: University of Cape Town.

Carruthers, J. (1993). 'Police boys' and poachers: Africans, wildlife protection and national parks, the Transvaal 1902 to 1950. *Koedoe*, 36(2), 11–22.

Couzens, E. (2003). The influence of English poaching laws on South African poaching laws. Retrieved from: http://reference.sabinet.co.za/webx/access/electronic_journals/funda/funda_n9_a4.pdf (last accessed 20 July 2021).

Craigie, F., Snijman, P., & Fourie, M. (2009). Environmental compliance and enforcement Institutions In A. Paterson, & L. J. Kotzé (Eds.), *Environmental compliance and enforcement in South Africa: Legal perspectives* (pp. 72–85). Cape Town: Juta Law.

Department of Forestry, Fisheries and the Environment. (2021). National Environmental Compliance & Enforcement Report 2020-21. Retrieved from https://www.dffe.gov.za/sites/default/files/reports/necer2020.2021report.pdf (last accessed 8 July 2022).

Dlamini, J. S. T. (2020). *Safari nation: A social history of the Kruger National Park*. Athens: Ohio University Press.

Dowie, M. (2009). *Conservation refugees: The hundred-year conflict between global conservation and native peoples*. Massachussetts: Massachusetts Institute of Technology.

Dudley, N., & Stolton, S. (2008). Defining protected areas: An international conference in Almeria, Spain. Gland, Switzerland: IUCN. Retrieved from: https://portals.iucn.org/library/sites/library/files/documents/2008-106.pdf (last accessed: 30 September 2021).

Duffy, R. (2014). Waging a war to save biodiversity: The rise of militarised conservation. *International Affairs*, 90(4), 819–834.

Duffy, R., St John, F. A. V., Büscher, B., & Brockington, D. (2016). Toward a new understanding of the links between poverty and illegal wildlife hunting. *Conservation Biology*, 30, 14–22.

Gobush, K.S., Edwards, C.T.T, Balfour, D., Wittemyer, G., Maisels, F. & Taylor, R.D. (2021). Loxodonta africana. The IUCN red list of threatened species 2021. Retrieved from: https://doi.org/10.2305/IUCN.UK.2021-1.RLTS. T181008073A181022663.en (last accessed 1 October 2021).

Hall, T. (2013). Geographies of the illicit: Globalisation and organised crime. *Progress in Human Geography*, 37(3), 366–385.

Henkeman, S. (2012). Restorative justice as a tool for peacebuilding: A South African study. Durban: University of KwaZulu-Natal. Retrieved from: http://hdl.handle.net/10413/11840 (last accessed 21 February 2021).

Herbig, F. J. W. (2008). Conservation crime: South African concerns and considerations from a criminological perspective. *Acta Criminologica: Southern African Journal of Criminology*, 21(3), 52–64.

Heyman, J. M. (2013). The study of illegality and legality: Which way forward? *PoLAR: Political and Legal Anthropology Review*, 36(2), 304–307.

Heyman, J. M., & Smart, A. (1999). States and illegal practices: An overview. In J. M. Heyman (Ed.), *States and illegal practices* (pp. 1–24). Oxford: Berg

Hübschle, A. (2016). A game of horns: Transnational flows of rhino horn. Cologne: International Max Planck Research School on the Social and Political Constitution of the Economy.

Hübschle, A. (2017a). The Groenewald criminal network: Background, legislative loopholes and recommendations. Bogotá, Colombia: The Global Observatory of Transnational Criminal Networks.

Hübschle, A. (2017b). Contested illegality: Processing the trade prohibition of rhino horn. In J. Beckert, & M. Dewey (Eds.), *The architecture of illegal markets* (pp. 177–197). Oxford: Oxford University Press.

Hübschle, A. (2019). Fluid interfaces between flows of rhino horn. In A. Amicelle, K. Côté-Boucher, B. Dupont, M. Mulone, C. Shearing, & S. Tanner (Eds.), *The policing of flows: Challenging contemporary criminology* (pp. 198–217). London: Routledge.

Hübschle, A., Dore, A., & Davies-Mostert, H. (2021). Focus on victims and the community: applying restorative justice principles to wildlife crime offences in South Africa. *The International Journal of Restorative Justice*, 4(1), 141–150.

Hübschle, A., & Jojo, B. (2021). *Community and community practitioners' attitudes, perspectives and perceptions of protected areas, conservation and community crime in the context of illegal wildlife trade.* White River, South Africa: USAID & WWF Khetha.

Hübschle, A., & Jooste, J. (2017). On the record: Interview with major general Johan Jooste (Retired), South African National Parks, Head of Special Projects. *South African Crime Quarterly*, 60 (June), 61–68.

Hübschle, A., & Shearing, C. (2018). *Ending wildlife trafficking: Local communities as change agents.* Retrieved from: https://globalinitiative.net/analysis/ending-wildlife-trafficking/ (last accessed 22 January 2022).

IPBES (2019). *Global assessment report on biodiversity and ecosystem services of the Intergovernmental Science-Policy Platform on Biodiversity and Ecosystem Services.* Bonn, Germany: IPBES Secretariat.

Johnstone, G., & Van Ness, D. (2007) *Handbook on restorative justice.* Portland: Willan.

Kahler, J. S., & Gore, M. L. (2012). Beyond the cooking pot and pocket book: Factors influencing noncompliance with wildlife poaching rules. *International Journal of Comparative and Applied Criminal Justice*, 36(2), 103–120.

Kidd, M. (2002). Alternatives to the criminal sanction in the enforcement of environmental law. *South African Journal of Environmental Law and Policy*, 9(1), 21–50.

Kidd, M. (2003). Sentencing environmental crimes. *South African Journal of Environmental Law and Policy*, 11, 53–79.

Kershen, L. (2019). Restorative approaches to environmental harm. In E. Biffi & B. Pali (Eds.), Environmental justice restoring the future: Towards a restorative environmental justice praxis (pp. 47–50). Leuven: European Union Forum for Restorative Justice.

Lederach, J. P. (2003). *The little book of conflict transformation.* New York: Good Books.

Lindsey, P. A., Roulet, P. A., & Romañach, S. S. (2007). Economic and conservation significance of the trophy hunting industry in sub-Saharan Africa. *Biological Conservation*, 134(4), 455–469.

Mackenzie, J. M. (1988). *The empire of nature: Hunting, conservation and British imperialism.* Manchester: Manchester University Press.

Meskell, L. (2012). *The nature of heritage: The new South Africa*. Oxford: Wiley-Blackwell.

Moneron, S., Armstrong, A., Newton, D. (2020). *The people beyond the poaching*. Cambridge, UK: TRAFFIC.

Mutongwizo, T., Holley, C., Shearing, C. D., & Simpson, N. P. (2021). Resilience policing: An emerging response to shifting harm landscapes and reshaping community policing. *Policing: A Journal of Policy and Practice*, 15(1), 606–621.

Paterson, A. (2011). *Bridging the gap between conservation and land reform: Communally-conserved areas as a tool for managing South Africa's natural commons*. Cape Town: University of Cape Town. Retrieved from: https://open.uct.ac.za/handle/11427/11498?show=full page 1 (last accessed 22 January 2022).

Paterson, A.R. (2007). Wandering about South Africa's new protected areas regime. *SA Public Law*, 1, 1–33.

Prinsloo, D., Riley-Smith, S., & Newton, D. (2021). *Trading years for wildlife: An investigation into wildlife crime from the perspectives of offenders in Namibia*. Cambridge, United Kingdom: TRAFFIC. Retrieved from: https://www.traffic.org/site/assets/files/13405/namibia-offender-survey-2021-web.pdf (last accessed 9 March 2021).

Rademeyer, J. (2016). *Tipping point: Transnational organised crime and the 'war' on poaching*. Geneva: Global Initiative against Transnational Organized Crime.

Republic of South Africa (1996). Constitution of the Republic of South Africa. Republic of South Africa. Retrieved from: http://www.thehda.co.za/uploads/images/unpan005172.pdf (last accessed 25 October 2014).

Roe, D., & Booker, F. (2019). Engaging local communities in tackling illegal wildlife trade: A synthesis of approaches and lessons for best practice. *Conservation Science and Practice*, 1(5), e26. https://doi.org/10.1111/csp2.26.

Schirch, L. (2004). *The little book of strategic peacebuilding*. New York: Good Books.

Stauffer, C. (2015). Formative Mennonite mythmaking in peacebuilding and restorative justice. In A. Klager & M. Gopin (Eds.), *From suffering to solidarity: The historical seeds of Mennonite interreligious, interethnic and international peacebuilding* (pp. 140–160). Eugene: Pickwick Publishers.

Thwala, W. (2006). Land and agrarian reform in South Africa. In P. Rosset, R. Patel, & M. Courville (Eds.), *Promised land: Competing visions of agrarian reform* (pp. 57–72). Oakland, California: Food First Books.

UNEP-WCMC (2021). Protected area profile for South Africa from the world database of protected areas, September 2021. Retrieved from: www.protectedplanet.net (last accessed 1 October 2021)

Van de Water, A., Henley, M., Bates, L., & Slotow, R. The value of elephants: a pluralist approach. *Ecosystem Services* (In Preparation).

Van Schendel, W., & Abraham, I. (2005). Introduction: The making of illicitness. In W. Van Schendel, & I. Abraham (Eds.), *Illicit flows and criminal things: States, borders, and the other side of globalisation* (pp. 1–37). Bloomington: Indiana University Press.

15

Exploring Environmental Restorative Philosophy for Victims: The Pollution and Life-World in Minamata, Japan

Orika Komatsubara

1   Introduction

Is restorative justice an attractive idea for victims of environmental harm? If the victims desire the restoration of their destroyed nature and communities, where does this desire originate from? Participants' spontaneity is vital for restorative justice, and it should not be externally imposed. This means that a restorative philosophy is required to emerge from within communities, rooted in the actual place that has been environmentally harmed. In this chapter, I will use case studies to illustrate the restorative philosophies developed by victims of environmental harm.

I focus on the severe case of Minamata disease (MD), which arose as a consequence of pollution in Japan (see also Komatsubara, 2021b). MD is a neurological disease that can cause death in acute cases. It originated in Minamata, when a large company, the Chisso corporation (hereafter Chisso), discharged factory wastewater containing methyl mercury into the sea. In 1956, the government first recognised MD and its victims.

O. Komatsubara (✉)
Japan Society for the Promotion of Science, Tokyo, Japan

And, in 1968, the government officially acknowledged the pollution that caused it. In 1969, the first group of victims filed a lawsuit against Chisso, seeking compensation for all the suffering they had endured due to MD, which they won in 1973. Later, a second and third group of victims also took Chisso to court. Since then, several compensation schemes for MD victims have been established. By 2020, 22,182 people had applied to be officially recognised as MD victims [水俣病認定患者][1] in Kumamoto and 1790 people received said certification. In addition, 7225 medical handbooks [医療手帳] and 41,516 MD victim handbooks [水俣病被害者手帳][2] have been issued to date as part of administrative compensation (Kumamoto Prefecture, Minamata Disease Screening Section and Minamata Disease Insurance Section, 2020).[3] In 1994, a new group, *Hongan-no-kai* [the Club of Original Vow本願の会], including victims of MD who had filed a lawsuit in 1969, became actively involved in the *Moyainaoshi movement* (the community-building project in Minamata). For members of *Hongan-no-kai*, the MD problem was not to end with the receipt of compensation through court action and administrative procedures but rather with a move towards community rebuilding—a kind of restorative justice (Komatsubara, 2021b). Clarification of the logic upon which MD victims based their participation in the community rebuilding project contributes to the study of environmental restorative philosophy.

I will focus my analysis on MD victims. The policy of the *Minamatabyo-o-kokuhatsusuru-kai* [水俣病を告発する会 Association for MD Legal Action], the support group for the first lawsuit, is a typical example of a victim-centred social movement. The slogan of the group was 'we help by

---

[1] Official recognition as an MD victim refers to those who have submitted an application to the prefectural governor and have been recognised by the 'Minamata Disease Examination Section [水俣病審査課]'. Those who qualify are entitled to temporary compensation, monthly allowances, and medical services, amongst other benefits.

[2] In 1995, people who were not officially recognised as MD victims were offered temporary compensation and certification in the form of a 'medical handbook [医療手帳]' to reduce their medical expenses as administrative compensation. In addition, in 2010, similar temporary compensation payments and certification in the form of an 'MD victim handbook [水俣病被害者手帳]' were offered to those who had not been covered by the compensation in 1995.

[3] The author refers to several Japanese sources. All quotations taken from the Japanese literature have been translated by the author.

morality [義によって助太刀いたす]', and they sometimes preferred the phrase 'to go along with [つきあう]' instead of 'to fight with [共闘する]' the victims (Watanabe, 2017). In 1970, MD victims, together with supporters of *Minamatabyo-o-kokuhatsusuru-kai*, attended a shareholders' meeting at Chisso. When they met the President of Chisso, they made themselves special costumes and sang a requiem. For them, this action was a memorial to the victims (Komatsubara, 2019). In other words, MD victims were not weak, one-sided recipients of support; instead, they were autonomous stakeholders in social movements, despite having suffered severe economic and physical damage from the pollution. Therefore, I believe this inner strength of the MD victims lies behind their philosophy that would eventually launch a social movement.

In recent years, a growing body of literature has reassessed the links between the thoughts of MD victims and social movements from the perspective of spirituality (Hagiwara, 2018; Komatsubara 2016; Yoneyama 2019). The key word here is 'life', which is translated as '*inochi* [命]' in Japanese. In 1993, a survey in Japan indicated that most people living in Japan held wide-ranging concepts of *inochi* (Morioka, 1993). It was viewed not only as a biological reaction but also as something that exists in connection with non-human beings or even beyond death. The philosopher Takashi Uchiyama (2007) observed the lives of people in Japanese villages and concluded that they live in a unique 'life-world'. That is, according to Uchiyama, in Japan, on the one hand, people consider that human beings live and die as individuals, on the basis of their biological individuality. On the other hand, people also believe that we all have a role to play in the whole life-world. Uchiyama explains the difference between the two ideas of life by referring to the metaphor of the tree and the forest. Each tree has its own individuality, its own process of sprouting, growing, and dying. At the same time, the forest as a whole is influenced by the surrounding trees. If one of its trees are cut down, it becomes difficult for the forest to sustain life. Therefore, the individual life of each tree can only exist if it is united with the whole life-world of the forest (Uchiyama, 2007, p. 111). In other words, people in the local villages of Japan recognise that their life is their own but that they also live in connection with non-human beings within the life-world. The life they envisage is not limited to plants and animals, but also includes mythical beings such as mythical foxes [化け狐] and the kappa [河童]. This is epitomised by the local people's narrative of 'being tricked by a fox'. However, Uchiyama (2007) observed that in the late 1960s, when Japan's industrial development was proceeding rapidly, the narrative of being tricked by a fox disappeared and the relationship with the life-world that envelops human beings became weaker. It was around this time that numerous pollution incidents occurred in Japan, and MD victims began to seek legal justice. Therefore, the environmental harm caused by pollution and the loss of the life-world that the local people had in their hearts are considered to be related.

In Minamata, most people, including MD victims, possess a similar concept of the life-world, including non-human beings, within their spirituality (Munakata, 1983a). The historian Daikichi Irokawa (1982, p. 203), who was the leader of *Shiranuikai-sogo-gakuzyutu-chosadan* [Shiranui Sea Integrated Academic Research Team 不知火海総合学術調査団], describes the life-world in Minamata:

> They [MD victims] have a strong belief that they have lived in the embrace of the Shiranui sea [不知火海] from their ancestors, listening to the sound of the wind that tells of spring and the tide that tells of summer, blended in and surrounded by the great life, and watched over by the ancestral spirits.

MD victims also lived in the midst of a life-connection that included non-human beings, such as the sea, wind, creatures, and spiritual beings of their region. I believe that their image of the life-world serves as the driving force behind their quest for restorative justice.

In this article, I propose a model of the process by which environmental restorative justice is formed, through adopting two theoretical frameworks: environmental embodied ethics and the philosophy of life. The former is an ethic that individuals acquire in their childhood from their natural environment. The latter is a philosophy that views environmental harm from a bird's eye view and linguistically explores the relationship between human and non-human beings. My hypothesis is that environmental restorative philosophies emerge in the process of splitting and integrating both empirical and intellectual knowledge. I apply the theoretical framework to analyse four cases in which the life-world was manifested in Minamata. Through the case studies, I will clarify how MD victims developed their own restorative philosophies and how it led them to be involved in the community rebuilding projects in Minamata.

2 Theoretical Framework

In this section, I propose two theoretical frameworks for analysing the environmental restorative philosophy in Minamata. The first framework is environmental embodied ethics. This framework was introduced in *Ethics Embodied: Rethinking Selfhood Through Continental: Japanese and Feminist Philosophies* by Erin McCarthy (2010). McCarthy (2010) developed embodied ethics by referring to the Japanese philosopher Tetsuro Watsuji (1889–1960). McCarthy (2010) argued that the mainstream of Western ethics considers, as in the philosophy of Plato, that 'the body is a betrayal of and a prison for the soul, reason, or mind' (p. 38), while, in the body-mind conception, as in the philosophy of Watsuji, 'the two work together and in fact, their integration must be cultivated rather than fled from or avoided' (p. 39). She also regarded Zen meditation as an example of cultivation and states that in enlightenment, '[t]heoretical or conceptual understanding is not enough—the body must learn as well' (McCarthy, 2010, p. 47). Thus, embodied ethics—as envisioned by McCarthy—harmonises bodily experiential knowledge with knowledge of the mind.

Droz (2020, p. 34) warned that in Watsuji's ethics, non-human beings are not 'pure nature', that is, others outside humans.[4] In other words, the relationship between man and nature is not an opposing binary, but an interactive one. For example, there is a concept of *satoyama* [village and mountain 里山] in Japan, where humans use nature for sustainable production through agriculture and forestry. Thus, in Watsuji's philosophy, the wisdom of local communities to co-exist and co-create, rather than to destroy and exploit nature, is also seen as ethical.

---

[4] An environmental ethic that assumes 'pure nature' as Others is deep ecology, with its emphasis on wilderness.

Synthesising the views of both McCarthy and Watsuji, I propose the environmental embodied ethics framework. Our environmental ethics are influenced by the relationships with non-human beings that we construct in our childhood. For example, the values of children born and raised near the beach are constructed partly by their physical experience of the cold and salty seawater, the fear of being swallowed by the waves, and the feeling of the seaweed that grows in the water. Through their observation of the environment, these children can also develop an ethical sense. For example, a person watching little crabs on the beach as they came out of their holes and ran around can come to feel that the crabs were interacting with each other in a non-verbal way and that they were to be loved (Onitsuka, 1986). Therefore, environmental embodied ethics is not only formed by logical thinking but also by the experiences we acquire in our daily lives.

The second framework is the philosophy of life. Morioka et al. (2008) proposed the philosophy of life as an academic genre[5] that fundamentally questions what life is and examines our way of life and society. They argued that in the philosophy of life, we should not only discuss pragmatic solutions to environmental problems but also question why we protect the environment. They hoped that 'we will reach a philosophical activity that considers the nature of the relationship between human life and non-human life' (Morioka et al., 2008, p. 4). Thus, it is by considering our relationship with non-human beings that we discover the meaning of life and how we should live.

I propose a model in which an environmental restorative philosophy emerges during the transition from environmental embodied ethics to a philosophy of life. The weakness of ethics constructed by embodied common experience is that it does not provide a clear path for interacting with others who hold different values. To move towards restorative justice, we require a comprehensive concept that allows us to resolve conflict, including the type of conflict that leads to environmental harm. In other words, beyond the embodied experience of the individual, we are confronted with the universal question: why should we protect the

---

[5] Morioka et al. (2008) clearly stated that this is different from the philosophy of life in the English-speaking world, which talks about a personal view of life.

environment? In considering a restorative philosophy, it is necessary to illuminate how victims of environmental harm move from seeing their problems personally towards seeing them from a broader social perspective.

3   Life-World in Minamata

I will illustrate four cases in which the life-world of MD victims has been made visible and analyse them within the two theoretical frameworks: environmental embodied ethics and the philosophy of life.

3.1   Literary Works by Ishimure Michiko [1968⁶–2018]

The first case concerns the literary works of Michiko Ishimure. As an artist, Ishimure provided the social movement of MD victims an image of their ideal new world (Komatsubara, 2021c). She was also a pioneer in recognising the environmental embodied ethics displayed by the MD victims. Ishimure's masterpiece, *Kugai Jodo: Waga-Minamatabyo (Paradise in the Sea of Sorrow: Our Minamata Disease) (Part 1)*, was published in 1968 and instantly became a sensation in Japan (Watanabe, 2013). She claimed to have written the book based on her interactions with the MD victims after listening to the voices in their hearts (Watanabe, 2013). In effect, Ishimure sensed and visualised both the human and non-human elements of the life-world of MD victims and translated it into her work (Komatsubara, 2021c). In a conversation with activist Eishin Ueno in 1973, Ishimure (2004, p. 523) emphasised the MD victims' respect for nature concerning their experience of fishing:

> They [MD victims] respect the [non-human], not as something subordinate or equal to themselves but as something different, as a divinity with an

---

⁶Although Ishimure published a prototype in a literary magazine before Kugai Jodo was commercially published, I have taken 1968 as the starting point for her work, when it started gaining public awareness.

immaculate rank that interacts with them. They live in the same world as the gods who play the pastoral songs, and they are entertained by them. (…) In the world inhabited by fish, foxes and other non-human beings, there is also a deep sense of compassion [憐れみ]. For example, speaking the word 'chikusho [non-human[7] 畜生]' is an expression of this, and it seems that gods [have such compassion].

Ishimure brought up compassion when she mentioned the MD victims who are fisherfolk.[8] The victims respect non-human beings, who have a deep sense of compassion, greater than that of human beings. Ishimure realised that these victims had an image of a life-world inhabited by non-human beings, including gods, that cared for all. Ishimure was aware of how the MD victims learned compassion for others through their relationships with non-human beings because she was also born and raised in Minamata. She was taught environmental embodied ethics by her grandmother in her interaction with non-human beings. For example, her grandmother was instructed to share the gifts of nature with crows, rabbits, and foxes while collecting fruit from the trees in the mountain (Ishimure, 2015). As a resident of Minamata, she understood that for them, the image of a life-world, including non-human beings in Minamata, was not just a myth but an ethical foundation. However, the logic of environmental embodied ethics in the philosophy of life is not readily apparent (at least not in Ishimure's early literary work).

3.2   Shiranuikai-sogo-gakuzyutu-chosadan

The second case arises from the work of the *Shiranuikai-sogo-gakuzyutu-chosadan* [Shiranui Sea Integrated Academic Research Team 不知火海総合学術調査団] (*chosadan*). *Chosadan* was an interdisciplinary research team consisting of researchers from the humanities, social sciences, and

---

[7] Although *chikusyo* [畜生] also has the meaning of 'animal', I have interpreted it to mean 'non-human' as used in Buddhist terminology, depending on the context. Even though *chikusho* [畜生] can also be used as an offensive word, in this context I believe it has a positive meaning.

[8] Many of the early victims of MD were fisherfolk, who lived in coastal areas and ate seafood and were, therefore, more vulnerable to marine pollution.

natural sciences, formed in 1976.[9] The research team was formed in response to Michiko Ishimure's request for an academic study of the inner world of MD victims (Irokawa, 1983).

Remarkable discoveries were made by Iwao Munakata, a sociologist of religion who conducted fieldwork on the daily lives of MD victims, with a special focus on fisherfolk. Munakata (1983a, p. 93) discovered that these fisherfolk possessed 'pure ethics', an ethic that he argued was underpinned by an invisible spiritual world. According to him, the spiritual world of MD victims is deeply connected to the natural environment of Minamata. He described Minamata's nature as follows:

Sea of Shiranui [不知火] are rarely wild and are often calm all year round ... The sea is a 'living nature', exposed to the tide, the wind, the sun, the moon, and the sunshine, creating a thousand different scenes. The hatching of fish and shellfish continues in Modowan [Modo bay 茂道湾] and the surrounding coastal areas of Fukuro [袋] and Yudo [湯堂], and the schools of young fish ripple on the calm surface of the sea, signalling the birth of new life. (Munakata, 1983a, p. 101)

Munakata (1983a) classified the spiritual world into three layers. According to his research, the first layer is formed by the relationship between humans and nature. The author explained that 'the nature of the Shiranui Sea tells the story through its symbolic beauty to the hearts of those who live in its embrace around the world' (p. 107). In other words, the beauty of the lives of non-human beings has touched the spirit of MD victims and enriched their inner world. Thus, they form their inner world, not through verbal thought but by their physical perception of nature. He relied on the narratives of the victims to describe this inner world:

The nature of the sea, the mother of all life, has nurtured a certain view of life and death in the minds of the fisherfolk. The relationship between the sea and all living things is close, and a unique view of reincarnation has developed between them. Human beings are born from the sea, the infinite

---

[9] The research group comprised social science, medical, and biological groups. Munakata, the focus of this article, belonged to the Social Science Group (12 researchers and administrative staff).

mother of life, and after their life in this world, they return to this beautiful natural sea to merge and become whole again. The spirits of the dead who have finished mourning will return to the spiritual world within the natural world, as the 50-year memorial service is still being held among the fisherfolk of Modo. (Munakata, 1983a, p. 182)

In the above citation, I believe Munakata is trying to convey the idea that the Minamata fisherfolk, through their relationship with non-human beings, acquire a sense of the life-world that involves coming from the sea and returning to the sea. Their ethical foundation is a sense of oneness with nature, which transcends their individual lives.

The second layer is centred on human relationships. For the fisherfolk in Minamata, ancestral memorial services are important for maintaining human relations. To them, the human community includes not only the living but also the dead. These relationships extend not only to blood relatives and ancestors but also to neighbours and even non-human beings (Munakata, 1983a). Thus, the fisherfolk have an image of a community in which human beings and non-human beings live together. The images in Munakata's examination (1983a) also share similarities with Ishimure's early literary works.

The third layer is centred on the human activity of exchanging goods (economic activities). Munakata (1983a) underscored that the industrial development that causes pollution occurs in the third layer. According to his theory, overexpansion of the third layer and downsizing of the first and second layers of the spiritual world in the Minamata area were the main causes of the massive pollution. Munakata (1983b) argued that the victim's spiritual world was unilaterally and dramatically destroyed. However, he also argued that the spiritual world of the fisherfolk was reacting to their destruction and trying to repair it. He formulated the following hypothesis:

When the existing order is destabilised by an external shock, the possibility of restoration emerges autonomously from its depth, and the original order is recovered. However, this process of restoration is not simply the recovery of the order that existed in the past in its original form but is a process of creative restoration based on the experience of this external shock and destabilisation of the order. (Munakata, 1983a, p. 180)

The positive force driving the creative restoration of the Minamata area was triggered by the negative event of pollution, which stimulated the spiritual world of the victims. Munakata's hypothesis gives us an idea of the process by which restorative philosophy emerges from environmental embodied ethics; the MD victims' inner environmental ethos activated the forces of restoration to push back on the impact of environmental harm. This overlaps with Nils Christie's (1977) restorative philosophy, based on which conflicts revitalise communities.

Munakata (1983a) put forward Teruo Kawamoto as an example of an MD victim who displayed a restorative philosophy. After Kawamoto's father died of MD, he began to confront the problem of pollution because of his and his family's suffering. He also began to work alone to support other victims in the community beyond his blood relations. Munakata (1983a) argued that there was an invisible spiritual world at the root of Kawamoto's ethos to help and restore solidarity amongst people in his community. His argument suggested the possibility that Kawamoto shifted from environmental embodied ethics (derived from his personal experience) to a philosophy that emphasises solidarity with the community and the other, but he provided neither evidence nor the logic behind the shift.

The results of Munakata's research were published in *chosadan* reports (which provided feedback to the MD victims). Ishimure agreed with Munakata's argument and further claimed that the remarks of outsiders were what triggered a re-evaluation of these values because Minamata insiders take the beauty of the Shiranui Sea for granted (Sakurai et al., 1983). As mentioned in the first case, Ishimure noticed and verbalised the environmental embodied ethics as a Minamata insider. However, when an outsider with different values tried to make sense of their inner world on a different scale, the insiders' ethics were observed in a wider context. In other words, outsiders promote the universalisation of the ethics of insiders.

Another small research group, *ikimono-o-daijinishiyou-kai* [Group for the care of all life 生き物を大事にしよう会]—consisting of MD victims and their supporters, former researchers of the *chosadan*, and Ishimure—was formed in 1979 (Komatsubara, 2018). In a conversation about fish and shellfish, when one member suggested 'going to the beach and eat shellfish right now', various responses were given. Someone asked how that was caring for life. Some claimed that you had to know what a creature tasted like to really understand them. Someone else suggested a meeting where they could 'have a meeting to taste and value life' (Komatsubara, 2018, p. 100). They were researching, free from academic methods, and were beginning to crystallise and having discussions about non-human beings. Thus, *chosadan*'s research work has not only verbalised and given value to environmental embodied ethics but also inspired MD victims to crystallise, observe, and think about their life-world. I propose that the influence of *chosadan* was a step in the transition of MD victims from environmental embodied ethics towards a philosophy of life.

3.3   Ogata, the restorative philosopher

The third case is that of Ogata's philosophy of life. He was one of the MD victims who led the *Moyainaoshi Movement* (the community rebuilding project) in Minamata in the 1990s (Komatsubara, 2021b). While he was indeed a restorative philosopher, he did not go to university; instead, he lived outside the academic world of scholars and explored his own philosophy in his daily life in Minamata. Ogata was born in Minamata in 1953 and lost his father, a fisherfolk leader, to MD when he was six years old. He witnessed his father suffering from MD and dying in a state of agony. He and his family also suffered from MD. He joined the MD movement because he resented the offending company Chisso and wanted to avenge his father (Ogata, 2001). In 1974, he applied for an official recognition as an MD victim at Kumamoto Prefecture, but this request was refused. He also became a vice-leader of the victims' group, and he was an active member of the social movement until 1985. After that, he left the organised social movement and began to work on his self-expression (Ogata, 2020).

It was a severe mental breakdown that triggered Ogata's transition from social movement to self-expression. He felt threatened by the fact that in the social movement, the process of seeking justice and compensation was led by lawyers and not the victims. He did not like that there was no opportunity for victims to confront their offenders face to face and express their feelings (Ogata, 2001). In 1985, he decided to withdraw his application for official recognition as an MD victim, believing that financial compensation was not the essential solution to the problem. However, he felt isolated and misunderstood by others. He began to wonder alone what the true source of the pollution was, but he suffered a mental breakdown. He was intensely fascinated by death. He became disgusted with all mechanical things, so he smashed his TV and crashed his car into a rock. Furthermore, he began talking to the plants and trees to understand the meaning of life from the blowing wind. Finally, he was taken to a psychiatric hospital by the supporters and recovered (Ogata, 2020). Ogata described this experience in a 2013 interview with the religious scholar Shoko Yoneyama (2019, p. 64):

> I feel a sense of dialogue in each individual working of life. For instance, birds singing, the sound of the wind, the sound of waves, all things in the universe … The relationality of the dialogue is felt as a sense of mercy from nature, a sense of ease of being watched over, or a sense of existence, which I think is something we all want … When I went crazy in 1985, I felt that I was pulled back to nature. I felt loved by nature … felt at ease … It is like playing in an infinite world.

He escaped his psychological crisis by regaining his link to the Minamata life-world. It is important to note that his mystical experience was based on his daily life, not on the ideological worship of nature. Ogata (2001, p. 174) believed that he 'received help from his childhood experiences (playing in nature), from plants, trees, fish, birds, and the sound of the wind'. In other words, he attributed his recovery to the embodied memory of his relationship with non-human beings that he acquired from his childhood in Minamata. Thus, his environmental embodied ethics helped him to recover from his mental breakdown.

During his mental breakdown, Ogata let go of his anger towards Chisso. In thinking about the essence of pollution, he realised that he too had bought a car, a television, a refrigerator, and even a plastic boat (Ogata, 2001). He considered the possibility that he might not have reported the outbreak of Minamata disease even if he had known about it, had he been in the offender's position of wanting to protect his own interests—an employee of Chisso. He could not say with conviction that he had played his part against the pollution by resisting the pressures of companies like Chisso around him (Ogata, 2001). Thus, he found himself a part of a modern society that was causing pollution. He gained a newfound awareness that 'Chisso was me'.[10] He also argued that we were all 'another Chisso' (Ogata, 2001, p. 49) in the sense that we lived in a society pursuing material wealth and that pollution was not a problem to be solved with financial compensation but rather by thinking about how we ourselves can 'break free of our own curse [of a modern society]' (Ogata, 2001, p. 49). Thus, Ogata stopped accusing Chisso as a collective entity in organised social movements and started questioning how we live in modern society.

Ogata's life history can be viewed as a transition from environmental embodied ethics to a philosophy of life. Ogata's environmental embodied ethics helped him to recover from his mental breakdown. He also gained awareness of a life-world living with non-human beings and began to explore a philosophy of life that questioned the essence of pollution, not just the MD that he and his family had suffered. In the process, he role-played the offenders in self-reflection and recognised that they also belonged to the same community as he did. In this way, he created a restorative philosophy that went beyond the usual focus on a victim-offender dichotomy.

In expressing a philosophy of life that questioned the essence of pollution, Ogata took tangible actions. For example, he began sit-ins in front of the Chisso factory in Minamata in December 1987. Before that, he had sent a *toikake-no-sho* [letter of questioning 問いかけの書] to the Chisso company. In the concluding part of the letter, Ogata (2020, pp. 150–151) wrote:

> As a witness to the history of Minamata disease, which denied the existence of both human beings and nature, I would like to ask the following two questions. I would like you to answer these questions.
>
> I want you to admit that you killed my father, made my mother and our family drink poisoned water, and tried to kill us.

---

[10] This expression is the title of Ogata's book (Ogata 2001).

I want you to confess that the Minamata disease case was a crime committed by Chisso, the government, and the prefecture, and that there is a 30-year history behind the crime.

If you sincerely acknowledge and give a written answer to these two questions, then I will be able to recognise you as a human being and forgive your sins.

Human beings do not violate nature.

Human beings are fostered by nature.

Human beings must live among other human beings.

After that, Ogata ordered a new wooden boat, and every week, he would depart from the harbour near his house up the river to the Chisso factory. The name of his boat was *tokoyo-no-fune* [the ship of the other world 常世の舟]. In Japanese, *tokoyo* refers to the world after death. For him, the spiritual state he reached after his mental breakdown was *tokoyo*, a state of 'peace' and 'selflessness'. He carried a photograph of his father and a mat on which he had written a message appealing to the people to answer *toikake-no-sho* and sat in front of the factory. Chisso staff came to where he was sitting and invited him to talk, but he refused as he believed that the gap between him and them could not be bridged by words. He called his sit-in 'exposing the body [身を晒す]' (Ogata, 2020, p. 170). In my interpretation, Ogata was not protesting against Chisso as a victim, but rather it was a serious questioning of pollution from one person towards Chisso. In other words, it was a non-verbal expression of his philosophy of life. This action was also a call for the staff of Chisso to confront pollution together and reflect on the future of modern society.

Ogata would later remember his solitary sit-in as a lonely struggle, but he also described it as a peaceful scene. For example, when the policeman normally in charge of policing social movements saw that he was sitting alone without making any demands, he cheered him on and wished him good luck (Ogata, 2020, p. 164). He was also interviewed by journalists and was offered donations by tourists visiting the spa, which he refused. He realised that it would be children who would take his message the most seriously, so he added some easy-to-read text to the display for them to read. During the sitting, he would drink *shochu* [traditional Japanese distilled spirit 焼酎] and grill fish to eat on the *shichirin* [Japanese small charcoal grill 七輪]. Attracted by the smell of the fish, a cat came to him.

He gave some of his fish to the cat, who stayed with him for more than an hour and made him very happy (Ogata, 2020). He stopped his sit-in after about six months because the sun and heat were too harsh on him (Ogata, 2020). His struggle was a part of his daily life and was influenced by non-human beings, such as the cat and the sun. In other words, although he was pondering on a profound philosophy of life, he remained grounded and firmly rooted in his physical self. His philosophy of life that started from environmental embodied ethics, and through self-reflection, returns to expressive activity through the body. The body as a vehicle is an integral part of the environmental restorative philosophy.

3.4   Hongan-no-kai

The fourth, and last, case that I analyse concerns the activities of *Hongan-no-kai* [the Club of Original Vow 本願の会]. *Hongan-no-kai* was founded in Minamata in 1994. *Hongan-no-kai* consisted of MD victims, their supporters, and activists. The club's members actively participated in the *Moyainaoshi movement* (community rebuilding project in Minamata) in the 1990s. In other words, the MD victims were the ones leading the restorative practice. From the time MD was first officially recognised in 1956 until the 1990s, the MD victims had split up into several groups with their own activists and supporters, each with different opinions and opposing arguments on the approach to lawsuits and compensation. However, in the founding statement, *Hongan-no-sho* [The statement of original vow 本願の書], there were 17 participants, which included the names of the MD victims beyond the boundaries of the groups. *Hongan-no-sho* declared in 1994:

> Minamata Bay was once a treasure box for fish. Migrating fish spawned here in schools, their young grew up here and became adults, and once they left, they came back. This place was similar to the mother womb for the fish. Around the current reclaimed area from *Hyakken* [百間] to *Myojin-misaki* [明神岬], sardines and *konoshiro* (white fish) glittered with their silvery scales, mullet leapt from the sea, and shrimps and crabs danced. We have lived on the shores of the sea, gathering shellfish and eating wakame seaweed and hijiki seaweed that sway in the waves. It is by giving us *inochi* [life 命] that we have been able to nourish ourselves.

However, how many creatures, from sea creatures to human beings, have been murdered by the poisonous water produced by industrial civilisation! This original sin is a historical fact that cannot be erased and must be inscribed in human history forever. In this sense, we would like to dedicate a number of stone statues (*nobotoke* [small Buddhas in a field] 野仏) in the land of the reclaimed bitter sea [苦海], and we would like to think deeply about the sinfulness of human beings and pray for the salvation of *tamashii* [souls 魂] together.[11]

The main author of this statement was Masato Ogata. He described the life-world of Minamata in correspondence with the sociologist Iwao Munakata's article in 1983. It is important to note that the statement clearly states that the human sin of genocide against non-human beings in Minamata must be recorded and conveyed to the next generation. As a way of explaining the history of the pollution, the members of *Hongan-no-kai* chose to sculpt a stone statue and offer a prayer. In 1997, Ogata (1998, p. 18) said:

Human beings may accept financial compensation, but I do not think it works for the sea. It does not work for fish, it does not work for cats, it does not work at all. How can [compensation] be accepted? Essentially, it does not work for the dead. So, I think it is a challenge for us to be aware of the sin of poisoning our seas and to confront it ... We organised a small group called *Hongan-no-kai* to sculpt a '*tamashii-ishi* [soul stone 魂石]'—*nobotoke*—and place it on the reclaimed land [in Minamata Bay]. This is also a sign of awareness.

Even though MD victims were divided by the politics of a social movement demanding economic compensation, they agreed on *Hongan-no-kai*'s activities to remember and pray for the genocide of non-human beings. The efforts of the MD victims could be seen as a call for everyone to work together to convey the memory of pollution in Minamata to the future rather than an attempt to accuse human beings of sin. Michiko Ishimure, who took part in *Hongan-no-kai*, wrote (Ishimure, 1997, p. 9):

---

[11] A copy of *Honngan-no-sho* is in the collection of the Minamata Disease Centre, Soshisha. The copy viewed by me had only 16 signatures, however, other documents and the *Hongan-no-kai*'s newsletter repeatedly state that 17 victims had signed *Hongan-no-sho*.

The patients [MD victims] who participate in *Hongan-no-kai* in Minamata clearly show that they do not hate human beings. No resentment. Rather, they are fascinated and compassionate towards human beings. Through the experience of sorrow, their compassion has become genuine. I think it is because of their hope for humanity that they are able to remain so, despite all the difficulties they have suffered. Hope does not come from hatred or disgust.

Ishimure's assertion that MD victims have compassion for human beings is consistent with the point she made in the first case, that MD victims in 1974 were driven by environmental embodied ethics they developed from the compassion they learned from non-human beings. When one member of *Hongan-no-kai* declared their intention to inscribe Minamata's history, activists wanted to sculpt a stone statue with a pained expression, but MD victims wanted to 'make something that is sweet and pretty and makes you want to gently stroke its head' (Kanasashi, 1998, p. 26). We can thus see that the sculpture and prayer of *Hongan-no-kai* were not intended to blame humans for their sins but rather for humans to face themselves and deepen their compassion for others. From my observation of the attitude and behaviour of MD victims participating in *Hongan-no-kai*, they, like Ogata, seem to have shifted from environmental embodied ethics to a philosophy of life.

In addition, the exploration of the philosophy of life in *Hongan-no-kai* is not based on language but the sculpture of stone. According to anthropologist Kentaro Shimoda (2017), one of the members who worked on the stone sculpture at *Hongan-no-kai* said that as he immersed himself in sculpturing, the image of the stone statue he wanted to design faded from his mind's eye and that all he could hear was the sound of the waves. He emphasised that, beyond his intentions, the statue had come out of stone (Shimoda, 2017). One of the other members was unable to start carving for nearly a year, but she wrote:

For me, I feel that this time is necessary. I thought that sculpting for a requiem, for nostalgia, for all the victims of MD (including non-human beings), to put on the reclamation site in the beginning. It was difficult for me to face stones. However, only when I felt uncertain emotion that I did not know what it was I was able to confront my stone. (Hirakida, 1997)

Another group member spoke about receiving messages from the stones while sculpting them, instead of entrusting them with their own messages. 'Hmm, sometimes I feel like [the stone] is trying to tell me an alternative prayer' (Shimoda, 2017, p. 100). While carving stone statues, some sculptors lost their previously linguistic thoughts and apologies towards non-human beings, and they entered a non-verbal world, facing the stone in silence. In an interview with the *Hongan-no-kai* newsletter, Ogata (2000, p. 8) said, 'Although much has been discussed about the MD, I feel that I want to put myself in a situation before the language. I want people to feel this place in a non-verbal way'. In addition, he stated, 'We have to create a story. It is the task of *Hongan-no-kai* to complete the story of life' (p. 9). Therefore, the stone sculptures of *Hongan-no-kai* aim to explore the meaning of life in the inner world without language and to express their philosophy of life in a non-verbal way. While the members of *Hongan-no-kai* participated in restorative practices, their environmental restorative philosophy was underpinned by non-verbal activities.

4 Discussion

By examining the above four cases, I have illustrated that the MD victims in Minamata have created and continue to create a restorative philosophy as they move from environmental embodied ethics to a philosophy of life. I will discuss two points regarding environmental restorative philosophy.

First, environmental restorative philosophy is generated through interaction with others. Referring to the second case (i.e., Munakata's (1983a) fisherfolk), the researchers who came from outside stimulated the curiosity of MD victims and their interest in the relationship with non-human beings. MD victims, influenced by the researcher who interviewed or observed them, also began to verbalise their environmental embodied ethics, and this verbalisation of their inner world always happens through dialogue with others. As mentioned earlier while presenting the third case, Masato Ogata is regarded as an environmental restorative philosopher who has pursued the philosophy of life through his own means, without reference to the literature of other famous philosophers. However, his description of environmental embodied ethics is similar to those of the authors mentioned in the first and second cases. Despite this, it is not clear whether he was aware of their literature. However, at least in Minamata, the basis for the linguistic expression of environmental embodied ethics had already been laid down by them. In addition, he organised *Hongan-no-kai*, with the members of which he explored the philosophy of life in a non-verbal way through stone sculptures. This work was the basis for the *Moyainaoshi movement* (community rebuilding project in Minamata) in the 1990s. In this context, environmental restorative philosophers, such as Ogata, actively developed their ideas in interaction with others.

Second, environmental restorative philosophy is created through the back and forth between verbal and non-verbal expressions. For Michiko Ishimure, the Minamata life-world was essentially impossible to verbalise (Komatsubara, 2021a). Ishimure's intuition of the inexpressibility of language is correct in that an individual's embodied experience cannot ultimately be shared with others. For example, we can imagine the pain of others, but we cannot directly perceive it with our bodies. In the case of MD victims, their bodies were in severe pain due to neurological disorders. When Ishimure called an MD victim one morning, she was told that the victim had not been able to sleep all night due to severe pain and that no painkillers had worked. She was shocked to hear the victim's voice, which was very different from the victim's usual voice (Ishimure, 2000). Given Ishimure's delicate sensitivity as an artist, she was able to realise the limits of verbalisation and the importance of other sensory input.

Unfortunately, the inner world of MD victims can only be expressed through language, even if it is an imperfect medium. Ishimure herself was aware that as an artist, she had to express herself in language, for it is only through language that the speakable and the unspeakable of the life-world in Minamata could be divided, and a new world of creation be opened between them (Tsurumi, 2002). She claimed she was trying to 'build a bridge' (Tsurumi, 2002, p. 176) between the inner world of MD victims and the world of her readers in a conversation with the sociologist Kazuko Tsurumi. The leap from environmental embodied ethics to the philosophy of life requires a similar verbalisation.

Masato Ogata too alternated between verbal and non-verbal expressions. His philosophical thinking was driven by questioning the relationship between human and non-human beings, about pollution as a concept, leaping from his own bodily experience. However, he expressed his philosophy by 'exposing his body'.

He explored the relationship between human and non-human beings in the philosophy of life and expressed what he found in the non-verbal approach of 'exposing the body'. One photographer, Shigemi Miyamato, who participated in *Hongan-no-kai*'s activities, called MD victims 'people who exposed the body' (Miyamoto, 2000, p. 23). As Miyamoto gazed at the victims praying, she realised that prayer is 'an acceptance of the unspeakable and the invisible' [言葉に出来ない、わからないことを、引き受ける] (Miyamoto, 2000, p. 24). Based on Miyamoto's interpretation, in the expression of Ogata's philosophy and the activities of *Hongan-no-kai*, the speakable and the unspeakable of something, once divided by language, are re-integrated. Thus, Ogata's environmentally restorative philosophy appears in the process of restoring the division between the verbal and the non-verbal, the mind and the body, the visible and the invisible, the knowable and the unknowable.

5   Conclusion

Through the case studies, I have illustrated the history of philosophy in Minamata up until MD victims acquired an environmentally restorative philosophy. MD victims were not taught an environmental restorative philosophy. None of the MD victims developed an environmental restorative philosophy alone, despite the presence of key persons, such as Masato Ogata. Their environmentally restorative philosophy was birthed from their interactions with artists, researchers, other victims, and the verbalisation of their embodied knowledge and experiences. I also identified a rich and creative thought process behind their participation in the

Moyainaoshi movement (the community rebuilding project in Minamata) in the 1990s.

Even though this is a theoretical model for analysing environmental restorative philosophies, it does not mean that the entire process that occurred in Minamata has been fully examined. In Minamata, I believe that many MD victims, residents, supporters, and activists have their own thoughts about pollution and that some members of the *Moyainaoshi movement* joined the project with different ideas. Through ongoing research, I will continue to further examine the theoretical model through an analysis of the *Jishu Kosho* [自主交渉, voluntary negotiations] that MD victims, including Teruo Kawamoto, had face to face with the President of Chisso between 1971 and 1973.

However, this theoretical model can already contribute to research that is exploring the possibility of restorative justice for environmental harm outside of Minamata. For example, in cases where there are yet to be victims participating in restorative justice, it is possible to consider that there is potentially a seed of restorative philosophy in their invisible inner world. Artists and researchers can also be involved in creating an environmentally restorative philosophy by dialoguing with the victims about their inner life-world. In other words, practitioners of restorative justice from outside do not teach victims a restorative philosophy but enable it to grow organically from their interaction with the victims. The best path moving forward is that we take the long road to the practice of restorative justice by seeking to uncover and share in the life-world of the victims.

References

Christie, N. (1977). Conflict as property. *The British Journal of Criminology*, 17 (1), 1–15.

Droz, L. (2020). *The milieu as common grounds for global environmental ethics* [PhD Thesis]. Kyoto: Kyoto University of Global Environmental Studies.

Hagiwara, S. (2018). Minamatabyojiken to "mouhitotsu-no-konoyo". *Gendai Shukyo 2018*, 112–132.

Hirakida, R. (1997). Tamashii ishi tono deai. *Hongan-no-kai* [The newsletter of *Hongan-no-kai*], 4, 5.

Irokawa, D. (1982) Shiranuikai sogo chosa 5 nen o hete: kankyo hakai to nigen sasei no dorama. *Dojidai heno chosen* (pp. 179–230). Tokyo: Chikuma Shobo.

Irokawa, D. (1983). Soron: shiranuikai sogo chosa no keika to mondaiten. In D. Irokawa (Ed.), *Minamata no keiji: shiranuikai sogo chosadan hokoku* (pp. 91–154). Tokyo: Chikuma Shobo.

Ishimure, M. (1997). Fukaku kagayaku tamashii no hikari. *Akebono, 2*, 8–9.

Ishimure, M. (2000). *Ishimure Michiko taidanshu: Tamashii no kotoba o tumugu.* Tokyo: Kawade syobo shinsha.

Ishimure, M. (2004). Kugai Jodo koshikata yukukata. In M. Ishimure (Ed.), *Ishimure Michiko Zenshu, 3* (pp. 511–531). Tokyo: Fujiwara Shoten.

Ishimure, M. (2015). *Kugai jodo.* Tokyo: Kawadeshobo shinsha.

Kanasashi, J.. (1998). Watashi no Minamata. *Tamashii utsure.* Minamata: Hongan-no-kai: 24–27.

Komatsubara, O. (2016). Minamata no inori to yurushi: 1990 nendai no Moyainaoshi jigyo o Kento suru. *Gendai seimei tetsugaku kenkyu, 5,* 51–73.

Komatsubara, O. (2018). 'Kogai mondai' kara "kankyo mondai" he: Minamata chiki niokeru "shiranuikai sogo gakujutsu chosadan" no katsudo o tegakarini. *Gendai seimei tetsugaku kenkyu, 7,* 74–106.

Komatsubara, O. (2019). 'Higaisha no jonen' kara "higaisha no hyogen" he: Minamatabyo "hitokabu undo" (1970) niokeru higaisha kagaisha taiwa o kento suru. *Gendai seimei tetsugaku kenkyu, 8,* 57–129.

Komatsubara, O. (2021a). 'Kitsune ni damasareru chikara' o torimodosu: Minamatabyo o toshita kankyo kyoiku no kanosei. *Gendai seimei tetsugaku kenkyu, 10,* 96–118.

Komatsubara, O. (2021b). Imagining a community that includes non-human beings: The 1990s Moyainaoshi Movement in Minamata, Japan. *International Journal of Restorative Justice,* 4(1), 123–140.

Komatsubara, O. (2021c). The role of literary artists in environmental movements: Minamata disease and Michiko Ishimure. *International Journal for Crime, Justice and Social Democracy,* 10(3). https://doi.org/10.5204/ijcjsd.1984.

Kumamoto Prefecture, Minamata disease screening section and Minamata disease insurance section. (2020). *Minamatabyo Kanren Tokei,* 31/08/2020.

McCarthy, E. (2010). *Ethics embodied: Rethinking selfhood through continental: Japanese, and feminist philosophies.* Plymouth: Lexington Books.

Miyamoto, S. (2000). 'Inoru' kotono muzukashisa. *Tamashii Utsure, 3,* 23–24.

Morioka, M., Inaga, M., & Yoshimoto, R. (2008). Seimei no tetsugaku no kochiku ni mukete (1): kihongainen, berukuson, yonasu. *Ningenkagaku*, 3, 3–68.

Morioka, M. (1993). The concept of Inochi: A philosophical perspective on the study of life. *Global Bioethics*, 6(1), 35–59.

Munakata, I. (1983a). Minamata no naitekisekai no kozo to henyo. In D. Irokawa (Ed.), *Minamata no keiji: shiranuikai sogo chosadan hokoku* (pp. 91–154). Tokyo: Chikuma Shobo.

Munakata, I. (1983b). Minamata mondai ni miru shukyo. In K. Kadowaki & K. Tsurumi, K. (Eds.), *Nihonjin no shukyo shin* (pp. 173–194). Tokyo: Kodansha.

Ogata, M. (1998). Tamashii no yukue. *Tamashii utsure*, 1, 15–18.

Ogata, M. (2000). Tamashii no monogatari o mezashite. *Tamashii Utsure*, 2, 5–9.

Ogata, M. (2001). *Chisso ha watashi de atta.* Fukuoka, Japan: Ashishobo.

Ogata, M. (2020). *Tokoyo no fune o kogite: jukuseiban.* Shimonoseki, Japan: Sokei Publishing.

Onitsuka, I. (1986). *Oruga Minamata.* Tokyo: Gendai shokan.

Sakurai, T., Munakata, I., Ishimure, M., Kawai, H., Yasue, R., & Tsurumi, K. (1983). Ishikisuru syukyo toha. In K. Kadowaki & K. Tsurumi (Eds.), *Nihonjin no shukyo shin* (pp. 173–194). Tokyo: Kodansha.

Shimoda, K. (2017). *Minamata no kioku o tsumugu: hibikiau mono to katari no rekishi jinruigaku.* Tokyo: Keiogijuku daigaku syuppannkai.

Tsurumi, K. (2002). *Taiwa mandara: ishimure Michiko no maki.* Tokyo: Fujiwara Shoten.

Uchiyama, T. (2007). *Nihonjin ha naze kitsune ni damasarenaku natta noka.* Tokyo: Kodansha.

Yoneyama, S. (2019). *Animism in contemporary Japan: Voices for the Anthropocene from post – Fukushima Japan.* London: Routledge.

Watanabe, K. ([1973]2013). 'Kugai Jodo' no sekai (the world of 'Kugai Jodo'). In K. Watanabe (Ed.), *Mou-Hitotsu-no-Konoyo: Ishimure Michiko no Sekai* (pp. 8–33). Fukuoka, Japan: Gen Shobo.

Watanabe, K. ([1972]2017). Shimin to nichijo. In K. Watanabe (Ed.), *Shimin to nitijo* (pp. 22–37). Fukuoka, Japan: Gen Shobo.

16

The Art of Repair: Bridging Artistic and Restorative Responses to Environmental Harm and Ecocide

Brunilda Pali, Maria Lucia Cruz Correia,
Marine Calmet, Vinny Jones, Lode Vranken,
Margarida Mendes, Evanne Nowak, and Mark Požlep

1 Introduction

I don't think art has a duty to be beautiful or uplifting, and some of the work I'm most drawn to refuses to traffic in either of those qualities. What I care about more ... are the ways in which it's concerned with resistance and repair.
—Olivia Laing, Funny Weather: Art in an Emergency (2020, p. 2)

Imagination is without doubt the human faculty that has the greatest capacity to enlarge the field of the possible. Enlarging the field of what's possible in responding to environmental harms is something that cannot be underestimated given the current equally demobilising trends of

B. Pali (✉)
Department of Social and Cultural Anthropology, Faculty of Social Sciences,
KU Leuven, Leuven, Belgium
e-mail: brunilda.pali@kuleuven.be

M. L. C. Correia • L. Vranken • M. Požlep
Gent, Belgium

B. Pali et al. (eds.), *The Palgrave Handbook of Environmental Restorative Justice*,
https://doi.org/10.1007/978-3-031-04223-2_16

385

catastrophism and ecological techno-utopianism (Varona, 2021). In this chapter, we elaborate and reflect on interdisciplinary collaborations involving responses to environmental harm that we have developed and the ways in which they have expanded fields of possibility. Even though we have come to the quest of thinking about responses to environmental harm from different perspectives—social sciences, arts, and law—we have met each other on the terrain of the imagination (Pali, 2020), a terrain where we co-cultivate our right, duty, and power to imagine a different way of living together with the Earth, with other humans, and with more-than-humans. In conducting durational projects with others (artists, community members, but also more-than-human beings and entities), whereby the intent of the inquiry is to 'create a different world, and to ask what kinds of futures are imaginable', we need to think about and do research differently (Springgay & Truman, 2018, p. 204).

Exciting and important alignments have been made between the movement to recognise ecocide and proponents for Earth jurisprudence, the Rights of Nature, the duty of care for the environment, and restorative justice (Wessels & Wijdekop, 2022). Largely inspired by the Ecocide Mock Trial which took place in 2011 in the UK Supreme Court and which tested international lawyer and environmental activist Polly Higgins' s idea of using restorative justice as part of the sentencing process for ecocide convictions (Kershen, 2019), in 2019, environmental

M. Calmet
Paris, France

V. Jones
Amsterdam, The Netherlands

M. Mendes
Centre for Research Architecture, Goldsmiths University of London, London, UK

E. Nowak
Utrecht, The Netherlands

artist Maria Lucia Cruz Correia[1] (herein referred to as Lucia) staged the *Voice of Nature: The Trial* as 'a restorative court unit' in the Old Court in Ghent, Belgium.[2] The project aimed to investigate ecocide and create a proposal for a new justice system, experimenting with ways in which law and justice can serve ecosystems by designing a new type of courtroom and a new form of 'restorative trial'. In this performative 'restorative trial', several attempts were made to hear the voice of Nature and to legally recognise its rights, and the human participants present in the trial declared their intentions to engage in collaboratively contracted actions of repair, by becoming guardians and custodians of Nature.

This project was the first of a series of ongoing collaborations through which we have placed the ethos and praxis of restorative justice at the centre of art-based interventions that imagine novel responses to ecocide. From the courtroom, the project developed into the *Voice of Nature Kinstitute*,[3] a utopian public service that engages with long-term projects dedicated to climate justice, mitigation of environmental harms and ecocide, while envisioning justice through artistic, juridical, ritualistic, and restorative practices. Our praxis in the *Voice of Nature Kinstitute* is based on six research pillars, which we describe in the following section. After describing the research pillars, we move on to elaborate first on the project *The Voice of Nature: The Trial* mentioned above and second on the project *Natural Contract Lab*,[4] a project that has been so far a central part of the Kinstitute's work and which aims to address harms made to several rivers in Belgium, Portugal, and Switzerland.

2   Research Pillars

Ecocide: The first pillar of research is Ecocide ('oikos', 'caedo'), which literally means 'killing the house'. Currently, there is no body of international law to prosecute those who are destroying the environment and

---

[1] See http://mluciacruzcorreia.com/.

[2] See http://mluciacruzcorreia.com/works/the-voice-of-natu; https://voiceofnaturekinstitute.org/projects/performance-trial/trailer-voice-of-nature.

[3] See https://voiceofnaturekinstitute.org/.

[4] See http://mluciacruzcorreia.com/works/walking-with-the-river-zenne.

ecosystems (Higgins et al., 2013), and consequently, no international institution is competent to judge these harms. There have been, as a result, concerted efforts advocating for the amendment of the Rome Statute—the treaty that established the International Criminal Court (ICC)—to add ecocide as the fifth crime against peace. The campaign to define in legal terms and criminalise ecocide was led, most prominently, by Polly Higgins. According to its campaigners, making ecocide a crime aims first and foremost at creating a cultural shift in how the world perceives acts of harm towards nature (Climate Academy, 2020). In December 2019, at the ICC in The Hague, the Vanuatuan ambassador to the European Union (EU), representing one of the many Small Island States in the South Pacific severely threatened by climate change and rising sea levels, suggested making the destruction of the environment a crime, thereby starting a collaboration with other states such as the Maldives and those in the EU. In 2021, legal experts from across the globe mandated by Stop Ecocide International,[5] and led by lawyer Philippe Sands, have drawn up a historic definition of ecocide, intended to be adopted by the ICC to prosecute the most serious offences against the environment, such as oil spills, deep-sea mining, industrial livestock farming, and tar sand extraction (Siddique, 2021). The draft law defines ecocide more specifically as 'unlawful or wanton acts committed with knowledge that there is a substantial likelihood of severe and widespread or long-term damage to the environment being caused by those acts' (Stop Ecocide Foundation, 2021). For Philippe Sands (cited in Siddique, 2021),

> The most important thing about this initiative is that it's part of that broader process of changing public consciousness, recognising that we are in a relationship with our environment, we are dependent for our wellbeing on the wellbeing of the environment and that we have to use various instruments, political, diplomatic but also legal to achieve the protection of the environment.

---

[5] See: https://www.stopecocide.earth.

Rights of Nature: The second research pillar focuses on the Rights of Nature. Recognising legal rights to more-than-humans acknowledges natural beings' legal equality with humans (and with legal entities such as companies). This constitutes a profound cultural and legal shift in terms of how we relate to and interact with the Earth's systems. Framing the broken relationship between humans and Nature and the commodification of Nature at the heart of the problem, advocates for the Rights of Nature argue that we need to move away from a self-entitled right to exploit towards a duty of care. Starting at the level of consciousness, this transformation can eventually translate into binding laws and ethics. First, they argue for a fundamental relational change between humans, more-than-humans, Nature, and the environment, where humans feel not as owners but as part of the Earth. Second, they argue that legal systems should be at the service of regulating this new balance and relation (see Wessels & Wijdekop, 2022). Recognising natural entities, such as rivers, for example, as legal subjects with appropriate rights, aims to compel individuals, societies, and governments to respect the Rights of Nature. It opens a new pathway to instituting new regulations to prevent the destruction of ecosystems and to demand reparation and compensation for their damage in the event of violation of their interests. Across the world, many states and local communities have recognised the legal personality and rights of ecosystems, encompassing such natural entities as rivers, mountains, forests, and glaciers. These new rights have been enshrined in law or recognised through landmark court decisions in New Zealand, Bangladesh, Bolivia, Colombia, Ecuador, India, Mexico City, Uganda, and the United States.[6] In our practice, we actively explore the possibilities of recognising the Rights of Nature for the rivers with which we engage. Marine Calmet, an environmental lawyer who is part of our team and co-author of this chapter, conducts research on each of the rivers, on which she bases her arguments to demonstrate the need to recognise river rights.

The guardian of Nature: The third research pillar focuses on the role of the guardian of Nature. Under this pillar, we investigate the legal procedures and the background that are necessary in order to become a

---

[6] See Global Alliance for the Rights of Nature website garn.org for the most updated state of the art.

guardian of Nature. In law, the concept of guardian implies the legal representation of other beings; to articulate a relational jurisprudence as a protector, guardian, and steward and mobilise the laws for articulating an equitable, reciprocal, and balanced relationship with the more-than-human world. Arguing that natural beings, such as rivers, forests, and trees, should have legal rights to make claims to protect against or to seek compensation and reparation for damage goes hand in hand with them voicing their claims through appointed legal spokespersons or guardians (Stone, 1972). In this scenario, the question of who has legitimacy and standing to advocate for a river and questions of relationality and positionality of the guardian become important. Through our research, we ask whether the guardian is necessarily a legal expert or whether she or he needs other skills and capacities, and if so, are those given and identifiable or can they be acquired?

Restorative justice: The fourth research pillar is restorative justice. Restorative justice is a global social movement and an alternative paradigm of justice, which aims at transforming the way in which our societies respond to crime and conflict (Johnstone & Van Ness, 2011). It prioritises repairing the harm caused over determining the laws that have been broken; it involves all affected stakeholders in the justice process instead of delegating the matter to the state justice system; and it searches for ways to promote repair and accountability instead of imposing punishment for punishment's sake (Pali, 2020). Environmental restorative justice (ERJ) can potentially respond to the whole spectrum of environmental crimes, harms, and conflicts. From this perspective, all harms and victims of environmental harm—human and more-than-human, individuals, communities, future generations, and the environment—could potentially be taken into account and given a space within restorative processes. The most valuable contribution restorative justice makes to environmental conflicts is creating safe and structured spaces for all stakeholders to exchange different views, tell their stories, and hold difficult and brave conversations around what has happened and what must happen in the future to ensure accountability and repair (Pali & Aertsen, 2021). The ERJ perspective is driven largely by the principles of harm reparation, restoration, and healing of communities, relationships, institutions, and ecosystems damaged by human action or inaction.

In cases of environmental harm, simply punishing the perpetrator of the pollution or destruction of natural resources is meaningless if there is no repair of the damage done and if there is no assurance that this type of behaviour will not happen again (Minguet, 2021). As a result of restorative processes, a plan of action or a restorative contract which contains suggestions to prevent or repair harms and damage can be drawn up, and because it is inclusive and participatory, such a contract and plan of action has the potential to be sustainable and transformative. Being held accountable for doing harm and for repairing harm is central to ERJ. Accountability and repair are also closely associated with the virtues of humility and restraint, which recognise the necessity of limits, limits to hubris and *desmesura*[7] (Varona, 2021). So, the transformation of cultural values is, in a way, a precondition for the practice of ERJ.

Ecological/climate grief: Our fifth pillar of research is climate or ecological grief. Ecological/climate grief is a psychological response to loss caused by environmental destruction or climate change. Climate grief is related both to changes that have already happened and to changes that are coming, or are in the process of happening, and can manifest itself as anxiety over loss or anticipated loss of a healthy planet.[8] Other words describing emotions connected to climate grief include *solastalgia*, a homesickness due to environmental changes, and *terrafurie*, a rage arising from the mindless destruction of Nature (Albrecht, 2020; Pihkala, 2020). The way forward towards healing is not denial or bypassing of grief but instead going through a grieving process. It can be profoundly relieving to admit and share feelings of grief and sadness in a safe setting, a setting which often takes time and requires trust to be created.

Artistic and sensorial practices: The last pillar of research important for our work is artistic and sensorial practices. The experimentation with environmentally engaged art and embodied sensory practices is chosen to actively change the perspective of the audience/participants from looking at, to 'being with' Nature and more-than-humans. To this end, we have continuously developed exercises and methodological practices that activate a specific sensory focus, be it through stimulating ways of looking,

---

[7] Wastefulness, squandering, or a lack of moderation.
[8] See https://www.psycom.net/anxiety/coping-climate-grief-anxiety (last accessed 28 January 2022).

listening, or sensing. This is aimed at widening the spectrum of perception, by provoking environmentally aware reactions, while potentially opening up the possibility of generating reciprocal relationships, as one grows aware of the ways in which we can affect and be affected (Springgay & Truman, 2019). Some of these artistic and sensorial practices are introduced via sculptural props to be utilised during our collective walks, as well as by listening exercises and environmental meditations, that are introduced to the audience and compiled and distributed later in toolkits and resource websites. In what follows, we describe the two projects *The Voice of Nature: The Trial* and *Natural Contract Lab* and the ways in which they incorporate the research pillars in their praxis.

3   Voice of Nature: The Trial

In the *Voice of Nature: The Trial*, the first artistic project on which we collaborated, Lucia was on a quest for radicalising the meaning of justice in relation to environmental harms or crimes. Some of the questions she posed were: What does justice mean in the aftermath of ecocide? What is the role of our laws and legal systems in serving and protecting the Earth? What is the contribution of science, Indigenous knowledge, art, and activism in such pursuits? After a long research and transformational journey in which she encountered many people with different forms of expertise, Indigenous communities, and scholarly and artistic resources that could answer these questions, she set about reimagining, with a multidisciplinary team,[9] a new form of a performative trial for investigating ecocide and proposing a different justice response (Cruz Correia, 2019, 2021).

The participatory and immersive performance of the *Voice of Nature: The Trial* that resulted from this multidisciplinary cooperation was enacted for the first time in the old court of Ghent in the context of the *Same Same But Different* festival with the support of the Vooruit Kunstcentrum. There

---

[9] The team included experts like lawyers Hendrik Schoukens and Juan Auz, dramaturge Ingrid Vranken, performer/artist Caroline Daish, sound designer Joao Bento, light designer Vinny Jones, and video maker Mark Požlep.

is a long worldwide tradition of using trials, tribunals, and truth commissions in a theatrical context (see Nellis, 2021), a tradition on which our team could rely and from which it could learn. The particularity of the *Voice of Nature: The Trial* was the experimentation around combining elements from a conventional court, restorative justice practices, as well as transformative rituals. Through enacting this speculative justice proposal, we not only redesigned the courtroom but also proposed an active role for a restorative approach to ecocide, by acknowledging multispecies victims and environmental harm and prioritising repair and accountability instead of punishment for punishment's sake. At the same time, we explored the difficulties and ambivalences of listening to the voice of Nature, of granting personhood to more-than-humans, and of transforming mountains and rivers into legal entities (Image 16.1).

Even though it was enacted in a courthouse, the performance of the *Voice of Nature: The Trial* is mobile and has a potentially hybrid architecture as the performance can be adapted to other formal and informal

Image 16.1   Voice of Nature: The Trial, Maria Lucia Cruz Correia, 2019 (Photo © Mark Požlep)

'spaces of justice', such as public squares, parliaments, and abandoned courts. Within the courtroom, the scenography reassembled to some degree the traditional set-up of the courtroom. All the familiar elements and roles were present in the performance but were transformed into new assemblages where everyone inside the room was implicated somehow in the ecocide crime, as victim, perpetrator, bystander, witness, or judge. The script assembled a fictional lawsuit for ecocide based on research material that Lucia had collected in Ecuador. Marked by an ecocidal legacy of oil extraction, given that in 1964, oil giant Texaco invaded the northeastern part of Ecuador and then—after 30 years of mining—left tons of chemical waste in the Amazon basin, Ecuador is also home to inspiring developments. Such developments include the incorporation of the Rights of Nature into its Constitution and the inspiring lawsuit won in 2012 by the Indigenous Sarayaku community against the Ecuadorian state, after Ecuador allowed a foreign oil company to conduct a seismic search for oil in Sarayaku territories.

Through the presentation of testimonies from the ecocide victims and communities encountered during Lucia's research, the emphasis in the *Voice of Nature: The Trial* was on the need to learn to see, hear, and feel the harm that has been inflicted on humans and more-than-humans, near and far away, in the past, present, and future, requesting from participants a presence that is at the same time intellectual, affective, and ethical. During the four hours of the performance, the courtroom transformed into an ecosystem and a regenerative space. Architecturally and performatively, attention was drawn to the centre by highlighting the many shapes in Nature, as well as the organicity and principles of interconnectedness to the whole. The space brought the audience into everchanging constellations, shifting lights, soundscapes, and materials. There were moments for sitting, walking, taking part in rituals, and signing a restorative contract for the Earth (Image 16.2).

During our collaboration in the *Voice of Nature: The Trial*, we tried to walk the difficult line between denouncing crimes so cruel and of such a massive scale and scope, while resisting a punitive impulse. Our ethos was based on our convictions that each of our actions leads to consequences, that we are all entangled in ecocide (even though this entanglement is unequally distributed), and that we each have our share of work to do. It

Image 16.2 Voice of Nature: The Trial, Maria Lucia Cruz Correia, 2019 (Photo © Mark Požlep)

is once that feeling sinks in and we are freed from the sentiment that the scale of things is simply too big or that things take place too far from us, that we can start to respond, each according to our own possibilities and resources. We thought of the term 'response-ability' quite literally: as our ability to respond as we 'stay with the trouble' of living in a wounded and vulnerable Earth (Haraway, 2016).

The ability to respond is profoundly damaged by demoralising trends such as catastrophising which leads to utter despair, or the other side of the coin, ecological techno-utopianism, characterised by naïve hope and fantasies of unlimited technological progress (Varona, 2021). Both trends are equally demotivating and impede social and moral accountability in relation to environmental harm.

In proposing a new type of 'verdict' in the form of a restorative contract, humans and more-than-humans come together to find a collective language, based on repair and accountability. Art thus is given an active role, as a possibility and space in between, to imagine and transform our relations. This prepares the audience to step into the path of understanding of what a duty of care could mean in relation to environmental harm, by opening their hearts to hear the voice of Nature in themselves (Cruz Correia, 2019, 2021).

4   Voice of Nature Kinstitute: Natural Contract Lab

From the court project, our collaboration developed in 2020 into a 'Kinstitute', a utopian public service acting as kin to more-than-humans. The Kinstitute aims at conducting long-term projects dedicated to environmental and climate justice, ecocide, and the possibilities of proposing justice through artistic, juridical, ritualistic, and restorative justice practices. We draw our thinking about kinship from Indigenous beliefs about their kinship relations to others, to the universe, to land and water and from Donna Haraway's (2016) concept of feminist kin-making and her ethical plea to attend to accountabilities and obligations to care across generations. The Kinstitute's work is based largely on the research pillars we describe above, and rather than being a real place, it is an always ongoing and open readiness to engage in collective thinking and acting around responding to environmental harms.

The first and most developed project the Kinstitute has engaged with so far is *Natural Contract Lab*, an artistic project that aims to address harms made to several rivers in Belgium, Portugal, and Switzerland. *Natural Contract Lab* designs a sustained protocol of repair for bodies of water under deep ecological transformation. In our project, we were influenced by the concept of *hydro-logics* (Neimanis, 2009), a concept attuned to the ways in which water flows through, across, and between human and more-than-human bodies politically, socially, and environmentally (Springgay & Truman, 2019, p. 47). Neimanis (2009) writes that 'our bodies of water open up to and intertwine with the other bodies of water with whom we share this planet—those bodies in which we bathe, from which we drink, into which we excrete, which grace our gardens and constitute our multitudinous companion species' (pp. 162–163).

The artistic format explored during the process of this project is hybrid and dynamic, as it unfolds as a different response to each place, situated as a social encounter, an event, or a collective laboratory. The team that calls itself the 'Body of Repair' is led by artist Maria Lucia Cruz Correia and constituted by environmental jurist, Marine Calmet; restorative justice scholar, Brunilda Pali; scenographer, Vinny Jones; climate grief expert, Evanne Nowak; architect, Lode Vranken; videographer, Mark Požlep; and sound researcher, Margarida Mendes. The Body of Repair is a transdisciplinary collective that questions the state of environmental justice through practices of walking, rituals of care, sensorial exercises, restorative processes, Rights of Nature pleas, landscape grieving, and guardian of Nature contracting.

In what follows, we will mainly focus on the practice we developed on the river Zenne (Belgium) which we call *Walking-with-Zenne*, which started at the beginning of 2021 and is ongoing. The river Zenne flows through the three regions of Belgium (Wallonia, Brussels Capital Region, and Flanders) and passes through the city of Brussels and about 30 villages. Starting from its source in Soignies, it flows into the river Dijle in Heffen from where it reaches the river Schelde and the North Sea. Nevertheless, it is not possible to follow Zenne's course throughout because it often flows underground or along so-called private property. Zenne is a river that has been substantially polluted, diverted, harmed, and forgotten throughout the centuries. For a long time, it was mainly used as an open sewer for the excrements and wastewater of thousands of families and for the chemicals of the adjacent factories. From around 1867 to the 1970s, Zenne was subjected to major urban interventions, which led to its canalisation, tubing, diversion, and vaulting (Holst, 2015). In Brussels, Zenne has been essentially treated simply as water to be managed, used, and governed through concealment and diversion, and yet its water still flows beneath the city, bubbling up, breathing.

Moved by the ecocidal history of the river and the relative silence surrounding this history and its impacts, at the first stage of the research, we designed a sustained protocol of repair that unfolds through a series of artistic 'walking-with' practices supported by our six research pillars. The sustained protocol of repair is essentially a step-by-step long-term process intended mainly as a roadmap of intentions and possibilities. The 'walking-with' practices are concrete ways to enact some of the intentions and actions conceived in the protocol of repair. In the following sub-sections, we will first describe the sustained protocol of repair and then more specifically the 'walking-with' Zenne practices.

4.1   A Sustained Protocol of Repair for Zenne

Even though our work is largely based on artistic processes, we find it important to move beyond performative and short-term practices that do not answer the needs for designing a long-term and sustained protocol of repair for damaged ecosystems. The sustained protocol of repair for Zenne which we have envisioned and designed could take place ideally in six steps: (1) Mapping river relationality; (2) Documenting harms to Zenne and its relations; (3) Documenting existing repair and resistance attempts and processes; (4) Engaging community in repair processes and elaborating commitments; (5) Concentrating and sustaining further repair processes; and (6) Expanding ripples of repair to include a larger community and/or other communities. We describe here the steps as generally conceived and will get more concrete and illustrative at the next sub-section.

Step 1. Mapping river relationality. Relationality is a central concept in our work. The concept refers to (a) being and living in the world (the fact of relationality, relationality as a given); (b) the damage done in the relational sphere by crimes, harms, and injustices; (c) the damage done to relationality by patriarchal, colonial, racist, and extractivist systems; and (d) an aspired relationality and attentiveness to the repair and transformation that is needed to achieve it. These points make clear that we do not praise relationality as a good in itself without ethically and politically engaging ourselves with undoing harmful relationality and acting and working towards an aspired relationality. Given the centrality of the concept of relationality both in terms of harm and in terms of repair, we set the 'mapping of relationality' as a first task for the research.

Mapping of relationality is essentially a temporally and spatially multidimensional process, where different artistic and sensorial techniques can be used such as drawing, writing, photography, film, poetry, storytelling, listening, observing, touching, and collecting. Translated to the reality of Zenne, this has basically meant trying to understand and imagine in which relations is Zenne entangled, how those relations are configured, and how could they be questioned, reconfigured, or transformed? This consisted in mapping Zenne's inhabitants and their activities and entanglements with the river. These included everything and everyone that uses, benefits from, or harms the river (e.g. municipality, corporations, farmers, citizens, infrastructure, planners) and their activities (e.g. walking, polluting, swimming, cleaning, benefiting), but also everyone who cares for the river (e.g. guardians, friends, protectors). It included all the river's other non-human animals, plants, soil, and water relations (e.g. farms, villages, other rivers).

Step 2. Documenting harms to Zenne and her relations. In addition to the mapping of relationality, we also planned to document and record the harms made to Zenne and its relations. These harms can be thought of as belonging to the past, present, and future. The documentation of harms can be done through a combination of material assessment (i.e. assessing the level of waste, chemicals, etc.) and through collecting testimonies and holding conversations with 'witnesses' and 'experts' (e.g. urban planners, environmental regulators, biologists, sociologists, historians, river community members, civil society actors, poets/artists, activists). Whereas both the material assessment and the stories can be collected at any time and place, they are best collected while engaging in 'walking-with' the river practices, where they can be documented as 'voices from the river'.

Step 3. Documenting existing repair and resistance attempts and processes. Another aim of the sustained repair protocol is to document not just harms, but also existing attempts and processes at repair and resistance. Despite the ecocidal heritage, or because of it, the river has guardians and protectors who 'care' for it. Their work, gestures, attempts, and processes of repair and resistance need to be honoured and learned from. They have all left a trace and a trail, and we often literally follow their footsteps in our own practices.

Step 4. Engage river relationality constellations in restorative processes. A central aim of the sustained protocol of repair is to engage all the participants that are involved either in harming or in protecting the river in processes of repair. These processes of repair can be short-term or long-term, modest or serious, personal or collective, so we conceive of them as layers of engagement, that can go from one-time engagements to sustained engagements. It is through a first cycle of community restorative circles and dialogues that the participants and temporary community involved in the 'walking-with' practices can reflect together on ways in which to repair the harms that were documented and recorded through the first steps in the protocol. They have to collectively prioritise harms and repair strategies that matter to them. Each of the circles and dialogues can lead further to collective commitments and restorative action plans. Commitments can be as 'light' as declaring an intention to meet again and continue the conversation or as 'serious' as engaging in a concrete collective action.

Step 5. Concentrated repair process with river relations. During this step, those who have committed to continue a sustained engagement with the repair and accountability process come together again to continue with their collective thinking and acting. In artistic processes, some members of the temporary communities that are created can also be engaging lightly or out of curiosity for the process rather than due to a commitment to the ecosystem in itself, so in designing another and more concentrated repair process, we acknowledge that some members of such communities will drop out in time, whereas others will stay, continue, and persist. As mentioned above, while both types of engagements are welcomed, we hope for and try to create more long-lasting collectives. And as Jolly et al. (2022) write, 'lasting solutions to problems are ones that grow out of, or can fit with, the knowledge, experiences, and desires of those most affected and there are capacities for growth, renewal, regeneration, healing and repair even in distressed networks including families, communities, and environments'.

Step 6. Expanding ripples of repair and river relationality. At the final stage, 'the ripple' of repair and accountability, and river relationality, gets expanded through broader civic engagement and public interventions. Whereas the desirable and available options need to originate organically from the participating communities themselves, some preliminary ideas we have come up are in the line of organising activities in public spaces (e.g. 'processions', 'parades'), collective statements (e.g. 'manifestos'), video ('films', 'documentaries'), or networks or institutions (e.g. 'River Solidarity Federation').

These steps constitute a roadmap which we have conceived as a long-term process of repair for Zenne, but they are not necessarily consecutive and often overlap. In what follows, we will describe more concretely the practice of 'walking-with' Zenne that we have developed where we illustrate the ideas described in some of these steps.

During 2021, besides the participation in several art festivals, we engaged in a total of five walks with Zenne (each time a different segment of the river), each time together with six/seven 'guest' participants in addition to the active team of five/six members. The participants differ every time and participate in the walks either due to their 'expertise' (i.e. being native to the river, having knowledge about it, or having been involved in some kind of project about the river), or simply out of curiosity for the project. Some of them were invited by us, as witnesses or experts, whereas others were free to participate and found the project announcements or invitations online or in various art forums. In the future, we plan to continue the 'walking-with' practices along Zenne by designing a rhythm of two-monthly walks, each walk together with a different community and along a different segment of the river's course, including an uninterrupted walk from its source to its end.

4.2 Walking-with Zenne

None of us is native of the Zenne River. During our engagement with it then, we had to carefully think about what modes of embodiment, attention, and imagination we would need in order to know this place well and create with it an ethical relationality (Myers, 2016, p. 3). In the words of Richard Louv, 'We cannot protect something we do not love, we cannot love what we do not know, and we cannot know what we do not see. Or touch. Or hear' (Louv, 2012, p. 104). From the early development of the work, it became clear that we had to find a methodology and a practice that differed from the more traditional scientific methods of research that tend to foreground rationality over relationality. It also became apparent that discursive-driven practices would not be enough and that our practice needed to be embodied and physical in order to reach other scopes of impact.

As mentioned above, the use of embodied sensory practices aiming at actively changing the perspective of the audience/participants from looking at to 'being-with' the river was quite important. This led to the group discovering that it wanted to pursue this research by 'walking-with' Zenne. The act of walking engages all the senses and creates an attitude of respect that 'folds body, self, other humans and non-humans, time-space and place together' (Edensor, 2010, p. 78). Walking is also a 'more literally grounded approach to perception … since it is surely through our feet, in contact with the ground (albeit mediated by footwear), that we are most fundamentally and continually "in touch with our surroundings"' (Ingold, 2004, p. 33).

We conceive of 'walking-with' not only as a method of research, but also as the basis of sharing our practice with the participants. It was through 'walking-with' that we could co-design our practices with the communities in, along, and around the river, through which we could map the socio-political impact of ecocide in relation to the hydro-commons of Zenne, and through which we could engage in meaningful and sustained processes of collective healing and repair. At the same time, therefore, 'walking-with' becomes a movement of collective thought not only with a plurality of others, but also a process of engaging with erased and forgotten histories of harms (Springgay & Truman, 2019), where we focus our attention to the presence (and often absence) of the Zenne. 'Walking-with' is for us also a methodology for thinking ethically and politically because it is in this process of thinking and 'walking-with' water that we can 'open up questions about human and nonhuman entanglements' (Springgay & Truman, 2019, p. 1; also Alaimo, 2010).

To this end, we continue to develop practices that activate a specific sensory focus, ways of looking and listening and feeling the river and its surroundings, aimed at opening up a reciprocal relationship with the river, and at being more aware of the ways in which we can affect and be affected by it. Such practice creates sensorially aware and relational experience that opens up a space and time for dialogue where embodied perceptions, knowledge, experience, memories, presence, and intentions can be shared. A large part of attention goes to the 'affective tonality' (Gendron-Blaise et al., 2016) of the walks and to how our 'attention would be distributed' (Ahmed, 2008, p. 30) during the walks. Affect, argues Sara Ahmed (2004), is contagious and circulates from one body to another. In that way 'walking-with' becomes a 'form of solidarity, unlearning, and critical engagement with situated knowledges' (Springgay & Truman, 2019, p. 11).

Our artistic experiments and the way in which they are scored within the walk are co-developed with our participants, and therefore, this research is ongoing and never finished. The work also remains a site-responsive and context-specific practice, responding to the situation of the river we are 'walking-with' and the group of people who are walking with us. However, by this stage of the research we can talk about some general principles that have emerged and about how we engage with the river through artistic and sensory means. We aim to give each walk a specific dramaturgy, choreographing the experience through the scoring of sensory practices at specific moments along the walk, combining with the restorative circle work. These practices are focused on the internal sensory experience of the participant, opening up to a reciprocal sensing with the river, and are practiced together, building up a sense of trust and intimacy within our temporary roaming community.

Our research practice is augmented by 'props' that we carry with us as we walk. Designed around the theme of mapping the river by engaging with practices of observing, listening, and collecting, the props were designed and made by artist/architect Lode Vranken, with the aim of being portable objects that could function as both a symbolic and a practical tool through which, and when combined with sensory practices, we could activate specific ways of being with the river. The props are simple wooden objects that have a distinct aesthetics, creating a strong image when carried by a group of people walking along the river. From within the group, they may be seen as generative of a sense of identity for our temporary community, while creating a curious image for the outside eye. They are designed with the idea of being ergonomic; however, they are also somehow contrastingly awkward to carry and necessitate a certain concentration while walking the river parcourse. These props necessitate a moment of collective work and activate the idea of sharing the load, as the objects are passed around the group. Simultaneously, the props embody a sense of physical care and appeal to humour. Through them, we harness the ability of the artistic practice to take the act of

Image 16.3 Natural Contract Lab, Maria Lucia Cruz Correia, 2021 (Photo © Mark Požlep)

walking to another level, opening up for ways of being, and being with, that might be difficult to be quickly achieved in everyday life. The props include chairs constructed with the purpose of collecting and sitting, a collecting basket, a water collecting object, a listening antennae, and a periscope (Image 16.3).

The call for group assembly for each walk is open, and its participants are always different. Hence, our walks include both participants we do not know and members of the river community who accept to be our witnesses and tell their stories about the ecocide or about the resistance and repair attempts. They usually live close to the river and have a strong affinity and bond with its waters. After welcoming our temporary community for the day, we briefly introduce ourselves, the project, and our props, which we then share with everyone. We start our walk by collecting water from the river, asking its permission to let us 'walk-with' it, water which we release at the end back to the river. This water carries the energy of our journey. We also use the water as a 'talking piece' or as an object to anchor the circles and try to remain connected to the river while sharing our stories, thoughts, and feelings. Starting the walk this way also conveys to the participants that we are present with Zenne with purpose and respect and sets the tone for the walk.

Sometime after the departure of the walk, we introduce an Observation Practice (see excerpt from the script below) to create focus and concentration, aiming to take participants out of everyday lives and into the practice of 'walking-with' the river. The idea of reciprocity is introduced in the practice: looking is not a one-way road—when we observe the natural world, it observes us back. We are both looking and being looked at. We are a part of the world and may not simply be constituted as outside observers (Image 16.4).

Observation Practice

Open your eyes.
Take some time to notice what you notice.
What is present here? What attracts your attention or makes you curious?
Continue to observe in silence and without judgement for several minutes as you expand your awareness of your surroundings.
What is present and entangled with the river? And what is absent?
Open your inner eye.

Image 16.4 Natural Contract Lab, Maria Lucia Cruz Correia, 2021 (Photo © Mark Požlep)

*Try to move your perspective from outsider observer to inside the environment,
as an entangled part of the landscape. Try to observe with your whole body and
not only your eyes. Look for the feeling your observations produce?*
[After some minutes]
While you are observing, notice that you are also being observed in return.
*Open your feelings to the presence and sentience of that which you are observ-
ing, observing you in return.*
[After some minutes]
*Take some time to notice what you notice. What has changed in the time you
have been observing?*

After some time, we sit together in the first restorative circle. This first
circle is usually facilitated by restorative justice expert, Brunilda Pali, and is
dedicated to creating a bond amongst the group, creating community with
the surrounding environment and the river. We use the 'talking piece' to
regulate the circle and ask the participants to share stories about their relation with the river or with other significant water bodies in their lives.
Besides the intention of 'creating community', our restorative circle practices also attune to a collective thinking in the 'presence of others' (Stengers,
2005, p. 996). Thinking in the presence of others, creates a space for hesita-
tion, resistance, and extended imagination that produces new modes of
relating. Collective thinking enables the development of a humble and a
critical reflex which demands that 'we don't consider ourselves authorised to
believe we possess the meaning of what we know' (Stengers, 2005, p. 995).

After two or three rounds of questions, we introduce the case of the
ecocide, which we call the Plea for the Rights of Nature, introduced by
the environmental lawyer Marine Calmet. We present here the Plea for
the Rights of Nature (Zenne River) in its full version:
Plea for the Rights of Nature

Ladies and gentlemen, I am speaking today to all entities,
human and non-human.
You who are part of this restorative justice circle,
If we are together here today
it is because,
as an advocate for the Zenne River, I tell you,
the situation is serious.
The fundamental and most basic rights of the River Zenne have been violated.

So to repair this injustice,
an awareness of the harm is needed.
It is not a question of carrying out a cold scientific analysis,
nor is it a matter of mimicking a political tirade
but of feeling deep inside ourselves
the pain, the loneliness, the distress
that the river has felt over the years.

Indeed,
the case that brings us together,
the story of the Zenne,
is the story of a destruction.
A destruction carried out by generations of human beings
who have taken advantage of the generosity of the river to move around, to feed
themselves, to drink...
but who have not given it any care in return.

So today, if we are here
it is to do her justice.

I have to tell you, when I am here,
so close to the river,
that this stream of water still flowing,
I feel like I'm at the bedside of a sick old woman.
In her youth, the Zenne was a carefree stream
wandering through a bucolic region.
Fish swam in her waters
The trees and grasses on her banks were abundant.
The Zenne made her way through the woods and marshes of the region.
On an island formed by one of the river's branches
the first village, called Bruocsella, was established.
The inhabitants living peacefully on the Zenne had a very different relationship
with nature than we do.
Proud of their beautiful home,
they chose the yellow iris as the town's emblem,
a flower symbolising wisdom,
which flourishes on the banks of the Zenne.

How ironic that this plant is also well known for its virtues of water purification.

What ingratitude!
Here we are
a few hundred years later,
totally unaware, amnesiacs.
Have we totally erased from our memories that,
it is only thanks to the presence of the river Zenne
that the little village has grown to become the nerve centre of Europe?

Who would have turned the water, paper and grain mills?
Who would have transported the merchants' boats?
Who would have supplied water to the workshops, the tanneries, the breweries,
the laundries until Brussels became this great trading place?

So maybe, you will say: but the Zenne was not always easy
It's a capricious river,
That is true!
Her frequent overflows caused many floods
But it was men who did this to her!
With the construction of water mills, bridges and houses along the riverbed, the
water flow was slowed down
and the ponds that used to dot the valleys of Brussels and that mitigated the rise
of the floods disappeared under the constructions.

Attempts were made to control and to tame the Zenne bassin,
At the end of the 19th century, a large number of its ponds were filled in,
and many marshlands were drained.

Today the Zenne is only a small stealthy stream, choked by the galloping urban-
ism of the city.
Torn from its bed,
diverted by the construction of the canal,
buried in rubble,
polluted by the discharges of the surrounding industries
and imprisoned in a long underground journey,
the fate that humans have reserved for the Zenne is a painful path.

The life-sustaining river that once supported a wonderful diversity of species and ecosystems,
feeding wetlands and other aquatic habitats with abundant water,
providing vital nutrients to the surrounding land,
transporting sediments to river deltas teeming with life,
performed vital ecological functions.

Humans have caused significant pollution of the Zenne River, which is now contaminated upstream,
including organic matter from sewage, plastic waste, pathogens and nutrients from agriculture, contaminants from industry, in addition to many other forms and sources of pollution.

These contaminations impact aquatic health and biodiversity.

Humans have caused large-scale physical changes to the river
through bridges, dams, locks, embankments and other infrastructure,
which impact the entire river,
resulting in habitat fragmentation,
reduced biodiversity, and endangered fish populations.
A threat now exacerbated by climate change.

Humans have created environmental laws that are largely inadequate to pro-tect the integral health of the Zenne River and its river basins.
These laws also fail to ensure that current and future generations of humans and other species and ecosystems have an adequate supply of water to meet their basic needs.

The fundamental rights of the river were violated!
the right to flow,
the right to perform essential functions within its ecosystem,
the right to be free from pollution,
the right to feed and be fed by sustainable aquifers,
the right to native biodiversity, and
the right to regeneration and restoration;

Why doesn't the human community living in the Zenne basin recognise these rights?

Elsewhere, other humans have already initiated legal recognition of the rights of the rivers.
A New Zealand treaty recognises the Whanganui River as an indivisible and living whole and a legal entity, with guardians appointed to represent the river's interests.
The Constitutional Court of Colombia has ruled that the Atrato River Basin has rights to protection, conservation, maintenance and restoration.

A decision of the Ecuadorian Provincial Court to condemn the state to respect the constitutional rights of the Vilcabamba River and request its restoration and rehabilitation.

The City of Curridabat in Costa Rica has recognised bees, but also humming-birds, butterflies and bats, as citizens like any other human.
The local administration has redesigned all urban planning to create ecological corridors allowing pollinators to freely move from one green space to another. It's a city designed for animal and plant species, in which humans can also live better.

The Zenne River should also obtain the recognition of its fundamental rights in order to favour the creation of a new legal and social paradigm based on living in harmony with Nature.

Today we are here to meet this new paradigm, to represent this new order in our society, an order including the Rights of Nature.

For this I will call my first witness; [here the witness invited for the day is introduced]

I call now my second witness: the river Zenne I ask you to listen carefully…

When Zenne is called as first-hand witness to ecocide, we usually introduce one of the props, which is a listening antennae. Underlying the idea that a river is a body with integral rights, we place the antennae in the circle. We connect a hydrophone (underwater microphone) to the antennae so that the 'voice' of the river can be presented in the circle and listen collectively to the river. This is a moment where the river can 'speak for itself'. After one or two rounds of questions, we conclude the first circle. Following this step of the walk, we engage in individual listening to the underwater sound and engage in a listening practice before starting to walk again, this time in silence to give the focus to listening to the river.

During the walk we engage with an ongoing practice of 'mapping and collecting'. These mapping tools aim to collect what we observe and to collect and remove harms that we encounter. When walkers encounter something that they feel is important to collect, be it an object, an observation, or the contact of a person or group connected to the river, they are asked to place it in the (collecting) chair. When walkers see a harm that can be removed, such as rubbish or invasive plants, we ask them to place it in the collecting basket as a gesture of care. Both objects cannot be operated by one person alone, as it is difficult to both carry and fill them while walking. The objects require that we work together. What is essential to highlight here is that through the props and the sensory

Image 16.5 Natural Contract Lab, Maria Lucia Cruz Correia, 2021 (Photo © Mark Požlep)

Image 16.6 Natural Contract Lab, Maria Lucia Cruz Correia, 2021 (Photo ©
Mark Požlep)

practices, participants are immersed in the doing of things, -observing,
listening, and so on-, with a particular attention and context (Image 16.5).

After some time walking, the second circle is organised. This circle is
focused on grieving the river and is led by Evanne Nowak, who has a
research background into climate apathy and currently facilitates climate
grieving processes (Image 16.6):

> Grieving of the river
> *My name is Evanne*
> *I live in Amsterdam*
> *I feel often paralysed because of the destruction of the planet. Climate change.*
> *Sea level rise. The uncountable crimes against Nature. The uncountable loss of*
> *species.*
> *The predictions of all these future losses.*
> *I'm on a quest to learn how do we understand, think, and feel the meaning of*
> *all the ecological loss? How to not become numb, anxious, depressed and apa-*
> *thetic of everything that is at stake?*

I'm here to guide you through 'grieving a river', or more specifically: to grieve Zenne.
However, I have never been here before. I've even never seen the Zenne.
To be honest: I have never mourned a river.
So I'm a bit uncomfortable.
Please bear with me.

We live on a very damaged planet. I know the IPCC reports frightening predictions on the future, the WWF Living Planet reports—but at the same time, there is a gap: between the numbers I read, the graphics I see of declining biodiversity ... and the world I experience around myself. I see water, here at Zenne, and I do not feel anything at all—I don't know if it is going well or badly ... I just enjoy it, as it is.
I read a text of philosopher Judith Butler that shook me up:
She states:
who and what we do not mourn tells us as much about ourselves as how and what we do mourn.
There are myriad deaths and losses that go unnoticed and unmourned by ourselves and others.

Rivers, minerals, landscapes, weather, snow, worms, fish, mountains, the sea... have you ever mourned them?
I haven't.

And yet, while we may not explicitly mourn, we are still shaped and affected by those losses,
consciously or unconsciously.
Philosophers and climate psychologists say that mourning our damaged landscapes might be a doorway to action. A pathway to learn to respond. To see with new eyes, to feel deeper, to learn to love the more-than-human. Mourning as a call to responsibility to engage with what was lost.
Mourning exposes our connections to others: human, animal, vegetable, mineral..., and provides an opportunity to connect ourselves and others through loss and shared vulnerability.
They say:
To be creatures who love, we must be creatures who despair at what we lose.
If we truly love what is beyond us, what is around us, what grounds us, and what creates us, we must be prepared to feel the full depth of loss, of grief, of despair, at that loss of non-human bodies.

So here we are today.
To explore:
what counts as a mournable body? Is a river a mournable body?
On an little exploration—to learn to mourn beyond the human.
how does grief help us live better with others? To live more gentle, caring,
responsible in
relation to our rivers, with the Zenne?

[Visualisation with closed eyes]:
Think of the rivers, the little streams, the canals, the large rivers you have
crossed in your life:
during your daily routines—to work, to study, to shop, to the train station,
while visiting friends and family, etc.
during holidays—when you were canoeing, sailing, hiking, swimming,
ice skating
the rivers where you have swum as a child
the rivers where you built little dams with stones
the rivers where you have been fishing
the rivers where you have swum naked
the rivers from which you have been drinking
the rivers that scared you, or maybe even harmed you
the rivers that were forbidden to swim in
the rivers you longed for
the rivers that cooled you during hot summer days and heatwaves.

[Writing: Describe a river that you have lost in your life]
think of a river you have lost; in a broad sense;
maybe the river is:
irreversibly changed;
forgotten; taken for granted;
polluted;
being dammed;
filled in/closed;
or its inhabitants might have gone extinct;
or maybe its banks are being transformed, harmed, polluted
maybe its surrounding landscape is being lost, changed irreversibly.
Describe this river's colour, smell, where it is/was streaming, its surrounding
landscape.

Describe what is changed, or lost...
Describe how you feel concerning this loss.

[Reading/sharing and listening]
[Writing: What touches you when listening to the others? What resonates and why?]
[Reading/sharing and listening]
[Writing: What does this mean, for your relationship with the Zenne?]
[Reading/sharing and listening]

After the grieving stage, which gets participants to dig further into their own memories and emotions, but also to realise both the affinity with and the difference of their own perspectives with those of others in the temporary community, we continue the walk further, until we reach the end of the walk, and sit together for the last circle. Our final step is dedicated to the co-creation of a guardian of Nature contract. This is based on our firm assumption that 'we are no more predisposed towards being destroyers of Nature than we are toward being stewards' (Loring, 2020, p. 20). To be a guardian means to uphold a view of people as part of Nature, neither separate from nor above it. A guardian should be able to articulate a relational jurisprudence and formulate arguments to defend laws for equitable, reciprocal, and balanced relationships with non-humans. As an outcome of this agreement, participants take the responsibility as guardians of the river, and together we sketch a proposal that envisions the protection of the vital cycle of the Zenne river. The contract also works as a process to clarify expectations; who will take what role, when, in what manner. As mentioned previously, while we turn our attention to accountability, we also shift the idea of accountability from being responsible for to a response-ability-with (Thiele, 2014) where to be accountable is eventually about 'making commitments and connections' (Barad, 2007, p. 329), so that we can enhance our collective abilities to respond to harms and imagine new ways of co-existence.

5   Conclusion

As we work to heal the earth, the earth heals us.
—Robin Wall Kimmerer, Braiding Sweetgrass (2013, p. 339)

Our research-led artistic practice is an example of the important role that imaginative processes can have in responding to environmental harms. The modality of its precedent opens a debate among practices and fields of knowledge, expanding on the crucial expertise that the arts, law, and the environmental humanities can have on this era of imminent planetary changes and climatic upheavals. Imagination-led tools and artistic practices have demonstrated that they can open a terrain where we co-cultivate our right, duty, and power to imagine different modes of living together with the Earth and plural forms of life that inhabit it.

Working through collaborative forms of exploratory address of intricate political topics such as ecocide, the Rights of Nature, and climate grief, our group has developed novel, situated, and transdisciplinary methodologies that use the grounds of restorative justice to sensitise and bridge new communities. Hence, it has opened itself to collaborative and interdisciplinary practices, resulting in exploratory and critically engaged formats.

Given the interdependency of our work with situated participants and locations, our research practice differs from traditionally conceived methodological approaches, whereby research processes can be designed as closed circuits and applicable formulas, fixed from the start or decided from outside our (temporary) collectives. Rather, it can be seen as an enmeshed practice, that dwells in particular contexts and encounters, taking place temporally and physically 'in the middle'. It is 'in the middle', where things always move, grow, expand, and pick-up speed (Deleuze & Guattari, 1987) and where 'modes of thinking-making-doing' emerge from the processes themselves and those engaged in them (Springgay & Truman, 2018, p. 206). While our six research pillars and intentions elaborated in the chapter have given us firm anchorage, 'the middle' can't be known in advance: one has 'to be "in it", situated and responsive' (ibid.), in order to be nurtured by the answers this brings.

Much like the tacit pact of allowing oneself to float on unknown waters to reach a state of presence within a milieu, one must dare to leave open space for questions, experimental methodologies, and collective intuitions, in order to find reasoning and reach accounts of the lived experience in each of the contexts we study. What we have seen emerging from the walks is not only a revived sense of care for the river and the more-than-human inhabitants of its ecosystem, but also a sense of reciprocal care that grows as we collectively open ourselves to the river, and witness a sensation of replenishment emerging, that is for many of us a form of healing that comes from 'walking-with' and 'being-with' a river.

References

Ahmed, S. (2004). *The cultural politics of emotion.* New York: Routledge.

Ahmed, S. (2008). Open forum imaginary prohibitions: Some preliminary remarks on the founding gestures of the 'new materialism'. *European Journal of Women's Studies,* 15 (1), 23–39.

Alaimo, S. (2010). *Bodily natures: Science, environment, and the material self.* Bloomington: Indiana University Press.

Albrecht, G. (2020). Negating solastalgia: An emotional revolution from the anthropocene to the symbiocene. American Imago, 77 (1), 9–30.

Barad, K. (2007). *Meeting the universe halfway: Quantum physics and the entanglement of matter and meaning.* Durham: Duke University Press.

Climate Academy (2020). This movement wants to make harming the planet an international crime. *The Guardian,* 16 September 2020. Retrieved from https://www.theguardian.com/climate-academy/2020/sep/16/ecocide-environment-destruction-international-crime (last accessed 18 January 2022).

Correia, M.L.C. (2019). Voice of Nature: the trial—re-storying environmental justice. *Newsletter of the European Forum for Restorative Justice,* 20(3), 6–8.

Correia, M.L.C. (2021). Voice of nature: the trial. An artistic response for environmental justice. The International Journal of Restorative Justice, 4(1), 166–171.

Deleuze, G., & Guattari, F. (1987). A thousand plateaus: Capitalism and schizophrenia. Minneapolis: University of Minnesota Press.

Edensor, T. (2010). Walking in rhythms: Place, regulation, style and the flow of experience. *Visual Studies,* 25 (1), 69–79.

Gendron-Blaise, H., Gil, D., & Mason, J. E. (2016). An introprocession. *Inflexions*, 9, i–vii.

Haraway, D. (2016). *Staying with the trouble: Making kin in the Chthulucene*. Durham: Duke University Press.

Holst, P. (2015). Transgression/transition: An exploration of Zenne and its surroundings. Retrieved from https://paolettaholst.info/research/transgression-transition-an-exploration-of-the-zenne-and-its-surroundings/ (last accessed 18 January 2022)

Higgins, P., Short, D., & South, N. (2013). Protecting the planet: A proposal for a law of ecocide. *Crime, Law and Social Change*, 59(3), 251–266.

Ingold, T. (2004). Culture on the ground. *Journal of Material Culture*, 9 (3), 315–340.

Johnstone, G., & Van Ness, D. W. (2011). The meaning of restorative justice. In G. Johnstone & D. W. Van Ness (Eds.), *Handbook of restorative justice* (pp. 5–23). London and New York: Routledge.

Jolly, R., Gehman, R., & Burford, G. (2022). Looking for the restoration in restorative justice's response to civil disobedience, this volume.

Kershen, L. (2019). Implementing restorative justice to environmental harm. In E. Biffi & B. Pali (Eds.), *Environmental justice: Restoring the future*. Leuven: European Forum for Restorative Justice.

Kimmerer, R. W. (2013). *Braiding sweetgrass: Indigenous wisdom, scientific knowledge and the teachings of plants*. Minneapolis: Milkweed Editions.

Laing, O. (2020). Funny weather: Art in an emergency. W. W. Norton & Company.

Loring, P. A. (2020). *Finding our niche: Toward a restorative human ecology*. Nova Scotia: Fernwood Publishing.

Louv, R. (2012). *The nature principle: Reconnecting with life in a virtual age*. Chapel Hill: Algonquin Books.

Minguet, A. (2021). Environmental justice movements and restorative justice. *The International Journal of Restorative Justice*, 4(1), 60–80.

Myers, N. (2016). Becoming sensor in sentient worlds: A more-than-natural history of a black oak savannah. In G. Bakke & M. Peterson (Eds.), *Anthropology of the arts*. New York: Bloomsbury.

Neimanis, A. (2009). Bodies of water, human rights and the hydrocommons. *Topia*, 21, 161–182.

Nellis, S. (2021). Enacting law: The dramaturgy of the courtroom on the contemporary stage. Journal of the Cultural Studies Association, 10.1, https://csalateral.org/issue/10-1/enacting-law-dramaturgy-courtroom-contemporary-stage-nellis/ (last accessed 28 January 2022).

Pali, B. (2020). Nourishing the restorative imagination. In G. Varona Martínez (Ed.), *Arte en prisión: Justicia restaurativa a través de proyectos artísticos y narrativo* (pp. 23–40). Valencia: Tirant lo blanch.

Pali, B., & Aertsen, I. (2021). Inhabiting a vulnerable and wounded earth: restoring response-ability. Editorial Special Issue Environmental Restorative Justice. *International Journal of Restorative Justice, 4*(1), 3–16.

Pihkala, P. (2020). Climate grief: How we mourn a changing planet. BBC Future, 3 April 2020. Retrieved from https://www.bbc.com/future/article/20200402-climate-grief-mourning-loss-due-to-climate-change (last accessed 18 January 2022)

Stone, C. D. (1972). *Should trees have standing?* Oxford: Oxford University Press.

Stop Ecocide Foundation (2021). Independent expert panel for the legal definition of ecocide: Commentary and core text. Retrieved from https://static1.squarespace.com/static/5ca2608ab914493c64ef1f6d/t/60d1e6e604fae2201d03407f/1624368879048/SE+Foundation+Commentary+and+core+text+rev+6.pdf (last accessed 18 January 2022)

Siddique, H. (2021). Legal experts worldwide draw up 'historic' definition of ecocide. The Guardian, 22 June 2021. Retrieved from https://www.theguardian.com/environment/2021/jun/22/legal-experts-worldwide-draw-up-historic-definition-of-ecocide (last accessed 18 January 2022)

Springgay, S., & Truman, S.E. (2018). On the need for methods beyond proceduralism: Speculative middles, (in) tensions, and response-ability in research. *Qualitative Inquiry, 24*(3), 203–214.

Springgay, S., & Truman, S.E. (2019). *Walking methodologies in a more-than-human world. Routledge Advances in Research Methods.* London: Routledge.

Stengers, I. (2005). The cosmopolitical proposal. In B. Latour & P. Weibell (Eds.), Making things: Public atmospheres of democracy (pp. 994–1003). Cambridge, MA: MIT Press.

Thiele, K. (2014). Ethos of diffraction: New paradigms for a (post)humanist ethics. *Parallax, 20*(3), 202–216.

Varona, G. (2021). Why an atmosphere of transhumanism undermines green restorative justice concepts and tenets. *The International Journal of Restorative Justice, 4*(1), 41–59.

Wessels, H., & Wijdekop, F. (2022). Restorative justice and Earth jurisprudence, this volume.

17

Harm to Knowledge: Criminalising Environmental Movements Speaking Up Against Megaprojects

Anna Di Ronco and Xenia Chiaramonte

1   Introduction

As Forsyth et al. (2021) eloquently put it, for a process of restorative justice to be initiated at all, someone has to *acknowledge* and *take responsibility* for the caused harms. We know that this has occurred in a few instances, even beyond the context of restorative justice conferences. For example, the mining company Rio Tinto recently took responsibility for the destruction of a sacred Indigenous site in the expansion of an iron ore mine in Western Australia. Although the company was obviously not able to bring back to life the 46,000-year-old rock shelters at Juukan Gorge, which it had destroyed, it replaced its leadership with a new executive team that seeks to rebuild the relationship with and trust of Traditional Owners by actively promoting the protection of Indigenous

A. Di Ronco (✉)
Department of Sociology, University of Essex, Colchester, UK
e-mail: a.dironco@essex.ac.uk

X. Chiaramonte
ICI Berlin, Berlin, Germany

cultural heritage (Wahlquist, 2021). Although this may sound like 'too little, too late' and a mere 'window-dressing' effort of the mining company—which was pressured by investors into acknowledging the disaster it caused (Butler et al., 2020)—what matters from a restorative justice perspective is that Rio Tinto eventually acknowledged that it had wronged Indigenous peoples with its unilateral and non-transparent decisions. The recognition of the harmful consequences of one's actions is, indeed, a necessary first step towards repairing and healing harms. Such a step is, however, also necessarily followed by a meaningful relational and dialogic engagement with the harmed subjectivities (Forsyth et al., 2021).

Usually, harms are not acknowledged by relevant corporations and by the state until it is too late, that is, when often irreparable harms have already been caused to humans, non-human animals, and plants, entire ecosystems, and biospheres (Hamilton, 2021). Harms, however, are often denounced by environmental activists and affected populations *in their making* and frequently even *before they are produced*, for example, when corporate projects are made public or when legislative changes—which are deemed detrimental to the environment—are announced by governments. These grievances are often ignored or downplayed by multinationals, states, and mainstream media alike, with environmental movements often being misrepresented as 'eco-terrorist' and ideological enemies impeding economic progress (Hasler et al., 2020)—and ultimately being criminalised.

This chapter focuses on two highly silenced and criminalised environmental movements in Italy: the No Tav and No Tap movements, which are currently fighting against the Turin-Lyon high-speed railway (TAV) and the Trans-Adriatic Pipeline (TAP), respectively. As we illustrate in this chapter, these two environmental movements have produced practical, technical, localised, and situated knowledge on the harms caused by the two megaprojects; however, despite the ability of such knowledge to avoid or mitigate harms, it has yet to be properly considered by relevant state and corporate actors. In the chapter, we discuss how this dismissed knowledge—and the criminalisation of the producers of such knowledge—constitutes a proper form of harm, what we call 'harm to knowledge'. The latter is especially insidious as it dismisses the counter-knowledge produced from below, ultimately frustrating its transformative potential.

This chapter is structured as follows. It starts by reviewing the concept of environmental harm from a green criminological perspective, and the recent green criminological literature that—often using innovative methodologies—captured environmental harms as perceived, experienced, represented, and expressed by the affected individuals and communities. After providing background information on the two megaprojects and the struggles against them, the chapter briefly discusses the authors' previous studies which inform this piece. In its central part, it explores some forms of environmental harms connected to the two megaprojects as emerging from activists' voices—voices which we collected through interviews, ethnographic fieldwork, and social media research. Through on-the-ground and virtual ethnographies we were also able to collect activists' artistic expressions such as songs, vignettes, videos, and street art pieces, which we also included in our analysis. We conclude this chapter by analysing environmental harms through the concept of 'harm to knowledge' and by arguing for the need to expand the scope of environmental restorative justice (ERJ) to include the resolution of present and future conflicts.

2 Environmental Harm in Green Criminology

A burgeoning perspective since the 1990s, green criminology focuses its attention on the study of crimes and harms affecting the environment, the planet, as well as human and non-human species inhabiting the Earth, and environmental (in)justice. It is seen as an evolving 'perspective' (South, 1998) grounded in critical criminology (Ugwudike, 2015), open to multi- and interdisciplinary approaches (Ruggiero & South, 2013) as well as to various theoretical orientations (see, e.g., White, 2013)—with the latter also including cultural criminology (Brisman & South, 2013, 2014) and the social harm perspective or zemiology (Beirne & South, 2013; Brisman & South, 2018).

Proponents of this perspective oppose environmental degradation, destruction, harms, and injustices and expose the negative impacts of such practices on humans, non-humans, and the planet as a whole (Ugwudike, 2015). They often do so in an effort to build what Goyes (2016, p. 508) called 'green criminological activism', which is a 'stance where [...] criminological knowledge and activity is placed at the service of those victimised on the basis of class, species, gender, sex, race, ethnicity or age'; in practice, this stance involves making 'a purposeful attempt to try to prevent such victimisation by making an impact in the social, political or cultural realms via research, teaching or service'.

Central to the perspective of green criminology is the concept of *harm*, which not only includes but also transcends criminal harm—or any harmful action or inaction that is punished by the criminal law. The idea is that all harms—including those that are *not* protected by the criminal law and are perhaps caused by entirely *legal* activities—deserve criminological attention for generating human and non-human suffering. The extension of the focus of green criminology to non-criminalised harms also matched developments within the field of green victimology (Hall, 2014), which in recent years has approached as victims not only individuals but also non-human species and environments affected by non-criminal harms (White, 2018)—all victims that have also been recognised a seat at the table of ERJ conferences (Forsyth et al., 2021; Varona, 2021). This is also in line with the recent socio-legal studies literature, which recognised the limitation of the human rights perspective to account for the complexity of environmental harms and victims (not only human but also non-human) (Altopiedi, 2020). To address this concern, in the Latin American tradition—in Ecuador and Bolivia, in particular—legal recognition has been given to the harms affecting non-human entities through the so-called rights of nature (Acosta & Martínez, 2011).

To study environmental harms, green criminological studies have been informed by at least one idea of justice that contributes to what White (2008) called an 'eco-justice approach': environmental justice (which addresses specifically the negative impacts of environmental harms on humans at the intersection of race, gender, class, and other systems of oppression), ecological justice and species justice (where the focus is on protecting the environment and animals, respectively) (White, 2013).

From the perspective of environmental justice, there has recently been a number of green criminological studies that have focused on capturing environmental *harms as perceived, experienced, and articulated* by the affected individuals and communities. Such studies have also experimented with innovative methodologies, which included itinerant soliloquies (Natali & de Nardin Budó, 2019), interviews with photo elicitation (Natali, 2016, 2019), involvement of 'peer' researchers from the Indigenous researched communities in the data collection (Goyes et al., 2021), and computational social science methods applied to social media material (Di Ronco et al., 2019; Di Ronco & Allen-Robertson, 2020). In many such studies, harms have emerged through interviews and participatory (mobile) methodologies, as well as through art—including memorials in natural landscapes (Varona, 2020)—and social media messages (Di Ronco et al., 2019; Di Ronco & Allen-Robertson, 2020). This chapter contributes to this emerging scholarship by examining how environmental harms have been articulated and represented by No Tav and No Tap activists in formal and informal interviews with us, and in activists' artistic expressions (such as songs, vignettes, videos, and street art pieces), which we collected during both ethnographic fieldwork and (only in the research on the No Tap movement) social media research. In the last sections of this chapter, we demonstrate how these articulations often constitute proper (counter-)knowledge on environmental harms produced from below and use the concept of 'harm to knowledge' to analyse relevant authorities' systematic dismissal and silencing of this knowledge.

3   Study Background: The Cases of No Tav and No Tap

Megaprojects have an inevitable impact on the territories where they are implemented and therefore need a careful analysis of their associated risks. Although there is an obvious need for information and evaluation of high-impact projects, the international history of megaprojects has often turned out to be a mix of underestimated environmental impacts and overestimated economic revenues (Flyvbjerg et al., 2003). Frequently, projects as such—their scope and concrete execution modalities—are not properly communicated to the public (Calafati, 2006) and by simply resorting to the general need for development and progress, these infrastructures are often defined as necessary and 'strategic'.[1] The lack of adequate information and meaningful public participation in the phases that lead to the approval and execution of the project often motivate local populations to study the projects on their own and organise counter-investigations, which can then lead to more or less lasting mobilisations. This opposition may—or may not—be accepted by the state and its law enforcement agencies, which can also rely on criminalisation to intimidate activists and ultimately suppress dissent. This path is shared by the No Tav and the No Tap movements, on whose histories we ground the following subsections.

3.1 No Tav

The Turin-Lyon high-speed railway project (TAV) dates back to 1989. Grassroots resistance to the megaproject began at the same time. Today, the project is run by TELT (Tunnel Euralpin Lyon Turin)—a French company owned 50 per cent by the French State and 50 per cent by the state-owned company managing the Italian railways; it is also co-financed by the European Union. In the late 1980s, the local population had received no information about TAV. The first attempts to probe the ground, however, attracted the attention of some inhabitants of the Susa Valley (or *valligiani*) who saw self-styled technicians operating without any authorisation on their land. In 1991, a 'tiny group' of *valligiani* called the first demonstration against the project (Dosio, 2006, p. 10).

Officially, the No Tav movement was born in December 1991, by a decision of *Habitat*, a committee of 60 inhabitants, university professors, and local administrators from the Susa Valley. Residents, academics, and administrators have constituted the backbone of the committee and, later, of what became known as the No Tav movement. Through the production of technical counter-knowledge on the megaproject, activists

---

[1] Usually, this definition has a practical consequence: it increases penalties for those who illegally access the construction site, which is defined as an "area of national strategic interest". In the No Tav case, penalties were increased through the 2012 *Legge di Stabilità* (art. 19).

have created and disseminated an expertise that has greatly contributed to the social struggle against the TAV project.

Essentially, No Tav activists have opposed the project for three main reasons. Firstly, they consider the project useless on the ground that existing infrastructures have not yet reached a level of saturation and that neither traffic by road nor rail has increased. Secondly, the project is believed to have excessive costs: the 2004 Paris agreement between France and Italy on TAV imposes most of the economic burden on Italy, requiring it to contribute to two-thirds of the overall TAV costs, in spite of the fact that only one-third of the route would cross Italian soil. Thirdly, the project is challenged for the harmfulness of substances contained in the mountains to be excavated—uranium, radon, and asbestos (Giunti et al., 2012).

To oppose the megaproject, No Tav activists have relied on a vast repertoire of protest practices, such as symbolic cuts of the construction site's protection fences, street art, marches, sit-ins, and festivals involving artistic performances and expressions. In 2005, No Tav activists succeeded in blocking the installation of the construction site, after having suffered acts of violence committed by the police (Chiaramonte, 2019a). As a sort of reward, the government set up an Observatory whose publicly stated goal was to negotiate and find an agreement between the residents and the company. However, those who made the most radical proposal, namely that of not building the megaproject at all, were expelled, their position being considered as incompatible with that of the Observatory.

In January 2010, one of the most famous anti-terrorism and anti-mafia Italian district's attorney offices instituted a specific pool to investigate the No Tav movement. Eventually, in 2011 the TAV construction site was installed and put under 24-hour police surveillance.

Soon after, criminalisation began and has not yet stopped: in fact, it has now gone on for more than 10 years involving over 1500 people (as both suspects and defendants), about 150 criminal proceedings, the use of pre-trial precautionary and preventive measures, and extremely serious criminal charges including that of terrorism (Chiaramonte, 2019a). Especially in the last decade, the movement has received solidarity, participation, and attention from a large number of social movements at the national and international levels. In recent years, it has also become the catalyst for an ecological proposal shared with the youngest and most global movements against climate breakdown, namely, Extinction Rebellion and Fridays for Future (see Chiaramonte, 2020).[2]

3.2   No Tap

The Trans-Adriatic Pipeline (TAP) has a more recent history than TAV. TAP is a state-authorised project that brings natural gas from Azerbaijan to Italy—and through it to Europe—through Turkey, Greece, and Albania. It is part of the Southern Gas Corridor—an EU project that is partly funded by the European Investment Bank (EIB). Its landing point in Italy has been identified in the Salento province of the southern-east Puglia region and, more specifically, in a part of the municipality of Melendugno called San Foca, which is famous for its marine (bathing beach) of San Basilio, its protected natural habitats, and its nature reserve of Torre Guaceto with century-old olive trees.

Opposition to the pipeline started in Melendugno and surrounding villages even before it was approved by the Italian Parliament in 2013: it dates back to 2010 and 2011, when the location of the pipeline's landing point started being discussed in the region (*No Tap*, n.d.). Since then, various No Tap groups and organisations have been established by residents and activists. Ever since the company received its state authorisation to start building the Pipeline Receiving Terminal in Melendugno in 2014—which did not involve any consultation with residents—these groups have challenged the pipeline for being 'illegal', 'useless', and 'harmful' (Di Ronco & Allen-Robertson, 2020). Since then, No Tap activists have also gathered, produced, and disseminated knowledge and information about the pipeline through local and international events and their use of social media.

The No Tap protest succeeded in reaching a wide public participation at protests especially during the spring and summer 2017, when TAP

---

[2] At the No Tav summer festival, called Alta Felicità ('High Happiness'—instead of 'high speed'), a round table on political ecology has been started in the last four years. Its aim is to reformulate the movement's political struggle in collaboration with the most recent global protests against the climate crisis.

uprooted and removed olive trees from the first construction site (*No Tap*, n.d.). The removal of olive trees sparked outrage in the local population, who spontaneously gathered around the construction site to express dissent. The images of those days of protesting show hundreds of people—including families with children and old people—peacefully protesting against the removal of olive trees (see Di Ronco et al., 2019). Following these protests, activists reported a substantial increase in police presence in the area, which involved its militarisation by the police, and also a fiercer criminalisation of activists (*No Tap*, n.d.; Papadia, 2018)—a criminalisation that is still ongoing and on which we will focus in the findings section.

4   Methodology

In this section, we review the methods for data collection that we utilised in this study to capture how No Tav and No Tap activists expressed and articulated the harms caused by the TAV and TAP megaprojects, respectively. In both case studies, these methods relied on ethnographic fieldwork and interviews, and—in the case of the research on the No Tap movement—also on social media research. Given the different resulting datasets, we analysed these two case studies separately. The thematic analysis of the collected material (which was carried out individually for each case study, see below under Sects. 5.1 and 5.2) allowed us to identify how No Tav and No Tap activists perceived, experienced, and articulated environmental harms.

4.1   No Tav

To illustrate the way No Tav activists expressed TAV-related environmental harms, we relied on data collected by the second author during (i) a two-year observant participation in the Susa Valley (2013–15), which was followed by in-depth interviews with 17 Tav activists (in 2015), and (ii) a two-year judicial ethnography at the bunker courtroom in the Turin's prison (2013–15) (see Chiaramonte, 2019a). This material served to illuminate the harms identified by activists in their narrations: in particular, these are the harms to human and non-human health. The data was also used to illuminate the main criminalisation techniques that have harmed the No Tav movement and its resources, which have however not discouraged No Tav activists from fighting the project; as an activist expressed it in an interview, the movement's criminalisation is 'a medal of honour' (Magno, 15 October 2015)—that is, an important recognition of its political relevance.

4.2 No Tap

To illuminate the environmental harms as described by No Tap activists, this chapter relied on visual social media and semi-structured and informal interviews with No Tap activists. Visual social media (images, pictures, and videos, in particular) were extracted from two of the first author's previous studies of #NOTAP activism on Twitter (Di Ronco et al., 2019; Di Ronco & Allen-Robertson, 2020). In these two studies, Twitter posts were computationally collected through a 'Listener' tool which utilised a Twitter streaming API and collected tweets in real time 24 hours a day during two different time-frames: from early June until the end of August 2017, in the first study (Di Ronco et al., 2019), and from October 2018 until June 2019, in the second (Di Ronco & Allen-Robertson, 2020). Formal and informal interviews with activists were conducted by the same author during on-the-ground ethnographic fieldwork in the municipality of Melendugno and surrounding areas in 2019 (see Di Ronco & Allen-Robertson, 2020). This data is also complemented with a more recent virtual ethnography of activists' posting of visual material on Facebook and Instagram during the COVID-19 pandemic. From the thematic analysis of visual social media material and of (formal and informal) on-the ground interviews, different harms emerged. For the purposes of this chapter, we only focused on the two most prominent harms emerging from activists' narratives and representations: harms to the land and olive trees, and harms of repression.

5 Speaking Up About Environmental Harms

In the following subsections, we focus on the main forms of environmental harms as emerging from our analyses. They include harm to human and non-human health (in the case of No Tav), harm to the land and olive trees (in the case of No Tap), and harms of criminalisation/repression (in both cases).

5.1 No Tav

Harms to Human Health and the Healthy Environment

No Tav activists contend that TAV is harmful to people's health and a healthy environment. As Magno—a local No Tav activist who is also affected by cancer—expressively said, 'I am obliged to believe in it in my eyes' (15 October 2015; see Chiaramonte, 2019a, pp. 78–81). What Magno meant was that the TAV-related harms suffered by the population come from their direct, bodily experience. As demonstrated by Fazio and Minnelli (2020), the megaproject significantly raises the already high carcinogenic levels in the area, in particular in light of the presence of asbestos and uranium in the excavated mountains.[3] Of course, this also negatively affects the people who work for TAV; as Angela (12 October 2015), a local resident and activist, put it: 'There are, I was told, 11 workers who take turns entering [the construction site]. They work amid asbestos and uranium so such a work kills and devastates' (Chiaramonte, 2019a, p. 346).

Health is conceived by activists not only as a human right or need but also as a duty to respect the environment and all its inhabitants (including not only humans but also animals and plants). For example, activists noted that the Susa Valley is a geographical area on which many other

---

[3] In particular, as Fazio and Minnelli (2020, p. 479) put it, 'In early 2000s, pleural and peritoneal malignancies were found in excess in some municipalities of Upper Susa Valley (Piedmont Region, Northern Italy), where tremolite asbestos in rocks surfaced by natural ground erosion or originating from construction activities'.

'developmental' infrastructures have already been built; they include two highways, a power line, factories, industries, tracking terminals, amongst others. Hence, the imposed high-speed railway project is considered by activists as yet another infrastructure that will harm this territory and therefore also its non-human and human entities (Chiaramonte, 2019a).

No Tav activists, together with experts such as qualified scientists, university professors, and engineers, have produced in-depth information and scientific research on the harms caused to human health and the 'healthy environment' by the TAV project. These included cost-benefit analyses that demonstrated the uselessness of the project (Mattone, 2014) and even counter-proposals,[4] which were, however, discredited and bypassed: in 2001, the Italian Parliament with the Law No 443 (*Legge Obiettivo*) simplified the procedures for the so-called strategic infrastructures, excluding citizens' democratic participation. To disseminate this rich knowledge, activists have also resorted to the so-called paper barricades, which included municipal resolutions against the megaproject, petitions related to the health risks connected to its realisation,[5] tireless collections of signatures, and appeals to EU institutions.

Despite this heroic if not tragic struggle for law, these activist practices have largely been ineffective and TAV has been approved through a typical 'infernal alternative' (Stengers & Pignarre, 2011)—health or work?—according to which otherwise jobs would be lost, the area would be left 'underdeveloped', and, in any case, penalties to pay for breach of contract would be too high.

Harms of Criminalisation

There have been two main trials against No Tav activists. The first one involved charges for terrorism and resulted in the acquittal of all activists from this accusation, who however had been subject to a year of

---

[4] For example, No Tav activists proposed the modernisation of the existing line. In particular, the transit of large freight trains would be allowed there simply by lowering the track level in the cross-border tunnel; with specific technological innovations, this would help increase the efficiency and existing traffic capacity at a much lower cost than that of building a new railway (Mattone, 2014).
[5] See https://www.europarl.europa.eu/meetdocs/2009_2014/documents/peti/cm/791/791817/791817it.pdf (last accessed 29 January 2022).

preventive detention. The second and most important trial is the so-called *maxi-trial*: it started in 2012 and involved 53 defendants, who were accused of resistance to and violence against the police during a two-day protest held between June and July 2011. Given the difficulty of proving who actually resisted the police, the public prosecutor decided to accuse all those who were present during the two-day protest of complicity in the crime (*concorso*). The first-instance judges supported this thesis and clarified that demonstrators who see a protest turning violent should leave it, otherwise they would become accomplices in the violent crimes committed by others. In other words, the individual nature of criminal liability was denied as well as the right to assembly (Chiaramonte, 2019b). It was only in April 2021 that these violations by the Court of First Instance were recognised by the Turin Court of Appeal, which overturned the decision.[6]

No Tav activists opposed the charges in multiple ways. For example, their protest repertoire involved the courtroom, where hundreds of activists followed the hearings for years; it also involved organising protests in the areas outside Turin's prison to support those activists who have preventively been detained in prison while awaiting trial. The harms of criminalisation, however, did not only involve the restriction of people's freedom. Some activists lost their jobs (Giordano, 2017), while others were convicted in non-related criminal trials on the basis of their participation in the No Tav movement (Chiaramonte, 2019a). Activists and their legal team also denounced police violence and especially the impunity guaranteed to law enforcers (no trials were initiated to investigate their behaviour) through a documentary called *Archiviato*, which means 'dismissed'.[7]

The media had also a central role in the criminalisation process. They provided news coverage that systematically stressed only the violent aspects of the No Tav protests as if they were structural elements of their protesting (Gitlin, 1980; Chiaramonte & Senaldi, 2015, 2017). Furthermore, a *diversionary reframing* (Freudenburg & Gramling, 1994; Gilbertson & Watterson, 2007) depicted the protest as unreasonable,

---

[6] Judgment No 470 of 21.01.2021.
[7] The documentary is available on YouTube at https://www.youtube.com/watch?v=Ev7Sa-cuz5w (last accessed 29 January 2022).

rebellious, and therefore against 'progress' (Freudenburg et al., 1998), paving the way for the judicial criminalisation that followed. With respect to the stages of the criminal trials, the media offered only an account of the 'official' documents, provided by 'authorised' sources, such as the police, prosecutors, and judges (Cohen & Young, 1973; Hall et al., 1978; McLeod & Hertog, 1988, 1992; McLeod, 2007). As a consequence, the grievances of the No Tav movement have been sternly silenced.

As a way to oppose this authoritarian treatment, the movement—in addition to not giving up, albeit with resources that have substantially been reduced by the high costs of various trials—has adopted many strategies and has sought visibility also through artistic expressions. An example is provided by the mural below (Image 17.1). It depicts the image of a tree that is threatened by excavators alternating with police batons.

The physical harm to people caused by the police—represented here only through the baton—is profoundly intertwined with the harm that excavators do to the land. But interestingly enough, there are no people

Image 17.1   Mural in Chiomonte. (Photo © Luca Perino)

portrayed. Centrality is given here to a large tree with the mountains behind it, the Alpine mountain range. Yet, the two harms—the one to the criminalised population and the other to the territory—seem entangled, as shown by the fact that the bulldozer arm becomes a baton.

5.2   No Tap

Harms to the Land and Olive Trees

Both in visual social media and interviews, No Tap activists focused on the harms caused by TAP on the land and olive trees. These harms are visually rendered in various ways by activists on social media. For example, numerous are the pictures posted by activists on Twitter of beautiful natural landscapes tainted by uprooted or chopped olive trees or by the presence of militarised construction sites defacing the territory. One of these photos, for instance, shows uprooted olive trees laid down on the ground carrying the caption 'assassins' on top of it. The message encoded in this image is reiterated in many other textual and visual materials posted by activists on Twitter, where they accuse TAP of committing 'violence', 'rape', and a 'massacre' against their land and olive trees in particular (see Di Ronco et al., 2019).

Also on Twitter, various artistic expressions have been posted that speak about the harms caused by the pipeline to the land, olive trees, and non-human animals. In addition to vignettes depicting olive trees coiled around and strangled by pipes, which have also been used as event flyers and protest banners (see, e.g., *La Repubblica Bari*, 2019), the collected sample of visual social media material also included pictures of No Tap-inspired street art. While one of these photos portrays the TAP pipeline as leaking into the sea, a second street art piece represents the pipeline as an evil-looking octopus killing fishes and destroying the marine ecosystem. A third example of street art represents the pipeline as a python coiling around its human prey and squeezing them. In a Facebook post, Nemo—the author of this last piece—says:

The TAP problem is very complex. I do not know if it's a right thing or a wrong one, but I know that these works are destroying this beautiful land and its nature. This piece was born listening to the inhabitants of these lands.[8]

Both in visual social media and informal interviews, activists constructed the land as 'home-space'. For example, this is evidenced by the many protest banners featuring the mantra 'defend your home', whose pictures were posted on Twitter, and by songs where nature was regarded as people's home. An example is the song titled 'Salento doesn't want TAP' (Treble Lu Professore, 2014), where the lyrics say: 'you [TAP company] want *our home* and pretend that who inhabits it doesn't ever rebel' (emphasis added). Nature is described as 'home' to be protected also in a song authored by Terroni Uniti (2017)—a group of singers that supported the protest of Sioux Indigenous people against the Dakota Access Pipeline (DAPL), which passes through their sacred land in North Dakota (US). The song expresses solidarity not only to the NODAPL protest ('Dakota are not alone / here we are all Sioux') but also to other protests happening in Italy, including No Tav and No Tap, arguing that 'we are all in the same boat'. The mantra 'one earth one fight' is recurrent also in the many images of protest banners posted on Twitter across time—including in an online 'flash mob' which No Tap activists organised on Facebook and Instagram to express solidarity to No Tav activists after an episode of police violence against them. Amongst the solidarity messages that No TAP activists posted on social media through selfies of themselves holding posters, there were the following: 'On the side of those who defend their land' and 'Proud to defend our land'.

In addition to conceptualising the land as their 'home', activists *humanised* olive trees, which were described in informal interviews as 'our history' and specifically as 'our family', for example, 'they are like our mother, our brother … they are part of our family'.[9]

---

[8] https://m.facebook.com/whoisnemos/photos/a.499960873372868.102273.49474 6737227615/1472231202812492/?type=3.

[9] Olive trees were humanised also by a famous Italian pop band, Negramaro, who in 2017 posted on social media images of olive trees with captions indicating human names, such as 'Gianfranco'. They also added the message 'The only possible way to treat nature is the absolute human one. They are trees with a name, [..] with a life, land with a large, infinite soul'. See Di Ronco et al. (2019).

Harms to the land and olive trees—in addition to other environmental harms—have been pinpointed by the movement through an expert commission, which has been established in 2013 by the municipality of Melendugno and has involved engineers, university professors, and lawyers, amongst others. The commission produced many important documents, including the project's counter-evaluations, which also assessed the legality of the TAP project against relevant legislation. In particular, the commission established that TAP failed to assess the *cumulative impacts* of the various tunnels[10] envisaged in the project on the conservation of natural habitats, which are protected by EU law.[11] Currently, TAP is facing multiple charges in court (Colluto, 2020; Gerebizza & Taglieri, 2020)—charges that are also based on evidence that the commission has actively contributed to gathering. It is noteworthy here that ever since the beginning of the constructions by TAP, activists have also closely monitored the company at its construction sites. Through their daily monitoring, which also involved crowdsourced counter-surveillance through social media and apps (see Di Ronco & Allen-Robertson, 2020), activists photographed and recorded TAP removing olive trees from the area of the first construction site without the necessary authorisations. This evidence helped the prosecutor's office build a case against TAP, which is currently being trialled also for this specific misconduct.[12]

Harms of Repression

Harms of repression have visually been represented on social media not only through images and videos capturing episodes of police violence (see

---

[10] The company allegedly only assessed the impacts on the environment of the micro-tunnel that enters the coast and lands at the Pipeline Receiving Terminal in Melendugno—*not* the additional impacts caused by the other connecting tunnels, which bring natural gas from Melendugno to Mesagne, a town 65 km north of Melendugno, where gas flows into the Italian natural gas grid. If confirmed, this accusation (which recently translated into a pressed charge, see, e.g., Colluto, 2020) would invalidate the environmental authorisation TAP received back in 2014, which allowed it to start working on the pipeline's landing point.

[11] See EU Council Directive 92/43/EEC of 21 May 1992 on the conservation of natural habitats and of wild fauna and flora.

[12] At the time of writing, the trial against TAP has not yet started. So far, it has been postponed twice because of the need of complying with COVID-19 regulations (the first time) and of replacing the judge who took unpaid leave (the second).

Di Ronco et al., 2019) but also through creative artistic expressions, such as vignettes depicting old people defending olive trees from the police, and various songs, including the iconic No Tap song titled 'And it rains' (Treble, Rocky and The Dangeroots, 2018). This song speaks about rain as water that 'thirsts those who resist and won't give up' their fight for justice, despite police use of 'shields and batons' against activists.

According to the formal and informal interviews with No Tap activists, police brutality and violence are, however, not the only tools used by the police to intimidate them and ultimately discourage their protesting and mobilising. Activists have also been issued onerous fines (up to EUR 4000), which proved difficult to pay by some of them and their families—with some families having to pay for more than one fine issued against two or more family members. Activists—specifically those who participated in non-authorised protests or publicly spoke at them—have also been cautioned by the police and banned from entering the town of Melendugno and, at times, also the city of Lecce—which is the region's capital city hosting many economic activities and hospitals, amongst others. As three activists put it in a recent online video, these bans have not only limited their mobility but also determined their dismissal from jobs they had in Melendugno's bathing beaches (*GoFundMe*, 2021). Other activists also reported feeling 'constrained' when travelling around Italy as they would systematically be stopped and harassed by the police (i.e. activists would be kept waiting for a long time without explanation and be asked many questions, including why a No Tap activist went to that particular area of Italy). As one activist said during fieldwork:

> I will have to go soon to Verona for work but I will only go there by flight or train. I will never again travel by car, otherwise the police will stop me and it will be the same old story.

Three activists—including a No Tap lawyer and a technical consultant of the municipality—also reported feeling constantly monitored by the police and 'afraid' for their safety when going back home at night or being home alone.

Finally, activists have also been charged with offences that—according to them—have been invented at best and ill-substantiated (i.e. notified on a wrong date) or unnecessarily exaggerated at worst (see Papadia, 2018). For example, activists who threw flowers and paint-filled eggs in the direction of the police or showed the sign of the horns (usually signifying cuckoldry) at one of their helicopters have been charged with insulting and resisting the police. They have also been charged with trespass for having crossed land that was not marked as private by signs or fences. This notwithstanding, recently a judge confirmed most charges against the activists, who were also sentenced to detention ranging from six months to over three years (V.Val, 2021). In some instances, the judge even increased the penalties requested by the public prosecutor—something that is not unusual in trials against 'public enemies' in Italy (Prison Break Project, 2017).

6   Discussion: Understanding Environmental Harms Through the Concept of 'Harm to Knowledge'

Our analysis of data collected through formal and informal interviews and visual and artistic expressions emerging from both ethnographic fieldwork and social media research (the latter only in the case of the No Tap research) has revealed various articulations of environmental harms as expressed and represented by No Tav and No Tap activists. Our aim was to contribute to the emerging green criminological literature that— also through innovative methodologies—examined how affected individuals and communities perceive, represent, express, and articulate environmental harms (see, e.g. Natali, 2016, 2019; Di Ronco et al., 2019; Natali & de Nardin Budó, 2019; Di Ronco & Allen-Robertson, 2020; Varona, 2020; Goyes et al., 2021).

As we highlighted in the chapter, environmental harms have mostly been associated by activists with harms to human health and the healthy environment (in the case of No Tav) and with harms to the land and olive trees (in the case of No Tap). Such harms, however, have also been described as having entangling negative repercussions on humans, non-human animals, plants, and entire ecosystems. This was represented well in a series of street art pieces, which showed, for example, pipelines strangling humans and olive trees (for No Tap) and excavators (with hands holding batons) menacingly approaching a tree (for No Tav). The interconnection between humans and non-humans is especially well rendered in the case of the No Tap protest, where the land has been described by activists as 'home-space' to be protected, and olive trees have been humanised and considered as family members.

All these harms have so far not been acknowledged by the relevant companies (TAV and TAP, respectively) nor—not until now at least—by the state, despite the many channels and tools used by activists to convey knowledge about them. Indeed, these two environmental movements have produced rich knowledge about the harms of these megaprojects—a knowledge that is practical, technical, localised, and situated. As Stengers (2015) underlines, the practical knowledge that spreads and might create 'citizens' juries'[13] is *local*; it is also the only possible source of knowledge that can dismantle the 'official' scientific knowledge that systematically bends to authority and the market. For example, No Tav activists provided evidence—and therefore generated knowledge—on the dangerous materials present in the mountains to be excavated by TAV. No Tap activists, on their part, have generated knowledge through their daily monitoring of the TAP company—a knowledge that has helped the public prosecutor's office to build a case against TAP that is currently ongoing (see Colluto, 2020; Gerebizza & Taglieri, 2020). In both the No Tav and No Tap cases, moreover, knowledge has been generated by qualified scientists, university professors, engineers, and other experts, who have produced in-depth information on the megaprojects, including cost-benefit analyses and (in the case of No Tav) counter-proposals.

Overall, this precious localised and technical knowledge has so far largely been ignored by relevant authorities, which have excluded activists from the decision-making process and also silenced them through the movements' systematic criminalisation, configuring what we called 'harm

---

[13] In general, by citizens' juries we mean the groups or assemblies that allow people to participate in shaping public decisions and to access the process of building scientific knowledge with respect to many topics, including climate change. In recent decades, scholars have also used the term Citizen Science to refer to these innovative experiments based on the joint participation of citizens and scientists (see, e.g. Bonney et al., 2009).

to knowledge'. In other words, needed and important counter-knowledge produced from below has been 'harmed' through its dismissal by relevant authorities and the criminalisation of its producers. As we illustrated above, the repression of No Tav and No Tap activists included enhanced police surveillance, cautions, place bans, administrative fines, charges for more or less serious offences (with the most serious indictment being that of terrorism for No Tav activists), and pre-trial detention. All these measures have had serious negative effects on the lives of many activists, who had their freedom restricted, have lost their jobs, and have had troubles paying the onerous fines they (and often their family members) have been issued. Criminalisation, however, has not only had negative effects on the lives of individual activists but also had adverse consequences for these movements, which have mostly been considered 'troublemakers' rather than credible knowledge-producers and counterparts in decision-making processes concerning the approval and execution of these mega-projects. Unfortunately, relevant authorities may never recognise the value and transformative potential of the knowledge produced by these movements and may continue to criminalise their members and exclude them from decision-making circles.[14]

7   Conclusion: For a *Proactive* Environmental Restorative Justice

As we argued in the introduction of this chapter, for a process of restorative justice to be initiated at all, someone has to *acknowledge* the harms caused (Forsyth et al., 2021). Without this acknowledgment, ERJ conferences may never occur—and the transformative potential of restorative justice may never be realised. This is the case of TAV and TAP in Italy, which are two state-authorised and 'strategic' projects whose harms have not been acknowledged by the state and the two corporations—and

---

[14] For the No Tav movement, an acknowledgement came from a decision of the Permanent Peoples' Tribunal (PPT), which in November 2015 recognised the 'systematic violation of the fundamental right of a community to be an indispensable and priority subject in the decision-making processes regarding its context and its present and future living conditions' (PPT, 2015). As it is known, however, this is only a tribunal of opinion whose decisions are not binding.

unlikely will be in the future. As discussed in this chapter, No Tav and No Tap activists have, however, produced rich localised, practical, and situated knowledge on the harms caused by these megaprojects on humans, non-human animals, plants, and ecosystems.

To valorise and not to waste the knowledge that activists and affected communities have produced on the harms of megaprojects, we conclude this chapter by arguing for the need to expand the scope of ERJ. In line with what Wilson (2016) called 'proactive restorative justice', which was applied by Hamilton (2021) to the context of ERJ, we make the case for a *proactive environmental restorative justice* that not only looks back at past offending but also resolves present and prevents future conflicts and harms. As Hamilton (2021) put it, to become 'proactive', ERJ should seek to (amongst other things) build a constructive dialogue and share knowledge with local communities in the various phases of a major project, including its assessment, approval, and implementation. An ERJ of this sort may not stop a (mega)project from happening; yet, increased people's participation in all the preparatory and execution phases of these projects may help scale down the level of conflict and reduce (or at least not excessively aggravate) the harms caused to humans and non-humans.

This obviously raises the question of how to ensure that dialogue and people's participation actually take place in the various project phases? There may be many ways, but these also include avoiding simplified procedures for the so-called strategic infrastructures, which *de facto* exclude the democratic participation of citizens—a practice that so often has occurred in Italy, as the cases of No Tav and No Tap in this chapter have demonstrated. Until this gets fixed, restorative justice scholars and practitioners could certainly be of help. For example, they could *proactively* use their skills and techniques to facilitate dialogue between activists and those in power positions (the state, corporations) during the various project's phases, in this way contributing to mitigating harms before they are actually generated. This may involve the setting up of ERJ conferences in the natural places that are—or soon will be—affected by megaprojects, offering structured and safe spaces for people to share their stories and knowledge, and where they can listen to and meaningfully engage with others in a constructive dialogue (Forsyth et al., 2021; Hamilton, 2021). The work of 'activist' green criminologists (Goyes, 2016) seems to us also paramount: through traditional as well as innovative methodologies, green criminologists should keep focusing on silenced and criminalised environmental movements and their perceptions and knowledge of environmental harms, also with a view to enhancing their visibility and exposure through sound analysis and dissemination. This is—we hope—also what this chapter contributes to doing.

References

Acosta, A., & Martínez, E. (2011). *La naturaleza con derechos. De la filosofía a la política*. Quito: Abya-Yala.

Altopiedi, A. (2020). Ambiente, giustizia e diritto(i). *Sociologia del Diritto*, 2. https://doi.org/10.3280/SD2020-002005.

Beirne, P., & South, N. (2013). *Issues in green criminology*. London: Routledge.

Bonney, R., Cooper, C.B., Dickinson, J., Kelling, S., Phillips, T., Rosenberg, K.V., & Shirk, J. (2009). Citizen science: A developing tool for expanding science knowledge and scientific literacy. *BioScience*, *59*(11), 977–984.

Brisman, A., & South, N. (2013). A green-cultural criminology: An exploratory outline. *Crime, Media, Culture*, *9*(2), 115–135.

Brisman, A., & South, N. (2014). *Green cultural criminology: Constructions of environmental harm, consumerism, and resistance to ecocide*. London: Routledge.

Brisman, A., & South, N. (2018). Green criminology, zemiology, and comparative and inter-relational justice in the Anthropocene era. In A. Boukli & J. Kotzé (Eds.), *Zemiology. Reconnecting crime and social harm* (pp. 203–221). Basingstoke: Palgrave Macmillan.

Butler, B., Allam, L., & Wahlquist, C. (2020, September 11). Rio Tinto CEO and senior executives resign from company after Juukan Gorge debacle. *The Guardian*. Retrieved from: https://www.theguardian.com/business/2020/sep/11/rio-tinto-ceo-senior-executives-resign-juukan-gorge-debacle-caves (last accessed 5 January 2021).

Calafati, A.G. (2006). *Dove sono le ragioni del sì? La Tav in Val di Susa nella società della conoscenza*. Turin: SEB.

Chiaramonte, X. (2019a). *Governare il conflitto. La criminalizzazione del movimento No Tav*. Milan: Meltemi.

Chiaramonte, X. (2019b). The right to freedom of assembly under the test of criminal evidence. An Italian case of political justice. *Droit & Philosophie*, 11, 169–187.

Chiaramonte, X. (2020). The struggle for law: Legal strategies, environmental struggles and climate actions in Italy. *Oñati Socio-legal Series*, 10(4), 932–954.

Chiaramonte, X., & Senaldi, A. (2015). Criminalizzare i movimenti: I No Tav fra etichettamento e resistenza. *Studi sulla Questione Criminale*, *1*, 105–144.

Chiaramonte, X., & Senaldi, A. (2017). Criminalizar los movimientos sociales: La resistencia a la construcción del tren de alta velocidad en el norte de Italia. *Revista de Derecho Penal y Criminología*, *1*, 55–81.

Cohen, S., & Young J. (1973). *The manufacture of news: Social problems, deviance and the mass media*. London: Constable.

Colluto, T. (2020, January 7). Tap, in 19 vanno a processo: ulivi espiantati, inquinamento falde e lavori senza permessi. Pm: "Illegittima l'autorizzazione ministeriale". *Il Fatto Quotidiano*. Retrieved from: https://www.ilfattoquotidiano.it/2020/01/07/tap-in-19-vanno-a-processo-ulivi-espiantati-inquinamento-falde-e-lavori-senza-permessi-pm-illegittima-lautorizzazione-ministeriale/5655478/ (last accessed 16 September 2021).

Di Ronco, A., & Allen-Robertson, J. (2020). Representations of environmental protest on the ground and in the cloud: The NOTAP protests in activist practice and social visual media. *Crime, Media, Culture*. https://doi.org/10.1177/1741659020953889.

Di Ronco, A., Allen-Robertson, J., & South, N. (2019). Representing environmental harm and resistance on Twitter: The case of the TAP pipeline in Italy. *Crime, Media, Culture*, 15(1), 143–168.

Dosio, N. (2006) *Intervista*. In Centro Sociale Askatasuna and Comitato di lotta popolare No Tav (Eds.), *No Tav. La valle che resiste*. Turin: Velleità alternative.

Fazio, L., & Minnelli, G. (2020). Early mortality from malignant mesothelioma in Italy as a proxy of environmental exposure to asbestos in children. *Annali dell'Istituto superiore di sanità*, *56*(4), 478–486.

Flyvbjerg, B., Bruzelius, N., & Rothengatter, W. (2003). *Megaprojects and risk: An anatomy of ambition*. Cambridge: Cambridge University Press.

Forsyth, M., Cleland, D., Tepper, F., Hollingworth, D., Soares, M., Nairn, A., & Wilkinson, K. (2021). A future agenda for environmental restorative justice? *The International Journal of Restorative Justice*, 4 (1), 17–40.

Freudenburg, W.R., Frickel, S., & Dwyer, R.E. (1998). Diversity and diversion: Higher superstition and the dangers of insularity in science and technology studies. *International Journal of Sociology and Social Policy*, 18, 6–34.

Freudenburg, W.R., & Gramling, R. (1994). *Oil in troubled waters: Perceptions, politics, and the battle over offshore oil*. Albany: Suny Press.

Gerebizza, E., & Taglieri, F. (2020, September 11). Gasdotto Tap in Puglia, al via il processo per disastro ambientale. *Il Manifesto*. Retrieved from: https://

ilmanifesto.it/gasdotto-tap-in-puglia-al-via-il-processo-per-disastro-ambientale/ (last accessed 16 September 2021).

Gilbertson, M., & Watterson, A.E. (2007). Diversionary reframing of the great lakes water quality agreement. *Journal of Public Health Policy, 28,* 201–215.

Giordano, A. (2017). *Non ho visto niente. Sul come essere No Tav comporti perdere il lavoro.* Rebibbia: Sensibili alle Foglie.

Gitlin, T. (1980). *The whole world is watching. Mass media in the making & unmaking of the new left.* Berkeley: University of California Press.

Giunti, L., Mercalli, L., Poggio, A., Ponti, M., Tartaglia, A., Ulgiati, S., & Zucchetti, M. (2012). Economic, environmental and energy assessment of the Turin-Lyon high-speed rail. *International Journal of Ecosystems and Ecology Sciences (IJEES), 2*(4), 361–368.

GoFundMe (2021). Cassa di Resistenza NoTap. Retrieved from: https://www.gofundme.com/f/no-tap?utm_campaign=p_cp_url&utm_medium=os&utm_source=customer (last accessed 16 September 2021).

Goyes, D.R. (2016). Green activist criminology and the epistemologies of the South. *Critical Criminology, 24*(4), 503–518.

Goyes, D.R., South, N., Abaibira, M.A., Baicué, P., Cuchimba, A., & Ñeñetofe, D.T.R. (2021). Genocide and ecocide in four Colombian Indigenous communities: The erosion of a way of life and memory. *The British Journal of Criminology.* https://doi.org/10.1093/bjc/azaa109.

Hall, M. (2014). Environmental harm and environmental victims: Scoping out a 'green victimology'. *International Review of Victimology, 20*(1), 129–143.

Hall, S., Critcher, C., Jefferson, T., Clarke, J., & Roberts, B. (1978). *Policing the crisis: Mugging, the state, and law and order.* Teaneck: Holmes & Meier.

Hamilton, M. (2021). Restorative justice conferencing in Australia and New Zealand: Application and potential in an environmental and Aboriginal cultural heritage protection context. *The International Journal of Restorative Justice,* 4 (1), 81–97.

Hasler, O., Walters, R., & White, R. (2020). In and against the state: The dynamics of environmental activism. *Critical Criminology, 28*(3), 517–531.

La Repubblica Bari (2019). In Salento corteo No Tap per celebrare i 'due anni di lotta'. Retrieved from: https://bari.repubblica.it/cronaca/2019/03/17/news/in_salento_corteo_no_tap_per_celebrare_due_anni_di_lotta_-221797046/ (last accessed 16 September 2021).

Mattone, P. (2014). *Tav e Valsusa: diritti alla ricerca di tutela.* Naples: Intra Moenia.

McLeod, D.M. (2007). News coverage and social protest: How the media's protect paradigm exacerbates social conflict. *Journal of Dispute Resolution,* 1, 185–194.

McLeod, D.M., & Hertog, J.K. (1988). *Anarchists wreak havoc in Downtown Minneapolis: A case study of media coverage of radical protest*. Retrieved from: https://files.eric.ed.gov/fulltext/ED295274.pdf (last accessed 16 September 2021).

McLeod, D.M., & Hertog, J.K. (1992). The manufacture of public opinion for reporters. Informal cues for public perceptions of protest groups. *Discourse and Society*, 3, 259–275.

Natali, L. (2016). *A visual approach for green criminology: Exploring the social perception of environmental harm*. London: Palgrave Macmillan.

Natali, L. (2019). Visually exploring social perceptions of environmental harm in global urban contexts. *Current Sociology*, 67(5), 650–668.

Natali, L., & de Nardin Budó, M. (2019). A sensory and visual approach for comprehending environmental victimisation by the asbestos industry in Casale Monferrato. *European Journal of Criminology*, 16(6), 708–727.

NoTap (n.d.). La storia. Retrieved from: https://www.notap.it/storia/. (last accessed 16 September 2021).

Papadia, E. (2018). Difendere i difensori dei diritti della Terra: un dossier sull'esperienza giudiziaria dei movimenti salentini. Retrieved from: https://ecor.network/userfiles/files/Difendere%20i%20difensori%20della%20terra.pdf (last accessed 16 September 2021).

Permanent Peoples' Tribunal (2015). *Fundamental rights, participation of local communities and mega projects from the Lyon-Turin high-speed rail to the global reality Turin-Almese (5–8 November 2015)*. Retrieved from: http://permanentpeoplestribunal.org/wp-content/uploads/2015/11/TPP_GRANDI-OPERE_EN.pdf. (last accessed 16 September 2021).

Prison Break Project (2017). *Costruire evasioni. Sguardi e saperi contro il diritto penale del nemico*. Lecce: Edizioni Bepress.

Ruggiero, V., & South, N. (2013). Toxic state-corporate crimes, neo-liberalism and green criminology: The hazards and legacies of the oil, chemical and mineral industries. *International Journal for Crime, Justice and Social Democracy*, 2(2), 12–26.

South, N. (1998). A green field for criminology? A proposal for a perspective. *Theoretical Criminology*, 2(2), 211–233.

Stengers, I. (2015). *In catastrophic times: Resisting the coming barbarism*. Retrieved from: http://openhumanitiespress.org/books/download/Stengers_2015_In-Catastrophic-Times.pdf (last accessed 16 September 2021).

Stengers, I., & Pignarre, P. (2011). *Capitalist sorcery: Breaking the spell*. New York: Palgrave Macmillan.

Terroni Uniti (2017). Simmo tutte Sioux. Retrieved from: https://www.openddb.com/music/simmo-tutte-sioux-terroni-uniti/ (last accessed 16 September 2021).

Treble Lu Professore (2014). Lu Salentu non vuole la Tap. Retrieved from: https://www.youtube.com/watch?v=fSSkog5ifY4 (last accessed 16 September 2021).

Treble, Rocky and The Dangeroots (2018). E piove. Retrieved from: https://www.youtube.com/watch?v=PacmKhMU4fc (last accessed 16 September 2021).

Ugwudike, P. (2015). *An introduction to critical criminology*. Bristol: Policy Press.

V.Val (2021, March 19). Scontri No-tap, arriva la sentenza in tre processi con novantadue imputati. *LeccePrima*. Retrieved from: https://www.lecceprima.it/cronaca/Scontri-no-tap-arriva-sentenza-in-tre-processi-con-92-imputati.html (last accessed 16 September 2021).

Varona, G. (2020). Restorative pathways after mass environmental victimisation: Walking in the landscapes of past ecocides. *Oñati Socio-Legal Series*, 10(3), 664–685.

Varona, G. (2021). Why an atmosphere of transhumanism undermines green restorative justice concepts and tenets. *The International Journal of Restorative Justice*, 4(1), 41–59.

Wahlquist, C. (2021, March 23). Rio Tinto pledges to protect cultural heritage after Juukan Gorge disaster. *The Guardian*. Retrieved from: https://www.theguardian.com/business/2021/mar/23/rio-tinto-pledges-to-protect-cultural-heritage-after-juukan-gorge-disaster. (last accessed 16 September 2021).

White, R. (2008). *Crimes against nature: Environmental criminology and ecological justice*. Devon: Willan Publishing.

White, R. (2013). The conceptual contours of green criminology. In R. Walters, D. Solomon Westerhuis & T. Wyatt (Eds.), *Emerging issues in green criminology: Exploring power, justice and harm* (pp. 17–33). Basingstoke: Palgrave Macmillan.

White, R. (2018). Green victimology and non-human victims. *International Review of Victimology*, 24(2), 239–255.

Wilson, C. (2016). Proactive restorative justice: A set of principles for enhancing public participation. *Environmental and Planning Law Journal*, 33(3), 252–263.

18

Looking for the Restoration in Restorative Justice's Response to Civil Disobedience

Rachel Jolly, Rachel Gehman, and Gale Burford

1   Introduction

When a civil disobedience case was first referred to a New England community justice centre for a restorative justice process, staff had to first determine who were the harmed and harming parties. From there, many more questions arose regarding systemic injustices, representation of the voiceless, and template models of restorative justice rooted in human-centric and individualistic worldviews. These questions became the basis to initiate research in the form of interviews, surveys, and literature review on the use of restorative justice in cases of civil disobedience. In this chapter, we share what our findings revealed about the problems of our current restorative approach to civil disobedience and strategies to

R. Jolly (✉)
Burlington Community Justice Center, Burlington, VT, USA
e-mail: rjolly@burlingtonvt.gov

R. Gehman • G. Burford
University of Vermont, Burlington, VT, USA
e-mail: Gale.Burford@uvm.edu

overcome these problems. We conclude that ongoing revisitation of the complex intersections of the multiple ecologies of regulation, crime, protest, race, gender, and environment is fundamental to breaking down silos between and amongst regulatory actors, counterbalancing the "creep" of criminalisation, and galvanising collective action at multiple levels to focus on solutions to complex problems.

Beyond crafting ethical treatment of activists, restorative approaches at their best might act as a platform to amplify and deepen environmentalist messages through public dialogue and building of collaborative partnerships amongst regulatory actors. At worst, restorative justice could undercut these messages by emphasising activist criminality and acting as a governmental instrument to silence protest and reinforce existing power structures. We offer perspectives from the small state of Vermont in hopes that our successes and shortcomings might inform other's search for ethical and generative restorative practice.

2   An Arresting Development

After their arrest in 2019, a small group of protestors was referred by a prosecuting attorney to a Community Justice Centre (CJC) where they would participate in a process of reparative-restorative sanctioning (referred to as either a restorative panel or a reparative board). Innovated and largely funded by the State of Vermont's Department of Corrections (DOC), this approach uses community volunteers to 'hold the responsible party accountable for the effects of their actions on others [and] discusses the circumstances and impact of the crime and ways the responsible party can avoid making similar mistakes in the future.'[1]

The lack of fit between the circle process used at the CJC and the nature of the case was apparent from the opening moments. A letter from a representative of the harmed party asking only that the protestors admit to having disrupted office routines was read. No one representing that office attended. The meeting facilitator used a standard script that

---

[1] See Restorative Justice Panels at the webpage of Burlington Community Justice Center https://www.burlingtoncjc.org/restorative-justice-panels (last accessed January 24, 2022).

included questions asking protestors to acknowledge the harms their behaviours caused, to hear a written statement from the harmed party, and to determine their role in repairing those harms. One protestor compared the process to 'doing penance,' adding, 'like you should have to do because you aren't sorry.' Another called the suggestion of an apology a 'childish kind of thing.'[2]

A representative of the police agency that made the arrests attended the meeting and shared his take on the role of law enforcement in such matters. Protesters acknowledged having intentionally broken a trespassing law and further acknowledged that accepting to participate in the restorative meeting did represent a way to avoid going to court and the possibility of ending up with a criminal record. While few believed the matter would go to court, even if they refused to participate in the restorative process, it was acknowledged within the group that some members would suffer greater impact than others. Out of respect for one another's different circumstances, the group agreed to practice solidarity and attend the meeting. Their case fit the definition of non-violent civil disobedience (NVCD): the refusal of a citizen to obey certain laws, demands, orders, or commands of a government. Importantly, at no time was there ever a suggestion that protestors had used or threatened to use violence or vandalism (Brouwer, 2019).

The complexities of the case helped frame our early questions as we began to design our research inquiry: What are the risks and potentials, including benefits, of invoking a restorative process, as opposed to prosecution, or to simply diverting matters without the expectation of

---

[2] There is precedence in the State of Vermont for using loosely defined restorative justice practices with non-violent civil disobedience (NVCD). Students engaged in non-violent civil disobedience were arrested by campus police 'when it became apparent they wanted to be arrested' (UPI Archives, 1987). The then local prosecuting attorney declined to take matters to court. Faced with considerable media attention, the university offered what the students understood as a 'restorative' alternative to being expelled (personal communications with students). Each individual student was offered the opportunity to write an essay addressing the topic 'What hill are you willing to die on?' in which they were to acknowledge, among other things, the harms they had inflicted on the university community. Students were required, under threat of expulsion, to sign an agreement that they would not disclose to anyone the details of the assignment or reveal the conditions under which it was undertaken. Some students saw the requirement for 'secrecy', the admonishment of any claim they had to morally challenge the university, and the threat of expulsion for failure to comply with the assignment, as most un-restorative.

'restoration'? How do we approach questions of harm and victimhood in cases of NVCD? What is lost when we focus on the incident at the expense of systemic problems? Who does and does not get access to the diversionary, reparative, and restorative options in Vermont?

3   Our Method: An Iterative Exploration

Because our questions have not yet been the subject of research, we kept the aims of our study modest and exploratory. Little administrative data is available that would shed light on our questions, especially those that might reveal suspected racial disparities in access to restorative practices.[3] In addition to exploring our primary questions, we also wished to investigate the potentials and challenges for state-sponsored organisations to foster space for protestors and other groups caught up in complex intersections of harm and conflict, aspiring that they be included as legitimate partners in restorative and regulatory processes.

We employed a mixed-methods approach, including a brief online survey, un-recorded video interviews (via Zoom), examination of sample agreements from restorative meetings, and a brief review of relevant media articles, literature, and policy documents. During the interviews we were prepared with sample questions but whenever possible we invited participants to guide the conversation and to pose their own questions.

The sample includes local, national, and international participants such as defense attorneys, prosecutors, police, restorative justice facilitators, community leaders, academics, and those who were referred to a restorative approach[4] either as persons who had caused harm or had been harmed. The online survey was carried out between March 2020 and May 2021, and the interviews were carried out between December 2020

---

[3] Many departments in the state, including the DOC and police agencies, have only recently begun to invest in data tracking systems that would shed light on racial disparities in enforcement, including traffic stops, referrals to court diversion, and other indicators of racialised practices (Davis, 2021; French, 2021; Petenko, 2022). At the same time, advocates remain alert to the biases in data collection (cf., Henne et al., 2021; Shelby et al., 2021).

[4] We use the term 'a restorative approach' as a blanket term for reparative, reintegrative, restorative, and transformative justice, while acknowledging their meaningful differences.

and May 2021.[5] The final sample consisted of 28 responses to the online survey and follow-up interviews with 24 of the respondents.[6]

4   Participant Demographics

Participants spoke to issues from the vantage point of ample experience and multiple identities. Of the 28 survey responses collected, 20 identified as female and 8 as male. Three identified as Black, one as Asian or Asian American, and one as multi-racial; the remaining 23 identified as Caucasian. Ten participants indicated they were between the ages of 26 and 44, 12 between 45 and 59, and 6 as 60 years or older. All participants indicated that they had completed an undergraduate degree, with the majority (n = 22) reporting some level of graduate training. Nineteen participants were local and five were national or international.

We sought diverse perspectives, including those of activists, prosecutors, educators, and facilitators. Nearly all participants indicated in the survey that they had more than one role connecting them to the study and could therefore offer multiple and nuanced perspectives. For example, three not only had current or previous experience as a prosecutor but also had experience in NVCD and/or restorative justice, two indicated they were primarily involved with law enforcement (both having experienced harm while policing NVCD), seven identified as having been involved in restorative processes as the person who caused the harm/disturbance, and four had experience as restorative justice facilitators. Two people with experience as panellists/facilitators also had experience being arrested and attending a reparative/restorative meeting as an arrested protester. Ten indicated 'other,' which included those who had ties to municipal/State government, programme administrators, facilitators of non-violent training, or others who engaged in advocacy, academic research/teaching, and training.

---

[5] Copies of the consent forms and procedures, the survey, and other protocol descriptions used in the study can be obtained directly from the corresponding author. This study was approved by the University of Vermont Institutional Review Board (IRB) (CHRBS #STUDY00001186).
[6] Fifteen of the interviews were with individuals. Three small group interviews involved a total of nine persons.

Participants had relevant training and education, perhaps indicating the extent to which restorative justice and related concepts and practices of conflict and dispute resolution have become infused in many disciplines, governance approaches, and popular culture. Most (*n* = 19) reported having some formal education/training in restorative justice, or related areas, which included college/university courses, workshops, panel discussions, or online courses. Experience with protesting was considerable, ranging from protests of local to national and international relevance. Eighteen reported participating in NVCD such as sit-ins, occupying geographical areas or sites, protesting police brutality, racial injustice (particularly with Black Lives Matter), unjust war, women's health issues, housing and homelessness, and a range of environmental injustices including resource extraction, bad faith negotiating on the part of politicians and bankers, clandestine corporate and university policies, historical harms, and failure to meet obligations on the parts of public officials. All but one reported having mainly favourable views towards using NVCD, seeing it as necessary to challenge sources of domination and exploitation.

Throughout the chapter we weave in quotes and paraphrased statements to help narrate our learning from the study process. The structure of the chapter, or lack of it, tracks our learning and questioning journey more than conforming to a standard outline. Finally, we have used pronouns "they, them, and their" to anonymise participants' identities.

5   Vermont as a Host for Environmental Restorative Justice Initiatives

Nestled in the heart of New England, Vermont is known for, among other things, its progressive politics, rural and agricultural landscape, strict environmental regulations, and 95% white population (Vermont Census, July 1 2019). Public image is one of pride in a political tradition of civic engagement and responsive government through its citizen legislature (Merrick & Newbury, 2020, p. 479), its historical use of town hall meetings as an expression of direct democracy, and national recognition for child and community wellbeing (Gallup Sharecare, 2018).

At the same time, the state is known for its legislatively sanctioned use of carceral practices including high rates of incarceration (Prison Policy Initiative, 2018), for its swings in the pendulum between child and youth removal and family support (Crist & Bech, 2018; Kenyon, 2015), its historical state-sanctioned genocidal practices towards Native Americans (Gallagher, 1999; Walters, 2020), and continuing explicit acts of racism towards Black and Brown Vermonters (Davis 2021; Galloway & Keays, 2021; Racial Disparities Advisory Panel, 2020).

In Vermont, the path to the development of the restorative panels/ boards was paved in part through the efforts that created an Agency of Human Services.[7] Driven by an ethic of support for mobilising community leadership through state-local partnerships, leaders in the DOC in the mid-1990s conducted a survey of Vermonters (Perry & Gorczyk, 1997) which identified the lack of victim voice and community engagement in the criminal-legal system as primary areas of dissatisfaction. What would become known as The Community and Restorative Justice Unit within the DOC was born, funding 20 CJCs at its height.[8] Much earlier, in the late 1970s, the Attorney General's Office (AGO) began funding diversion programmes to divert minors out of the court system. This later expanded to adults, and in 2016, began using restorative justice principles and practices in many of its programmes. The AGO funds Court Diversion and Pretrial Service Programs in all 14 counties of Vermont, through 12 agencies, 7 of which are co-located at CJCs.

While enjoying high levels of public support, Vermont's programmes, and the restraints they find themselves under, are coming increasingly under scrutiny. They are under pressure from conflicting allegations that they utilise practices that are 'soft on crime,' that they are window dressing that serves to direct attention away from the defunding of social services and lack of adequate stable housing, and that they increase scrutiny of certain groups, to name a few. Once characterised as engaging with

---

[7] Created in 1969, the Vermont Agency of Human Services oversees the Departments for Children and Families and of Corrections, Disabilities, Aging and Independent Living, Health, Mental Health and Vermont Health Access.

[8] That number is currently being reduced through amalgamations in some districts.

'minor crime in a quaint setting' (Karp & Drakulich, 2004, p. 304), the approach developed in the State and tied to its CJCs has been prevented over many years from taking complex cases such as sexual and domestic abuse (Vermont Statutes Title 24) and is opposed by a conservative legal community who have seen restorative justice, among other things, as a threat to plea-bargaining (Dembenski, 2004). By extension, CJCs face concerns about 'net-widening,' that is, accepting cases that would not stand up in court, or, as one participant in this study put it, being 'an off-ramp for cases prosecutors don't want to deal with.' In the face of reductions in social services in Vermont, CJCs have seen increases in referrals involving mental health and substance use problems for which they lack adequate training and resources.

Still, the State's DOC and the Attorney General's Office have been widely recognised for pioneering and visionary work, including for mobilising community volunteers to sit on youth and adult Diversion Reparative Panels (Bazemore & Umbreit, 2001; Boyes-Watson, 1999; Karp & Drakulich, 2004), Reparative Probation Boards (Humphrey et al., 2012), Community Reentry panels/boards (Fox 2010, 2012), the use of Circles of Support and Accountability (Fox 2013; Wilson and Fox 2019), and for the unique embeddedness of state-funded restorative justice centres throughout the state. So successful were their efforts that ultimately legislation was passed requiring that principles of restorative justice be included "in shaping how the criminal justice system responds to persons charged with or convicted of criminal offenses, and how the State responds to persons who are in contempt of child support orders" (Vermont Statutes Online 28 V.S.A. § 2a).[9]

---

[9] While Section § 2a of the law sets out policy objectives and details the requirements of offenders who participate in restorative justice, its impact has been limited by its location in statutes that applied only to the DOC instead of the wider State Criminal Justice legislation (Personal Communication with John Perry November 26, 2021). This narrowed the work of CJCs whose main source of funding comes from the DOC.

6 Findings and Discussion: Harm, Victimhood, and State Control in Cases of NVCD

In this section, we draw together what we learned and trace links to the theme of this volume: how a restorative justice approach can further our efforts to combat the biggest and most challenging threats to the planet, that is, those wrought by human domination of nature. We reasoned that exploring the issues around protest, including those in the tradition of non-violent protest and defiance of laws, should shed light on links between local incidents of offending and larger environmental harms that impact everyone, albeit disproportionately. We worried that if restorative justice does not position itself to contribute to these larger issues, that it will remain on the margins, vulnerable to capture and even corruption by the very forces that benefit from suppressing protest.

Nearly all participants believed that taking a restorative approach in matters of NVCD could or should be made available. Their perspective is built largely from a rejection of the gross inequities and harms wrought of the conventional criminal justice system. At the same time, most participants acknowledged how restorative justice narrowly conceived for use as an extension of criminal justice system logic risks amplifying the very harms it seeks to reduce.

Along these lines, several themes emerged from our synthesis of participant surveys and interviews. Broadly, these include (1) matters originating in NVCD challenge restorative justice to think outside its standard boxes; growing the conversation away from the incidental offense and towards the social/ecological issues that it represents. (2) Restorative justice should look beyond the crime for defining 'harm' (e.g., naming environmental harms alongside the harms of inconvenienced bystanders at a demonstration) and, by extension, releasing the singular focus on who is the 'harmer' versus the 'harmed.' (3) The main restoration outcome from the conference should target the social/ecological harm rather than the incidental harm. This will be easier if authorities such as politicians and executives are present at the conference, along with the civilly disobedient, to make commitments. (4) State sponsorship limits partnership with activists and the malleability of the restorative justice format. We should aspire towards more diverse funding structures and partnerships. (5) Restorative justice should take action to serve diverse communities with representative facilitators, particularly given the racial disparities in arrests in cases of NVCD.

Some protestors feel that publicly breaking the law and the subsequent visibility that prosecution brings is the central purpose of their actions, as it facilitates wide coverage of their message. Their willingness to endure and even challenge state punishment is a testament to the importance of their cause. No matter how well intended diversion from the justice system might be in such cases, the consequence can be to defuse the efforts of protesters to gain visibility for their issue. At the same time, nearly all participants held the view that court processes are overused, rarely lead to clear solutions to complex problems, and put certain groups who are historically disadvantaged in the access to legal protections at risk for greater penalties. The following sentiment was common amongst participants with legal backgrounds:

> I don't see coming to court as any progress for a cause. I think there's a big misunderstanding of how much you get to "speak your peace" during your day in court. You have a few minutes and then months later, you might get some resolution to your case. (Study participant/Justice official)

For the protestors who do not see prosecution as furthering their message, the conventional court system is also undesirable because most cases of NVDC do not involve a concrete harm. One participant with prior experience as a prosecutor and protester unpacked some of this nuance:

> When we think about civil disobedience coming into our criminal justice system, we're typically thinking about law breaking, not really harm. Maybe there's been a crime (obstructing justice by failing to disperse, etc.). That's not really crime in the spirit of why we even have a penal code. Generally civil disobedience is about higher ideals of how we can be a more civilised society. I don't see that as criminal behaviour. If we're talking about civil disobedience in the context of wanting to be heard/unrest, restorative

justice is a preferred response than arresting and prosecuting. It is different for me, when there is harm. (Study participant/Former prosecutor)

Other participants shared this view, including in some instances when damage to property related to protest or impositions to the rights of others have occurred. As long as restorative justice referrals are sourced from lawbreaking individuals, the responsible party in cases of NVCD will inevitably be the activist. The default focus is then on the law or rule that is broken rather than on the wider context in which protest has occurred or the necessity for collective responsibility taking. One experienced restorative justice facilitator in Vermont summed up the dilemma succinctly:

> I think the way restorative justice is practiced via VT state-supported programs is a mismatch for any possible restorative approaches to NVCD. Even though they are restorative to an extent, such restorative justice programs still take an individualistic view of crime because of how cases are referred: there is an assumption that the individual who broke the law created and must repair harm caused; there are others who were impacted. I just don't believe that lens is a fit for responding to people breaking the law as part of an intentional, collective, effort to challenge an unjust law or practice. I do think a transformative justice lens makes sense; with that lens, a restorative justice process would also focus on the harms of the injustice that people are protesting against. A circle process where all people in a community have a stake and responsibility in the situation, whatever their role, makes slightly more sense to me than someone who "offended" through NVCD being sent to a restorative justice process. But the latter actually makes me cringe! Can you imagine MLK being asked or told to engage in a restorative process because he led a boycott? Yikes! (Study participant/Vermont restorative justice facilitator)

As this participant points out, basing restorative justice in legal definitions of harm lays bare all the ways in which legality is not synonymous with harm. In most cases of NVCD, the harm caused is difficult to measure. How does one measure the 'harm' caused by one experienced protestor, who told of participating in a human chain that blocked a roadway, inconveniencing untold numbers of drivers? It caused stress and anxiety for those who were late for work and appointments, yet many may support the protestors' actions even so. Others not. Even when the magnitude of the harm *is* clear, some actions are illegal but not harmful and others are harmful but not illegal. Restorative justice is championed as overcoming the shortcomings of the legal system. However, its superiority can only be realised if it is not limited to legal definitions of harm.

Equally as tricky as 'harm' is the question of who the 'victim' is in cases of NVCD. Some protestors saw themselves as the victim even as they were treated as the responsible party in the restorative justice conference. On the other hand, focusing on the activist as victim can neglect the significant distress NVCD can cause citizens. Another participant offered that:

> In the ICE (U.S. Customs and Immigration Enforcement) protest case from last year, I later got a call from someone who self-described as a 'victim' of the protest that closed off the street where a woman was late to work and then was fired. These are powerful stories that should be heard. (Study participant/Protester)

This focus on the fair acknowledgement of individual harm and victimhood is core to restorative justice but risks drawing attention away from what activists and restorative practitioners truly want: to target the systemic injustices underlying the incidental harms that pass through panels. In undertaking this study, we became more concerned that opportunities to foster deeper understanding of causes and to draw links to wider ecological and systemic connections in victim-offender-focused restorative interventions are either mostly lost or, when they do happen, are serendipitous. Politicians who could leverage policy change are largely absent from restorative conferences, including those involving NVCD. The question of 'who is the victim?' was aptly complicated by one participant who was arrested during a protest:

> I felt I was the victim. (After literally years of trying…), we never had the opportunity to have conversation to try and determine the harm on me and the wider community. Never had the opportunity to hold (elected official) to account. We wanted to have a conversation. So many systems,

layers of bureaucracy, in place to keep them from having to interact with us. Conversation needs to happen in formats like this (research interview), built into things. (Study participant/Protester)

In our view, the state's implementation of restorative justice through its panels and boards represents missed opportunities to draw links between individual incidents of offending behaviour and the multiple historical and ecological contexts in which they occur. The individualised practice does not accommodate fully restorative dialogic, inclusive processes, what McCold (2000) characterised as fully restorative processes, that would focus on the inseparability of environmental problems from social injustices such as poverty, racism, sexism, unemployment, and poor health and educational outcomes (Brisman, 2007; Brisman & South, 2020). One participant went further in helping us frame our central question about the fit of restorative justice and civil disobedience:

It isn't just in cases of Civil Disobedience where we are confronted with these failures, the way we do restorative justice, when we pretend it is adequate/possible to just look at "the incident" and allows us to decontextualize incidents from their context and allows us to ignore racism, sexism, etc. We have to look at who is responsible for harms at multiple levels. (Study participant/Researcher, academic)

This theme surfaced in participants' attributing outsized attention to incidents of harm at the expense of holistic needs and attention to wider contextual complexity:

We keep obsessing about victim participation in restorative justice and I think that in doing so we do not ask why this is a goal worth pursuing … so this idea that restorative justice in response to criminal harms (which are by definition public harms and not merely interpersonal) requires individual victims to participate in a process with individual offenders to address the harms they experienced. Often this is not the nature of the harms experienced by victims, and restorative responses require something broader. [...] Part of what makes them more fully restorative in my book is that they disrupt individual notions of responsibilities and harm that the current justice system trades on … in doing so they better meet the need of victims—those harmed—because they can conceive of those harms beyond the victim-offender relationship. [...] We get it wrong when we measure restorative justice by the participation of victims as defined by the criminal justice system and participation in processes about the offender instead of measuring the justice on offer to meet their needs. (Study participant/ Educator, researcher)

In Vermont, state funding for restorative justice enables its wide scale and reach. Simultaneously, state sponsorship holds it to legal definitions of harm that risk rendering it as an instrument of regulatory action without the ability to attend to the individualised nuances of neighbourhood dilemmas, including those that might spark NVCD. We learned from participants both in Vermont and beyond that when facilitators take on more fully restorative initiatives like community or family conferences or restorative inquiries, it is most often done on top of the work they are paid for or at least that the time it takes to carry these processes out with integrity is not fully compensated. Furthermore, state sponsorship limits panel's autonomy and contributes to their lack of connections with environmental activists, interpersonal and family violence advocates, other regulatory agencies, and vice versa. At an extreme, the state could support restorative justice insofar as it is used to chastise activists, exaggerate any harm they caused and require they publicly apologise for and repair those harms. Indeed, strong doubts were raised that groups whose primary source of funding comes from agencies vested in the criminal justice system can prevent domination by state interests:

The issue in VT is that restorative justice exists at the behest of state entities. It is funded by the state, relies almost exclusively on referrals from state agencies, and is regulated by different state actors with different foci and goals. This means that restorative justice as an alternative to the formal system in VT requires permission from the formal system to exist. It can thus be critiqued as a means to expand state control of people (via NVCD) under the guise of restorative justice, while failing to define, operate, or adhere to restorative justice principles. (Study participant/Educator, researcher)

The issues raised here are indicative of the tug-of-war over the purpose, scope, and definition of restorative justice. Does restorative justice exist to repair individual-community conflict or existential threats to the national-global stage? Are harms defined by the state, the neighbourhood, or the activist? One participant referred to restorative justice as 'soup de jour, everybody wants to talk about it, there is no one definition.' Another similarly asked, 'How can we move forward together absent some agreement on the core ideas, the lack of agreement about what is restorative justice and what it looks like in Vermont?' On the other hand, we have seen the risks of reductionism that come with fixating on one definition of restorative justice that works across-the-board at the expense of hybrid arrangements that enable more flexible expression of restorative values (Braithwaite, 2020). One presenter called attention during a regular meeting of the Vermont Restorative Justice Consortium to the issue this way: 'There are seven tables of people talking about restorative justice in the State. None of them are talking with each other and none trust each other.' These questions raise familiar concerns about competition to 'own' or put their own brand on restorative justice processes while losing sight of its fundamental purpose.

The difficulties of fitting NVCD into the Vermont DOC model point to the necessity of expanding our perception and uses of restorative justice. Restorative justice needs to be understood as more than a method, technical application, or routine imposed on a given crime or charge. The future of restorative justice rests in finding new spaces in which to actualise restorative values and principles. Ones that embrace, in Ommer's (2021) view, stewardship that is historically and culturally informed. At the same time, those involved in NVCD are looking for what John Lewis would call 'good trouble,' trouble that ignites tension against power in a way that captures public attention. For restorative justice to not take the wind out of activists' sails, it needs to engage with good trouble and project it to a wide audience.

Throughout the study period, media attention to protest and to restorative justice has increased locally, nationally, and internationally. Critics have used media to advance what are arguably dangerous misrepresentations (Barker et al., 2021), such as claims that #BLM protestors employed violence. These claims have persisted despite evidence to the contrary

(Geoff & McCarthy, 2021). For example, the Guardian reported that over 90% of those arrested at Black Lives Matter protests were never charged, suggesting that arrests were part of a strategy to cast protestors in a violent light (Beckett, 2020). Others have highlighted restorative justice in the face of large-scale arrests during protests as a positive way to divert matters from court (Melamed, 2021). We found few studies that depict public views of restorative justice in the media aside from Vaandering and Reimer (2019), who affirm that representations in the media can be profoundly different than how restorative justice advocates intend. We would go further to suggest that research is urgently needed to compare views of restorative justice between and among judges, police, restorative facilitators, politicians, lay public, and beyond. Through this inquiry we can better understand the nuanced ways that racialised, gendered, and stigmatising images are reproduced over time.

Study participants reported that many of these protestors were more likely to be the victim of violence rather than the perpetrator of it, describing cases of civil disobedience where BIPOC protestors experienced more aggressive or violent treatment than White protestors. One arrested protestor/environmental justice advocate offered that in making calculations of risk and collateral consequences before a protest, the risks are magnified by those who hold marginalised identities. They shared that arresting officers and jailers gave them preferential treatment over Indigenous and Black co-protestors. What's more, they described how White arrested protestors were able to use their privilege and influence to help a Black colleague from suffering further misfortune. One of the protestors involved observed that:

> We're all White. They [police] treated us very respectfully, kind, even … It was almost a pleasant experience, and I hadn't expected that. This plan impacts BIPOC people at a disproportionate rate, but we can speak up because we don't run the same risk as being arrested as BIPOC folks. (Study participant/Protester)

Clearly, the peacefulness with which police disrupt or end demonstrations depends on who you ask and the presence of unanticipated developments that happen including intimidating counter-protest activity.

Such concerns were born out in other interviews with White subjects who acknowledged feelings of responsibility to speak up, believing their BIPOC counterparts would face more damaging consequences for the same actions.

> There was a large White, middle-class presence and orgs like the Sierra Club, Patagonia were also there and supportive. So many tribal members went to Standing Rock from other countries, and other tribes around the US. There were more arrests and prosecutions of Indigenous protestors because the state of North Dakota really cracked down on criminal arrests/prosecutions after Standing Rock. The fear of the criminal justice system may be a deterrent for BIPOC protestors who would worry about getting detained, have their kids taken away, what happens to Indigenous people when they're inside the system. They know it's not good. (Study participant/Educator, environmental justice advocate)

As an example of the vulnerability of restorative justice programmes to be captured by White privilege, one prosecutor, educator, and researcher observed that restorative justice pilot projects typically get located in areas with the highest property values. In thinking about the use of restorative justice in the face of protest, they explained:

> Getting a restorative justice program off the ground in neighbourhoods that are over-policed is challenging; it's way easier to get a restorative justice program off the ground in an affluent neighbourhood ... The mind doesn't even go there for communities of colour. It's seen as a slap on the wrist for BIPOC/poor folks, and fair for White/affluent folks. Amy Cooper[10] had to do 5 counselling sessions and cultural sensitivity trainings. This was not a restorative approach, even though it was billed this way. She said she was forced to go and might want to sue because the counselling did her harm. If it's an affluent person that's the responsible party, of course, restorative justice is the default. That's not the case with BIPOC folks. [...] The ones who are loudest are often the ones with the most access and are also inclined to want others to live in a way that's similar to theirs. We might have to

---

[10] Amy Cooper was arrested for calling 911 on a Black birdwatcher in New York's Central Park. Her criminal case was dismissed after completing a diversionary counseling program that prosecutors said was meant to educate her on the harm of her actions (Sisak, 2021).

rethink who has the right to call law enforcement? What laws are we enforcing? My premise is that LE is there to enforce laws and … many of our laws (are) unfair or unjust, so examine the laws we are looking to enforce, who they're protecting and why. (Study participant/Educator, community leader, former prosecutor)

Beyond police presence, the use of volunteers as facilitators can preclude authentic community representation and consequently contribute to social and racial stratification. As our participants made clear, restorative justice will have to take active measures to combat the tendency to serve already-privileged communities with a set of already-privileged facilitators. Yet again, this theme is exacerbated in cases of NVCD, as BIPOC individuals are at a heightened risk of arrest and poor treatment by law enforcement. This further underscores the need to position legal definitions of harmed and harming parties as but one of many ways of understanding harm, responsibility, and accountability.

7   Final Thoughts

In contrast to criminal offenders who seek to conceal their identity and evidence of their culpability, NVCD is generally an act of intentional communication to make visible the principled and conscientious reasons for defiance of the law (Delmas & Brownlee, 2021). The civilly disobedient intentionally break laws they perceive as unjust as an act of indirect (e.g., stopping traffic to call attention to an injustice) or direct protest, as in the case of Rosa Parks, who refused to give up a seat on a bus reserved for White people in defiance of discriminatory policy. The uniquely socially intentioned lawbreaking of civil disobedience shares a spirit with restorative justice, in which transparency is fundamental to building trust as it lays the groundwork for reciprocity, mutual aid, and common cause. The legal facts are considered in light of their social and environmental context. Such transparency and commitment to participatory practices adhere to the notion that "sunlight is the best disinfectant" to eke out the misleading claims and denial of harm used by corporations to avoid being held responsible for environmental harms (cf., McGreal, 2022).

The discussion-cantered and community-oriented approach of restorative justice is in direct contrast to this secrecy. Restorative conferences even make space for conversations around civil disobedience related to vaccine hesitancy, environmental degradation, or more complete versions of Indigenous people's history, for example. Broaching these topics in a restorative justice format might avoid the partisan divisiveness raised by the criminal punishment of these cases.

The entrance of NVCD into restorative justice offers a unique opportunity to use restorative conferences as a platform from which to amplify discussion of and solutions to environmental and systemic inequities that both activist and restorative practitioners seek to combat. However, as our study indicates, Vermont's CJCs and Diversion agencies have operated within a climate of legal, resource, and social restraints that prevent them from developing a more fully restorative menu of offerings or from partnering with social movement groups. Study participants emphasised the need to clarify the purpose of Diversion and Community Panels, noting problematic conflation between their original purpose, unrelated programming, and restorative practices. Such concerns have long been raised in other venues (Clamp, 2014; Richards, 2014). In building Vermont's restorative justice story, we should remain alert to the dangers of the single story (Ngozi Adichie, 2009) of restorative justice that is regulated by a single actor, in this case, the State. Non-state actors may have more agility to adapt their protocols to align process outcomes with the intentions of civilly disobedient participants. When confined to legal, State-defined definitions of harm, the restoration in restorative justice can end up being directed disproportionately towards the minor and incidental harms of activist lawbreaking instead of social and environmental causes. To realise a restorative vision means transcending, but not abandoning, the useful micro-practices and technical applications of restorative justice that are so well known and replicated at the interface of practice.

If we embrace both a subjective and an objective ownership of the charge to 'make the public whole' (as dictated by the Vermont's 2017 Natural Resources Damages Assessment and Restoration Plan) and continue to bring the voices of those most impacted by harm, individually, socially, and systemically, to our collective table, we might be emboldened to redefine 'responsible parties' and go deeper when constructing contracts for repair or restoration. As recent events in the US and internationally have shown,[11] protest is growing more and more dangerous for BIPOC persons and their allies. The need for expanding restorative imagination and the need to support protest grow in parallel.

Examples of 'bright spots' in Vermont include the use of 'upstream' preventative practices in the schools, apologies and dialogue with Abenaki leaders and community members (Gokee, 2021; Walters, 2020), interrogating racialised histories of institutions in Vermont (e.g., Loftus, 2018), a restorative inquiry of abuse and neglect in a Catholic orphanage (Stand Up Resources, 2020), to name a few. Moving forward, the National Centre for Restorative Justice, at the Vermont Law School where graduate and certificate programmes in restorative justice have begun, including a recently announced training development between Vermont Law School and the State Department for Children and Families (Vermontbiz, 2022), and a State Environmental Conservation Department looking to put more and more focus on 'relational regulation'[12] are potentially galvanising efforts to widen the restorative imagination and cultivate partnerships amongst state and non-state actors. Of particular interest to us was the connection with state environmental personnel who succinctly assert in their written guide for assessment and restoration: 'Each alternative shall be designed so that, as a package of one or more restoration actions, the alternative(s) would make the environment and public whole' (Vermont Agency of Natural Resources, 2017). Importantly, the guidelines offer a menu of support and enforcement possibilities that support compliance with laws and regulations. The achievement of such aims across regulatory institutions requires developing comfort with and knowledge of the importance of regulation as a field, or craft, as Sparrow (2020) calls it.

We can start to imagine what restorative spaces look like through investigating positive results of restorative meetings that emerged in surprising ways. One such example was offered by a participant who told of a demonstration organised to protest local officials' cooperation with

---

[11] For example, Joseph Rosenbaum, Anthony Huber, and Gaige Grosskreutz in the US and Breiner David Cucuñame in Columbia.

[12] Examples of definitions can be found in Braithwaite (2002), Burford et al. (2019), and Llewellyn and Morrison (2018).

Immigration and Customs Enforcement (ICE) in which protesters blocked streets, causing gridlock traffic. The protest resulted in arrests, during which "several people were dragged away by police, so it did get 'ugly.'" This same participant observed that 'protestors felt victimised in the process of the protest and after-effects,' during which time protestors experienced 'severe online threatening' related to racial issues. In this example, police and other community members attended restorative meetings after which arrested demonstrators reported having been 'surprised how curious they [police] were and open-minded about the motivations rather than just one-sided view of them as law-breakers.' Restorative justice is well positioned to foster more of these generative conversations when the roles of 'harmed' and 'harmer' are loosened.

In another successful example of NVCD in restorative justice, one study participant located in a large urban area described an inspirational effort that engaged many arrested protestors in community conversations through the prosecutor's office. The conversations focused less on identifying offender and victim than on the impact of the issues they were protesting, the impact on their communities and families, and the issues that tied them to one another, to the arresting officers, and to the larger community. Unsurprisingly, the conversations focused on wider issues that built what Haraway (2016) might accept as kin connections, rather than on their differences. Importantly, the work to contact people, prepare them, and to ensure that meetings were safe and centred on dialogue was on top of the facilitator's regular paid work. What's more, the efforts were backed by local prosecutors, police, and others who saw value in bringing people together.

Like so many restorative pilots and one-time efforts that have contributed significantly to restorative theory and practice, the efforts are often not sustainable in the sense of continuous development. Hotspots of good work show up here, then there. Champions leave. Corners are cut in scaling up and out that inevitably step away from the most important principles of the work (Adams, 2019; Adams & Chandler, 2004). At worst, CJCs may be seen as coming late to the party, so to speak, by other groups more rooted and experienced in grassroots community relations than restorative panels and boards have achieved. At the same time, the CJCs hold a somewhat unique position of being a responsive option in the network of regulatory institutions. In the fall of 2021, the VT Joint Justice Oversight Committee of the Legislature engaged in an 'RJ Study' to reexamine the landscape of state-funded restorative justice, the various relevant statutes, and work toward recommendations to the full Legislature. This has needed to involve some education regarding the current make-up of the two dozen state-funded RJ agencies and their funding sources (Showalter, 2022). Perhaps the CJC's greatest potential to expand lies in cultivating relationships with other regulatory institutions committed to the principles of strengthening democratic participation and reducing domination in the service of tackling complex problems. The need for harmonisation of regulations and rule-making processes between regulatory institutions is a fundamental step the State could take to clear the way for growing restorative values and supporting community initiatives and activities.

As the contemporary restorative justice movement works, ever so slowly, towards a more holistic form of justice, we can try and incorporate core environmental principles of interconnection and valuing all forms of life into our work. Boyes-Watson and Pranis (2020) mention the concept of interconnection in their Seven Core Assumptions of Restorative Work, but in our experience in the Vermont restorative justice field, we are still highly human-centred, and ripple-effects on the natural world and other species are rarely, if ever, topics of conversation. Civil disobedience on behalf of the Earth presents more explicit challenges for our human-centred paradigm and opportunities to innovate when considering how to, and who could, represent 'impacted parties.'

Martin Luther King, Jr., famously said, 'In the final analysis, a riot is the language of the unheard. And what is it that America has failed to hear?'[13] What if we cultivated CJCs as 'epicentres of dialogue'—which provided restorative spaces for community building, understanding, repair, and de-escalation of conflict? Imagine a rotating series of themes for a given community to come together to problem-solve controversial topics such as school mascots, public art, gentrification, public safety,

---

[13] Martin Luther King Jr. "The Other America" speech, March 10, 1968. Retrieved from: 24/01/2022, https://www.beaconbroadside.com/broadside/2018/03/martin-luther-king-jrs-the--other-america-still-radical-50-years-later.html (last accessed January 25, 2022).

environmental harms, or the role of government, to name just a few. This is an image of restorative justice that befriends the civilly disobedient, magnifying their capacity to make a meaningful difference beyond the limits of the justice system.

Acknowledgements We thank the people who volunteered their time to respond to the survey and who generously contributed to our inquiry through interviews, consultations, and guidance. We thank the slate of international and national researchers, policy leaders, administrators, and practitioners who have given so freely of their time, often contributing their own travel costs, to support restorative justice development efforts in Vermont over the past 20+ years to share knowledge from places as far flung as AU, NZ, EU, UK, CA. We are happy to share these experiences from Vermont and hope they contribute to the international and global dialogue.

References

Adams, P. (2019). Families and farmworkers: Social justice in responsive and restorative practices. In G. Burford, J. Braithwaite, & V. Braithwaite (Eds.), *Restorative and responsive human services* (pp. 91–99). New York: Routledge.

Adams, P., & Chandler, S. M. (2004). Responsive regulation in child welfare: Systematic challenges to mainstreaming the family group conference. *Journal of Sociology and Social Welfare, 31*(1), 93–116.

Barker, K., Baker, M., & Watkins, A. (March, 2021, updated June, 2021). *In city after city, police mishandled black lives matter protests.* The New York Times. Retrieved from: https://www.nytimes.com/2021/03/20/us/protests-policing-george-floyd.html (last accessed 16 November 2021).

Bazemore, G., & Umbreit, M. (2001, February). A comparison of four restorative conferencing models. *OJJDP Juvenile Justice Bulletin.* Retrieved from: https://www.ojp.gov/pdffiles1/ojjdp/184738.pdf (last accessed 16 November 2021).

Beckett, L. (2020, September 5). *Nearly all Black Lives Matter protests are peaceful despite Trump narrative, report finds.* The Guardian. Retrieved from: https://www.theguardian.com/world/2020/sep/05/nearly-all-black-lives-matter-protests-are-peaceful-despite-trump-narrative-report-finds (last accessed 16 November 2021).

Boyes-Watson, C. (1999). In the belly of the beast? Exploring dilemmas of state-sponsored restorative Justice. *Contemporary Justice Review* 2(3): 261–281.

Boyes-Watson, C., & Pranis, K. (2020). *Circle forward: Building a restorative school community* (Revised Edition). St. Paul: Living Justice Press.

Braithwaite, J. (2002). *Restorative justice and responsive regulation.* Oxford: Oxford University Press.

Braithwaite, J. (2020). Restorative justice and reintegrative shaming. In C. Chouhy, J. C. Cochran, & C. L. Johnson (Eds.), *Criminal justice theory, volume 26: Explanations and effects* (pp. 281–308). New York, Routledge.

Brisman, A. (2007). Crime-environment relationships and environmental justice. *Seattle Journal for Social Justice*, 6(2), 727–825.

Brisman, A. & South, N. (2020). *Routledge International Handbook of Green Criminology* (2nd edition). New York: Routledge.

Brouwer, D. (2019, September 9). *F-35 opponents arrested during sit-in at Leahy's Burlington office.* Seven Days. Retrieved from: https://www.sevendaysvt.com/OffMessage/archives/2019/09/09/f-35-opponents-arrested-during-sit-in-at-leahys-burlington-office (last accessed 16 November 2021).

Burford, G., Braithwaite, J., & Braithwaite, V. (2019). *Restorative and responsive human services.* New York: Routledge.

Clamp, K. (2014). A 'local' response to community problems? A critique of community justice panels. *British Journal of Community Justice*, 12(2): 21–34.

Crist, L., & Bech, T. (2018). Bending the curve to improve our child protection system: A multiyear analysis of Vermont's child protection system & recommendations for improvement. *Vermont: Vermont Parent Representation Center.* Retrieved from: https://vtprc.org/wp-content/uploads/2018/11/BTC-11-09-18-FINAL.pdf (last accessed 16 November 2021).

Davis, X. R. (2021). *Report of the executive director of racial equity.* General Assembly of Vermont. Retrieved from: https://legislature.vermont.gov/assets/Legislative-Reports/EDRE-Report-to-GA-2021.pdf (last accessed 16 November 2021).

Delmas, C. & Brownlee, K. (2021, Winter). Civil disobedience. In E. N. Zalta (Ed.), *The Stanford Encyclopedia of Philosophy.* Retrieved from: https://plato.stanford.edu/archives/win2021/entries/civil-disobedience/ (last accessed 27 December 2021).

Dembenski, J. (2004). Restorative justice in Vermont: Part two. *Vermont Bar Journal*, 30(1), 49–53.

Fox, K. J. (2013). *Circles of support & accountability: Qualitative evaluation.* Vermont: State of Vermont Department of Corrections.

Fox, K. J. (2012). Redeeming communities: Restorative offender reentry in a risk society. *Victims & Offenders*, 7(1), 97–120.

Fox, K. J. (2010). Second chances: A comparison of civic engagement models for offender reentry programs. *Criminal Justice Review*, 35(3), 335–353.

French, E. (2021, April 8). *Legislators want new data system to understand where racial disparities begin*. VTDigger. Retrieved from: https://vtdigger. org/2021/04/08/legislators-want-new-data-system-to-understand-where-racial-disparities-begin/ (last accessed 25 January 2022).

Gallagher, N. (1999). *Breeding better Vermonters: The eugenics project in the green mountain state (Revisiting New England: The new regionalism)*. UPNE.

Galloway, A., & Keays, A. J. (2021, June 23). *Panel: Troopers discriminated against Clemmons Family Farm director*. VTDigger. Retrieved from: https:// vtdigger.org/2021/06/23/panel-troopers-discriminated-against-clemmons-family-farm-director/ (last accessed 16 November 2021).

Gallup Sharecare. (2018). *Well Being Index: 2017 State well-being rankings*. Sharecare. Retrieved from: https://wellbeingindex.sharecare.com/wp-content/uploads/2018/02/Gallup-Sharecare-State-of-American-Well-Being_2017-State-Rankings_FINAL.pdf.

Geoff, K., & McCarthy, J. D. (2021, October). *Critics claim BLM protests were more violent than 1960s civil rights ones. That's just not true.* The Washington Post. Retrieved from: https://www.washingtonpost.com/politics/2021/10/12/critics-claim-blm-was-more-violent-than-1960s-civil-rights-protests-thats-just-not-true/ (last accessed 29 December 2021).

Gokee, A. (2021, February 21). *90 years after Vermont eugenics survey, lawmakers propose apology to those affected*. VTDigger. Retrieved from: https://vtdigger. org/2021/02/21/90-years-after-vermont-eugenics-survey-lawmakers-propose-apology-to-those-affected/ (last accessed 29 December 2021).

Haraway, D. J. (2016). *Staying with the trouble: Making kin in the chthulucene*. Durham, NC: Duke University Press.

Henne, K., Shelby, R. & Harb, J. (2021). The datafication of #MeToo: Whiteness, Racial Capitalism, and Anti-Violence Technologies. *Big Data & Society*, 1–14. https://doi.org/10.1177/20539517211055898.

Humphrey, J. A., Burford, G., & Dye, M. H. (2012). A longitudinal analysis of reparative probation and recidivism. *Criminal Justice Studies*, 25(2), 117–130.

Karp, D., & Drakulich, K. M. (2004). Minor crime in a quaint setting: Practices, outcomes, and limits of Vermont reparative probation boards. *Criminology & Public Policy*, 3(4), 655–686.

Kenyon, J. (2015). *The brutal 1981 crime that spawned woodside*. Valley News. Retrieved from. https://www.vnews.com/Archives/2015/02/Woodside sidebar-jsk-vn-022215 (last accessed 16 November 2021).

Llewellyn, J., & Morrison, B. (2018). Deepening the relational ecology of restorative justice. *The International Journal of Restorative Justice*, 1(3), 343–355.

Loftus, S. (2018). *Proposal to remove Guy Bailey's name from library open to public comment*. The Vermont Cynic. Retrieved from: https://vtcynic.com/news/proposal-to-remove-guy-baileys-name-from-library-open-to-public-comment/ (last accessed 29 December 2021).

McCold, P. (2000). Toward a holistic vision of restorative juvenile justice: A reply to the Maximalist model. *Contemporary Justice Review*, 3(4), 357–414.

McGreal, C. (2022, January 18). *How Exxon is using an unusual law to intimidate critics over its climate denial*. The Guardian. Retrieved from: https://www.theguardian.com/environment/2022/jan/18/exxon-texas-courts-critics-climate-crimes?CMP=Share_iOSApp_Other (last accessed 18 January 2022).

Melamed, S. (2021, March 27). *Hundreds arrested in Philly uprisings may avoid prosecution through restorative justice*. Philadelphia Inquirer. Retrieved from: https://www.inquirer.com/news/philadelphia-unrest-restorative-justice-george-floyd-larry-krasner-20210326.html (last accessed 29 December 2021).

Merrick, G., & Newbury, A. (2020). Proactively protecting Vermont's participatory democracy: Reforms to election structure, campaign finance, and voter engagement. *Vermont Law Review*, 45, 481–562.

Ngozi Adichie, C. (2009, July). *The danger of the single story*. [Video]. TED Conferences. Retrieved from: https://www.ted.com/talks/chimamanda_ngozi_adichie_the_danger_of_a_single_story (last accessed 16 November 2021).

Ommer, R. (2021, September 10). The importance of scale complexities in fisheries research. *University of British Columbia Institute of Fisheries*. Seminar presented to the Institute of Oceans and Fisheries, University of British Columbia. Retrieved from: https://www.youtube.com/watch?v=FJWcs_f-ksc (last accessed 16 November 2021).

Perry, J. G., & Gorczyk, J. F. (1997). Restructuring corrections: Using market research in Vermont. *Corrections Management Quarterly* 1, 26–35.

Petenko, E. (2022, January 9). The pandemic tanked traffic stops in 2020—but racial disparities remained, data shows. *Vermont Digger*. Retrieved from: https://vtdigger.org/2022/01/09/the-pandemic-tanked-traffic-stops-in-2020-but-racial-disparities-remained-data-   (last   accessed   29 December 2021).

Prison Policy Initiative. (2018). *Vermont Incarceration Pie Chart 2018*. Retrieved from: https://www.prisonpolicy.org/profiles/VT.html#:~:text=Vermont%20 has%20an%20incarceration%20rate%20of%20288%20per,in%20 Vermont%20and%20why.%20Jump%20to%20COVID-19%20data   (last accessed 16 November 2021).

Racial Disparities Advisory Panel. (2020). *Report of the racial disparities in the criminal and juvenile justice systems advisory panel concerning Section 19, Act 148 –An act concerning justice reinvestment*. Retrieved from: https://legislature.vermont.gov/assets/Legislative-Reports/RDAPAct148Report-FINIS. pdf (last accessed 16 November 2021).

Richards, K. (2014). Blurred lines: reconsidering the concept of 'diversion' in youth justice systems in Australia. *Youth Justice*, 14(2), 122–139.

Shelby, R., Harb, J., & Henne, K. (2021). Whiteness in and through Data Protection: An Intersectional Approach to Anti-Violence Apps and #MeToo Bots. *Internet Policy Review*, 10(4), 1–25.

Showalter, P. (2022). State-funded restorative justice in Vermont. *The future of structure, funding, and flow*. IRP Spring.

Sisak, M. (2021, February 16). *Case dropped after woman in racist NYC run-in gets therapy*. AP News. Retrieved from: https://apnews.com/article/amy-cooper-case-dropped-ca04b20d80580837645641480303b558 (last accessed 29 December 2021).

Sparrow, M. (2020, July 30). *Fundamentals of regulatory design*. Kindle Direct Publishing. Retrieved from: https://scholar.harvard.edu/msparrow/ fundamentals-of-regulatory-design (last accessed 29 December 2021).

Stand Up Resources. (2020). Saint Joseph's Orphanage Restorative Inquiry. Retrieved from: https://www.stjosephsrjinquiry.com/ (last accessed 29 December 2021).

UPI Archives. (1987, October 29). Police arrest 17 students in CIA protest. Retrieved from: https://www.upi.com/Archives/1987/10/29/Police-arrest-17-students-in-CIA-protest/7092562482000/ (last accessed 2 January 2022).

Vaandering, D., & Reimer, K. (2019). Listening deeply to public perceptions of restorative justice: What can researchers and practitioners learn? *The International Journal of Restorative Justice*, 2(2), 186–208.

Vermont Agency of Natural Resources. (2017, November 21). *Natural resource damage assessment and restoration rules*. Retrieved from: https://dec.vermont.

gov/sites/dec/files/2017_11_29_NRDA_Rule_Ch_36A_ADOPTED_
CLEAN.PDF (last accessed 16 November 2021).

Vermontbiz. (2022, January 20). Vermont law school and DCF to infuse state
work with restorative justice principles. Retrieved from: https://vermontbiz.
com/news/2022/january/20/vermont-law-school-and-dcf-infuse-state-work-
restorative-justice-principles (last accessed 22 January 2022).

Vermont Census. (2019, July 1). Retrieved from: https://www.census.gov/
quickfacts/fact/table/VT/PST045219 (last accessed 16 November 2021).

Walters, J. (2020, February 20). Final reading: Abenaki tribe supports state-
house eugenics apology. *VtDigger.* Retrieved from: https://vtdigger.
org/2020/02/20/final-reading-abenaki-tribe-supports-statehouse-eugenics-
apology/ (last accessed 16 November 2021).

Wilson, R.J., & Fox, K.J. (2019). Why do we exclude the community in "community safety". In G. Burford, V. Braithwaite, & J. Braithwaite (Eds.),
Restorative and responsive human services (pp. 195–209). New York: Routledge.

19

Environmental Restorative Justice in the Philippines: The Innovations and Unfinished Business in Waterways Rehabilitation

Jennifer Marie S. Amparo, Ana Christina M. Bibal, Deborah Cleland, Ma. Catriona E. Devanadera, Aaron M. Lecciones, Maria Emilinda T. Mendoza, and Emerson M. Sanchez

1 Introduction

But how do you apologise to a river?

It's better if you have not done the wrong thing in the first place. But time, like the river, does not flow backward. So you go forward, like the river. You try to do better.

—Eliza Victoria, Filipina author from her short story *Down the River* in the anthology Multispecies Cities: Solarpunk Urban Futures, 2021, World Weaver Press.

J. M. S. Amparo
Department of Social Development Services, College of Human Ecology, University of the Philippines Los Baños, Los Baños, Philippines
e-mail: jsamparo@up.edu.ph

A. C. M. Bibal
University of the Philippines Los Baños, Los Baños, Philippines
e-mail: ambibal@up.edu.ph

B. Pali et al. (eds.), *The Palgrave Handbook of Environmental Restorative Justice*,
https://doi.org/10.1007/978-3-031-04223-2_19

477

The rivers, oceans, and watercourses of the Philippines have suffered much harm over the course of colonisation and industrialisation, such that *doing the good work* is an urgent imperative. Across the country, 75 per cent of mangroves and 78 per cent of wetlands have been lost (Friess et al., 2019), and one of its river systems, the subject of a case study in this chapter, was named as one of the most polluted in the world in a landmark study (Blacksmith Institute, 2009). Perhaps because of this legacy, and perhaps in spite of it, the Philippines is the site of some truly remarkable legal and institutional innovations that have enormous potential to aid in the quest for environmental restorative justice (ERJ).

By enabling the voices of those affected by environmental harm to be heard through the courts and through organisational structures, the Philippines has made it possible to hold both governments and corporations accountable for harm, while creating a healing orientation. At the

D. Cleland (✉)
Australian National University, Canberra, ACT, Australia
e-mail: deborah.cleland@anu.edu.au

M. C. E. Devanadera
Department of Community and Environmental Resource Planning, College of Human Ecology, University of the Philippines Los Baños,
Los Baños, Philippines
e-mail: medevanadera@up.edu.ph

A. M. Lecciones
College of Architecture, University of the Philippines Diliman,
Quezon City, Philippines
e-mail: amlecciones@up.edu.ph

M. E. T. Mendoza
Department of Social Services, College of Human Ecology, University of the Philippines Los Baños, Los Baños, Philippines
e-mail: mtmendoza2@up.edu.ph

E. M. Sanchez
Crawford School of Public Policy, Australian National University,
Canberra, ACT, Australia
e-mail: emerson.sanchez@anu.edu.au

same time, applied case studies show that it can be very tricky to identify what harm clearly and definitively can and should be repaired and who are the victims, offenders, regulators, and broader community actors, when these roles are blurred and shifting across time and space.

This chapter introduces and explores the innovations of the Philippines in ERJ and then looks at three examples showing how while national regulatory and legal frameworks may make ERJ possible, even sustained coordination and activism have not yet guaranteed happy endings for the waterways in question. Three common threads unite the case studies: (1) questions of accountability across international borders; (2) complex trade-offs between different livelihoods (which sustains life for the humans engaged) and more-than-human ecologies; and (3) the complexities about what relationships, places, bodies, and activities should be restored—and who gets to make those decisions.

2   Frameworks for ERJ in the Philippines

This section briefly introduces four key pillars of law and policy that enable ERJ for waterways in the Philippines—the constitutional right to a healthy environment, protection and recognition of intergenerational environmental justice, specialist environmental courts, and integrated water management through Water Quality Management Areas.

The Philippines was the first nation to enshrine the right to a healthy environment, with Section 16 of Article II of the 1987 Constitution of Philippines (1987) proclaiming: 'The State shall protect and advance the right of the people to a balanced and healthful ecology in accord with the rhythm and harmony of nature'. Since this right is protected, many environmental harms are recognised in the eyes of the law, meaning that robust and enforceable mechanisms for restitution and repair are possible—if not always realised.

Accompanying this ground-breaking constitutional right was growing awareness of the limits to development in the face of environmental destruction. With the creation of the Philippine Council for Sustainable Development in 1992, the first Asian country to do so after the Rio Summit (Reyes, 2014), the environmental movement was growing in the consciousness of Filipinos (Magno, 1999). The confluence of civil society, environmental, and democracy movements in the latter half of the century served as building blocks of paradigm shifts that recognised the inseparability of social justice and environmental restoration. This awareness spread to the courts, where decisions and opinions on the matter strengthened Philippine environmental jurisprudence. In 1993, through an opinion by Justice Davide in *Oposa v. Factoran*, the constitutional right to a sound environment was decided by the Supreme Court as enforceable and thus 'grant[ed] standing to [the plaintiffs'] children in the present generation to represent both their own interests and those of future generations' (Oposa v. Factoran, 1993). Once again, the Philippines was the first country to recognise what is now widely accepted as inter-generational justice (Allen, 1993; Gatmaytan, 2002): 'every generation has a responsibility to the next to preserve that rhythm and harmony [of nature] for the full enjoyment of a balanced and healthful ecology' (Oposa v. Factoran, 1993).

The Minors Oposa doctrine came into play at the helm of civil society environmental justice movements' clamour to usher post-Martial Law reforms and curb massive denudation and deforestation of watersheds brought by the elite capture and transnational corporate control of large-scale logging concessions and extractive industries. Though deemed a novel innovation in the legal realm, the doctrine echoes a core principle of ecological spirituality and environmental worldview, common amongst the diverse Indigenous societies of the archipelago. The Minors Oposa doctrine proved potent in highlighting the primary mandate of the State to uphold peoples' welfare over profit as duty-bearers of environmental protection, conservation, and restoration (La Viña, 1994).

More than a decade later, this landmark case fed into the establishment of specialised environmental courts[1] in 2008 (Davide & Vinson, 2010), such that fully one-third of the world's environmental courts are in the Philippines (Preston, 2014; Pring & Pring, 2009). This too followed the Rio Summit and accompanying declaration around access to justice for environmental

---

[1] The Administrative Order No. 23-2008 *Designation of Special Courts to hear, try, and decide on Environmental Cases* designated 117 courts as environmental courses for 'efficient administration of justice and pursuant to Section 23 of BP Bldg 129 *An Act Reorganising the Judiciary, Appropriating Funds therefor, and for other purposes'*.

harms. Specifically, Principle 10 of the Rio Declaration on Environment and Society declares that environmental issues are best handled with 'appropriate access to information', 'the opportunity to participate in decision-making processes', and 'effective access to judicial and administrative proceedings, including redress and remedy' for 'all concerned citizens' (United Nations Conference on Environment and Development, 1992). The environmental courts swiftly established rules of procedures which cemented intergenerational justice as a core consideration of the court, explicitly giving 'future generations' 'standing'—or the right to sue.

The rules of procedure for the specialists' courts also include the *Writ of Continuing Mandamus*, again a globally rare and future-orientated power for the court (Poddar & Nahar, 2017). A continuing mandamus is an extensive, persistent, and continuing order of the court to implement an action plan, meaning that those responsible for causing harm are held legally responsible until they have satisfactorily demonstrated repair and restoration to the court.

Finally, the Water Quality Management Area (WQMA) specified under the Clean Water Act facilitates cross-institutional responsibility and action for rehabilitating waterways and associated ecosystems (Arcala Hall et al., 2018). The key mechanism available to WQMA is the governing body, which brings together stakeholders from across sectors to decide together what should and can be done to restore the river. Where other jurisdictions have catchment management authorities, that do not necessarily have specific or integrated roles for business or citizens reference groups that may respond to the specific environmental issues of a particular firm, the WQMA's flexible and inclusive structure allows for representatives from across all those responsible for harm, affected by the harm, and with the power to stop the harm being done.

The following three case studies incorporate aspects of these regulatory innovations and critically examine how these innovations have worked in practice. Firstly, we turn to a wetland that evolved from land reclamation and is now protected, while still being accessed by community. Secondly, we examine one of the world's most polluted river systems and the mixed success of rehabilitation efforts. Finally, we turn to the tragic tale of the Marcopper Mining Disaster, where no amount of national legal innovation has forced the international company involved to account.

3   Case Studies

3.1   Case Study 1: The Las Piñas Parañaque Wetland Park (LPPWP)

Cognizant of the profound impact of human activities on all components of the natural environment particularly the effect of increasing population, resource exploitation and industrial advancement, and recognising the critical importance of protecting and maintaining the natural, biological, and physical diversities of the environment notably on areas with biologically unique features to sustain human life and development, as well as plant and animal life, it is hereby declared the policy of the State to secure for the Filipino people of present for future generations, the perpetual existence of all native plants and animals through the establishment of a comprehensive system of integrated protected areas.
—Preamble of the E-NIPAS Act (2018), Republic of the Philippines

In 2018, the Las Piñas Parañaque Wetland Park (LPPWP) was declared over an area of 181 hectares in southern Manila Bay, following decades of conflict over land use and development. This finally gave legislative backing to decades-old fights to access and conserve the area by activists and local communities. This case study showcases three features that are common in the Philippines and relevant more broadly. Firstly, the area includes 'novel ecosystems'—that is, ecosystems resulting from dramatic human-induced landscape-level changes—including the small islands known locally as 'Pulo', remnants of a failed reclamation in the 1970s (Coronel & Tordesillas, 1998). As conservationists let go of the idea of 'untouched wilderness', these 'disturbed' places are increasingly recognised as critical interfaces for biodiversity. Secondly, the case study demonstrates how legally orientated decisions supporting ERJ were in themselves not enough without grassroots activism. Thirdly, the 'traditional' roles of restorative justice—victim, offender, regulator, and community—are porous. The main 'offender' causing environmental harm in this case was successive national governments, through publicly funded projects to extend human habitats through land reclamation and urban expansion. In turn, activist communities have held the government to account, thereby playing the role of regulator.

Historically, fisherfolk communities used the cultural ecosystem services of the LPPWP area for recreation and the provisioning ecosystem services of the adjacent bay for their food and livelihoods (Lecciones & Devanadera, 2021; Estoque & Murayama, 2015). However, rapid urbanisation and development have excluded and displaced fisherfolk since the 1990s, while Metro Manila's urban area grew around 5 per cent per year (Estoque & Murayama, 2015). Huge development projects transformed the rural landscape at the urban fringe. One example is Boulevard City, which was part of an older plan to reclaim 3000 hectares from Manila Bay (van den Muijzenberg & van Naerssen, 2005). While this project brought thousands of jobs and considerable tax revenue to the local government, it also resulted in the contested relocation of thousands of families and the displacement of fisherfolk (Gabasan, 2006; Rodolfo, 2016). Another land reclamation project undertaken in the late 1990s was the construction of the Manila-Cavite Expressway—a government project meant to alleviate local traffic congestion and to connect the emerging municipalities of Parañaque, Las Piñas, and Bacoor with Metro Manila (Estoque & Murayama, 2015), but which again separated fisherfolk from their traditional catch areas.

In response to the declining water quality caused by unchecked development and population growth, in 1999 concerned residents filed a complaint with the Regional Trial Court in Imus Cavite against several government agencies for the clean-up, rehabilitation, and protection of Manila Bay. The complaint alleged that water quality in the bay had fallen below standards set by the Philippine Environmental Code and a host of other Philippine laws (Supreme Court of the Philippines, 2019). In 2002, the Regional Trial Court found merit in the complaint and ordered the defendants—various government agencies—to rehabilitate and restore the bay.

A year later in 2003, after years of investigation into corrupt business dealings, the Supreme Court also voided the joint venture to reclaim the three islands comprising the Pulo, declaring the project unconstitutional and stopping the completion of the last section of the Boulevard reclamation project. The Pulo was then an incomplete reclamation site whose unfinished foundations were slowly washed away by the tide, leaving what remains today as two large islands. By this time in the mid-2000s, many of the displaced fisherfolk returned to the area and resumed their lives, albeit much changed by the expressway and the slowly degrading reclamation site.

At the same time, growing environmental awareness mobilised communities into blocking efforts to revive old reclamation projects with new proposals in the area. The issue reached a turning point with the Supreme Court's landmark ruling in 2008, following appeals by defendants against the Manila Bay decision of the Regional Trial Court. Anchored on the concept of intergenerational rights established in *Oposa Minors* (Ristroph, 2012), the ruling upheld the issuance of a *Writ of Continuing Mandamus*, as introduced above, directing 13 government agencies to restore and rehabilitate Manila Bay within 10 years (Metropolitan Manila Development Authority v Concerned Residents of Manila Bay, 2008).

The Writ provides a legal and moral imperative for ongoing action and accountability. However, as precedent would predict, despite the favourable rulings in these environmental courts, actual progress in restoring and rehabilitating Manila Bay has not been realised (La Viña, 2014). Yet, even by the time of the Supreme Court ruling, a coalition of groups concerned for human rights and environmental justice was working to keep the two islands of *Pulo* a space for nature amongst the urban sprawl. By this time, the islands were covered in mangroves, mixed beach forest, mudflats, and lagoons, with wetlands habitat for more than 80 species of migrant and resident wild birds. The civil society sector and local politicians from communities around Manila Bay banded with environmental groups from across the country that were dismayed at the disfiguring of nature for economic progress. They united against threats to the environment of Manila Bay, including habitat and biodiversity loss, pollution, and degraded foreshore ecology, and the plight of displaced fisherfolk, including the disabling of social networks, the loss of livelihood and physical capital (Gabasan, 2006). Thanks to this grassroots advocacy, the area was first protected by Presidential Proclamation 1412 and 1412A and designated as the Las Piñas Parañaque Critical Habitat and Ecotourism Area (LPPCHEA) in 2007 (DENR, 2012).

As more groups joined the environmental activism, the wider the reach and more diverse the activities they conducted became. The group working to protect the wetlands expanded to include the Department of

Environment and Natural Resources—which recognised the natural value of the wetlands and thus had the mandate to protect them; The Wild Bird Club of the Philippines—which documented the bird species of the area; Villar Sipag Foundation—which conducted environmental awareness activities on behalf of Senator Cynthia A. Villar, a long-time resident in the area; and the Society for the Conservation of Philippines Wetlands—whose sole mission was to conserve wetlands in the country and whose Vice-President was also the Ramsar Communication, Education, Participation and Awareness (CEPA) Focal Point. Through their efforts, the LPPCHEA in 2015 was declared a Ramsar Wetland of International Importance. In 2018, through long-time ally Senator Cynthia A. Villar, the area was officially placed under legislative protection in the Expanded National Integrated Protected Areas Systems Act of 2018 (Republic Act 11038) under the name the Las Piñas Parañaque Wetland Park.

The work towards protecting the wetlands through legislation was done through a conscious effort by the partners to act in strategic and inclusive ways to maximise their strengths (Gera, 2016), whether in political clout, technical expertise, grassroots activism, or environmental governance. Some activities to strengthen horizontal collaboration included regular programmes and activities such as coastal clean-ups, bird watching, field visits, and other communication, education, participation, and awareness activities. Key local government actors, local government units of Las Piñas and Paranaque Cities, Bakawan Warriors, nearby communities, and local and national media participated in these activities, which coincided with the celebration of Earth Day, World Wetlands Day, World Water Day, World Migratory Day, and Urban October, amongst others. The partners also leveraged international agreements, including the Ramsar Convention on Wetlands, Convention on Migratory Species, the Convention on Biological Diversity, the East Asian-Australasian Flyway Partnership, to increase the impact of conservation activities at LPPWP. In the future, the coalition is hoping to build a wetland centre, which will be an avenue for restoring the relationship lost between the community and the wetlands by providing a secure way to access the protected area and by engaging the community with biodiversity friendly livelihood activities. Raising the awareness of the community and empowering them to become sustainable stewards of wetland ecosystems pave the way towards inclusive ERJ in LPPWP.

It took more than a decade, from the 2007 Presidential Proclamation, the 2015 Ramsar Site declaration, to the inclusion of wetlands in the 2018 E-NIPAS legislation, for LPPWP to be designated as a 'Protected Area'. Without the support of key actors, environmental justice is not possible. The story of the Pulo islands does not end here. Work still needs to be done to fully heal the physical and relational harm done through decades of missteps. The fisherfolk still do not have access to the Pulo and their livelihoods remain in danger, while the concerned residents of Metro Manila continue to be wary of large reclamation projects that destroy the natural patrimony of Filipinos in Manila Bay. However, there is hope. With the legislative protection provided by the E-NIPAS Act, the LPPWP is now secure—it no longer has the imminent threat of reclamation through an incompatible change of land use. For the fisherfolk and coastal communities, the work to establish a wetland centre is an initial step in restoring with the natural environment and addressing old grievances for these communities. The eventual story of the Las Piñas Parañaque Wetland Park will depend on the willingness of stakeholders to work together towards a collective vision of conservation and wise use.

3.2   Case Study 2: Marilao-Meycauayan-Obando River System

There are lobsters here with big claws. The river then was so clear that every fiesta of St. John we would go swim and bathe in the river. You could see the river floor of gravel and sand.
—Fish farm caretaker, Meycauayan City, Philippines

Similar to the first case study, the Marilao-Meycauayan-Obando River System (MMORS) rehabilitation effort is a story of use conflict, novel institutional arrangements which have the potential to facilitate repair and dialogue and widespread delays and challenges for municipal governments and citizen activists in implementing the mandates of courts and laws.

MMORS has been identified as one of the dirtiest watercourses in the world, exceeding allowable levels of heavy metals and failing in other water quality and biota standards (Blacksmith Institute, 2009; Prudente et al.,

1994). Around 20 kilometres away from Manila, and one of the major tributaries of Manila Bay, MMORS covers 130 kilometres with a total length of 52 kilometres running through 7 municipalities (David, 2011). Despite widespread heavy metal pollution, the river is officially used for recreation, fisheries, and industrial needs (Amparo, 2021). The river is home to freshwater and brackish water fishes, marine fish, and shellfish species found near Manila Bay and hundreds of fish farms along the river system (BFAR Region 3, 2013). The system supports around 1500 hectares of brackish water ponds that grow milkfish (*Chanos chanos*), tilapia (*Oreochromis niloticus*), and prawns (*Penaeus monodon*) (BFAR Region 3, 2013). It is surrounded by numerous industries and settlements, including 794 firms representing 21 industry types (David, 2011).

The upstream municipalities are known for both legacy industries, such as tanneries and gold smelting, and highly industrialised areas. These legacy industries work with toxic chemicals and heavy metals, such as chromium for tanneries and copper and mercury for gold smelting. They are mostly small- and medium-scale enterprises commonly operating inconspicuously along residences in the upstream towns. In addition, the largest used lead acid battery (ULAB) plants formerly operated in one of the upstream towns. This means the primary pressure to clean up lies in upstream communities.

In contrast, fish farms are located mid- and downstream (Amparo, 2021). Some seafood like milkfish, mussels, and oysters had elevated levels of heavy metals based on the 2008 biota sampling done by Blacksmith Institute (2009). The estimated average daily consumption of milkfish and tilapia in the two coastal areas along the river system is double compared to the national average of daily fish consumption (Amparo et al., 2017b). Thus, the resulting pollution of the river poses significant direct health risks to the fish-consuming public (Amparo et al., 2017a).

Further, the river is peppered, not only with industrial, agricultural, and commercial establishments but also with settlements. Coastal barangays do not have regular solid waste collection systems, and burning of wastes is considered illegal. Thus, most of the solid waste eventually finds its way to the river. This led one fisherfolk leader to claim, 'we also throw waste in the river but not as much compared to upstream cities' (Personal communications, December 4, 2015, cited from Amparo, 2021).

Many different actors therefore contribute to environmental harm to the river, which in turn harms the health and livelihoods of others. These harms largely come about from a myriad of economic activities, contrasting with the government-directed urban development of the previous case study. Here, economic and environmental trade-offs are diffused and entwined across the landscape, with no easy answers as to how to restore health to the waterways and look after the basic needs of the people.

Even so, Manila Bay and its tributaries, such as MMORS, have been the focus of regular national restoration efforts since the 1990s (Velasco, 2009). The local government unit (LGU)-led clean-up initiatives, which started in the early 2000s, include the Heal the Meycauayan River (2003) and Marilao River Council (2004). Early clean-up initiatives, we would argue, were mostly fragmented and reactive given the limited water quality data and resources afforded to local stakeholders. In 2005, different LGUs and regional regulatory agencies created the MMORS Stakeholders' group to consolidate river rehabilitation efforts.

International environmental NGOs put a public spotlight on MMORS that helped fast track rehabilitation. Reports from international environmental NGOs like Greenpeace and Blacksmith Institute revealed the degrading state of water quality in the area and, hence, the need for urgent, collaborative, and holistic action (Greenpeace, 2007; Blacksmith Institute, 2007). The rehabilitation efforts grabbed national and international attention. The Worst Polluted Places Report published by Blacksmith Institute (now Pure Earth), for instance, contained a global list of severely polluted waterways that posed a significant public health threat (Blacksmith Institute, 2007).

As a tributary of Manila Bay, the system is subject to the 10-year restoration and rehabilitation orders from the Supreme Court's 2008 decision introduced in the previous case study for the relevant government bodies to comply with (Metropolitan Manila Development Authority v Concerned Residents of Manila Bay, 2008). The confluence of all these earlier initiatives, together with the public attention to the river system, led to the MMORS to be declared a Water Quality Management Area (WQMA) in 2008 under the Philippine Clean Water Act of 2004 (Republic Act 9275), as its poor water quality required immediate management intervention across jurisdictions.

Four key players act in the rehabilitation efforts for MMORS—government, civil society, industries, and researchers. The Philippine Clean Water Act mandates the government, through its agencies and local government units, to lead in the rehabilitation of water resources. Industries are expected to comply with regulations and contribute to the rehabilitation efforts. Civil society organisations serve as representatives of the affected communities like fisherfolk organisations or by providing technical assistance and support. In addition, researchers from both national and local universities contribute through research and education for river rehabilitation. As canvassed in the introduction, all these stakeholders became part of the governing body, enabling them to have a say in the direction and scope of restorative activities.

Apart from these internal stakeholders, external local and international organisations play a role in the rehabilitation planning and initiatives. Various international development agencies such as Japan International Cooperation Agency (JICA), Asian Development Bank (ADB), United States Agency for International Development (USAID) through the Department of Environment and Natural Resources, and corporate foundations such as Coca-Cola Foundation through Blacksmith Institute (now Pure Earth) provided grant support to conduct baseline studies, stakeholders' meetings and workshops that were needed to complete the process of designating MMORS as a WQMA (Malenab et al., 2016).

The WQMA governance model provided the platform for different sectors with understandably divergent work cultures but a similar goal to restore the river in a way that was inclusive, accountable, non-confrontational, and based on dialogue. This is attuned to the Filipino relationship-oriented virtue ethics that values preservation of human relationships (Reyes, 2015, p. 148) through dialogue or *talakayan* and *tulungan* or collective support and cooperation, important values for Filipinos. Confrontational tactics may not entirely be effective in this context, as it could discourage industries or governments to sit down in meaningful conversations to discuss rehabilitation efforts. Through the WQMA Board, the ten-year action plan (2010–2020) was crafted for the MMORS rehabilitation. The action plan exemplified the united stakeholder backing based on data, without which efforts would be largely fragmented, misaligned, and wasteful trial and error. The ten-year action plan focused on controlling pollution at source and on point sources of chemical as well as solid waste pollution. It paved the way for both local and international organisations to provide technical assistance to the river rehabilitation efforts.

Approaching clean-up of the river through legalistic means alone could promote a reactionary and defensive stance rather than a cooperative one. For instance, the Department of Environment and Natural Resources (DENR) has issued closure orders to several smelting firms, tanneries, packaging plants, and even fishponds to control direct discharge of untreated wastewater to the river (Brozas, 2019; Reyes-Estrope & Enano, 2019). All are significant sources of local livelihoods to the coastal communities along the river system, with those working in the industries likely to have difficulty finding other work. And with more than 700 businesses, closing the odd point-source polluter is not likely to solve the systemic issues of pollution for the waterways. The WQMA opened up venues for pilot testing of cost-effective water treatment solutions with members of the Tannery Association of the Philippines, and some gold smelting facilities have persuaded some of their members and operators to adopt these technologies and process changes (see Alfafara et al., 2012; Pleto, 2015; Coronado et al., 2016; Vivas et al., 2019). On the other hand, closure of some non-compliant tanneries and small-scale gold smelters only forced them to continue operating at different sites or to become 'fly-by-night' operators.

Water governance requires fluidity to adapt to the local socio-political dynamics. To facilitate and command action from the different local government units and regional regulatory agencies, the MMORS WQMA Governing Board invited the Province of Bulacan Governor, who is a political authority over its constituent local government units, to co-chair the board together with the Environmental Management Bureau Region III and National Capital Region (NCR) Directors. This shows that a credible and an authoritative figure, like the Governor, could help mediate conflict, inspire and lead the members to act in a coordinated way to achieve mutual self-help. The Governor also provided a macro-context that prevents the 'blame' game amongst different LGUs and agencies, allowing instead for a focus on resolutions to dominate the discussions.

This is not to say there have not been issues, setbacks, and delays. For example, the crafting of the ten-year action plan (2010–2020) for the MMORS rehabilitation has been viewed as technically driven and as a means to help secure external funding rather than an endeavour/initiative emanating from local stakeholders (Malenab et al., 2016). The action planning, for instance, was facilitated through Japanese aid funding and most pilot testing of point-source pollution prevention and water remediation technologies was done by external organisations. Although these initiatives were conducted together with local governments and MMO WQMA members, external actors remain the lead actors in these endeavours. Another limitation is that the mainstreaming of these technologies remains incomplete due to a lack of sufficient incentives, local support systems, and enforcement needed to encourage these industries and sectors to adopt pollution control technologies and processes.

The ten-year action plan focused on controlling pollution at source and regular monitoring of both organic and heavy metals levels of the river water to guide rehabilitation interventions. Rehabilitation programmes for nature-based livelihoods like fisheries and fish farming were not explicitly discussed in this action plan. Nevertheless, mobilisation of the fishery and fish farming sector for river rehabilitation were moderated through existing government programmes (i.e. Philippine National Aquasilviculture Program) (Dieta & Dieta, 2015) and through organisational development projects of civil society organisations like Pure Earth Philippines (Malenab et al., 2016).

And so, the restoration of the Meycauayan-Marilao-Obando River remains incomplete. We argue that the MMORS WQMA and the *Writ of Continuing Mandamus* have set the platform for two out of four attributes of ERJ (Forsyth et al., 2021). First, the water governance structure and court case set the legal and organisational platforms for different stakeholders to effect change in the river rehabilitation—to work together with industries and civil society. Second, it has amplified the voices of the local stakeholders, whether government, civil society, or the public, on the need to focus on this Manila Bay tributary. However, it has yet to strengthen the space for storytelling, dialogue, and authentic listening to nature-based livelihoods of farmers and fishers, the community members, and even the small-scale legacy industry operators. Lastly, holding the government, industries, and even communities accountable for their waste remains a challenge. The MMORS saga could potentially learn from ecological restorative justice as it forges a stronger and clearer vision for a 'harmonious, balanced and restorative' human ecological relationship in this critical river system.

3.3   Case Study 3: The Marcopper Mining Disaster

I was washing clothes when this dam caved in, so the water was a bit like milk... now the shrimps were all jumping around me ... We did not eat them since they were poisoned
—A washer-woman describing the immediate after-effects of the toxic spill from the Marcopper Mine (Macdonald & Southall, 2005, p. 29).

The final case study goes to the heart of global institutional weakness of the ERJ movement. That is, ERJ is not possible wherever and whenever legal responsibility can be divorced from local consequences of environmental harm and moral responsibility holds no sway. Without cross-scale accountability enforced by states and judiciaries far from the original sites of damage, ERJ has limited reach.

Many corporate disasters demonstrate this point, but in the Philippines none more so than the 1996 Marcopper Mining Disaster. To date, virtually no meaningful rehabilitation has been undertaken. To understand why, we must ask: who decides what restoration takes place? Who is held accountable and why? Who decides what is done and how?

On 24 March 1996, Marcopper Mining Corporation's drainage tunnel burst, unleashing millions of cubic metres of mine waste into the Boac and Makulapnit river system, on the Island of Marinduque. Flash floods rapidly isolated some villages and thousands of residents. An entire village was buried, displacing hundreds. The toxic spill inundated a large area and caused huge crop and fisheries losses. Most residents in the area depended on this major water system for livelihood and domestic needs—in other words, for their daily survival. The national government briskly responded by suspending the mine's permit to operate. Only three days after the spill, then President Fidel Ramos declared a 'State of Calamity' for affected areas (Cabalda et al., 2002). The declaration acknowledged the severity of impact to nature and to people and guaranteed immediate and substantial assistance. Affected residents were also given financial support by the mining company (Macdonald & Southall, 2005).

With such widespread and toxic damage, containment and rehabilitation were an immediate concern. Placer Dome Technical Services (Philippines) (PDTS), a subsidiary of the Canadian part-owner Placer Dome, Inc, embarked on environmental and social assessments of affected areas. Their 1997 report concluded that submarine placement was the most suitable rehabilitation method, claiming that land-based disposal was not preferred, mainly due to long-term environmental liabilities.

In 1997, Marcopper applied for a permit for submarine tailings disposal (STD) from the DENR. Marcopper proposed to dispose of the tailings in the waters of Tablas Strait, located south of Marinduque Island. On October 30, 1997, DENR rejected the permit application, primarily for legal reasons. The then DENR Secretary Victor Ramos explained that 'under current laws and regulations, all the offshore and submarine areas of the country are considered to be Environmentally Critical Areas … Hence, your application for the submarine placement of redredged channel tailings materials is hereby denied' (Coumans, 2002, p. 3). Ramos also expressed concern that there might have been potential harm to the abundant fish areas near the strait and, thus, land-based disposal would be preferable.

Marcopper appealed the government's rejection of its first permit application. Then in 1998, DENR responded to the appeal by granting Marcopper the right to prepare an Environmental Impact Assessment (EIA). Although submarine disposal could be one of the options considered in the EIA, the Secretary Ramos emphasised that it should include land-based disposal alternatives for spilled tailings and a full rehabilitation proposal (Coumans, 2002).

PDTS hired Woodward Clyde (Philippines) as their main EIA consultants for tailings disposal options. Besides commissioning an EIA, PDTS also brought in Vancouver-based Derek Ellis of Rescan Environmental Services Ltd. to give presentations on submarine tailings disposal to government officials (Coumans, 2002), suggesting efforts to influence these officials' perception on the tailings disposal. When Woodward Clyde (Philippines) completed the EIA, Marcopper applied for their second permit application for submarine tailings disposal. Soon after, it was again rejected, mainly due to the 'absence of social acceptability' as evidenced by resolutions documenting the opposition of civil society groups and local government (Cerilles, 1999, para. 2–4). DENR further instructed Marcopper to submit a final clean-up and rehabilitation programme within 30 days to be implemented immediately by PDTS. Instead, Marcopper launched a social acceptability campaign amongst Marinduque residents and submitted their third permit application to dispose of the tailings in the ocean.

When the disaster happened in 1996, the environmental movement in Marinduque, led by the Marinduque Council for Environmental Concerns (MACEC), was already highly critical of environmental regulation in mining. The environmental movement had reason to believe that the experts contracted by the mine operators were submarine tailings disposal advocates. In 2000, after repeated calls to deny the submarine tailings disposal proposal, MACEC made the latter issue explicit in a letter sent to then DENR MGB [Mines and Geosciences Bureau] Director Horacio Ramos: 'We are really wondering why MGB/DENR is still entertaining the overtures of Marcopper/PDTS/Placer Dome to go into STD when this is already outlawed in Canada since 1977 … and the US Environmental Protection Agency has also held firm to its ban on the practice' (Mapalad, 2000, para. 5). Because of the movement's distrust of the mining company and government regulators who were not ruling out STD, the movement supported the search for an independent assessment team who could provide reliable rehabilitation strategies for Marinduque.

The mobilisation efforts of MACEC led to the unprecedented hiring of an independent assessment team led by the US Geological Survey (USGS), paid for by the national government and chosen by the provincial government. The findings of the team led by the USGS, among other studies, were used as evidence for the class suit filed in Nevada, USA, by the Marinduque provincial government against Placer Dome in 2005. In 2015, the Nevada State Supreme court junked the case for *forum non conveniens*, which means the US is not the proper jurisdiction to hear the case. The petitioners are now planning to file the case in Canada, the home country of Barrick Gold that acquired PDI in 2006 (Cinco, 2016).

Meanwhile, the people of Marinduque remain waiting. After 25 years, all attempts to hold the company to account have not been successful. Indeed, newspaper reports from late 2020 claimed that, on top of the legacy issues from the spill and abandoned site, 100 barrels of radioactive waste materials had been found during a government inspection, even as Marcopper seeks to resume mining activities on the Island (Cinco, 2020).

Despite these setbacks, benefits for the local environmental movement indicate the importance of supporting such organising efforts. There have been limited clean-up efforts financed and led by the different levels of government. There are some recent signs that authorities are beginning to commit to serious investment in rehabilitating the site and the broader aquatic ecosystem. In 2020, the national government pledged 5 million Philippine pesos to rehabilitation projects, including dredging the affected rivers—through which a private company is hoping to recover mineral resources (Mayuga, 2020). Beyond Marinduque, the disaster's aftermath, in the short term, resulted in broad positive responses from different actors. For example, the government reformed environmental policies to strengthen environmental monitoring of mining operations and to create a funding requirement for mining companies to be used for covering the cost of prospective damage of such high-risk projects. Government regulation of mining became stricter in the years right after the mining disaster (De la Cruz, 2017). The mining industry body committed to responsible mining practices (Cabalda et al., 2002). Civil society groups strengthened their organising activities (Nem Singh & Camba, 2016) and have been building capacity in scientific processes to counter the findings of industry experts (Camba, 2016). The latter civil society points to pathways for making environmental decision-making more inclusive, even for scientific ecological issues.

In the long term, however, mining disasters continued to happen, indicating the limits of the short-term gains. For example, in 2005, Lafayette Mining Limited's mine tailings spillage on Rapu-rapu Island resulted in one of the worst mining-related fish kills in the country. More recently in 2012, mine waste from Philex Mining Corporation's operations discharged into the Balog Creek and Agno River in Benguet. This disaster unleashed the biggest volume of tailings spill in Philippine's history to date.

This case study may seem like the antithesis of ERJ and instead just another example of business as usual. Certainly, we have included it after discussions between the Filipino and non-Filipino authors of the chapter—on the one side, the desire to showcase the Philippines' undercelebrated innovations and achievement in ERJ, and on the other, to expose the challenges and messiness in achieving real and substantive progress when corporations have the power to just walk away, and the resources to hold them to account are virtually non-existent.

It remains, perhaps, as a parable for ERJ more broadly. That we can have strong frameworks and principles, but without reorganising power and control, ERJ may remain better in theory than in practice. As the case demonstrated, steadfast organising and action by civil society groups is instrumental in this transformative work to move government and industry.

4   Conclusion

The three case studies present how ERJ in the Philippines unfolds with common threads of experiences that unveil the distinct complexities of waterway restoration. Spatially, the multiple facets of environmental harm, whether within urban wetlands, transboundary river systems, and small-island's tributary networks, literally and figuratively flow outside human-designated boundaries and political jurisdictions. The legal, technical, political, and logistical resources and processes revolving ERJ hence demand participation not only from upstream to downstream communities of residential, agricultural, commercial, and industrial sectors but also transnational actors including scientific and environmental monitoring bodies. The case studies likewise reveal that pursuing ERJ entails navigating a messy web of macro- and micro-political actors with competing or complementing resource interests, development worldviews, motivation, and capacities.

Temporally, multiple harms from visible catastrophes (e.g. mining spills, toxic contamination) occur as sensational immediate events, but its long-term repercussions are experienced across generations of ecologies and communities. The Minor Oposa doctrine for intergenerational justice now embedded in the specialist environmental courts aids as a legal tool that complements the temporal dimensions of ERJ. The features of harm and the pathways towards redress in the case studies reveal a broad temporal spectrum wherein long-term and multidimensional impacts are sometimes gradual or latently manifested, yet embedded in the lives across generations of families and communities. Amongst these impacts are the intangible realities of cultural fragmentation, psychological suffering, intergenerational trauma, and ecological death of rivers as consequences of ruptured environments. Along this line, the ERJ attribute of storytelling and dialogue is crucial, as exhibited by the continuing legal battle and 25th year of commemorating the Marcopper Mining Disaster. Unrelenting media coverage, scientific documentation, art expressions, and community narratives serve as a crusade of the aggrieved against the currents of impunity and a historical struggle against forgetting.

In the language of justice, it is important to recognise the skewed power dynamics, corporate influence, and the internal structural factors that lead to social and environmental harm. An important component of ERJ is not only strengthening society's collective agency towards healing environmental harm to people and nature but dismantling the systemic barriers to genuine healing. Examining whose labour is deployed in attempting to ensure accountability and whose actions are catalysts for on-ground change makes it clear that legal protections have only been the first step in what is likely to be a very long road to justice. The brunt of environmental harm, the burden of proof, and quest for justice have often been shouldered by marginalised grassroots communities. Though filled with optimistic ideals, realities show that ERJ cannot bank on the altruism of the perpetrators of harm, particularly when regulatory mechanisms and corporate actors circumvent rather than acknowledge culpability. As evidenced by local histories, the persistent collaborations of civil society movements composed of grassroots activism, legal advocates, environmental journalists, and engaged academics are potent forces in crafting, accessing, and employing the legal and institutional innovations for ERJ, as exhibited by the *Writ of Continuing Mandamus.*

The Philippine experience shows that despite barriers, a broad spectrum of social actors collaboratively advanced an arsenal of progressive environmental laws, conflict management strategies (e.g. dialogue, mediation, negotiation) and grounded community development processes (e.g. capacity building, advocacy campaigns) that are vital tools of ERJ. Likewise, continuing cultural featuring of strong community relations and persisting spiritual and moral connections to the environment serve as pillars in ERJ.

Finally, the ecosystems at play in the case studies explored in this chapter are highly modified socio-ecosystems that are continuing to co-evolve through human and more-than-human activities. Any pretence of returning to a pristine or 'untouched' state is moot when the area in question is home to large communities engaging in ongoing actions which can all be thought of as on a subjective spectrum between 'harmful', 'harmless', and 'restorative'. Whose activities are accepted, who is rejected, who is tolerated, and which activities will have consequences inevitably reflect the underlying structural, political, and material inequalities that characterise not only the Philippines but every country on Earth. ERJ will mean redistribution as well as repair. Filipino innovation in ERJ demonstrates the hopeful potential of forward-thinking laws and regulations, while Filipino realities show the practical difficulty of truly realising the ideal.

Acknowledgements The authors would like to thank Professor Valerie Braithwaite for her valuable comments and suggestions on the earlier draft. Also, we are grateful for Ms. Amy M. Lecciones from the Society for Conservation of Philippine Wetlands (SCPW) and Ms. Larah Ortega Ibanez and Ms. Marife Dapito from Pure Earth Philippines and University of the Philippines Los Baños for their inputs and comments on the LPPWP and MMORS case studies, respectively. Furthermore, the authors are thankful to the Marinduque Center for Environmental Concerns for giving access to their document collection. Some data for the Marcopper Mining Disaster case study were collected by one of the authors as a visiting fellow in 2016 at the Ateneo School of Government. Also, some data and discussions on MMORS were based on the PhD thesis in 2021 of one of the authors at Fenner School of Environment and Society at Australian National University under the Australia Awards Scholarship.

References

Alfafara, C., Maguyon, M. C., Laurio, M. V., Migo, V., Trinidad, L., Ompad, E., Amparo J. M. S. & Mendoza, M. (2012). Scale-up and operating factors for electrolytic silver recovery from effluents of artisanal used-gold-jewelry smelting plants in the Philippines. *Journal of Health and Pollution*, *2*(3), 32–42.

Allen, T. (1993). The Philippine children's case: Recognising legal standing for future generations. *Geo. Int'l Envtl. L. Rev.*, *6*, 713.

Amparo, J. (2021). Dynamics of social-ecological traps: The case of small-scale fisheries in the Philippines (Dissertation/Thesis). Retrieved from: https://openresearch-repository.anu.edu.au/handle/1885/219982 (last accessed 27 January 2022).

Amparo, J. M. S., Geges, D. B., Malenab, M. C. T., Visco, E. S., Mendoza, M. E. T., Jimena, C. E. G., Saguiguit, S. L. C., Mendoza, M. D., & Ibanez, L. O. (2017a). A balancing act: Managing multiple pressures to fisheries and fish farming in the Marilao-Meycauayan-Obando river system, Philippines. In P. Guillotreau, A. Bundy, & R. I. Perry (Eds.), *Global change in marine systems* (pp. 157–170). London: Routledge.

Amparo, J. M. S., Talavera, M. T. M., Barrion, A. S. A., Mendoza, M. E. T., & Dapito, M. B. (2017b). Assessment of fish and shellfish consumption of coastal barangays along the Marilao-Meycauayan-Obando River System (MMORS), Philippines. *Malaysian Journal of Nutrition*, *23*(2), 263–277.

Arcala Hall, R., Abansi, C. L., & Lizada, J. C. (2018). Laws, institutional arrangements, and policy instruments. In A. Rola, J. Pulhin, & R. Arcala Hall (Eds.), *Water policy in the Philippines: Global issues in water policy, Vol. 8* (pp. 41–64). Springer, Cham.

BFAR Region 3 (2013). Bulacan Fisheries Profile. Presented at the HSBC Ecoday Camp Conference, Bureau of Fisheries and Aquatic Resources (BFAR), Manila Philippines.

Blacksmith Institute. (2007). The World's Worst Polluted Places, Blacksmith Institute, New York. Available from: http://www.google.com.au/url?sa=t&rct=j&q=&esrc=s&source=web&cd=2&ved=0CCYQFjAB&url=http%3A%2F%2Fwww.worstpolluted.org%2Freports%2Ffile%2F2007%2520Report%2520updated%25202009.pdf&ei=dVrCUHEK8ukkwXpoYGYBQ&usg=AFQjCNFYRjIW1PI52mEU289e0Q6Dj3d8Jg&sig2=JPa_3j8UE2V20BNbj7zb1w&bvm=bv.70810081,d.dGI (Accessed 7 December 2014).

Blacksmith Institute (2009). *The world's worst polluted places*. New York: Blacksmith Institute. Retrieved from: http://www.worstpolluted.org (last accessed 1 September 2021).

Brozas, B. (2019, January 30). 15 pabrika ng leather sa Bulacan ipinasara ng DENR. *Philippine Daily Inquirer*. Retrieved from: https://radyo.inquirer.net/160790/15-pabrika-ng-leather-sa-bulacan-ipanasara-ng-denr/ (last accessed 16 November 2021).

Cabalda, M. V., Banaag, M. A., Tidalgo, P. N. T., & Garces, R. B. (2002). *Sustainable development in the Philippine minerals industry: A baseline study* (Report No. 184; Mining, Mineral and Sustainable Development Project of the International Institute for Environment and Development). International Institute for Environment and Development, World Business Council for Sustainable Development. Retrieved from: https://pubs.iied.org/sites/default/files/pdfs/migrate/G00614.pdf (last accessed 27 January 2022).

Camba, A. A. (2016). Philippine mining capitalism: The changing terrains of struggle in the neoliberal mining regime. *ASEAS—Austrian Journal of South-East Asian Studies, 9*(1), 71–88.

Cerilles, A. H. (1999, February 16). (Letter to Teodoro C. Gabor). Photocopy in possession of MACEC.

Cinco, M. (2016, November 9). Groups support refiling of suit vs Marcopper. *Philippine Daily Inquirer*. Retrieved from: https://newsinfo.inquirer.net/842402/groups-support-refiling-of-suit-vs-marcopper (last accessed 5 March 2017).

Cinco, M. (2020, November 26). Toxic substances found in Marinduque mine site. *Philippine Daily Inquirer*. Retrieved from: https://newsinfo.inquirer.net/1364952/toxic-substances-found-in-marinduque-mine-site (last accessed 14 April 2021).

Constitution of the Philippines (1987). Retrieved from: https://www.officialgazette.gov.ph/constitutions/1987-constitution/ (last accessed 27 January 2022).

Coronado, F. F., Unciano, N. M., Cabacang, R. M., & Hernandez, J. T. (2016). Removal of heavy metal compounds from industrial wastes using a novel locally-isolated Vanrija sp. HMAT2. *Philippine Journal of Science, 145*(4), 327–338.

Coronel, S. S., & Tordesillas, E. (1998, March 18). *The grandmother of all scams*. Philippine Center for Investigative Journalism. Retrieved from: https://old.pcij.org/stories/the-grandmother-of-all-scams/ (last accessed 16 April 2021).

Coumans, C., (2002, April). The successful struggle against submarine tailings disposal in Marinduque, Philippines. Marinduque Council for Environmental

Concerns. Retrieved from: https://miningwatch.ca/sites/default/files/marinduque_std_struggle_0.pdf (last accessed 21 November 2021).

David, C.P.C. (2011). Pollution loading in the Marilao-Meycauayan-Obando River System, *EAS Conference*, Manila. Retrieved from: http://pemsea.org/eascongress/international-conference/presentation_t6-2_david.pdf (last accessed 7 December 2014).

Davide Jr, H. G., & Vinson, S. (2010). Green courts initiative in the Philippines. *J. Ct. Innovation*, *3*, 121.

De la Cruz, G. (2017, March 25). *Look back: The 1996 Marcopper mining disaster* [News site]. Rappler. Retrieved from: http://www.rappler.com/move-ph/issues/disasters/165051-look-back-1996-marcopper-mining-disaster (last accessed 27 March 2017).

DENR (2012). *Saving the last coastal frontier: Framework plan for the coastal lagoons of Las Piñas and Parañaque.* Quezon City: Department of Environment and Natural Resources.

Dieta, R. E., & Dieta, F. C. (2015). The Philippine national aquasilviculture program. In *Resource enhancement and sustainable aquaculture practices in Southeast Asia: Challenges in responsible production of Aquatic Species. Proceedings of the International Workshop on Resource Enhancement and Sustainable Aquaculture Practices in Southeast Asia 2014 (RESA)* (pp. 77–83). Aquaculture Department, Southeast Asian Fisheries Development Center.

Estoque, R. C., & Murayama, Y. (2015). Classification and change detection of built-up lands from Landsat-7 ETM+ and Landsat-8 OLI/TIRS imageries: A comparative assessment of various spectral indices. *Ecological Indicators*, *56*, 205–217.

Expanded National Integrated Protected Areas System (E-NIPAS) Act (2018). Retrieved from: https://www.officialgazette.gov.ph/downloads/2018/06jun/20180622-RA-11038-RRD.pdf (last accessed 27 January 2022).

Forsyth, M., Cleland, D., Tepper, F., Hollingworth, D., Soares, M., Nairn, A., & Wilkinson, C. A (2021). A future agenda for environmental restorative justice? *The International Journal of Restorative Justice*, *4*(1), 17–40.

Friess, D. A., Rogers, K., Lovelock, C. E., Krauss, K. W., Hamilton, S. E., Lee, S. Y., Lucas, R. Jurgenne, P., Rajkaran, A., & Shi, S. (2019). The state of the world's mangrove forests: past, present, and future. *Annual Review of Environment and Resources*, *44*, 89–115.

Gabasan, R. (2006). *Policy paper on fisherfolk settlement.* Quezon City, Philippines.: Ateneo de Manila University.

Gatmaytan, D. B. (2002). The illusion of intergenerational equity: Oposa v. Factoran as Pyrrhic victory. *Geo. Int'l Envtl. L. Rev.*, *15*, 457.

Gera, W. (2016). Public participation in environmental governance in the Philippines: The challenge of consolidation in engaging the state. *Land Use Policy*, *52*, 501–510.

Greenpeace. (2007). *The state of water resources in the Philippines*. Quezon City, Philippines: Greenpeace Southeast Asia.

La Viña, A. G. (1994). The right to a sound environment in the Philippines: The significance of the Minors Oposa Case. *Rev. Eur. Comp. & Int'l Envtl. L.*, *3*, 246.

La Viña, A. G. (2014). After more than 100 years of environmental law, what's next for the Philippines? *Philippine Law Journal Centennial*, 195–239.

Lecciones, A. M., & Devanadera, M. E. (2021). *Rapid assessment of Wetland Ecosystem Services of the Las Pinas Paranaque Wetland Park*. Unpublished Manuscript.

Macdonald, I., & Southall, K. (2005). *Mining ombudsman case report: Marinduque Island*. Melbourne: Oxfam Australia.

Magno, F. A. (1999). Environmental movements in the Philippines. In Y. Lee & A. So (Eds.), *Asia's environmental movements: Comparative perspectives* (pp. 143–173). New York: M.E. Sharp.

Malenab, M. C., Visco, E., Geges, D., Amparo, J. M., Torio, D., & Jimena, C. E. (2016). Analysis of the integrated water resource management in a water quality management area in the Philippines: The case of Meycauayan-Marilao-Obando river system. *Journal of Environmental Science and Management*, 19(2), 84–98.

Mapalad, A. L. (2000, February 28). (Letter to Horacio C. Ramos). Photocopy in possession of MACEC.

Mayuga, J. L. (2020, March 4). DENR moves to rehab rivers ruined by Marcopper mining disaster. *Business Mirror*. Retrieved from: https://businessmirror.com.ph/2020/03/04/denr-moves-to-rehab-rivers-ruined-by-marcopper-mining-disaster/ (last accessed 15 April 2021).

Metropolitan Manila Development Authority v. Concerned Residents of Manila Bay, G.R. Nos. 171947-48, December 18, 2008, Philippine Supreme Court. Retrieved from: https://lawphil.net/judjuris/juri2008/dec2008/gr_171947_2008.html (last accessed 27 January 2022).

Nem Singh, J. T., & Camba, A. A. (2016). Neoliberalism, resource governance, and the everyday politics of protest in the Philippines. In J. Elias & L. Rethel (Eds.), *The Everyday Political Economy of Southeast Asia* (pp. 49–71). Cambridge: Cambridge University Press.

Oposa v. Factoran (1993). G.R. No. 101083, July 30, 1993. Retrieved from: https://www.lawphil.net/judjuris/juri1993/jul1993/gr_101083_1993.html (last accessed 15 April 2021).

Pleto, J. V. R. (2015). Assessment of different bioremediation strategies on the environmental quality of the aquaculture ponds of Brgy.[Village] Nagbalon, Marilao and Brgy. Liputan, Meycauayan, Bulacan, Philippines.

Poddar, M., & Nahar, B. (2017). Continuing Mandamus: A judicial innovation to bridge the right-remedy gap. *NUJS L. Rev., 10*, 555.

Preston, B. J. (2014). Characteristics of successful environmental courts and tribunals. *Journal of Environmental law, 26*(3), 365–393.

Pring, G. R., & Pring, C. K. (2009). Greening justice: Creating and improving environmental courts and tribunals. World Resources Institute. Retrieved from: https://www.law.du.edu/documents/ect-study/greening-justice-book.pdf (last accessed 27 January 2022).

Prudente, M. S., Ichihashi, H., & Tatsukawa, R. (1994). Heavy metal concentrations in sediments from Manila Bay, Philippines and inflowing rivers. *Environmental Pollution, 86*(1), 83–88.

Reyes, J. A. L. (2014). Environmental attitudes and behaviors in the Philippines. *Journal of Educational and Social Research, 4*(6), 87. Retrieved from https://www.mcser.org/journal/index.php/jesr/article/view/4067 (last accessed 27 January 2022).

Reyes, J. (2015). Loob and kapwa: An introduction to a Filipino virtue ethics. *Asian Philosophy, 25*(2), 148–171.

Reyes-Estrope, C., & Enano, J. O. (2019, January 28). 3 restos face closure for polluting Manila Bay. *Philippine Daily Inquirer.* Retrieved from: https://newsinfo.inquirer.net/1078263/3-restos-face-closure-for-polluting-manila-bay (last accessed 15 April 2021).

Ristroph, E. B. (2012). The role of Philippine courts in establishing the environmental rule of law. *Envtl. L. Rep. News & Analysis, 42*, 10866.

Rodolfo, K. S. (2016). *Dangerous aspects of reclamation along Manila Bay and Laguna de Bay: NAST policy discussion on the hazards risks and profits of reclamation.* Manila: National Academy of Science and Technology Philippines.

Supreme Court of the Philippines. (2019). *G.R. Nos. 171947-48, December 18, 2008.* Retrieved from: https://elibrary.judiciary.gov.ph/thebookshelf/showdocs/1/48335 (last accessed 15 April 2021).

United Nations Conference on Environment and Development (1992). *Agenda 21, Rio Declaration, Forest Principles.* New York: United Nations.

Van den Muijzenberg, O., & Van Naerssen, T. (2005). Metro Manila. *Directors of urban change in Asia*, 126–147.

Velasco Jr, P. J. (2009). Manila Bay: A daunting challenge in environmental rehabilitation and protection. *Or. Rev. Int'l L.*, *11*, 441.

Victoria, E. (2021). Down the river. In C. Rupprecht, D. Cleland, N. Tamura, & R. Chaudhuri (Eds.), *Multispecies cities: Solarpunk urban futures* (pp. 169–179). Alburquerque: World Weaver Press.

Vivas, E. L., Alfafara, C. G., Migo, V. P., Cho, K., Detras, M. C. M., Trinidad, L. C., Mendoza, M. D. & Lee, S. (2019). Comparative evaluation of alkali precipitation and electrodeposition for copper removal in artisanal gold smelting wastewater in the Philippines. *Desalination and Water Treatment*, *150*, 396–405.

20

Restoring Justice and Environmental Knowledge in Sámi Reindeer Husbandry?

Ida Hydle and Jan Erik Henriksen

1   Introduction

Reindeer herding, with its multiple traditions and knowledge of animal handling, lands, climate, grasslands, and so on, is generally acknowledged as a main historical basis for Sámi languages, culture, traditions, cosmology, and communication, in addition to being a central source of survival and income, namely, as a way of life. It is regarded as the main reason for the survival of the Sámi languages in Norway during the area of forced Norwegianisation from 1850 onwards. That is why the following 'Sara case'[1] is of crucial importance: a recent lawsuit from the Norwegian government concerning an increase in numbers of reindeer, which was won

---

[1] Yle Sapmi (2018). State of Norway wants reindeer herder to slaughter though the case has not been decided by UN. *The Barents Observer*. Retrieved from: https://thebarentsobserver.com/en/life-and-public/2018/12/state-norway-wants-reindeer-herder-slaughter-though-case-has-not-been (last accessed 22 January 2022).

I. Hydle (✉) • J. E. Henriksen
Norwegian Arctic University, Tromsø, Norway
e-mail: ida.hydle@uit.no; jan.e.henriksen@uit.no

at two court levels by a young Sámi reindeer owner. However, the state appealed the case to the Supreme Court of Norway, where the state won. The young Sámi reindeer herder has appealed the case to the European Court of Human Rights. The Norwegian Sámi Parliament supports him and requires that the authorities initiate a revision of the Reindeer Husbandry Act 2007.[2] The Norwegian State's handling of the case clearly shows an example of how neo-colonisation takes place and combines with or disguises itself in new public management (NPM). Countless examples across *Sápmi*, that is, the indigenous Sámi populations across Russia, Finland, Sweden and Norway, through the last three centuries disclose how state governing is not built upon Sámi herding expertise and legal traditional use and rights to the land, but on racism and illegal ignoring of civil rights[3] (Ravna, 2020), and invasion of lands, rivers, and fjords. Although the general state policy seemingly aims at a growing Sámi autonomy due to the Sámi Parliament and the Finnmark Act,[4] we question this double-bind governance. We also ask if this double-bind governance over time has led to most of the conflicts in the reindeer husbandry field that we have witnessed and maintain the importance of taking a broad view of this field of conflict. There is a growing need to find sustainable solutions, particularly to respond to the handling of the state's new public management. New public management policy privatises conflicts that are part of past and present state governance, shifting public conflict into so-called private civil legal cases, handled in court. These cases are often extremely costly 'solutions' for Sámi reindeer owners, their families, and local communities—rendering the state governance invisible and free from responsibility. In addition, our data show that only rarely do court decisions solve these conflicts at the personal or community level.

---

[2] *The Reindeer Husbandry Act* (2007). Retrieved from: https://www.pileosapmi.com/wp-content/uploads/2017/11/reindeer-husbandry-act-english.pdf (last accessed 22 January 2022).

[3] Lee Swepston and Lars Norberg (2017). A conversation on the Sami's rights as Indigenous people in their Homeland. Retrieved from: https://www.sametinget.se/11144 (last accessed 13 November 2017).

[4] The Norwegian State established the Finnmark Act in 2005. This decision reversed the management of the 45 000 km² area of Finnmark to the population of Finnmark. The Finnmark Act sets up a board where half of this board is appointed by the Sámi Parliament and the other half by the Finnmark County Council. One of the representatives appointed by the Sámi Parliament must be a reindeer herder.

In this chapter, we will describe a course of events that happened as we suggested restorative justice ways of handling these conflicts, through informing and linking the conflicting parties with an experiment in the Norwegian National Mediation Service.[5] We follow anthropologist Phillip G. Gulliver's (Gulliver, 1979, 1988, p. 249) approach to dispute management: 'Negotiations do not occur in a socio-political vacuum. They are intricately enmeshed in ongoing, wider social processes that constitute their essential environment and to which negotiations themselves contribute and often modify'. The negotiation cases presented must thus be understood as extended social processes, not as separate entities.

Sámi from Swedish, Finnish, and Norwegian sides[6] have demanded that their governments establish a truth commission concerning past colonisation of Sámi heritage, culture, and language. The Finnish and the Norwegian governments both established truth and reconciliation commissions which will investigate the consequences of the assimilation of the Sámi as an Indigenous People and the Kven people as a national minority.[7] Reindeer herding and husbandry have a crucial position in Sámi cultural and language traditions. We therefore maintain that truth commissions must include examination of the consequences of the past and present state governance of Sámi reindeer husbandry, such governance having caused serious damage to shared traditional knowledge and herding practices. The question includes how this governance is taking on new forms as well as continuing to harm indigenous ecologically based

---

[5] The National Mediation Service (Konfliktrådet) is a service for people in conflict or after an offence.

NMS have about 550 volunteer mediators throughout the country. The service is free of charge. All residents in Norway may contact their local Mediation Service with a request for a possible mediation meeting (restorative process). See https://www.regjeringen.no/en/dep/jd/organisation/underliggende-etater/national-mediation-service/id426406/ (last accessed 22 January 2022).

[6] One does not speak about Finnish, Swedish, Russian, or Norwegian Sámi; the Sámi people are considered as an Indigenous people resident in four nation-states. Thus, no matter where in Fennoscandia, the Sámi are one people.

[7] The Kven people are of Balto-Finnic ethnic origin. They descended from Finnish peasants and fishermen who emigrated from the northern parts of Finland and Sweden, to Northern Norway in the eighteenth and nineteenth centuries. They were granted minority status in Norway in 1996, and the Kven language was recognised as a minority language in 2005 in Norway.

practices and traditions. Our work for trying to find new ways to ensure justice and autonomy in reindeer husbandry conflicts may serve as an empirical example for similar fields of conflict.

2   Background to the Conflicts

The particular conflict areas in which we are involved comprise Indigenous, that is, Sámi groups; basic differences between Indigenous and neo-colonial knowledge fields and values, that is, epistemologies; and the various stages of pastoralism within reindeer husbandry and ecological concerns, that is, sustainability (Henriksen & Hydle, 2016). These diverse areas cross several lines of conflict (e.g., between reindeer owners and state or municipal developers or international firms, supported by the state or municipality) and involve both public and private interests, locally as well as internationally.

Anthropologists, biologists, and other scholars have described the development of these conflict lines over the past decades (Bjørklund, 2004; Bjørklund & Marin, 2015; Dahlstrøn Nilsson, 2003; Elenius et al., 2016; Ingold, 2000; Riseth & Lie, 2016; Riseth et al., 2016). Currently, Norwegian legal scholars, local and national politicians, and the Norwegian Reindeer Herding Organisation (NRL) are debating why the Norwegian governmental regulation of Sámi reindeer herding has gone so wrong in terms of conflicts over land and grazing areas. Not only are the traditional ways of organising herding and needed semi-nomadism destroyed, but also the ecological and economic basis of herding equilibrium has been threatened. The consequences of these conflicts impact upon people, animals, and grazing areas in 6 of Norway's 11 counties, in addition to wide areas in the Arctic parts of Sweden, Finland, and Russia. The failure of these four Arctic State's attempts at 'modern regulation' is widely known and partly acknowledged by both central and local authorities. Even so, governmental attempts to restore or improve conditions, in large part due to political pressure from Sámi bodies in the Nordic part of Sápmi, during the last decade have remained unsuccessful in the years after and continue to be unsuccessful to this day.

One example is the Programme for Change for Inner Finnmark, on which the Norwegian State spent NOK 330 million in the period 1993–2000, with the aim to retrain reindeer owners (to help them move off reindeer husbandry whether or not they desired this) and to improve education, competence, and welfare in reindeer husbandry for those who remained in it. This was prompted by the difficulties being experienced in lack of resources, poor economic conditions, and an increasingly challenging social situation within reindeer husbandry. The results were disappointing, both for industry and people's and animal welfare (Angell et al., 2003; Bergland, 2005). Today, in 2022, our research experience from our ongoing fieldwork reveals a great insecurity and lack of clarity, in addition to the external pressures on the grazing areas and the contested ownerships, both of land and reindeer. The insecurity and unclarity cause feelings of symbolic violence, resulting in both considerable stress for Sámi People and high numbers of suicides in the herding population (Kaiser & Salander Renberg, 2012; Kaiser et al., 2010; Møllersen et al., 2016; Stoor, 2016). A rising level of internal as well as external conflicts is measurable by the numbers of court cases. The media, as well as non-Sámi residents, describe reindeer husbandry in negative terms, alleging 'backwardness', harm, and violence (Henriksen, 2008).

Bringing an increasing number of conflicts to court does not solve the multi-layered content of the conflicts. The conflicts continue after court decisions and are shown to be both socially and economically disastrous for owners, families, siidas,[8] and communities. Attending lawyers are the only ones who get the benefit of increasing their income, a fact much discussed and regretted by both Sámi and public authorities, thus well known to all locally; for example, one reindeer owner told us: 'the authorities escape the responsibility for keeping and upholding the state of law, and push us over to private civil legal solutions that cost us millions of NOK'. Even if reindeer owners and public reindeer authorities regret this development and see the detrimental effects, they still seem to accept its occurrence as a 'matter of fact'. As participant researchers, we have

---

[8] The siida (in Northern Sámi, sijte in Southern Sámi, etc.) is a Sámi reindeer herding collective or community that has existed from time immemorial, mostly based on family hereditary relationship.

followed these legal as well as social and political processes closely in Finnmark County over the last few years, along with some examples from Troms, Nordland, and Trøndelag counties, as well as some from Northern Sweden with *samebyar* (reindeer herding collectives) that have, for example, summer grazing areas on the Norwegian side of the nation-state border. These case studies underpin our analysis in this chapter.

3   Methodologies

We are two scholars inherently comprising a wide variety of representations of importance for the research upon which this chapter is written. A short excerpt of our biographies is an inescapable part of the empirical basis for the ontology and epistemology of this chapter, as underscored by the American anthropologist Renato Rosaldo (2017). Although we are both employed as professors at the same university department, we also build a bridge between what separates Norwegian citizens in general concerning ethnicity, cultural heritage, geographical home, age, and gender. JE is a Sea Sámi and grew up on a small farm with berry picking, fishing, and hunting as important parts of the household's income. He graduated as a social worker, married into a Sámi reindeer herding family with reindeer herding as a considerable income source. JE is currently working as professor of social work at the Arctic University of Norway. He speaks Sámi and continues the practices of fishing, hunting, gathering, and taking part in herding as part of his leisure time. IH grew up in the capital, Oslo, in a non-Sámi Norwegian family, far away from Sámi language and traditions, graduated as a medical doctor and a social anthropologist, and is currently a professor at the same department as JE, within the same fields of restorative justice, Indigenous studies, and marginalisation. These personal background stories explain how JE functioned as a gate opener to interviewees whom IH would never otherwise have been able to approach. Our continuous and common research planning, participant observations, and analyses in the various and multi-sited fields of Sápmi and its nation-states' context are results of this mixture of personal, professional, and disciplinary breadth. Thus, together we not only cover the research areas as listed, but also Sámi language and practical reindeer herding experience, ecology, and sustainability. As academics and socially engaged professionals, we share extended knowledge and practical experiences within the field of restorative practices, also as mediators within social work, the Norwegian Mediation Service, and Red Cross Street Mediation. We are both keen to convey these experiences between the system of governance and the reindeer-based world of living. As 'wordwarriors' (Turner, 2006), we are both concerned to ensure that knowledge will contribute to change.

The interviewees in our fields of particular interest include (1) Sámi reindeer herders, old and young, on the Norwegian and Swedish sides, representing a multitude of traditions, modernisation, and economic systems. The scale goes from hunting, gathering, and herding economy with a small amount of monetary income to a full monetary economy; (2) The man–reindeer relationship, from having few reindeer as subsistence (40–200 reindeer) and a personal caring connection with the animals to an industrialised relationship with a herd of more than 5000, and the authorities speaking about animal welfare in terms of *biomass*, that is, meat quality and slaughter weight; (3) Representatives of the NRL, its leaders, and their governance; (4) The Norwegian State representatives of the responsible ministry of agriculture, in Oslo as well as the four abovementioned counties, the members of the Sámi Parliament, the special legal authorities for the governance of the Finnmark county;[9] (5) Representatives of the Norwegian Mediation Service locally and its secretariat coordinators in the ministry of justice in Oslo; and (6) Researchers from other scientific fields than ours, for example, geographers, lawyers, environmental managers, and literature theorists. In addition, we include public documents, laws (legislation and regulation), court decisions in reindeer herding conflicts, and local and national newspapers. The differences between interviewees' contexts are multitude, ranging from sterile glass and concrete offices in high-rise buildings surrounded by heavy traffic, hardly knowing anything about reindeer or Sámi culture, to the Arctic

---

[9] For details on the Finnmark Act (FEFO) see: http://www.fefo.no/en/Sider/AboutFeFo.aspx (last accessed 22 January 2022). For details on the Utmarksdomstolen (Finnmark Land Tribunal) see: https://www.domstol.no/en/Enkelt-domstol/Utmarkskommmisjonen/about-the-tribunal/ (last accessed 22 January 2022).

tundra in minus 30 degrees Celsius, on the snowmobile with the herding dog on the backseat, surrounded by reindeer looking for food.

The mixture of our personal origins led us to a customised construction of a particular network and a reindeer herding Nordic multiple sites' fieldwork focus. The anthropologists Biehl and Petryna (2013) describe the crucial role of ethnography in global health research and argue for a comprehensive, people-centred approach. Our ways into this fieldwork are through meeting people in their everyday locales, often outdoors, whether with or without reindeer around them. But if not literally present, reindeer were 'there' on screens,[10] in pictures on walls, or as food being offered and shared. We were following Biehl and Petryna (2013), and also Geertz (1973), who favoured 'thinking-in-cases' or, perhaps more aptly phrased, thinking *within* cases. There is a shortage of both anthropological and social scientific studies in general of present-day reindeer herding. For example, a broad range of scholars from a variety of disciplines have stated the lack of basic knowledge on reindeer herding.[11] Thus, offering ethnographic case studies as an approach to the field should make an important contribution. Through the analyses of our data, we identify and to a certain extent evaluate policies that supposedly 'work' or rather, as we show, do *not* 'work'. Our explanatory model should be possible to replicate or scale up 'across a range of often widely divergent social contexts and geographic locations' (Biehl & Petryna, 2013, p. 12). Following this, we agree with Biehl and Petryna's thoughts on ethnographic case studies, which 'brings granular ethnographic evidence to the forefront of analysis and enables analogic thinking'. We use their advice to pay 'close attention to particular realities on the ground and to the metrics in which they are cast' (Biehl & Petryna, 2013, p. 13), in order to highlight the productive and uneasy coexistence between the Nordic neoliberal systems' design of reindeer herding polity and the alternative models Sámi people craft for engaging the real world.

This approach coincides and works well within Indigenous methodology. It is seen as a critical methodology applied in marginalised research

---

[10] Microchips and GPS-collars are fitted on a selection of female reindeer so that herders can, from their sofa at home, watch at anytime where the plurality of the animals is on the tundra.

[11] See, for example, https://www.nmbu.no/en/projects/reign (last accessed 22 January 2022).

areas that often are excluded from mainstream research issues. It has, for example, the aim of changing perspectives, turning viewpoints into new positions, and challenging academic thinking and publishing (Drugge, 2016).

4 Bases for Our Knowledge and Approaches

Reindeer herding played a decisive role in what became Finnmark and the basis for the Finnmark population's existence. Until recently, the reindeer was the crucial source of meat, skin for clothing, antlers and bones for tools and jewellery, as well as for transport and traffic (Riseth & Lie, 2016; Sara, 1992, 2013). Finnmark County plays a significant role for understanding general conditions in reindeer herding development, applicable to the other five Norwegian counties as well as parts of northern Sweden that also have Sámi reindeer herding. We claim that our research, to a large extent, is applicable to other reindeer herding landscapes: according to The International Centre for Reindeer Husbandry (ICR), established by the Norwegian Government in 2005 in Kautokeino (Finnmark), there are '24 different Indigenous Peoples which base their lives on herding domesticated reindeer today. These nomadic reindeer herders live across the entire circumpolar Arctic and Sub-Arctic region, including areas in Sweden, Finland, Norway, Russia, China, Mongolia, US/Alaska, Canada, Greenland and Scotland. These groups currently herd around 2,5 million domesticated reindeer, and are altogether close to 100 000 people, all included'.[12] Considering the numbers of Indigenous Peoples in reindeer herding areas globally, these traditions in man-animal and herding relationships represent a considerable multitude of Indigenous Peoples' knowledge and traditions. Interestingly, veterinarians, climatologists, ecologists, biologists, and so on now see the reindeer as a signpost of global climate change, the Arctic serving as the most

---

[12] International Centre for Reinderr Husbandry, https://reindeerherding.org/projects/6wrhc/ (last accessed 22 January 2022).

vulnerable and overt area for studies of rising CO_2 emissions and effects, and for recording the impacts of rapid temperature increases.[13]

Conflicts between private and state-based enterprises and those involved in reindeer husbandry have increased since World War II, due to industrial growth in hydroelectric and wind power development, mining, oil and gas extraction, national and NATO military geopolitical strategic use of the territories, as well as tourism.[14] Some of the most comprehensive conflicts between nature conservationists/Sámi and the Norwegian authorities after World War II have taken place due to reindeer husbandry, such as during the dam building of Alta-Kautokeino River in the 1980s (Dalland, 1983). The reindeer, in addition to the salmon, is not only the signal animal for the Sámi, but Rudolf as Santa Claus's reindeer is known worldwide for bringing Christmas gifts to expectant children everywhere. However, the reindeer also represents an important carrier of Sámi traditions, culture, language, identity, and income that is now met by the NPM requirements, based solely on economic growth. Cultural, social, and symbolic power are hidden beneath the sole focus on economic growth or are overlooked; this is occurring at the same time as the reindeer herders face more hindrances, particularly in the form of an almost impossible demand from state authorities for participation in hearings on societal changes. These claims for participation in hearings based on the state's perception of societal development or even a perceived 'protection of this vulnerable Indigenous way of life' take up almost all of their time, with too little time left for herding.

To fight 'piece by piece' colonisation and destruction of grazing land for electric or wind power, mining, cabins, or tourism, the siidas have started a so-called complementary branding practice. Each time there are proposed destructive activities (state agencies and to a large extent, also municipal authorities, always use the term 'development') within the

---

[13] See, for example, the current centre of excellence ReiGN at its website: https://www.nmbu.no/en/faculty/biovit/about/department/iha/research/reign (last accessed 22 January 2022).

[14] The reindeer graze about 40 per cent of Norwegian land (as in Finland [36 per cent]). In contrast to other land-based primary sectors, where crops can be grown and planted on new land, the grazing and resource areas for reindeer are diminishing due to different kinds of encroachment, such as transportation corridors. See https://www.fylkesmannen.no/nb/Nordland/Landbruk-og-mat/Reindrift/Fakta-om-reindrift/Reindrift-i-Nordland/?id=96214 (last accessed 22 January 2022).

pastures of the siidas, the siidas in turn point out former destructive activities that have been completed. Prior to this more holistic response, comments were given only in relation to the specific case, often resulting in the grazing land being destroyed 'piece by piece'. Moreover, this approach to reminding of past developmental impacts helps to highlight that the state's development projects cause cultural and environmental harm that never gets addressed or remedied. Through using complementary branding practices, reindeer owners hope to reduce the burden of destruction and convey a better understanding of why they often say no to 'development'.

Thus, it can be said that the reindeer and their owners are victims of current, ongoing conflicts that arise due to the state's unquestioned implementation of several kinds of 'reforms', 'developments', and 'modernisations', often looked upon as parts of neo-colonial mechanisms which, for example, assume that all forms of development are neutral or even good in their outcomes, despite evidence often to the contrary. As we will show, the Fennoscandian States[15] use various direct or indirect economic, political, cultural, or other means to exploit or suppress Sámi cultures, traditions, and ownership to lands, soil, and fjords, in spite of Norway's ratification of the ILO convention article 169 on Indigenous rights and the United Nations Declaration of the Rights of Indigenous Peoples,[16] amongst other treaties and international statements that support Indigenous rights. One of the most disastrous consequences of this victimisation has been the rise of internal conflicts in or between siidas, a pattern typical of NPM. The responsibility for harm and suffering is left for and to the lowest part of the governance chain to deal with as best it can—unaided. In turn, the contexts, the structures, and processes leading to the harm and suffering are invisibilised. The book 'Samisk Reindrift—norske myter' (in English: 'Sámi reindeer herding: Norwegian myths') (Benjaminsen et al., 2016) is an attempt to unveil Norwegian authorities' arguments of the unsustainability of Sámi reindeer herding that implies

---

[15] Fennoscandia, Fenno-Scandinavia, or the Fennoscandian Peninsula, is the geographical peninsula of the Nordic region comprising the Scandinavian Peninsula, Finland, Karelia, and the Kola Peninsula.

[16] See http://www.un.org/esa/socdev/unpfii/documents/DRIPS_en.pdf (last accessed 22 January 2022).

there is the need for applying NPM to this most ancient of practices. If reindeer herders do not acknowledge this unspoken rationale lying behind the government's arguments, they will continue to be forced to live with ongoing conflicts and lean on and pay for private lawyers to help them solve state-produced conflicts as private civil court cases.

Although the Reindeer Husbandry Act of 2007 contains a mediation article, this has been barely used; this is due to the lack of regulations having been made for the mediation article by the Ministry of Agriculture and Food, as promised in the Act. When we ask why nothing has been done during the subsequent years, we get deflective answers, such as that there have been so many other challenging tasks related to reduction in numbers of reindeer, a need to progress the market and increase demand for reindeer meat, the reorganisation of the reindeer authorities, increased self-determination within the reindeer husbandry, and so on. This state failure to provide regulatory support for the mediation rights of Sámi has clearly increased the level of conflict in the Sámi reindeer husbandry.

As part of the neo-colonial Norwegianisation and NPM system's evolution, we see a clear manifestation of the expansion of juridical power, juridification, and the growing importance of law, legal discourse, and legal institutions in Sámi society, as well as in general social arenas. This aligns with what the anthropologist Bartoszko (2018) finds in studying, for example, Norwegian health policy. Juridification is defined as 'the mounting degree to which social relations, formerly left to autonomous and/or informal regulation, are being textured by formal legal rules' (Sieder et al., 2016). This increase in legal intervention relates exactly to the transition from the autonomous reindeer herding collectives that were present in the whole of Fennoscandia up until the end of the nineteenth century. Slowly, these collectives were destroyed due to the changing border policies between the Fennoscandian countries internally and towards Russia as an external enemy. To the Sámi people, these changing national borders were and still are tearing them apart and cause huge problems due to the reindeer needing to graze their homeland which, all of a sudden, was barred with fences. Judicialisation[17] is another concept

---

[17] Juridification and judicialisation are used differently in various contexts but we rely upon both terms here to cover the entire process we perceive as having occurred and continuing to occur with respect to the Sámi People and their way of life.

used to describe how social claims are pursued through the courts and court-like structures, for instance, the Norwegian county board for reindeer herding or the similar Swedish Länstinget.[18] In Sweden, the government transferred reindeer husbandry tasks from government agencies to the Sámi Parliament in 2007 against the Sámi Parliament's will, due to lack of resources. This was able to happen because the Swedish Sámi Parliament is more or less a part of the Swedish national authorities (Mørkenstam et al., 2016). Thus, the conflicts between the state and the Sámi herding collectives (Samebyar in Swedish) are seemingly 'internal conflicts', the state as actor is invisible.

In Norway, this took another direction, although with the same results, 'giving' the power to so-called civil legal solutions. Vallinder and Tate (1995, p. 5) contrast the 'judicialisation of politics' and a second, distinct phenomenon, namely, the 'less dramatic instance of expansion of judicial power, or judicialisation, [that is] the domination of nonjudicial negotiating and decision-making arenas by quasi-judicial (legalistic) procedures'.

We see juridification as close to Habermas' idea of colonisation of the lifeworld (Habermas, 1981)[19] and thus as a negative option for the Sámi. However, the same processes have affected the whole Fennoscandian population within different crucial areas of daily life, such as health and social fields (Bartoszko, 2018).

Several recent and ongoing law cases at all three administrative levels manifest the assumption and reality of superiority of national legislation over Sámi traditions, as well as over ecological or social facts and lived experiences. However, we do acknowledge that there might be a pitfall in oversimplifying the negative force of national law as a paradigm of power and domination. Law may be used not only as a matter of negative or suppressive social control or governmental power. Law may also have the

---

[18] This term means the county government.

[19] The *lifeworld* in Habermas' interpretation is based upon the philosopher Edmund Husserl's term, applied into a sociological area of interpretation and understanding as to how the world is experienced through communication with language and actions, both formal and informal understandings and mutual adaption, grounded in culture.

capacity to provide a meaningful map or a model for the world (Bartoszko, 2018). Law has, to a certain extent, been shown to act as an instrument for social change, such as in creating and supporting the Indigenous rights of the Sámi in Norway. Even though these legal changes were preceded by a nationwide and partly illegal demonstration and serious hunger strikes, imprisonment, high fines, and social and personal suffering in the years 1979–1982, the resulting legal changes led to the establishment of the Sámi Parliament with consultation rights to the national government. This illustrates the legal structures as landscapes of possibilities for disadvantaged groups, as Bartoszko (2018) found in her thorough study of patient groups fighting for their legal rights.

Sámi reindeer herders now see themselves as fighting against 'the system', but they also use the tools provided by that very 'system'. To some extent 'the system' also gives them tools to undermine the 'system', in this case the Norwegian NPM of reindeer herding, as we will show in our data presentation. Indigenous rights, for example, when or if forced through, may thus be a powerful instrument for social change.[20] In that context, the law may be seen as facilitating social cohesion. In other words, the cases that we present below are not only about outcomes of legal disputes or restorative justice efforts, whether successful or not. Single outcomes may not be crucial in the long run. Gulliver (1988) recognises how a majority of social scientists studying negotiations search for an explanation of outcomes. Our work has been led by the ambition to find explanatory powers as well as tracing as much as possible 'all the serious interactions which precedes, affects and sets up the end-game' (Gulliver, 1988, p. 251). They are personal stories, experiences, Fennoscandian, in particular Norwegian, reindeer herding and Sámi policy and history, climate and areal changes, histories of significant localities—aspects that are not only a background for negotiations and outcomes, but significant players. They are all actors in restorative justice efforts, possibly to come for the future.

---

[20] The question of a Sámi court of appeal, particularly in lawsuits concerning reindeer herding, has been raised and will perhaps be prompted once more as a result of the Truth Commission.

5   Results from Our Fieldwork

5.1   First Case: Cartography

In a meeting with scholars at the Sámi University College in Kautokeino, several of the scholars were mentioning the problem with cartography:

> There are many examples of this, meetings where the authorities have taken control of the local siida management, demanding of the siida owners: draw the border of your siida! Many lines here or there, without sense, the authorities behave as if they have no knowledge and the reindeer herders have the impression that they themselves have no knowledge of their own, too. ... Borders must be there, but those borders are not drawn on a piece of paper, they are verbally justified. An example is that "this coombe has always functioned as a fence". Traditionally, people did not use maps, this is a new medium that contributes to increase the level of misunderstandings and conflicts ... A map only functions if the reindeer choose to walk there, then a map is of little use, because there is a big disparity between a reality and a map. (Research summary)

We have observed the overall and consistent use of map as method in the authorities' governance of reindeer herding throughout our fieldwork. They use the map, as for them, it is an easy, straightforward, and neutral method in the following-up of their rules and regulations. Reindeer owners and herders do not recognise themselves in the authorities' maps and descriptions of the terrain. Their understanding is based upon the migration of the reindeer, climate, snow, ice, wind, rivers, creeks, and lakes, varying from year to year, season to season. While the authorities have four seasons, reindeer herders have eight. This is impossible to draw on a two-dimensional piece of paper—on maps. Maps can be used to conquer and thus govern peoples' opinions, actions, and understandings of a terrain, a landscape, and a policy. It has been a method for governing and administering colonised land in all European acquisition of former and, to some extent, remaining colonies. Cartography is the science of such mapping, a Western scientific way of reading a terrain. But a terrain may be read in at least three different ways.

A map can be a cognitive system, different from culture to culture (Ben-Ze'ev, 2012). As we met this problem, we saw how two different systems for understanding the terrain were incongruent. A map can also be an expression for material culture. We saw this when the authorities named a mountain or a particular space, which the Sámi held as holy, due to particular religious, historical experiences, or burial sites. A map is also a social construction, that is, socially created, in that someone once agreed upon the interpretation of drawings, lines, colours, and how to read them (Hunt & Stevenson, 2016; Woodward & Lewis, 1998).

5.2   Second Case: Cross-border Reindeer Herding

Due to the gradual nationalisation of Norwegian, Swedish, Finnish, and Russian lands into nation-states, nomadic reindeer herding has become more and more trapped by state borders. Particularly the case during the 1800s, the border lines still today cause considerable problems to human, animal, and state authority. The shifting and various closures of borders between the Fennoscandian nation-states followed as a result of several wars. Thus, Sámi reindeer herding that traditionally defined grazing land within Sápmi, that is, Fennoscandia as a whole, was gradually locked in and defined as either Norwegian, Swedish, Finnish, or Russian. The Skolt Sámi group was divided into two as they were forced to choose whether they should stay in Finland or Norway. Herders found themselves surveyed by state authorities, trapped between nation-states' differences in legal acceptance of Sámi civil rights, culminating in the state's veritable theft of these rights on each side of the borders.

We went to Sámi reindeer owners and villagers on both sides of the Swedish-Norwegian border, their grazing land having been kept intact due to time-immemorial use, now defined as either being situated on the Norwegian or Swedish sides of the respective nation-state's border. The old reindeer herding agreement between Norway and Sweden, the Lappkodicill, an addendum to the Border Treaty 1751, ceased to be in effect after 2005, which has meant that the reindeer herders can no longer pasture their traditional land.[21] We found several and· different consequences flowed from this during our fieldwork. Sámi reindeer owners in several counties were fined for herding in areas that now either municipalities, state or private industries, farmers (who had been given the land from the state), the armed forces, and so on, claimed as *their* property. Each claim for levelling a fine must be answered by the Sámi herders with the help of costly lawyers and court cases. 'Two of my brothers have committed suicide during one of these court cases', one herder on the Swedish side said desperately. 'We don't know how far this will go. The Norwegian municipality have fined us for more than 8 million NOK, despite us winning the Norwegian Supreme Court case against them, years ago'. The Swedish political scientist Anette Löf (2016, p. 426) underscores similarly in her PhD study that the governing of the Swedish Sámi reindeer husbandry remains mainly hierarchical and is characterised by inconsistencies:

> In contrast to well-established narratives of increased participation and indigenous peoples' right to self-determination ... there are large differences in understanding between key actors and, over time, only marginal change in governing structures and meta-images. Thus, reindeer husbandry actors appear to be locked out of essential governing functions and locked into a system that is proving hard to change.

The Swedish head negotiator for a new reindeer herding treaty between the two states, Lars Norberg, was a skilled and very respected diplomat, having been the head negotiator from Sweden during the Balkan Wars. He wrote an outstanding book on his extremely negative experiences with the reindeer authorities on both sides: 'The Swedish State has been trapped and thereby given away the rights of the Swedish Sámi (for grazing on the Norwegian side). And in that trap it fastened' (Udtja-Lasse, 2007, p. 52).[22] His book from the negotiations symbolically contains part of the same title as 'Bury my heart at Wounded Knee' by the Native

---

[21] See https://www.loc.gov/item/global-legal-monitor/2021-07-28/norway-supreme-court-defines-extent-of-swedish-sami-reindeer-herder-rights/ (last accessed 22 January 2022).

[22] See also a conversation between Norberg and Lee Swepston, the former senior advisor for human rights at the ILO in Geneva: https://www.sametinget.se/11144 (last accessed 22 January 2022).

American Dee Brown (Brown & Brown, 2007), bearing the very same message of state governance as neo-colonisation of the Sámi people by both Norwegian and Swedish State authorities. Norberg thereby compares the negotiation and its background history with the American government's continuing effort to destroy the culture, religion, and way of life of Native American peoples. Although the Nordic history does not contain the same kind of violence, killings, and rapes as the American one, the principal way of exploitation of land is the same. The consequences for people, animals, culture, and way of life may also be the same.

5.3   Third Case: When a Fence Turns Friends into Enemies

One of the cases we studied (text analysis, interviewing, and by observing in court) deals with an area that is used both for summer grazing for one siida and for moving the reindeer herd in the other siida. These two neighbouring siidas were friends and had a joint secretary only 15 years ago. However, the authorities have interfered in the Sámi traditional way of land use on at least two occasions in recent years, causing pressure and conflicts to erupt in other neighbouring areas. Some of the siidas managed to handle this through cooperation, but others responded by building fences against their neighbouring siida.

In this case, siida A set up a long and costly fence, doing so illegally. Siida B protested and claimed that the fence prevented their traditional moving pattern for the reindeer. Siida A was forced to tear down the fence. Both siidas engaged lawyers and expert historians, bringing resolution of the conflict to the court. Siida B won the case, but siida A appealed and won in the court of appeal. Siida B who lost the case now says that this is a case of injustice. After having had close and friendly relationships for at least three generations, both siidas are now bitter enemies. In addition, both parties have spent millions of Norwegian kroner to pay for lawyers and expert witnesses. Everyone who has a little knowledge about this case, claims that there are no winners.

The case should never have been brought to the court; instead, it should have been resolved by mediation. This would have kept relationships intact and prevented the creation of enmity. But, as noted earlier, the Norwegian mediation system lacks Sámi language and cultural knowledge as well as mediation regulations, and the reindeer herding authorities have yet to establish the reindeer mediation system promised way back in 2007. Our data and analysis show that a legally enforceable judgement does not solve the conflicts but rather amplifies conflict. One of the representatives from the county reindeer herding authorities said:

> The case is dead now. At the same time, I have not heard that they have a reasonable dialogue between them. This often ends up with people not talking to each other but sending sarcastic e-mails.

This outcome concerns congenial, sharing, and knowledgeable people from both siidas. Several people in the community have said that they do not support the judgement, some are angry and will never come to a reconciliation. In addition, the overall illegitimacy of the whole scenario of such civil lawsuits and cases between the siidas is the responsibility of the Norwegian State that through its impugnable governance of Sámi reindeer herding brings siidas into impossible deadlocks. The state makes itself invisible, the local authorities have no means to adjust, intervene, or give a helping hand, and some of the herders told us about their helplessness and sadness for being placed into such a squeezed position.

5.4   Fourth Case: The Jovsset Ánte Sara Case[23]

This case has recently received international media coverage. Sara, a Sámi herder, refused to comply with the demands from state authorities to reduce his number of reindeer from 116 animals to 75. This demand is part of the authorities' long-term struggle to reduce the numbers of reindeer, as previously described. The state decides a static maximum number for each district. In the 2007 Reindeer Husbandry Act, a clause was introduced that if all reindeer owners in a district did not agree to the

[23] See https://www.theguardian.com/world/2017/dec/22/norway-court-clears-way-for-contro versial-reindeer-slaughter (last accessed 22 January 2022).

reductions, the government would make an equal percentage reduction amongst all siida shareholders. Sara won his case both in the District Court and in the Court of Appeal. The District Court said that the state's demand for reduction violated the property protection according to the European Convention on Human Rights, P 1-1.[24] The Court of Appeal discarded the appeal of the state according to the International Covenant on Civil and Political Rights, article 27.[25] The state's decision on reduction in reindeer would imply that the reindeer owner no longer had a basis for economic surplus. Even if the Reindeer Herding Act's provision on reindeer numbers and reductions had a reasonable and objective ground, seen from the aim of the Act, there is a threshold for when an intervention can be seen as legitimate. This also refers to the state's reindeer herding governance. The threshold for illegal intervention was exceeded in this case. The state appealed to the Supreme Court, which upheld the plaintiff's (the state's) claim on reduction. Sara has since appealed the case to the European Court of Human Rights, with support from the Sámi Parliament.[26]

The case caused demonstrations in front of the National Assembly and received considerable support from many different groups within the Norwegian population, in particular artists. The arguments from the state, represented by the ministry of agriculture, were based upon arguable assumptions drawing on ecological, economic, and cultural sustainability rationales and purported concern with the animals' welfare. However, these arguments were counteracted by professors with expertise in environmental sciences, who saw the state response as a policy that was arrogant, lacking in verifiable knowledge and that bore no relation to the facts.

---

[24] Particular rights, such as reindeer herding, are seen as protected by the European Human Rights Convention, article 1-1.

[25] The central conventions that are relevant for protecting traditional reindeer herding are in the International Covenant on Civil and Political Rights, article 27, implemented in Norwegian law through the Act of Human Rights and the ILO Convention on Indigenous and Tribal Peoples 1989 (No. 169), which the Norwegian State has ratified, but only made directly current for Finnmark through the Finnmark Act.

[26] See, for example, https://inews.co.uk/news/reindeer-fight-government-109315 (last accessed 22 January 2022).

In this case, we do not only experience how the ministry of agriculture reproduces the 'Norwegianisation' of Sámi reindeer herding, as a prolongation of the Norwegianisation of the Sámi People, its language and culture. We also see how the independent Norwegian courts at all three levels exert a diffuse and confused approach to judgements of state policy when adjudicating Sámi reindeer herding conflicts, particularly when the state is a litigant.

6 Conclusion

Where does or does not restorative justice have a role to play in reindeer herding conflicts in Fennoscandia? If we stay with the term neo-colonisation, most of our state representative informants do their best as part of a civilising mission to educate, convert, and assimilate the Other, that is, to include Sámi and their reindeer herding traditions into the empire of the state and the NPM of agricultural governance as an indirect part of the Norwegianisation of the Sámi. In other words, they try to convince Sámi that they will benefit as standard Norwegian 'farmers'—one of 'us'—if they accept their conditions concerning boundarying, herding, grazing, slaughtering, marketing, and so on. The mechanisms of this kind of neo-colonisation today seem to be state subsidies, taxation rules, changing legal provisions and regulations, distribution of part-governance to Sámi institutions like the Sámi Parliament (albeit without legislative power), the Norwegian reindeer Herding Association, NRL, locally and nationally, and to the reindeer herding districts and siidas—the collectives, as reminiscent of the pre-colonial time. Cartography serves as an example of the state's NPM and as a neo-colonising tool—reindeer owners mapping borders of reindeer tracking as part of governmental neoliberal agricultural and homogenising policy.

The idea of the colonisers was to not only dominate but also 'civilise' and 'save'. Civilising in this context also means 'modernised agricultural reasoning and praxis', although this reasoning and praxis has an unsurpassed power to ruin climate, soil, forests, animals, humans, and human-animal relationships. Thus, there are in all, several categorical mistakes in play: firstly, to adjust herding of nomadic animals to a non-nomadic herding praxis; secondly, to force Norwegian (or Swedish or Finnish) language, social, cultural, and religious traditions upon the Sámi population, a policy in force until the late 1960s. With this two-pronged approach to decimating traditions, the Sámi cultural ground was broken and turned into shame, disaster, and suicides.

The latest Norwegian Reindeer Herding Act from 2007 may be regarded as a kind of reform, aimed at regulating both humans—Sámi owners—and animals—reindeer, the relationship between them, and the political and social changes needed to achieve these aims. Such an NPM-inspired 'reform' has led to so-called internal conflicts that originate mostly from state decrees and paternalistic involvement, leading to private civil court cases where citizens are obliged to meet the expenses of fixing problems that have been directly caused by government. Strangely enough, even if the National Mediation Service as a fully restorative justice practice exists as a state service, organised by the Ministry of Justice and Public Security, the Norwegian State has a regrettable absence of restorative justice as a concept or practice in its own conflicts with its own inhabitants, individuals, or groups.

To handle all these increasing and spreading conflicts between the state and the Sámi, the Norwegian State has appointed a Truth and Reconciliation Commission, without any reference to restorative justice—and with too few resources to work through all the stages needed in a restorative justice process. Thus, the Norwegian government avoids or evades getting its house in order to restore the power, oversight and ongoing relational aspects back to the Sámi People. We consider that perhaps there is not even enough for the stage of recognition—so what then can be said of achieving reconciliation and repair? Our analysis of political, social, and cultural reform cannot be limited to the organisational changes studied from a perspective of the Norwegian Reindeer Herding Act institutions and governmental bodies. It must include the individual, the subjectivity, and the narrative—including the transnationality. In that way, we may bring critical insights into the questions of how policies translate into the lives of the people that they aim to address (Bartoszko, 2018).

Our aim was to approach the Sámi people's understanding of themselves as 'owners' with 'Sámi rights' and explore how both reindeer owners and reindeer bureaucrats actively engage with the various socio-political categories and labels, to see whether they were using them to interpret or cope with the social and political culture of suspicion that we have shown in the cases described above. By looking and listening to their everyday practices and ways of dealing with traditions, laws, rules, and mistrust, we have focused on reciprocal relationships between social categories, herding behaviour, and identity. Our final analysis investigated the transformative power of bureaucracy and law, by also questioning our own assumptions as researchers from our different backgrounds.

Could our quest for suggesting the use of a restorative justice approach, given the present political situation, transform into a legal instrument for the state to continue the neo-colonising of Sámi land and reindeer herding? If that be the case, 'restorative justice' risks being turned into the opposite of its aim and converted into a tool for disguising not only Sámi civil rights and their national and international citizenship, but also destroying life-saving Sámi environmental knowledge. Bearing this risk in mind, it is a Sámi version of restorative justice that the state authorities would need to ask for. The authorities would need to acknowledge that it carries within it Indigenous ways of approaching conflict and that it should be wielded by both the Sámi Parliament and individual Sámi who have passed their claims for recognition and apology. It needs to be done in ways that resonate with Sámi traditionally and could be a benefit both to them and to the government agencies, with the latter's leaders to be invited to sit, listen, and learn.

References

Angell, E., Karlstad, S., & Nygaard, V. (2003). *Samiske samfunn i omstilling. Sluttevaluering av Omstillingsprogrammet for Indre Finnmark*. Retrieved from: https://www.nb.no/nbsok/nb/6234ce5c6071d2aaf9acf27d8e8522a9?lang=no#0 (last accessed 16 January 2022).

Bartoszko, A. (2018). The pharmaceutical other. Negotiating drugs, rights, and lives in substitution treatment of heroin addiction in Norway. Oslo: Oslomet (Oslo Metropolitan University).

Ben-Ze'ev, E. (2012). Mental maps and spatial perceptions: The fragmentation of Israel-Palestine. *Mapping Cultures: Place, Practice, Performance*, 237–259.

Benjaminsen, T. A., Eira, I. M. G., & Sara, M. N. (Eds.). (2016). *Samisk rein-drift—norske myter*. Bergen: Fagbokforlaget.

Bergland, E. (2005). *Reindrift, omstilling og identitet*. Guovdageaidnu: Sámi Instituhtta.

Biehl, J., & Petryna, A. (2013). *When people come first: Critical studies in global health*. Princeton: Princeton University Press.

Bjørklund, I. (2004). Saami pastoral society in northern Norway: The national integration of an Indigenous management system. In D. G. Anderson & M. Nuttall (Eds.), *Cultivating arctic landscapes: Knowing and managing animals in the circumpolar north* (pp. 124–135). New York: Berghahn.

Bjørklund, I., & Marin, A. (2015). Er Finnmarksvidda en allmenning? In T. A. Benjaminsen, I. M. Gaup Eira, & M. N. Sara (Eds.), *Samisk reindrift, Norske myter* (pp. 106–126). Bergen: Fagbokforlaget.

Brown, D. A., & Brown, D. (2007). *Bury my heart at Wounded Knee: An Indian history of the American West*. New York: Macmillan.

Dahlstrøn Nilsson, Å. (2003). *Negotiating wilderness in a cultural landscape: Predators and Saami reindeer herding in the Laponian world heritage area* (Phd). Uppsala: Uppsala University.

Dalland, O. (1983) The Alta case: Learning from the errors made in a human ecological conflict in Norway. *Geoforum*, 4(2), 193–203.

Drugge, A.-L. (2016). *Ethics in Indigenous research: Past experiences, future challenges*. Umeå: Umeå University, Faculty of Arts, Centre for Sami Research.

Elenius, L., Allard, C., & Sandstrøm, C. (Eds.). (2016). *Indigenous rights in modern landscapes* (1st ed.). London: Taylor & Francis Group.

Geertz, C. (1973). *The interpretation of cultures / selected essays*. New York: Basic Books.

Gulliver, P. (1979). *Disputes & negotiations: A cross-cultural perspective*. Cambridge: Academic Press.

Gulliver, P. (1988). Anthropological contributions to the study of negotiations. *Negotiation Journal*, 247–255.

Habermas, J. (1981). *Theorie des kommunikativen Handelns* (Vol. 2). Suhrkamp Frankfurt.

Henriksen, J. E. (2008). They can never take my dignity: Social and cultural capital among reindeer herders in Finnmark. In P. Huse (Ed.), *Northern imaginary* (pp. 147–151). Lillehammer: Delta Press.

Henriksen, J. E., & Hydle, I. (2016). Participatory handling of conflicts in Sámi areas. *International Social Work, 59*(5), 627–639.

Hunt, D., & Stevenson, S. A. (2016). Decolonising geographies of power: Indigenous digital counter-mapping practices on turtle Island. *Settler Colonial Studies, 7* (3), 372–392.

Ingold, T. (2000). *The perception of the environment: Essays on livelihood, dwelling and skill.* London: Routledge.

Kaiser, N., & Salander Renberg, E. (2012). Suicidal expressions among the Swedish reindeer-herding Sami population. *Suicidology Online, 3*, 102–113.

Kaiser, N., Sjølander, P., Edin Liljegren, A., & Jacobsson, L. (2010). Depression and anxiety in the reindeer-herding Sami population of Sweden. *International Journal of Circumpolar Health, 69*(4), 383–393.

Löf, A. (2016). Locking in and locking out: A critical analysis of the governance of reindeer husbandry in Sweden. *Critical Policy Studies, 10*(4), 426–447.

Møllersen, S., Stordahl, V., Eira-Åhren, I. M., & Tørres, G. (2016). Reindriftas hverdag. Interne og eksterne forhold som påvirker reineiere. Karasjok: SANKS—Samisk nasjonal kompetansetjeneste for psykisk helsevern og rus.

Mørkenstam, U., Josefsen, E., & Nilsson, R. (2016). The Nordic Sámediggis and the limits of Indigenous self-determination. *Gáldu Čála—Journal of Indigenous Peoples Rights* (1), 4–46.

Ravna, Ø. (2020). The duty to consult the Sámi in Norwegian Law. *Arctic Review, 11*, 233–255.

Riseth, J. Å., & Lie, I. (2016). Reindrifta i Finnmarks betydning for næringsutvikling og samfunnsutvikling. In E. Angel, S. Eikeland, & P. Selle (Eds.), *Nordområdene i endring—Urfolkspolitikk og utvikling* (pp. 182–207). Oslo: Gyldendahl Akademisk.

Riseth, J. Å., Tømmervik, H., & Bjerke, J. W. (2016). 175 years of adaptation: North Scandinavian Sámi reindeer herding between government policies and winter climate variability (1835–2010). *Journal of Forest Economics, 24*, 186–204.

Rosaldo, R. (2017). Grief and a headhunter's rage. In A. C. G. M. Robben (Ed.), *Death, mourning, and burial: A cross-cultural reader* (2nd ed., pp. 156–166). London: Wiley-Blackwell.

Sara, M. N. (1992). Boazudoallulahki ja boazudoalupolitihkka. *Dieđut, 2.* Guovdageaidnu: Nordisk Samisk Institutt.

Sara, M. N. (2013). *Siida ja siiddastallan. Å være en siida—om forholdet mellom siidatradisjoner og videreføringen av siidasystemet/Being siida—on the relationship between siida tradition and continuation of the siida system* (Philosophiae Doctor). Sámi Allaskuvlla, Guovdageaidnu.

Sieder, R., Schjolden, L., & Angell, A. (2016). *The judicialisation of politics in Latin America*. Berlin: Springer.

Stoor, P. (2016). *Kunskapssammanställning om samers psykosociala ohälsa*. Kiruna: Sametinget.

Turner, D. (2006). *This is not a Peace Pipe: Towards a critical Indigenous philosophy*. Toronto: University of Toronto Press.

Udtja-Lasse (2007). *Begrav mitt hjärta vid Udtjajaure*. Stockholm: Emma Publishing.

Vallinder, T., & Tate, C. N. (Eds.). (1995). *The global expansion of judicial power*. New York: New York University Press.

Woodward, D., & Lewis, G. M. (1998). *Cartography in the traditional African, American, Arctic, Australian, and Pacific societies*. Totowa: Humana Press.

21

Restor(y)ing the Past to Envision an 'Other' Future: A Decolonial Environmental Restorative Justice Perspective

Iokiñe Rodriguez

1   Introduction

In Latin America, Indigenous peoples, along with other marginalised groups such as Afro-descent communities and women, have been at the forefront of contemporary environmental justice struggles for more than two decades (Escobar, 2016; Leff, 2015). This is not fortuitous. In this, the new direction that capitalism has taken upon entering a new economic phase, coined by the British geographer David Harvey (2004) as *accumulation by dispossession*, has been decisive. As pointed out by Harvey, a significant percentage of the world's capital is currently being used to deprive people of their natural wealth (waters, forests, minerals, fauna) and their ancestral knowledge—associated with use of the commons—as part of processes of globalisation and the commodification and privatisation of land and natural resources. We are in the presence of, according to Harvey, a new colonialism, more rapacious than the one suffered by

I. Rodriguez (✉)
School of International Development, University of East Anglia, Norwich, UK
e-mail: i.rodriguez-fernandez@uea.ac.uk

the Indigenous peoples of Latin America between the fifteenth and eighteenth centuries. Indigenous people know this, and that is why they are trying to free themselves from it (Escobar, 2011; Leff, 2001).

The struggles for greater environmental justice of the Indigenous peoples of Latin America have given rise to a regional movement that has taken up a stance against the economic rationality driving the dominant model of development and the global project of modernity. This has involved, amongst other things, fighting for new culturally differentiated forms of decision-making in nation-state models to acknowledge their rights to their own forms of development, self-determination and political autonomy, the property of their territories and, most significantly, the preservation of their cultural integrity.

However, the project of modernity is a pervasive one. In Latin America, modern nation-state building has historically been premised on narratives of national identity and modernity that have sought to 'assimilate' Indigenous peoples into the wider society rather acknowledging their 'difference'. This trend has continued even within emerging pluri-cultural nation-state models, such as those currently favoured in Venezuela, Bolivia and Ecuador, where the economic rationality of development and the imperative for economic growth has remained as intact as it is in the rest of the region. Thus, environmental conflicts continue to rise.

In many cases, the lack of success in achieving greater justice and restoring environmental wrongs is linked to complex processes of cultural and social erosion created by coloniality and the ongoing project of modernity. In many Indigenous communities and territories, Indigenous youth are experiencing increasing disconnection from nature and the local environment because of rapid processes of cultural change and decades of assimilation policies (Pilgrim & Pretty, 2010). This gives rise to intra-community and intergenerational tensions and conflicts over the use of the environment, which, more often than not, limit the clarity, consistency and cohesive response with which Indigenous peoples respond to external threats and pressures to their culture and territories.

Thus, to do environmental restorative justice, in many parts of the world, endogenous processes of cultural revitalisation are needed to strengthen Indigenous peoples' own knowledge systems and cultural identities. In other words, actions are also needed to overcome a more invisible form of violence experienced by Indigenous people, which is seldom talked about in environmental justice literature: *cognitive or epistemic violence*.

Such processes of cultural revitalisation are what, in 1997, Shiv Visvanathan coined as *cognitive justice* and others refer to as *epistemic justice*. Epistemic justice, as Catherine Walsh (2005) argues, entails creating new knowledge in a way that confronts existing relations of domination in hegemonic paradigms and also helps to strengthen what the people themselves understand and reconstruct as 'theirs', in relation to identities, differences and knowledge. This emphasis on reconstructing, recovering and revaluing local knowledge is central in Latin American decolonial environmental justice theory and key to achieving justice in environmental struggles.

In some parts of the world, local revitalisation projects are well underway in areas such as traditional foods, economies, education, language, cultural practices and rights (Pilgrim & Pretty, 2010), but these are not necessarily conceptualised as environmental restorative justice processes, nor follow a decolonial knowledge production paradigm. I argue that an environmental restorative justice approach can be enriched by getting in touch with Latin American decolonial thinking and praxis and incorporating it into its disciplinary practice.

To do so, I first discuss some main propositions of Latin American decolonial environmental justice theory, which ground my approach to restorative justice. Secondly, drawing on John Paul Lederach's (2008) long-term peacebuilding perspective, I define how I understand a restorative environmental approach from a cultural revitalisation perspective. Thirdly, to exemplify what such an approach looks like in practice, I discuss two case studies: one in Canaima National Park, Venezuela, and another in the Indigenous Territory of Lomerio, in Bolivia, in which I have been involved in environmental restorative justice process using participatory action research for cultural revitalisation with Indigenous peoples in contexts of prolonged environmental conflicts. Fourth and finally, I discuss some key lessons from the two key studies with the hope that they can guide the efforts at conceptualising and doing environmental restorative justice in Latin America and beyond.

2   Main Propositions of Latin American Decolonial Environmental Justice Theory

In Latin America, in contrast to other parts of the world, environmental justice thinking has largely developed alongside decolonial thought, which explains social and environmental injustices as arising from modernity and the ongoing expansion of European cultural values and worldviews. The decolonisation of knowledge and social relations is highlighted as one of the key challenges to overcoming the history of violent oppression and marginalisation in development and conservation practice in the region. Arturo Escobar (2003, 2008, 2011, 2018) and Enrique Leff (2001, 2004) have been pioneers in positioning an environmental justice theory in the region with a 'decolonial turn'[1] (Castro-Gómez and Grosfoguel 2007). More recent additions to this body of knowledge from an environmental justice perspective include Alberto Acosta (2013), Eduardo Gudynas (2010, 2011) and Boaventura de Sousa Santos (2008), who, although not strictly Latin American, collaborates closely with Latin American decolonial scholars.

Indigenous peoples' contemporary struggles for social and environmental justice have laid important empirical and conceptual foundations for the emergence of decolonial theory in Latin America (Lander, 2000; Quijano, 2000; Escobar, 2003; Walsh, 2007; Mignolo, 2009). Decolonial thought is distinct from other post-colonial critical theory through its focus on the Global South and for identifying mechanisms of subordination and marginalisation in the project of modernity and the continual reproduction of European cultural values. Proponents of this school of thought are largely from Latin America (Quijano, 2000; Lander, 2000; Leff, 2001; Castro-Gómez & Grosfoguel, 2007; Escobar, 2003; Walsh, 2007; Mignolo, 2009), but important contributions have also come from India (Visvanathan, 1997), Portugal (Santos et al., 2008; Santos, 2010) and New Zealand (Smith, 1999), amongst others.

---

[1] A 'decolonial turn' refers to the task of decolonising or freeing oneself and society from the legacy of colonialism and *coloniality* in its different forms and manifestations. Section 3 explains this in detail.

Building on Indigenous peoples' own anti-modernity agenda, Latin American environmental justice thinkers have developed a series of core ideas that are central to how environmental justice is currently approached and perceived in the region.

2.1   Justice Beyond Recognition: The Need for the Construction of 'Otherness'

According to decolonial theory, 'colonialism' ended with political independence in the Global South, but coloniality persists through dominant Eurocentric colonial/modern values and worldviews that are institutionalised and disseminated through education, the media, state-sanctioned languages and behavioural norms. Thus, coloniality is a form of power that creates structural oppression over marginalised sectors of society, such as Indigenous peoples, whose alternative worldviews become devalued, sidelined and stigmatised in development and environmental management practice. From this perspective, coloniality is a particular mechanism and form of misrecognition that must be confronted in order to achieve emancipation and social/environmental justice.

Decolonial scholars argue that modernity leads to profound psychological harm for Indigenous peoples and other subaltern sectors of society because it erodes vital conditions for their wellbeing, including their cultural identity, freedom of choice and self-respect. It also has tangible impacts on the status and participation of Indigenous peoples in development and environmental management by disregarding local notions of authority and territory, frequently resulting in displacement or enforced changes to livelihoods. Structural oppression is perpetuated through a matrix operating at three levels: (a) power (political and economic); (b) knowledge (epistemic, philosophical and scientific) and (c) the self or ways of being (subjective, individual and collective identities).

The *coloniality of power* is exercised through two primary mechanisms: first is the codification of racial difference between Europeans and non-Europeans aimed at making the latter appear naturally inferior. This finds expression in normative rules such as definitions of development/progress. The second is the use of Western/modern institutional forms of power (like the nation-state) in non-Western societies to organise and control labour, its resources and products (Quijano, 2000). Hence, although coloniality continues to be intrinsically linked to global capitalism, it cannot be reduced to economics, as it also involves other invisible cultural mechanisms of domination.

The cultural and normative dimension of the coloniality of power also finds expression in the *coloniality of knowledge* (explained below), through the dominance of European knowledge and symbolic systems over non-European ones. Furthermore, the coloniality of power and knowledge impacts on the individual through the *coloniality of the being*, via mechanisms of subjectivation on the life, body, and mind of the 'colonised' or marginalised people, to the point of stripping them of their very essence and soul.

Thus, responses to coloniality necessarily involve decolonising power, knowledge and the being.[2] This involves moving away from unitary models of citizenship and civilisation to one that respects different local economies, politics, cultures, epistemologies and forms of knowledge, while also forging new categories of thought, constructing new subjectivities and creating new modes of being and becoming that can lead to emancipation.

This focus on the decolonisation of power, knowledge and the being marks an important divergence in environmental thinking from that of the Global North. From a Latin American decolonial perspective, environmental justice entails developing a politics of difference that is not simply based on the search for recognition or inclusion in dominant structures, such as the liberal nation-state or global economic systems, but focused rather on the construction of 'otherness':

> an "other" process of knowledge construction, an "other" political practice, an "other" social (and State) power and an "other" society; an "other" way to think and act in relation to, and against, modernity and colonialism. (Walsh, 2007, p. 57)

---

[2] Broadly speaking 'the being' is defined as the soul and essence of a person.

Bolivia and Ecuador serve as good examples of the construction of such 'otherness', through their recent shift towards becoming plurinational nation-states, their acknowledgement of differentiated cultural rights for Indigenous peoples and the institutionalisation of alternative concepts of development such as *Buen Vivir* in their national constitutions. In both cases, such changes represented an important 'decolonial turn' and a moment of epistemic rupture with modernity, which was greatly inspired and influenced by Indigenous peoples' life projects. However, as mentioned above, they also serve as good examples of the forces at play in the project of modernity that resist the 'decolonial turn'. Decolonising of power cannot be achieved solely by producing changes in the political or social spheres, while maintaining the dominant economic rationality. Ultimately, a shift in values systems (knowledge) and ways of being is needed.

2.2   There Is No Global Justice Without Cognitive Justice

A significant contribution of the decolonial environmental justice perspective is its focus on the epistemological dimension of oppression and domination. It highlights the need to engage with the invisible and extremely subtle ways in which violence is meted out in environmental justice struggles: through the imposition of particular ways of knowing the world at the expense of oppressing others, in other words, through *epistemic violence*, which refers to 'the different ways in which violence is exercised in relation to the production, circulation and recognition of knowledge: the denial of epistemic agency for certain subjects, the unacknowledged exploitation of their epistemic resources, their objectification, among many others' (Perez, 2019).

As Latin American environmental justice thinkers argue, the battle of Indigenous peoples and other socio-environmental movements in Latin America is not for the re-distribution of harms and benefits in the use of the environment, as stressed in environmental justice movements in other parts of the world. Rather, their struggle is for the right to live well, in accordance with their own identities, cultural imaginings and ways of knowing the world (Leff, 2015). Therefore, as suggested by Walsh (2005) and Santos (2008), the biggest challenge for emancipation from a decolonial perspective is to move towards a situation of greater *cognitive justice* in the world, learning from, and making visible, alternative forms of knowledge and being. Cognitive justice, as Visvanathan (2009) says,

> demands recognition of knowledges, not only as methods but as ways of life. This presupposes that knowledge is embedded in an ecology of knowledges, where each knowledge has its place, its claim to a cosmology, its sense as a form of life. In this sense knowledge is not something to be abstracted from a culture as a life form; it is connected to a livelihood, a life cycle, a lifestyle; it determines life chances.

According to this perspective, greater recognition of alternative knowledges in development requires changing the conditions of dialogue between knowledge systems to achieve a situation in which traditionally excluded actors, such as Indigenous peoples, do not have to fit in with the structures and standards of Western knowledge or worldviews. Far from it: research and development must be able to respond to the social, cultural, political, economic and environmental imperatives of the agendas of local and Indigenous peoples (Smith, 1999).

2.3   The Academic–Activist Nexus

As can be inferred from the above, from a decolonial perspective, academia has an important role to play in the making of the intercultural dialogues needed for emancipation and environmental justice. In fact, one of the distinctive features of Latin American environmental justice thinkers has been their commitment to understanding reality in order to transform it. In contrast to many environmental justice thinkers from the Global North, who largely engage with environmental conflicts and injustices as objects of study, Latin American environmental justice scholars have been conspicuous for taking a positive stand and active role against environmental injustices. They do so by unpacking the dominant rationality of modernisation, by entering into dialogue with local movements exploring their discursive techniques and strategies of struggle and, most significantly, by using research as a vehicle to transform power asymmetries in the dominant paradigms of knowledge production and development (Escobar, 2008; Gudynas & Acosta, 2011; Alimonda, 2011). Thus, behind decolonial environmental justice thinking, there is a political intention as much as an academic one. The long tradition of participatory action research in Latin America has been an important source of inspiration and influence in this trend (Fals Borda, 1986; Fals Borda & Brandao, 1986).

Another important aspect of the positionality of Latin American decolonial environmental justice theory is the growing acceptance of the theoretic production that takes place outside academia, specifically in activist circles and as a result of the interaction between academics and activists. Concepts such as *Buen Vivir*, which have been incorporated into the decolonial agenda and discourse, are an expression of this academic–activist interface, as is the theoretical and historical commitment of environmental justice academics to the construction of sustainable futures and other possible 'worlds' (Leff, 2015).

2.4 The Intercultural Challenge: The Core of a Decolonial Praxis

Decolonial thinkers propose the ecology of knowledge (Santos, 2008), also termed dialogues of knowledge/wisdoms (Leff, 2004) or the construction of interculturality (Walsh, 2005), as the core of a decolonial praxis.

But interculturality here is radically different from other more widely used functional definitions. Decolonial thinkers approach interculturality from a *critical perspective* (Tubino, 2005; Walsh, 2005, 2007; Santos, 2010). The term 'interculturality' should not be understood as a simple contact, but as an exchange that takes place in conditions of equality, mutual legitimacy, equity and symmetry. This encounter of cultures is a permanent and dynamic vehicle for communication and mutual learning. It is not just an exchange between individuals, but also a meeting of knowledge, wisdoms and practices that develop a new sense of co-existence in their difference.

As suggested by Viaña (2009), to achieve this, it is necessary to change the conditions of intercultural dialogue, to ensure that the conversation is not about the right of inclusion in the dominant culture, but about the historical and structural factors that limit a real exchange between cultures in each country. Only this can help create the conditions for more symmetrical conversations about the model of development needed for *Buen Vivir*, the type of solidary economy needed for life and the participatory political system needed for the consolidation of autonomies, territories and regions that seek different forms of government and self-governance.

Thus, the 'inter' space becomes an arena of negotiation where social, economic and political inequalities are not kept hidden but are made visible and confronted. Therefore, a starting point for such intercultural practice is to develop a politics of knowledge that helps strengthen Indigenous peoples' own initiatives and agendas of cultural revitalisation and knowledge production.

3   The Role of Cultural Revitalisation in Environmental Restorative Justice

Repeatedly, in my work with Indigenous peoples in Latin America over the last 20 years, I have heard my Indigenous colleagues and collaborators saying: 'We have to unearth our own history', 'We have to rescue/revitalise our traditional rules and norms of governing nature' and 'We want to teach public officials who we are: Our history, Our ways of life, Our forms of government'. These desires are an expression of the deeper and more invisible layers of environmental conflicts. Most of the Indigenous peoples I have worked with know they must produce changes in the dominant structures and social relations that reproduce social exclusion and marginalisation in development and environmental conservation. But they are very much aware too that domination is also expressed through processes of cultural violence that erase knowledge and displace identities. Thus, resisting such processes through cultural reaffirmation is key.

Furthermore, they know that the possibilities of developing symmetrical intercultural dialogues of knowledge with other actors in conflicts, to teach them about who they are and thus repair any harms that have been done to them, is dependent on them developing first their own internal dialogues: their own processes of knowledge revitalisation to reconstitute their identities, reconnect with their sense of place and restore their place in history.

In his book *The Moral Imagination: The Art and Soul of Building Peace*, John Paul Lederach (2005 original, 2008 Spanish translation) captured very well the key role of re-stor(y)ing in restorative justice. Contemporary environmental conflict analysis and interventions have a strong present time bias, which can contribute to erasing collective identities. A much longer term timeframe that gives a voice to history and counter-narratives is essential for preventing and reverting this process (see Fig. 21.1):

From the perspective of indigenous peoples, original violence can be better understood by viewing it as the disintegration, and on too many occa-

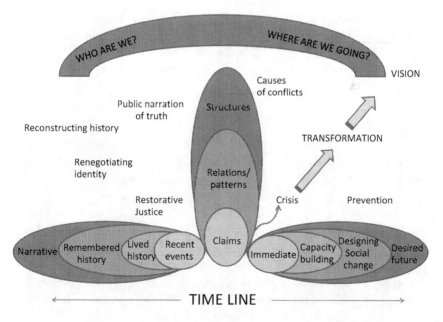

Fig. 21.1 Peacebuilding timeframe. (Adapted from Lederach (2008))

sions—the direct destruction, of the history of a people. These patterns are found on all continents and in the history of all aboriginal groups … You cannot go back and redo history. But that does not mean that history is static and dead. History is alive and requires recognition and attention. The challenge is how today can the people re-history … The narrative has the capacity to create, even to heal, but its voice has been frustrated. Narrative needs to be given a place and a voice again. (Lederach, 2008, p. 214)

As also pointed out by Lederach in an older text, 'identity shapes and moves the manifestation of conflicts. At the deepest level, identity is housed in narratives that describe how people see themselves, who they are, where they come from, and what they fear to be' (Lederach, 2003, p. 6). Thus, it is not possible for Indigenous people to think of an 'other' future where the *culture of life* and *many worlds are possible*, without reconnecting with their identity. This requires going beyond recent events as experienced in conflict episodes or crisis, where claims are generally made, and digging into deeper layers of the past, which involve touching upon the lived history, remembered history and even the wider past narrative, such as the myth of origin of Indigenous peoples. As said to me by an Indigenous leader in Venezuela once: 'we can't know where are going if we don't even have clarity of who we are'.

Such a longer peacebuilding timeframe involves paying attention to not only to the restorative process that can be put in place to undo wrongs caused to individuals and groups in environmental conflicts (such as compensations, new laws, institutional arrangements, programmes) but also to processes of historical reconstruction and public narration of the truth that are key for renegotiation of the identity of a social group and which can give it a stronger footing for imaging and crafting a more just future. Such a process requires a long-term commitment

to create multiple access points and a repetitive examination in order to address the issue of identity … Working with identity is not an immediate or instantaneous decision-making process, but a process of continuous learning about oneself and the other. This requires an insistent platform to detect and address identity concerns, within a broader framework of constructive change. (Lederach, 2003, p. 6)

In Latin America, there are valuable experiences in recovering the historical memory of Indigenous peoples, made by the protagonists themselves, as part of strategies aimed at confronting the dominant development model and its tendency to erode and erase the identity of entire peoples. A case in point was the project to recover the historical memory of the Talamaqueño people in Costa Rica, led by the American historian Paula Palmer in the 1980s (Palmer, 1994). The project sought to document the socio-economic changes experienced by the people in the region as well as conflicts with the state as lived and experienced by the Talamaqueño peoples themselves (Quezada, 1990). Recently in Colombia, the Muinane Indigenous People underwent a similar process, which culminated in a self-authored history book (Ancianos del Pueblo Fééneminaa, 2017).

Many Indigenous peoples in Latin America are making links between their past, present and future through the definition of their Life Plans (Planes de Vida), helping them to look ahead by reconnecting first with their past and their identity (Cabildo de Guambia, 1994; Jansasoy & Perez-Vera, 2006; COINPA, 2008; Espinosa, 2014), a perspective that holds great similarities with John Paul Lederach's long-term framework for conflict transformation (Box 21.1).

Box 21.1 Life Plan Definition

The 'plan de vida' is a plan made by Indigenous organisations and communities in an effort to survive and to maintain traditions, customs and the hope of having a society with its own identity based on the traditional knowledge of its people. It is a means of guaranteeing better conditions and a better quality of life for Indigenous communities. However, it is also a document to be used in negotiations with both the regional and national government. It includes the issues of health, education, territory, the environment, natural resources, the economy and production, government, justice, youth and women's and gender issues, amongst others (Perez, 2009).

In the case of socio-environmental conflicts, reconstruction of local stories is also key in helping clarify disputes over environmental and landscape changes, which are often and simplistically attributed to local practices (see, e.g. Rodriguez et al., 2014). Thus, affirming history from the local perspective can play an important role in developing environmental counter-narratives and counter-histories, which, in turn, by helping to change the collective way of thinking and seeing the environment, can help revalue and revitalise local knowledge and identities more broadly.

4   Doing Environmental Restorative Justice with a 'Decolonial Turn': Two Examples

My work with Indigenous peoples in Latin America on environmental justice and conflict transformation over the past 30 years has been largely driven by the imperative of my Indigenous colleagues to help revitalise the culture of their people. I share two examples in which we have used collaborative action research to respond to this need as part of a long-term collaboration framework.

4.1   Who Are We, Where Do We Come From and Where Are We Going? Developing a Pemon Life Plan

The first example is set in Canaima National Park (CNP) and UNESCO Natural World Heritage site, located in the south-eastern Venezuela. An area of 30,000 km of exceptional natural beauty, home to the Pemon Indigenous people, CNP has a long history of socio-environmental conflicts, which spring from the fact the park was originally created without prior consultation with the Pemon, thus disregarding their authority, historical presence and knowledge systems.

I started working in CNP in 1995, coordinating a conflict resolution project for a national non-governmental organisation (NGO). My frustration at seeing only minor progress in fostering dialogue between the Pemon and the park managers forced me to take a step back in 1997 and try to understand conflicts better before attempting to engage with them. Later, in 1999, I went back to CNP to carry out my PhD fieldwork.

When I approached the village chief of Kumarakapay, Juvencio Gomez, to seek permission to base myself in his village to study and analyse conflicts in the National Park, his response was:

> You can stay as long as you help me and my village reflect about who we are and who we want to be in the future. We—the Pemon—don't know where we are going because of all the projects that are being imposed on our territory; we are totally disoriented. We need to be clear about who we are and who we want to be as a People.

I accepted the challenge of working together to design and develop this process of self-reflection, alongside my wider research objectives.

As an Indigenous leader, Juvencio Gomez had participated in several international Indigenous rights forums and heard that Indigenous peoples in Colombia were conceptualising development in their own terms through the construction of Life Plans (Planes de Vida). Thus, the Colombian experience became an inspiration and a path to follow in the search for a self-defined society and future. It was Juvencio Gomez's view that the starting point for conceptualising this local development pathway was to confront themselves with their identity, reconnecting with their past and analysing their current situation. Promoting and constructing this process of personal and collective confrontation with cultural identity became our joint endeavour.

The result was a year-long participatory process of self-reflection which we conducted in conjunction with a group of approximately 30 people (about half elders and half youth). We focused on discussing and researching the following topics:

* Who are we? The origin of the Pemon according to mythological beliefs.
* Where do we come from? Historical and ancestral settlement areas.
* Community history: important historical figures, events, foundation processes of the village.
* How has our community and territory changed over time? Discussion of social and environmental change.
* Things that we need to solve to improve our living conditions and environment.
* Views of development.
* Good and bad things of the past and of the present.
* Vision of a desired future.

Community meetings and workshops were used to adapt the process of reflexivity and enquiry to the deliberative, oral-based decision-making structure of Pemon society (Thomas, 1980). Different participatory tools were used in these meetings and workshops, including oral testimonies, timelines, territory and community mapping, matrices, brainstorming, group and plenary discussions. When necessary, assemblies were held to validate information and have the community's consent to continue carrying out the process of self-enquiry.

Reflexivity about the origin, past, identity, changes and current situation of the Pemon was crucial for grounding a view of the desired future. The discussion about the future addressed three main questions: (a) How should the Pemon from Kumarakapay be in the future?; (b) What has to change to achieve this? and (c) What can we count on in order to achieve this change? These three questions were then used as a base to collectively construct a vision of the ideal type of society that the Pemon from Kumarakapay wanted to have. The result of this discussion can be found in Box 21.2.

Box 21.2 The Type of Society That the Pemon of Kumarakapay Want to Have (Source: Roroimökok-Damük (2010))

- A Pemon society with awareness of who we are and with a sense of identity and of belonging.
- Knowledgeable about our history, culture, tradition and language.
- Owners of our land—territory, knowledge, culture and destiny.
- A society educated with ancestral and modern knowledge.
- A society that values its wise people (parents and grandparents).
- A respectful, hard-working, obedient, kind, courteous, cheerful, generous, harmonious, understanding society where there is love.
- A productive, autonomous society.
- A society that defends its rights and is ready to confront pressures from Venezuelan society.

The varied reflexive processes carried out had different types of impacts. First, reconstruction of Pemon historical roots (in terms of both mythological and factual history) made both elders and youth revalue the importance of their cultural heritage. Important aspects of their past that were starting to be erased from the community's oral memory were discussed and made visible through this process. As a result, the elders issued a request to the younger generations to put this history into writing in order to make it known to the younger Pemon and wider Venezuelan society. This request effectively materialised in 2010 (see below). Secondly, reflections on their situation and socio-environmental changes led the Pemon from Kumarakapay to consider their livelihood potential and options for the future. A critical issue they face is food security. The shift in the settlement pattern experienced since 1950, from semi-nomadic to a permanent village, is depleting their farming land. Taking this situation into consideration, in 2000 the Pemon of Kumarakapay decided to focus on becoming a tourist community. Since then, they have been training more actively in this activity.

Under their own initiative, the inhabitants of Kumarakapay started undertaking a series of activities to revalue their identity, such as reconstructing the Pemon calendar, carrying out educational workshops, cultural activities, fairs of the Pemon culinary culture and native sports competitions.

With further external assistance between 2000 and 2004, they implemented a project for self-demarcation of the Pemon territory in the eastern part of the CNP (Sletto, 2009). This has now been issued as part of a formal claim for territorial property rights, which is still pending. Following the request of the elders made back in 1999, in 2008 we started work updating the material compiled in 1999 and 2010, the life plan team published a book entitled 'The History of the Pemon of Kumarakapay' (see Image 21.1), which is used in schools and other communities of the *Gran Sabana* as a guide for developing Life Plans (Roroimökok-Damük, 2010). This book expresses the need to reconstruct the past and revalue the Pemon identity to be able to visualise a desired future.

By putting their history in writing, the Pemon of Kumarakapay became more visible, showing the wider society that they exist as a People, with their own knowledge, language, culture and traditions, and developed local commitment to collectively start building their desired future. As a result of this publication, the community began a new series of cultural reassertion activities, such as workshops and seminars on community philosophy, with the guidance of elders (grandmothers and grandfathers), in order to orient their development, education and organisational-building.

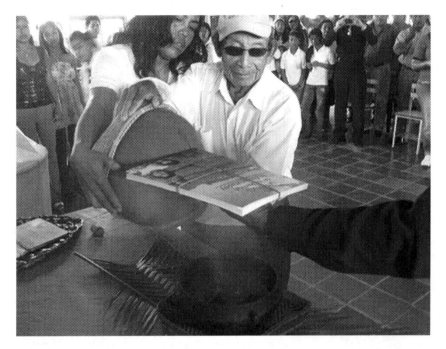

Image 21.1   The *History of the Pemon from Kumarakapay* book launch December 2010, Kumarakapay village, Canaima Nationa Park. (Photo © Iokiñe Rodriguez)

The Pemon performed a Chiuka ritual when launching their book in Kumarakapay in 2010 in a gathering with community members, national authorities and friends. The Chiuka ritual is performed to protect the newborn babies, ideas, projects or objects from evil spirits and to ensure their mission in life is fulfilled.

With a stronger and clearer vision of development and the future, in the years that followed, the Pemon from Kumarakapay became more assertive in their relations with national authorities and in negotiating projects for the area and enhanced their capacity for public deliberation with environmental managers regarding pressing issues of CNP management, such as the use of fire, a highly contested local practice (Rodriguez et al., 2013).

The process of cultural reaffirmation experienced in Kumarakapay was also key in reasserting Pemon cultural identity at a wider level, particularly with regard to the need to advance the construction of a Pemon Life Plan. Between the years 2000 and 2015, the Life Plan gradually came to be acknowledged as a platform for intercultural dialogue with external actors about the current and future wellbeing of the Pemon, as well as for articulating different institutional agendas (Rodriguez, 2016).

More recently, however, tensions with the national authorities have escalated as an important faction of Kumarakapay started to distance itself from national government and mark its opposition and resistance to national projects and plans that pose great threats to their physical and cultural survival (these include expansion of mining in southern Venezuela, but also other time- and context-specific policies, such as blocking the entrance of humanitarian aid to Venezuela in 2018–2019 during a period of great food and medicine scarcity). This has led to recent wave of physical violence and criminalisation against the Pemon, the most violent of which was a military attack by the National Guard in February 2019 against a group of Pemon who were protesting the blocking of humanitarian aid by the national government at the Brazilian border. The attack led to the death and injury of 6 and 13 village members, respectively. This attack, along with subsequent acts of violence and repression, ended up displacing 900 Pemon to Brazil, many of whom were the authors of the *History of the Pemon from Kumarakapay*, including Chief Juvencio Gomez and his family.

Chief Juvencio and his team, however, have not given up in their effort to keep their culture, sense of place and identity alive. In exile, Juvencio Gomez and his wife, Yraida Fernandez, are now working on a revised, second edition of the *History of the Pemon of Kumarakapay* in collaboration with the INDIS (Indigenous Interactions for Sustainable Development: https://indisproject.org/) Project from the School of International Development at the University of East Anglia, UK (where I now work), which is being written as a reminder to the Pemon youth, to always keep fighting to defend their right to their culture and territory. They are also part of a video production called *Pemon*, led by Apropos Productions Ltd[3] with Funding from Doc Society[4] and as a part of a col-

---

[3] See https://www.aproposltd.net/ (last accessed 22 January 2022).
[4] See https://docsociety.org/ (last accessed 22 January 2022).

laboration with DEV/UEA, which seeks to help raise international awareness of the destruction of the Pemon culture. Furthermore, a year after of having been displaced to Brazil, a portion of the Pemon from Kumarakapay living in exile established the Association of Indigenous Migrants of Roraima, which aims to promote, preserve and celebrate the Pemon Indigenous culture, no matter the country in case, by producing and selling their traditional crafts.

4.2   'For the First Time We Are the Protagonists of Our Own History': Using Cameras to Re-reconstruct the Monkoxi History

The second example is that of the Monkoxi people of the Indigenous Territory of Lomerio, Bolivia, with whom my Bolivian colleague (Mirna Inturias) and I have an ongoing collaboration since 2013.

The communal Indigenous territory (TCO) of Lomerio is an area of 256,000 hectares located in the department of Santa Cruz, in the lowlands of Bolivia, legally owned and managed by the Monkoxi Indigenous peoples since 2006.[5] Despite the success in obtaining territorial property rights, the collective management of this vast territory is complex due to the diversity of actors and public policies pressuring for access and use of its rich natural resources.

For the last decade, the Indigenous Organisation of the Native Communities of Lomerio (CICOL), the legal authority over the territory, has been experiencing great difficulty in regulating natural resource management in the area. Despite the recent shift to a Pluricultural National State, public policies in Bolivia to support community environmental management models are still weak. Some communities enter into very unfavourable negotiations with timber companies for forestry activities without following CICOL's recommendations. The government on its side is promoting mining and incentivising cattle grazing without

---

[5] In 2009, Bolivia changed its national constitution to become a 'Plurinational Nation-State' that acknowledges differentiated rights for Indigenous peoples (Article 2). TCOs are legally owned Indigenous territories that resulted from this intense period of mobilisations. There are currently 190 TCOs in Bolivia, covering 20.7 million hectares.

undergoing free, prior and informed consent procedures or involving CICOL in local territorial planning. What is perhaps more worrying to the Monkoxi leaders and elders is that, despite the success in obtaining legal rights over their territories and forests, the younger Monkox generations are oblivious of how the conquest of the land took place and know very little about their own history. The Monkoxi elders and leaders feel an urgent need to help remember the past, in order to revitalise their identity, and have a clear view of a desired future amongst the younger generations.

In 2013, the Universidad NUR from Santa Cruz, Bolivia, and the School of International Development (DEV) from the University of East Anglia (UEA) started a collaboration with CICOL to assess and initiate a community-wide reflectivity process about community tensions over the management of their territory.

Participatory videos were used in conjunction with other ethnographic methods, such as interviews, participant observation and secondary data collection, as a tool to examine local notions of environmental (in)justice in community forestry and to help give public visibility to environmental justice concerns. CICOL chose four of its members (three young women and a member of the council of elders) to work as community researchers in the project for a year, hand-in-hand with two external researchers/facilitators (the authors of this paper).

Participatory videos were made entirely by village members through a process that involved:

* Learning by practising: overcoming the fear of cameras through games.
* Participatory analysis (through the use of a variety of participatory assessment tools used to help creating the story, e.g. time lines, community mapping, problem trees, Venn diagrams, thematic pictures).
* Creating a storyline.
* Filming, screening and editing.

Three participatory videos were made. The first one, titled 'On the road to freedom: the History of the Monkox People', focuses on reconstructing the long struggle of the Monkox people of Lomerio to obtain territorial rights over their lands. It was carried out by a team of ten people from CICOL. The remaining two videos, titled 'The forest is our life, our home' and 'Our forest, our development', were carried out by community members from two small villages (Todos Santos and Santo Rosario, respectively) and focus on experiences of (in)justice in community forest management.[6]

The making of the history video in particular revealed significant concerns with regard to the loss of identity amongst the Monkoxi youth. As mentioned before, one of the biggest concerns of many Monkox leaders and elders is that current young generations do not value the heritage they have received from those who fought for a very long time to obtain legal rights over their territories and natural resources. They perceive that younger generations are responsible for a generalised lack of local governance in community forest management and, worse of all, that they are being left out of benefit sharing from community forestry.

It is with this concern in mind that the research team, in conjunction with CICOL's Board of Directors, decided to devote the participatory video to reconstructing the long struggle for liberation and territorial rights in Lomerio. The first half of this video captures, through the living testimonies of the elders, the long history of oppression experienced by the Monkox ancestors since the establishment of Jesuit Missions in the Bolivian lowlands (seventeenth century), followed by the rubber boom of the late nineteenth and twentieth centuries, when the Monkox were brutally exploited as forced labourers on plantations. Elders describe their own parents as having been slaves and subjected to exploitation by large landowners, which continued into the late twentieth century, even after they had escaped the mission towns and (re)established in their territory.

The second half of the video recounts the process of liberation experienced by the Monkox people in this last century, as initiated first through the agrarian and educational reforms in the late 1950s, and more recently since the 1990s, through new legislation and structural political reforms that acknowledge Indigenous peoples' differentiated rights, including amongst others, property rights to their collective territories through new

---

[6] Links to the videos can be found here: 'On the Road to Freedom' https://www.youtube.com/watch?v=AdeWZXFqcWQ, 'Our forest, our development' (Santo Rosario Community) https://www.youtube.com/watch?v=tTBIgbx3wkQ, 'The forest is our life, our home' (Todos Santos Community) https://www.youtube.com/watch?v=tuUzTfFH6fM.

figures like the TCOs. In this latter part of the video, attention is paid to explaining the contribution that the lowland Indigenous movements, including CICOL, had in making these political reforms possible. The video production team also devoted considerable time to reconstructing the different processes that CICOL had to undergo to obtain the territorial rights for the Lomerio TCO.

Once finished, the final version of the video was shown and discussed in general and with community assemblies and is being used by CICOL as a dissemination tool in meetings with other Bolivian Indigenous peoples related to claims for territorial autonomy. It has also been posted on YouTube for wider dissemination.

Shortly after finishing the video, CICOL decided to put the history narrated in the video into writing thorough a community authored book that is now used as part of their communication strategy to advance their claim for territorial autonomy (Peña et al., 2016). In 2018, the book was translated into English and is now used as part of CICOL's international dissemination strategy.[7] In 2020, as part of a new NUR, UEA and CICOL research collaboration through the INDIS project, we initiated work to continue developing strategies to revitalise the Monkoxi identity. In 2021, the book was translated into Besiro language and is going to be used in local schools as part of a wider cultural revitalisation strategy. We also initiated activities to help the youth reconnect with their territory and identity through photography. This time, we used the Photovoice method to develop Youth Stories about autonomy, identity, health and climate change action.[8]

Thus, work in identity revitalisation with CICOL has been continual since 2013. Similar to the Pemon, the Monkoxi know that knowledge and identity revitalisation must be an ongoing endeavour. The history video and book in particular have been very valuable strengthening the self-esteem and dignity of the Monkoxi, particularly of the CICOL members, who found a way of making their story of struggle for liberation and Indigenous rights known to the younger generations and the

[7] Retrieved from: https://indisproject.org/the-history-of-the-monkoxi-nation-in-lomerio-bolivia-book-now-available-in-english/ (last accessed 22 January 2022).
[8] A book that brings together the result of the process can be accessed at: Fotovoz_Reconexion_Monkoxi.pdf (uea.ac.uk).

general public. Recounting and making public the long and arduous process behind gaining territorial rights have been important for CICOL's legitimacy as territorial authority of the TCO. Most importantly, as said by the General Chief of CICOL, Anacleto Peña, 'for the first time we are the protagonist of our own history, and we have been able to tell the story ourselves, not someone from the outside. That is why "we" are the narrators and not some external person talking about "them"' (Rodriguez & Inturias, 2016, p. 44).

On a different level, the experience of participatory video and their mobilising potential within the community has prompted CICOL to start using participatory videos as an education tool in schools and social media to revitalise and document knowledge and skills of the communities in their everyday life as present in their oral history, language, practices and relationship with nature. As the Monkoxi people are predominantly an oral culture, they require tools like the videos and cameras to document their knowledge and cultural heritage. Due to their accessibility, videos and cameras can be used by any community member regardless of his/her level of education (see Image 21.2), allowing to rescue the voices and different histories of elders, women and youth, thus decolonising knowledge (Quijano, 2000). These local experiences are being shared with other communities, Indigenous peoples, public policymakers and officials, and the 'scientific community' at different levels: local, national and global, playing a role in the construction of intercultural dialogues.

5   Final Reflections and Lessons

Four key lessons can be drawn from these case studies in terms of the role of cultural revitalisation and action research in environmental restorative justice. The first one is that an environmental justice approach that places a focus on restor(y)ing history and identity, on its own, is no panacea. Doing environmental justice under the conditions of ongoing uncertainties, continual pressure and huge power asymmetries faced by Indigenous people in their territories is a mammoth task. Indigenous peoples have the arduous challenge of having to develop, simultaneously, strategies to

Image 21.2 Members of the CICOL Council of Elders filming the History Video, Lomerio, 2015. (Photo ©Iokiñe Rodriguez)

decolonise economic and political structures, human relations, dominant narratives, values and worldviews and the self (Rodríguez & Inturias, 2018), to succeed crafting an 'other' future where the *culture of life* and *many worlds are possible.* The Pemon case study, in particular, shows the fragility of processes of cultural revitalisation when the structural relations of power remain unchanged.

What such a process can help with, however, is in awakening and restoring the sense of dignity that is necessary to imagine and raise aspirations about change and an 'other' future. In turn, this can help redefine social consensus over norms and behaviour in a social group that are necessary to reshape conditions behind decision-making, both internally and in their relationships with others. Eventually, when the conditions are right, this can lead to changes in legal and political frameworks that are necessary for restoring justice at a more structural level.

In both case studies discussed, I have been able to witness over time how the efforts to revitalise cultural identity have had clear effects at a personal level for decolonising the self, strengthening the self-esteem and dignity and increasing the deliberation and leadership capacity over controversial environmental issues of those involved in participatory research processes. At the interpersonal level, I have also seen how the deliberation and reflective processes carried out have contributed to strengthening intergenerational bonds at the community level and helped to clarify inter-community conflicts. In the relationship with external actors, they have opened important opportunities for the articulation of new knowledge networks with sectors of academia that have been key in continuing to strengthen local knowledge systems or to legitimise them. In the case of the Pemon, the later has led to the emergence of environmental counter-narratives or counter-histories, which have started to change how the Pemon are referred to, or perceived, by environmental policymakers in CNP. In the case of the Monkoxi, the increased visibility gained by unearthing and telling their history has helped to build stronger links and networks with policymakers that have been key in pushing forward structural reforms to advance their new claim for political autonomy. Thus, although the revitalisation of culture does not instantaneously restore justice at all levels, it can be an important catalyst for change at multiple levels.

The second lesson is that, despite the fact Indigenous people have great expertise developing resistance and mobilisation strategies in environmental justice struggles to impact structural power, the same cannot be said about strategies to revitalise cultural identity, particularly when intergenerational conflicts are the norm. If anything, the two case studies show that Indigenous peoples have a great need for external collaboration to help revitalise identity and knowledge. The methodological and facilitation know-how developed through the action research processes created opportunities for reflectivity, joint-learning and knowledge revitalisation amongst the Pemon and Monkoxi that would not have emerged otherwise.

The third lesson is that taking part in this process as external collaborators requires a particular positionality as researcher. On the one hand, it requires us to take a back-stage role, one in which rather than leading the research process, we help others to do their own research. On the other hand, given that, as said by Lederach (2003, p. 6), 'working with identity is not an immediate or instantaneous decision-making process, but a process of continuous learning about oneself and the other', it requires from us researchers a long-term commitment to revisit and revitalise identity over time with our Indigenous collaborators. In both cases, having a sustained collaboration with my Indigenous partners has been key not only in developing different avenues to explore the tensions around identity, but also in developing as we go, new methodological approaches or outputs that can be of local relevance, such as the production of community authored books, videos, films or photographic exhibitions.

The fourth lesson is that, beyond the contributions at the community level restoring identity, action research can also play an important part helping to build intercultural relations with others as part of a long-term agenda for transformation of conflicts. But this requires taking interculturality seriously. As Catherine Walsh (2005, p. 45) says, building intercultural relationships 'is not solely a matter of acknowledging, discovering or tolerating the other or cultural difference. It is neither about making identities static. It is about actively promoting processes of exchange that allow building spaces of encounter among different beings, knowledge, logics and practices'. This requires an openness of mind on the part of academics to critically consider the what, why and what for, of knowledge production, in order to ensure its local relevance. This involves paying much more attention to creating the conditions for dialogue than is normally acknowledged and to giving visibility to identities and knowledge that have been made invisible by ongoing processes of epistemic violence. In both cases, the commitment to producing research outputs and organising engagement activities that can help give visibility to the local history, identity and knowledge as part of the research process has been key in levelling power relations with other actors, and in some cases in producing important structural changes, such as the progress made by the Monkoxi in recent years in their claim for territorial autonomy.

I. Rodriguez

References

Acosta, A. (2013). *El Buen Vivir: Sumak Kawsay, una oportunidad para imaginar otros mundo*. Barcelona: Icaria editorial.

Alimonda, H. (2011). La colonialidad de la naturaleza: Una aproximación desde la Ecología Política Latinoamericana. In H. Alimonda (Ed.), *La naturaleza colonizada: Minería y ecología política en América Latina* (pp. 21–58). Buenos Aires: CLACSO.

Ancianos del Pueblo Fééneminaa (2017). *Ancianos que caminan y cuentan historias*. Colombia: Consejo Regional Indígena del Medio Amazonas (CRIMA) y Forest Peoples Programme (FPP).

Cabildo de Guambia (1994). *Plan de Vida del Pueblo Guambiano*. Popayan: Colombia.

COINPA (2008). *Plan de Vida Pueblos Huitoto e Inga*. Documento de avance. Colombia: Consejo Indígena de Puerto Alegría (COINPA).

Castro-Gómez, S., & R. Grosfoguel (Eds.) (2007). El giro decolonial *Reflexiones para una diversidad epistémica más allá del capitalismo global*. Bogotá: Siglo del Hombre Editores.

Escobar, A. (2003). "Mundos y conocimientos de otro modo": El programa de investigación de modernidad/colonialidad Latinoamericano. *Tabula Rasa. Bogotá – Colombia*, 1, 51–86.

Escobar, A. (2008). *Territories of difference: Place, movements, life, networks*. Durham and London: Duke University Press.

Escobar, A. (2011). América Latina en una encrucijada ¿Modernizaciones alternativas, postliberalismo o postdesarrollo? *REVISTA CONTROVERSIA*, (197), 9–61. Retrieved from: https://doi.org/10.54118/controver.vi197.789 (last accessed January 25, 2022).

Escobar, A. (2016). Desde abajo, por la izquierda y con la tierra: La diferencia de Abya Yala/Afro/Latino/America. *Intervenciones en estudios culturales*, 2016, (3), 117–113.

Escobar, A. (2018). *Designs for the pluriverse: Radical interdependence, autonomy, and the making of worlds*. Durham, North Carolina: Duke University Press.

Espinosa, O. (2014). Los planes de vida y la política indígena en la Amazonía peruana. *Anthropologica* [online], 32 (32), 87–114.

Fals Borda, O., & Brandao, C. (1986). *Investigación participativa*. Montevideo: Instituto del Hombre.

Fals Borda, O. (1986). *La investigación-acción participativa: Política y epistemología*. In G. Camacho (Ed.), *La Colombia de hoy* (pp. 21–38). Bogotá: Cerec.

Gudynas, E. (2010). La senda biocéntrica: Valores intrínsecos, derechos de la naturaleza y justicia ecológica. *Tabula Rasa. Bogotá – Colombia, 13*, 45–71.

Gudynas, E. (2011). Desarrollo y sustentabilidad ambiental: diversidad de posturas, tensiones persistentes. In A. Matarán & F. López (Eds.), *La Tierra no es muda: diálogos entre el desarrollo sostenible y el postdesarrollo* (pp. 69–96). Granada: Universidad de Granada.

Gudynas, E., & Acosta, A. (2011). La renovacion de la crítica al desarrollo y el buen vivir como alternativa. *Utopia y Praxis Latonoamericana* 16 (53), 71–83.

Harvey, D. (2004). The 'new' imperialism: Accumulation by dispossession. *Socialist Register* 40, 63–87.

Jansasoy, J., & Perez-Vera, A. (2006). *Plan de Vida: Propuesta para la supervivencia cultural, territorial y ambiental de los Pueblos Indigenas*. Washington: The World Bank Environment Department. Retrieved from: https://acervo. socioambiental.org/sites/default/files/documents/C2D00026.pdf.   (last accessed 25 January 2022).

Lander, E. (ed.). (2000). *La colonialidad del saber: Eurocentrismo y ciencias sociales*. Buenos Aires: CLACSO.

Lederach, J. P. (2003). *"Conflict Transformation". Beyond Intractability*. Retrieved from: https://www.beyondintractability.org/essay/transformation   (last accessed 12 January 2022).

Lederach, J. P. (2008). *La imaginación moral: El arte y el alma de construir la paz*. Buenos Aires: Ediciones Norma.

Leff, E. (Ed.). (2001). *Justicia ambiental: Construcción y defensa de los nuevos derechos ambientales culturales y colectivos en América Latina*. México: UNEP.

Leff, E. (2004). Racionalidad ambiental y diálogo de saberes: Significancia y sentido en la construcción de un futuro sustentable. *Polis (Revista de la Universidad Bolivariana, Santiago de Chile)*, 2 (007), 0–28.

Leff, E. (2015). The power-full distribution of knowledge in political ecology: a view from the South. In T. Perreault, G. Bridge, & J. McCarthy (Eds), *Routledge Handbook of Political Ecology* (pp. 64–75). Routledge: London & New York.

Mignolo, W. (2009). Epistemic disobedience and the decolonial option: A manifesto. *Theory, Culture & Society*, 26(7–8), 1–23.

Palmer, P. (1994). Self-history and self-identity in Talamanca. In C. D. Kleymeyer (Ed.), *Cultural expression and grassroots development. Cases from Latin America and the Caribbean* (pp. 39–55). Colorado: Lynne Riener Publishers, Inc.

Peña, A., Tubari, P., Chuve, L., Chore, M., & Ipi, C. (2016). *Historia de Lomerio: El camino hacia la libertad*. Universidad de East Anglia, Universida NUR, CICOL.

Perez, C. (2009). "El Plan de Vida"—What is a life plan. *Acción Colombia*. Colombia Support Network, Spring 2009, 3–4. Retrieved from: http://colombiasupport.net//wp-content/uploads/2012/02/CSN_Spring_09_Newsletter1.pdfm (last accessed 21 January 2021).

Perez, M. (2019). Epistemic violence: Reflections between the invisible and the ignorable. *El lugar sin límites*, 1 (1), 81–98.

Pilgrim, S., & Pretty, J. (Eds.) (2010). *Nature and culture: Rebuilding lost connections*. Oxford: Earthscan.

Quezada, J. R. (1990). Historia oral en Costa Rica: Génesis y estado actual. *Estudios sobre las culturas contemporáneas*, 3 (9), 173–197.

Quijano, A. (2000). Coloniality of power and Eurocentrism in Latin America. *International Sociology*, 15, 215–232.

Rodriguez, I., Bjørn, S., Bilbao, B., Sanchez-Rose, I., & Leal, A. (2013). Speaking of fire: Reflexive governance in landscapes of social change and shifting local identities. *Journal of Environmental Policy Making and Planning*. DOI:https://doi.org/10.1080/1523908X.2013.766579.

Rodriguez, I. (2016). Historical reconstruction and cultural identity building as a local pathway to "Living Well" amongst the Pemon of Venezuela. In S. Whit, & C. Blackmore (Eds.), *Cultures of Wellbeing. Method, Place and Policy* (pp. 260–280). Hampshire: Palgrave Macmillan.

Rodriguez, I., Gasson, R., Butt-Colson, A., Leal, A., & Bilbao, B. (2014). Ecología histórica de la Gran Sabana (Estado Bolívar, Venezuela) entre los siglos XVIII y XX. In S. Rostain (Ed.), *Antes de Orellana* (pp. 113–121). Actas del III Encuentro Internacional de Arqueología Amazónica, Instituto Francés de Estudios Andinos, Facultad Latinoamericana de Ciencias Sociales, Embajada de EEUU: Quito, Ecuador.

Rodriguez, I., & Inturias, M. (2016). Cameras to the people: Reclaiming local histories and restoring environmental justice in community-based forest management through participatory video. *Alternautas* 3(1), 32–49.

Rodriguez, I., & Inturias, M. (2018). Conflict transformation in indigenous peoples' territories: Doing environmental justice with a 'decolonial turn'. *Development Studies Research*, 5 (1), 90–105.

Roroimökok-Damük, (2010). La historia de los Pemon de Kumarakapay. In I. Rodriguez, J. Gómez, & Y. Fernánde (Eds.), Instituto Venezolano de Investigaciones Científicas, Fundación Futuro Latinoamericano, Inwent y Forest Peoples Programme. Caracas, Venezuela: Ediciones IVIC.

Santos, B. de S. (2008). *Another knowledge is possible: Beyond northern epistemologies*. Verso: London.

Santos, B. de S., Arriscado, J., & Meneses, M.P. (2008). *Introduction: Opening up the canon of knowledge and recognition of difference.* In B. De Sousa Santos (Ed.), *Another knowledge is possible: Beyond northern epistemologies* (pp. vx–ixii). London: Verso.

Santos, B. de S. (2010). *Descolonizar el saber reinventar el poder.* Montevideo: Ediciones Trilce.

Sletto, B. (2009). Indigenous people don't have boundaries: Reborderings, fire management, and productions of authenticities in Indigenous landscapes. *Cultural Geographies* 16 (2), 253–277.

Smith, L. (1999). *Decolonising methodologies: Research and Indigenous peoples.* London: Zed books.

Thomas, D. J. (1980). Los Pemon. In R. Lizarralde, & H. Seijas (Eds.), *Los aborigenes de Venezuela: Etnologia contemporanea* (pp. 302–379). Caracas: Fundación La Salle de Ciencias Naturales.

Tubino, F. (2005). La interculturalidad crítica como proyecto ético-político. In *Encuentro continental de educadores agustinos.* Lima, 24–28 de enero de 2005. Retrieved from: https://oala.villanova.edu/congresos/educacion/lima-ponen-02.html (last accessed 12 January 2022).

Viaña, J. (2009). *La interculturalidad como herramienta de emancipación.* Bolivia: Instituto Internacional de Integración/Convenio Andrés Bello.

Visvanathan, S. (1997). *A carnival for science: Essays on science, technology and development.* London: Oxford University Press.

Visvanathan, S. (2009). The search for cognitive justice. *Seminar web edition.* (597). New Delhi. Retrieved from: https://www.india-seminar.com/2009/597/597_shiv_visvanathan.htm (last accessed 13 January 2022).

Walsh, C. (2005). Interculturalidad, conocimientos y decolonialidad. *Signo y pensamiento. Perspectivas y Convergencia*, 46 (24), 31–50.

Walsh, C. (2007). Interculturalidad y colonialidad del poder: Un pensamiento y posicionamiento "otro" desde la diferencia colonial. In S. Castro-Gómez, & R. Grosfoguel (Eds.), *El giro decolonial: Reflexiones para una diversidad epistémica más allá del capitalismo global* (pp. 47–62). Bogotá: Siglo del Hombre Editores.

22

Socio-environmental Harms in Chile Under the Restorative Justice Lens: The Role of the State

Daniela Bolívar, Liliana Guerra, and Felipe Martínez

1   Introduction

Could restorative justice be used to deal with environmental issues in Chile? Answering this question is a difficult task when looking at the broad picture: the economic, political, legal, and social context in which environmental harm takes place, which in the case of Chile—as in many countries of the Global South—is extractivism. Extractivism is a multi-scale process, involving the mobilisation of a significant amount and volume of natural resources, usually unprocessed (UN, 2015), and the specialisation of areas or territories towards producing one single type of product (Romero-Toledo, 2019). Extractivism relates to activities such as oil and mining, but currently the term refers to different industrial

D. Bolívar (✉) • L. Guerra
School of Social Work, Pontifical Catholic University of Chile, Santiago, Chile
e-mail: dbolivar@uc.cl; ldguerra@uc.cl

F. Martínez
Pontifical Catholic University of Chile, Santiago, Chile
e-mail: famartinez@uc.cl

activities in which natural resources, such as minerals, agriculture, or sea products, are massively exploited (UN, 2015). Unfortunately, as we review in this chapter, extractivism is a reality and a well-established policy not only in Chile but also in several countries of the Global South, making environmental harm and economic growth quite an inseparable duo. The fact that extractivism is promoted by governments, locates the state in a complex position when environmental conflicts arise.

This chapter has the goal, then, to discuss the role of the state in the context of extractivist policies when approaching environmental harm from a restorative justice perspective. Unfortunately, we cannot offer here clear and straightforward solutions, as the problem is highly complex. We offer instead some reflections and ideas based on what we believe could be useful dimensions to understand the state's role from a restorative justice perspective.

We first explain the notion of extractivism and its main characteristics, then move on to discussing elements of the Chilean context (legal and economic) in which environmental harm takes place. Later, we discuss the Chilean case viewed from premises of the restorative justice literature. We finally discuss what the role of the state could be, from a restorative perspective, in addressing environmental harms.

2   Setting the Context

2.1   Extractivist Policy in the Global South

It has been argued that extractivism or, more precisely, neo-extractivism is a 'comprehensive development project' (Burchardt & Dietz, 2014, p. 470) that has guided governments of different political colours in the Latin American region. Instead of international corporations obtaining the profits from exploitation of raw materials—as used to be the case decades ago—the current tendency is that governments, by nationalising goods or regulating the intervention of foreign companies, use the profits from such activities to encourage economic growth and social development (Burchardt & Dietz, 2014). The assumption is that extractivism could be key to promoting social development (Romero-Toledo, 2019), by increasing foreign investment and thus reducing poverty (Burchardt & Dietz, 2014). Economic benefits, in turn, offer social and political legitimacy to extractivist activities (Burchardt & Dietz, 2014; Romero-Toledo, 2019).

However, extractivism can be a harmful policy. Different studies have shown the interfering impacts of these policies on local communities, especially by excluding communities that are already marginalised (Romero-Toledo, 2019) and through causing significant environmental and social harm, as most of the time extractivism takes place where vulnerable populations and fragile environments are located (McNeish, 2012). Additionally, extractivism 'interrupts territorial integrity, disrupts local economies, destroys environments, and undermines local decision making' (Farthing & Fabricant, 2018, p. 10) affecting mainly Indigenous and Campesino habitants (Farthing & Fabricant, 2018). It increases poverty, marginalisation and inequality in the Latin American region (Schmink & Jouve-Martín, 2011).

This phenomenon grows as global demand increases, not only for raw materials and consumer goods (Svampa, 2012), but also for goods supporting a so-called sustainable life in industrialised countries. For instance, the quest for healthy diets increases demand for superfoods, which in turn fosters major production of monocultures. A concrete example: production of avocados in Latin America has started to take over forested areas and, in the case of Chile, abuse of water resources in an area where soils do not retain rainwater (Magrach & Sanz, 2020). Likewise, the demand for green energy in the Global North has contributed to environmental devastation in the Global South. In the case of lithium, a material required for electromobility and extracted in the north of Chile, extraction requires significant amounts of water resources (Jerez et al., 2021).

Chile's main productive activities are taking place within this context of an international ecology of production and economic exchanges. In the following section, we explain Chile's environmental legal context and then describe how extractivist policy is affecting Chile in particular. We finish with a concrete example on what happens when both the legal framework and extractivist policies come together.

2.2   Chilean Legal Context

Chile has ratified various international treaties related to environmental protection. Recent examples include the Kyoto Protocol (2002), the Stockholm Convention (2005), the Rotterdam Convention (2005), the Treaty N° 169 about Aboriginal Peoples and Tribal and Independent Communities (2010), the Paris Agreement (2017), and the Minamata Convention on Mercury (2018). Chile, at the time of writing, has not signed the Escazu Agreement,[1] even though it took a leading role in its preparation.

Considering all the normatives mentioned above, The Treaty N° 169 about Aboriginal Peoples and Tribal and Independent Communities has particular significance. This treaty has two basic principles. First, the importance of the right of Indigenous peoples to keep and strengthen their cultures, ways of life, and own institutions, and second, their right to participate effectively in decisions affecting them. The treaty also emphasises the right of Indigenous communities to define their own priorities in terms of development. This treaty obliges all (public or private entities) who make decisions that might have an impact on an Indigenous community to carry out consultation with such communities.

The most important national law regarding environmental issues is the *Law 19.300 Sobre Bases Generales del Medio Ambiente*. Promulgated in 1994, this law regulates the right to live in an environment free of pollution, as guaranteed in the Chilean constitution. Amended in 2010 by the law 20.417, this law created new institutions, such as *Superintendencia del Medio Ambiente* (Superintendent of Environment) and the *Ministerio del Medio Ambiente* (Ministry of Environment).[2] Thanks to the law 204.17, the environment is recognised as something the state must protect; therefore, punitive and reparative actions should be undertaken if

---

[1] This was the first regional agreement signed by 24 Latin American countries and the Caribbean. It was oriented to guarantee the right of all individuals to have timely access to information, participate in decisions that could affect their lives and their environments, and have access to justice when their rights are violated. The agreement resulted from the United Nations Conference on Sustainable Development (UNCSD) (Rio de Janeiro 2012) and executed by the United Nations Economic Commission for Latin America and the Caribbean (ECLAC).

[2] Environmental tribunals were created later, by the Law 20.600.

the environment is harmed (Corral, 2004). This law understands environmental harm as 'all type of loss, reduction, detriment or damage caused to the environment or to one or more components'.[3]

All these treaties and laws seem to indicate that Chile is making real and important efforts in dealing with environmental issues. Nevertheless, in practice there are significant problems and gaps.

First, the law establishes types of projects that require environmental assessment if they cause (or might cause) environmental harm (including risks to people's health). If a project type is not in this list, projects are required to present an environmental declaration, to state that no harm will be caused, accompanied by approval certificates from the sectors involved and sometimes supported by scientific advice. Both procedures have proved to be imperfect. On the one hand, environmental evaluation must include a citizens' consultation, which is expected to be a source of community participation. However, this mechanism occurs late in the process (Ocampo-Melgar et al., 2019), is perceived as not fully transparent by community members (Barría Meneses, 2019), and does not ensure citizens' opinions are seriously considered during the process (Höhl, 2020). Bigger problems appear when the citizens included in the consultation are Indigenous people. Some commentators have observed that consultation design does not consider cultural aspects of the participating communities (Barría Meneses, 2019), and 'the formal decision-making process is fragmented in such a way that Indigenous actors are excluded from the final decision' (Höhl et al., 2021, p. 764). On the other hand, environmental declarations have been used as a tool to avoid environmental evaluation. One way to do so is by splitting big projects that as a whole might have negative impact on a specific territory into smaller projects, presenting a declaration for every single one. This has been especially the case during the COVID pandemic 2020–2021. During this period, projects with citizen participation were frozen by the authority, increasing the number of projects solely using declarations. This has been seen by the Latin American Observatory of Environmental Conflicts

---

[3] Article 2 letter (e), Law 19.300.

(Observatorio Latinoamericano de Conflictos Ambientales, OLCA) as an alliance between corporations and government.[4]

Second, the General Law 19300 as well as other Special Laws that could be applied to specific environmental issues are oriented to offer compensation to the specific individuals harmed. This implies two specific problems: first, it does not consider harm against nature and it does not consider types of harms which are more difficult to quantify, for example, cultural loss of local communities.

Third, when a specific harm has been identified, a complaint can be made by individuals, municipalities, or the state through an institution called *Council for State Defence* (*Consejo de Defensa del Estado* (CDE)), thereby mixing public and private dimensions. The CDE, however, has limited tools to assess and understand environmental issues as its main role is to defend the general interests of the Chilean State, which are several and diverse in nature.

Fourth, the General Law and Special Laws, since they act under the logic of private law and, at the same time, deal with environmental issues from a compensatory perspective present one relevant difficulty to determine reparations: certainty.[5] The Chilean norm requires certainty of causality, namely, that a specific company has caused a specific effect. This matters when harm to nature is observed months or even years after a company's action, as some effects might be visible only after the passage of time. Let's imagine a river contaminated with substances from three companies, where such contamination develops slowly over time, making effects evident only after several years, through sickness that inhabitants of the region start developing. What happens if those effects can be observed only when the deadline has passed the status of limitations? When does that status 'start'? How can harm be repaired if most harm is already irreparable? How can harm be quantified? These are some of the gaps our legislation cannot answer satisfactorily.

---

[4] See report about abusive rise of projects in the system of environmental assessment in Chile during the pandemic: https://olca.cl/articulo/nota.php?id=107913 (last accessed 22 January 2022).

[5] Such certainty is a notion from civil law. It establishes legal civil responsibility when there is a clear and direct cause of a specific verifiable harm.

2.3 Extractivism and Environmental Issues in Chile

Significant extractive activities in Chile include mining, industrial fishing, aquaculture of salmonids, lithium extraction, hydroelectricity, agricultural monocultures, and forestry (Romero-Toledo, 2019; Mora-Motta et al., 2020). These activities have underpinned the Chilean economy (Chile Sustentable, 2016), especially since the end of the dictatorship (Montoya, 2014). Already in 2016, the Organisation for Economic Co-operation and Development (OECD) was warning that the Chilean economic model based on exploitation of natural resources was reaching a limit (Chile Sustentable, 2016).[6]

Chile has not been an exception in the region concerning the negative effects of extractivism generating pollution in urban areas, water shortages or water contamination, and increased vulnerability to climate change (Chile Sustentable, 2016; Farthing & Fabricant, 2018), as well as significant social problems. For example, the lithium industry in the *Salar de Atacama* (Chile's largest salt flat) has not only seriously reduced availability of water for the local population but has also caused the disappearance of native and vulnerable species, loss of cultural heritage of Indigenous population, and diminution of biodiversity. The companies involved are committed to compensating economically for these losses, but this does fail to compensate for the extent of the harm caused because it is incommensurable, traverses varying levels of harm and has ongoing impacts into the future (Garcés & Alvarez, 2020).

In addition to extractivist policies, a second phenomenon that has occurred in the country is the installation of (usually more than one) polluting industries in specific zones. These are called 'sacrificial zones', alluding to the fact that areas (environment and citizens) are sacrificed to favour the benefits of development. As this usually affects poor and marginalised areas with no political power, 'sacrificial zones' have been criticised as a discriminatory policy. 'Sacrificial zones' imply that specific communities have to deal with very high proportions of pollution and contaminating industrial waste, in their water and land, affecting their

---

[6] See http://www.chilesustentable.net/2016/07/modelo-economico-chileno-basado-en-recursos-naturales-se-agota-advirtio-la-ocde/ (last accessed 22 January 2022).

ecosystems, making people sick and increasing marginalisation and poverty even more.[7] In Chile, activists have identified five 'sacrificial zones' installed in the north, centre, and south of the country: Mejillones, Tocopilla, Huasco, Quintero-Puchuncaví, and Coronel.

2.4   Sacrificial Zone Meets Chilean Law: A Concrete Example

A good example that shows both how 'sacrificial zones' work and the deficiencies of Chilean legislation is what occurs in the industrial area of Quintero-Puchuncaví, region of Valparaiso. In this area, different industries (including a copper refinery and carbon thermal power stations) were installed from the 1960s on. The environmental conflict arose in 2018 when the inhabitants of these two towns—including several children who attended school in the area—started to present symptoms, such as nausea and vomiting. After initial investigations, the presence of high levels of sulphur dioxide, methyl chloroform, nitrobenzene, and toluene was observed in the area. Unfortunately, the limitations of the legislation became evident during subsequent legal procedures. Not only was it impossible to establish specific responsibilities, but also was evident the absurdity of the regulation making industries responsible for monitoring their own emissions.[8] At the end of the legal processes,[9] the supreme court established the negligence of the Chilean State (delayed and reactive reactions) and the limitation of Chilean regulations to protect the environment and citizen health. In addition, the court ordered the companies to execute a decontamination plan, which has still not been implemented. No other sanctions were made. In this case, companies actually never acted against the law. They caused harm, however,

---

[7] See https://chile.oceana.org/zonas-de-sacrificio-0 (last accessed January 2022).

[8] See https://www.indh.cl/empresas-y-ddhh-zonas-de-sacrificio-y-conflictos-socioambientales-vulneratorios-2/ (last accessed 22 January 2022).

[9] Due to the toxic gases produced in the area, different individuals, NGOs, municipalities, and other entities presented 12 protection remedies against all factories installed in the neighborhood: ENAP Refinerías S.A.; de Enel Generación Chile S.A.; de Copec S.A.; de Epoxa S.A.; de GNL Quintero S.A.; de Oxiquim S.A.; de Gasmar S.A.; de Codelco Chile División Ventanas; de Cementos Bío Bío S.A.; de Puerto Ventanas S.A.; de Aes Gener S.A.; de Asfaltos Chilenos S.A.

because Chilean environmental law is inadequate, and external control mechanisms are non-existent. Finally, communities' access to litigation for environmental harm is difficult, as retaining a legal representative specialist in the field is extremely expensive, thus often unaffordable.

This conflict is just one of several. According to the *Instituto Nacional de Derechos Humanos/National Institute of Human Rights*,[10] Chile has had a total of 127 socio-environmental conflicts to date, located in different regions of Chile, affecting different dimensions of human rights. Out of these, 35 per cent affect Indigenous population, and 28 per cent the poorest population, with energy and mining-related conflicts being the most common.

3   Communities Speak: Perceptions of Two Affected Local Groups

In this section, it is our intention to share the perspective of those affected by environmental harm in Chile. We describe findings based largely on two focus groups that took place in 2019 in the north of Chile, at the Region of Tarapacá. Both focus groups were carried out in the context of a project carried out by a multidisciplinary team of researchers, aimed at exploring the potential of community panels to solve environmental conflicts (Guerra et al., 2019). The two focus groups involved communities affected by environmental harm, one caused by industrial fishing (Group 1) and one by mining (Group 2). The first case involved a group composed of 12 fishermen, all of them inhabitants of a small town located in the cost, in which artisanal fishing was the main way of living. The second group was composed of eight community representatives from a village from the Andean Plateau, who were concerned about their water supply as a mining company was about to be installed next to them. Most of these participants were actively involved in the fight against the companies affecting them.

Participation in these focus groups was strictly voluntary, so we cannot ensure that their perceptions actually represent the view of the

---

[10] See https://mapaconflictos.indh.cl/#/ (last accessed 22 January 2022).

communities they belong to, as several members of the community rejected being part of this research. Finally, we need to make clear that we cannot offer more detailed information on the conflict they were involved in, as any additional material could easily help to identify which communities (and companies) we are referring to. Participants were very strict about emphasising the relevance of both their individual anonymity and of that of their villages.

In these focus groups, the opinions, experiences, and concerns of members of these communities were explored, in relation to environmental conflicts they were experiencing. In this section, we refer to two main issues: environmental harm experienced and implications for the affected communities. We warn the reader, however, that we can generalise this content neither to all members of the communities involved in this research nor to other communities experiencing specific environmental conflicts. Instead, the following paragraphs represent the specific views of those who formed part of these focus groups.

3.1   Environmental Harm and Implications for Communities' Lives

Members of Group 1 were affected by industrial fishing in their zone. This activity, in their view, was abusing natural resources through fishing without catch limits, depleting marine resources and affecting biodiversity. Such economic activity, they said, is disrespectful of nature and unfair to those communities who have lived off fishing since ancestral times. In their experience, since the industrial fishing started, their own work has been affected, as they no longer catch the same amounts of fish as before. This has put at risk their main source of livelihood and a tradition dating back generations. Participants see in industrial fishing the main cause of serious negative economic and cultural impacts. However, they also see that there is no acknowledgement whatsoever about what is happening, neither from authorities nor from fishing companies.

Group 2 were members of a community who were anticipating water-related issues as a mining company wanted to expand its installations and move closer to them. Their fear was based on what they had witnessed: a town located nearby which was closer to the mining company had already run completely out of water. In that case, inhabitants were required to migrate and the settlement became a 'ghost town'. At the time of the focus group, participants said that the 'ghost town' was later taken over illegally by drug dealers, transforming it into an area of criminality, encouraged by the abandonment and wilful blindness of judicial and police authorities. Participants told how they, as a community, in addition to feeling scared and concerned because of the imminent arrival of the mining installations, have witnessed the rise of a problem they have never experienced before: drug trafficking. For that reason, the community wanted to organise and protest to prevent the mining company from moving into their surroundings. Community leaders took an important role in empowering their community and creating awareness about the dangers they were facing.

3.2 Implications for Affected Communities

According to the participants' narratives, three dimensions largely shaped their experiences of harm: distrust, views on development, and lack of involvement and participation.

Distrust. Distrust towards corporations and government has a central place in both groups' narratives. On the one hand, there is a colossal power imbalance between the companies and communities. Participants believe that companies are only interested in their own profit, not considering the harm they are causing to both communities and nature. Companies do not fully inform communities about their projects and possible implications, or if they do, they do it using jargon that rural inhabitants do not understand. Participants are also under the impression that companies know they lack information about administrative and legal resources that could be used to make their voice heard and their rights respected. They also feel powerless because they lack economic resources to pay for legal advice or representation.

Corporations set up in such sites make promises about possible benefits to communities that either give rise to expectations that are not fulfilled or that are too insignificant compared to the harm caused. This goes together with the perception of bad practices on the part of companies. They divided both communities by giving material incentives (money or other material benefits) to some of their members to encourage them to convince others about the project, affecting relationships amongst community members.

All of these issues have reinforced the communities' perception about corporations that usurp ancestral lands, water, sources of work, and cultural heritage. Companies do not understand their ways of living, cultural background, cosmovision and relationship with nature, creating difficulties in communicating with companies, and promoting the prejudiced view of local communities as being rigid and hostile.

The government does not help either. In participants' experience, governmental authorities are allied with corporations, increasing power imbalances. State officials do not create room for dialogue with communities, nor do they create a mechanism through which communities can be heard, oriented, or helped. Focus groups' participants related numerous relevant frustrating experiences of visiting governmental regional or local offices for information and orientation that increased their feelings of disempowerment and abandonment. In practice, this is expressed by evasive or incomplete answers and bureaucratic procedures that do not lead anywhere. In their view, instead of protecting citizens, the state is a collaborator of those who cause the harm.

Different visions on development. Participants of the focus groups had the perception that their worldview about progress and development differs significantly from the views of the state and those of the companies. Companies tend to highlight all kinds of benefit that their intervention could bring to communities. However, they felt that companies emphasise economic benefits (e.g. increase of employment options) because they are poor, pressuring them to accept the project. That reflects a limited understanding of development, they argue, as for them cultural and environmental aspects are more important than material benefits or opportunities—and they see no interest whatsoever from the state or companies in seeking to understand their culture and cosmovision. This disrespect and disregard for their worldview leads to companies always winning; in their perception, projects succeed in being implemented despite communities' resistance.

Lack of participation. During the focus groups, it became clear that community members are unevenly involved in environmental fights. Not all representatives or leaders felt supported or legitimated by their communities, as they assume this role on a voluntary basis. Consequently, achieving goals at the local level is seen as a highly complex task, for it is demanding, stressful, and frustrating. Representatives face difficulties in achieving agreements across the community, collecting information from all its members or getting their approval when involved in negotiations with counterparts. Community involvement, then, is not always possible, even when essential issues are at stake, such as water resources conservation.

Several reasons might be behind such uneven participation. As mentioned above, division is likely an important one, as some community members feel motivated by economic or other material benefits offered by corporations. Another reason might be fear. In the Latin American context, activists have faced death threats, intimidation, and harassment, receiving hardly any protection from authorities. In fact, during 2019, a total of 212 environmental activists were killed worldwide, out of whom 67 per cent were in Latin America (Global Witness, 2020).

4   The Chilean Scenario Under the Restorative Justice Lens

So far, we have analysed the context in which environmental harm takes place in Chile. We have showed that such harm occurs in the context of a general extractivist policy promoted by the state, helped by weak and inadequate legislation. We have also described the perceptions of some individuals belonging to two communities affected by extractivism, emphasising how, in their view, government and companies are on the same side, making communities more vulnerable and marginalised.

In this section, we analyse the main features of environmental harm in Chile from a restorative justice perspective. In this exercise, we take different key statements of the restorative justice approach and discuss these according to how we see their expression in the Chilean case.

4.1   Restorative Justice Focuses on Harm That Has Been Caused

Walgrave has stated that restorative justice 'is primarily oriented towards repairing the individual, relational, and social harm caused by that offence' (Walgrave, 2008, p. 21). Crime leads to needs and obligations (Zehr, 2002). In that sense, restorative justice is 'any action that repairs the harm caused by crime' (Bazemore & Walgrave, 1999, pp. 47–48).

When talking about 'harm', we need to discuss how harm can be understood and who 'the victim' is in the context of environmental issues. Harm can be understood as a multi-layered harm, which includes social, physical harm of individuals and communities and also harm of nature (Forsyth et al., 2021). However, in the example discussed earlier, we could see that harm to people and environment can be silent and invisible until the passage of time reveals its seriousness. Harm can be, thus, difficult to identify, as sometimes effects on environment are indirect and long-lasting (Aertsen, 2018). In addition to the serious harm caused to vulnerable and fragile ecosystems, our interviewees showed us that harm encompasses negative effects at cultural, social, and wellbeing level, with great risk of affecting future generations as well.

Regarding the victims' perspective, from the previous section we can conclude that victims of environmental harm in Chile are usually collectives and, most of the time, are communities that share specific characteristics. The victims usually belong to Indigenous communities and therefore share a cultural background that differs from that of the dominant culture. In addition, they live in areas historically segregated or marginalised from the economic and cultural activity of the country.[11] The latter factor sits hand-in-hand with low visibility of the harm caused. Participants of the earlier discussed focus groups expressed frustration and strong feelings of powerlessness given the lack of visibility their

---

[11] See the map of socio-environmental conflicts in Chile: https://mapaconflictos.indh.cl/#/ (last accessed 22 January 2022). We must be aware, however, that this map shows conflicts, not environmental harms. To become a conflict, the harm has to be visible and identified by a community, which mobilises itself to deal with such a problem. Given the nature of environmental harm in our country, it would not be surprising if several harms are still taking place but have gone unnoticed or noticed without action having been taken.

communities have. Only one community felt that its leaders could be heard, thanks to the fact that a president of Chile had visited the area once, and they protested to draw his attention to the problem. However, they saw this fact as a random and rare occurrence that does not represent broader reality.

We also mentioned above that 'victims of environmental harm' is not a homogenous group. Sometimes specific members of the affected community do not see the harm that is being caused. According to the testimonies, not all community members are engaged with the struggle. As argued above, corporations' tactics create internal divisions and conflicts in order to receive the support from community members. Such initiatives might be informal or formal. In the latter case, they might take the form of Corporate Social Responsibility (CSR) activities. These are activities aimed at offering benefits to local communities, usually oriented to protect the environment (Slack, 2012) but that, combined with lack of monitoring and consultation, could create a situation in which companies become a 'sort of de facto government' (Hilson, 2012, p.132). For example, a mining company mentions on its website that during one year, it invested US$ 7.9 million on education, local development, and employment in the territory where it operated.[12] These kinds of 'precalculated reparations' seem to have malicious effects (Slack, 2012), as they do not necessarily address the environmental harm caused or, even worse, they are not enough to address the environmental and social impacts of mining in the developing world. In the views of the participants of this research, these strategies are a means to convince or pressure communities, by offering support or resources in areas of actual need.

In addition to human beings, individual and/or collective, we cannot forget another actor when defining the victim perspective in environmental harm: nature. This includes ecosystems, animals, and biodiversity. Green criminology representatives have already stated that justice systems require abandoning the traditional anthropocentric perspective in the criminal justice arena, to replace this with approaches that allow for protection of and redress for nature (Nurse, 2017).

---

[12] See https://www.collahuasi.cl/comunidades/inversion-social/ (last accessed 24 January 2022).

4.2   Restorative Justice Requires an Offender Willing to Take Responsibility

An offender that does not deny his/her responsibility in the offence is considered a requirement for holding restorative justice processes (UNODC, 2020). However, acknowledgment of the harm caused is not sufficient. In addition to offenders' capacity to express responsibility and remorse, they should also be able to make amends (Wood & Suzuki, 2020). This active responsibilisation implies that actions are taken to repair the harm. Interestingly, for several victims, 'repairing the harm' also requires that the offender takes actions not only to prevent further harm to that particular victim but also to others (Bolívar, 2019).

While this scheme is useful and relevant in the context of ordinary crimes, the situation differs in the context of environmental harm. On the one hand, different levels of responsibility are involved. Minguet (2021) has argued that environmental harm has a strong social justice dimension, as tensions are usually rooted in historically defined social structures, involve problems of a global character, imply long-term consequences, and reveal parties' different values regarding nature. The author highlights these features after analysing two cases from Nigeria and Ecuador. Minguet's description seems especially true for environmental harm that takes place in the Global South, as the postcolonial context is the general framework that has defined the relationship of these countries with their natural resources and ex-colonialising countries. Minguet's idea emphasises the strong historical, international, and macroeconomic dimension of environmental harm.

On the other hand, we cannot identify a single or a group of 'offenders', as several individuals and organisations often seem to be involved. Because of that, in the restorative justice literature some have replaced the word 'offender' by 'those taking responsibility for harm' (Forsyth et al., 2021, p. 33) or the 'offender dimension' (Aertsen, 2018, p. 247). Those who are called to take responsibility might include those morally and legally responsible for causing harm, who might have diverse motivations, time frames, and positionalities (Forsyth et al., 2021). In the Chilean case, there is a moral responsibility, as corporations cause environmental harm without necessarily breaking the law, encouraged by both a general national policy of extractivism and lax implementation of already weak legal frameworks.

A second issue requiring discussion is the meaning of 'full active responsibility'. According to Chapman,[13] responsibility encompasses different stages: admitting participation in the facts of the crime committed, recognising the harm caused, recognising that what was done was wrong, making amends, and, finally, preventing it from happening again. In transitional justice and human rights-related issues, the guarantee of non-repetition is a fundamental aspect. In environmental crimes, therefore, taking responsibility implies not only compensation for the concrete (short and long term) damage caused, by 'restoring biodiversity, ecosystem health, access to or safety of places that have been damaged and restoring or revivifying care of place, taking into account the particular histories, lore, values, inhabitants and potentialities of each site' (Forsyth et al., 2021, p. 30). It should also involve guarantees of non-repetition and protection against further damage, by, for example, stronger and protective legislation, robust mechanisms of control and accountability, stabilised and permanent (not temporal) recognition of the value of Indigenous cultures and heritage, to mention but a few.

4.3   Restorative Justice Works Under the Principle of Non-domination, Giving Voice to All Stakeholders

For Braithwaite (2002, p. 20), restorative justice practitioners should secure 'freedom as non-domination through repair, transformation, empowerment with others, and limits on the exercise of power over others'. This principle means that the restorative justice process should ensure that parties involved have the 'same weight' in terms of the decision-making process. Despite the fact that power imbalances are part of the nature of crime (Bolívar, 2019), imbalances around vulnerability should be of high concern (Pali & Aertsen, 2021), as in cases of gendered

---

[13] Personal communication.

violence (Daly & Stubbs, 2006; Keenan & Zinsstag, 2014). When victims are vulnerable or emotionally traumatised, restorative justice 'should also guarantee a minimum level of victim support' (Braithwaite, 2002, p. 20), so that counterbalance mechanisms are put in place.

However, in the context of environmental harm that takes place in the Global South, 'we enter into a world of systemic injustices, extreme power imbalances and high victim vulnerability' (Aertsen, 2018, p. 245). Corporations have economic power and governments have political power while communities lack power of that sort. On the contrary, they feel unprepared to understand all the important elements inherent in the conflict, given the low access to information and their (usually) low educational levels. In this regard, Pali and Aertsen (2021) argue that restorative justice should be conceived 'as distance-reducing and power-sharing mechanisms', making the 'offender more vulnerable and empowering the victim' (p. 6). How could that proposal materialise in the Latin American context, in which power imbalances are structural and a deeply ingrained heritage of colonial times? How could this be done when the complete macroeconomic system is based on goods that the Global South holds more of in proportion to the Global North? Power imbalances then replicate at the international level, leaving developing countries at a disadvantage. Just as we need an ecological view to understand the whole context involved in case of an ordinary crime, we need also an international perspective to realise the complete picture that surrounds environmental harm in Chile and other Latin American countries.

4.4   Restorative Justice Processes Are Guided by an Impartial Facilitator

Impartiality has long been considered an important value in restorative justice processes (Choi et al., 2010). Impartiality means to treat people fairly and respectfully without any discrimination (UNODC, 2020) and requires taking care of and communicating to all involved on an equal basis, without controlling or imposing ideas or conceptions (De Mesmaecker, 2013). When the principle of impartiality is affected, the success of the procedure is threatened. Impartiality does not imply being 'neutral' to the conflict, because in the restorative justice field it is important to assume that a harm has been done, some have been hurt, and others need to be held accountable (Bolívar, 2019; De Mesmaecker, 2013).

Impartiality has different levels. On a macro level, restorative justice programmes are located within a certain institutional or public policy context, which might facilitate or hinder the perception of a 'neutral' programme. Accessibility of the restorative justice programme is another important dimension. For example, it has been observed that victims who participate in diversionary schemes describe the restorative justice programme as oriented to 'help the offender', due to the legal benefits obtainable as a result (Bolívar, 2014, p. 21). Lack of accessibility to information about the possibility to access restorative justice could also be interpreted as lack of neutrality (Laxminarayan, 2014). On a micro level, biases of the facilitator might hinder the neutrality of the process, affecting participants' experiences, leading to feelings of dissatisfaction and, eventually, injustice. To deal with this problem, it is usually recommended that restorative justice practices are located within neutral services, that function outside the criminal justice system. Some countries have followed this mandate and installed their services in municipalities or NGOs which, despite being located outside the criminal justice system, are validated enough to communicate to the justice system what parties have agreed.

In the environmental harm arena, we wonder what could be, at an institutional level, a neutral position, especially when the state is actively involved in the conflict. This seems not to be an easy issue to solve, as Chilean NGOs and other local organisations strongly depend on governmental funding—for which they must reapply on a periodical basis.

5   The Role of the State

Throughout this chapter, we have analysed the Chilean case, identifying some features of the context in which environmental harm occurs. From a South American perspective, we cannot see environmental harm as something separated from the colonial and therefore the social and economic development mechanisms that Chile as well as other Latin American countries have been historically inserted in.

In the last section, we provided some suggestions in terms of actions the state would need to take, when addressing environmental harm from the restorative justice perspective.

The starting point is to assume that in the context of environmental harm, the state is playing two roles: it is on the 'offender's side', as it has collaborated in causing harm by promoting extractivism and neglecting communities and ecosystems, while, at the same time, it is a key actor in solving the conflict and holding corporations accountable for their actions. In the context of 'ordinary' crimes, members of the community who play these two roles[14] might, during a restorative justice process, take active responsibility in recognising their contribution to the conflict while also offering support to victims and offenders so as to promote the victim's reparation and offender's social reintegration (Beck et al., 2017). We see both roles being played by the state, as its main mandate is to protect and contribute to the wellbeing of all citizens, while promoting human rights. This implies a need to design and implement policies that are not only sustainable and environmentally friendly but also that take care of communities and cultural heritage neglected since colonial times.

5.1 Creating a Third Party

Minguet (2021) states that, from a legal point of view, a local court's jurisdiction cannot deal efficiently with multinational corporations that might affect different territories and/or countries, particularly when political aspects are involved where state and corporations operate as allies. Minguet (2021, p. 71) proposes that, in order to face legally environmental harm, it would be necessary to count on 'the help of courts from countries where the environmental damage did not occur, such as the court of the country where the polluter has its headquarters'. This suggestion could help to solve the legal aspect of environmental harm,

---

[14] This role is especially evident in cases of sexual abuse of children in which the non-offender parent has acted with negligence but has, at the same time, the important duty of supporting the victim and/or the offender in their recovery or reintegration process, respectively (Beck et al., 2017).

but not yet the issue of the third party that could intervene as a neutral actor in facilitating a restorative justice procedure, as third parties need to be in the territory and get involved directly with stakeholders.

We believe that lessons learnt from the transitional justice field are relevant here. Being aware of the differences, we identify at least one common issue: transitional justice is used when the state has been actively involved in human rights violations or has been negligent or unable to stop civil conflicts. In both instances, the state plays an important role in terms of its responsibility towards both having caused and solving the conflict. One of the main mechanisms used in the context of transitional justice is truth commissions. Truth commissions 'are temporary official bodies implemented to disclose the truth regarding the human rights violations of the past' (Ferrara, 2014, p. 4). Truth commissions focus on understanding past violence and its causes, which can include focusing on ongoing events (Sarkin, 2018). Truth commissions usually investigate patterns of violence occurring over a period of time, with the goal of writing a report with recommendations (Sarkin, 2018; Hayner, 2011). These mechanisms are 'officially authorised, empowered or sanctioned by the State, but may be established by an intergovernmental organisation or United Nations where the State is unwilling or unable to do so' (Sarkin, 2018, p. 354). In addition to truth-telling and recommendations, truth commissions have had as a main goal providing evidence for subsequent prosecution, reconciliation, reforms, reparations, and historical clarification (Kochanski, 2020), for which systemic and ongoing harms could not only be addressed and understood, they could also become preventive as reforms are initiated. Finally, some see these mechanisms as based on restorative justice principles, because they have the potential for doing justice restoratively (Llewellyn, 2007).

Truth commissions' goals seem to be consistent with the needs of addressing environmental harm from a restorative justice perspective. In particular, it could help to create a neutral third party, either composed by individuals of the same country from different sectors and groups of civil society—including scientific organisations—or by international organisations that could help the state to assume its responsibility (recognising the harm done) and care role (preventing further harm) in the context of systemic harm.

5.2   Hearing Victims

Being heard is one of the most important needs of victims of crime. According to Zehr (2005), giving victims the opportunity to express their emotions is important, as victimisation can cause feelings of anger, fear, and pain. However, not just any form of 'being heard' is sufficient. Victims need to be heard in a meaningful context, in which their accounts can be considered within the decision-making process (Daly, 2017). They also need to feel safe. In the Latin American region, environmental activists have long received threats and some even been killed, so it is not surprising when community members are afraid of voicing their opinions.

Any hearing should also occur in culturally informed settings, in which affected communities' views and cosmovisions are understood and respected. Designing a meaningful context in terms of both recognition of the cultural aspects and the relevance for the decision-making process is a first step in offering validation and voice to communities that have been historically (and are still now) marginalised in the Chilean society. However, community members might not feel evenly affected, which could in part be explained by any economic incentives offered to them to encourage their acceptance of corporations' projects. In other words, we must accept that hearing communities implies needing to hear different voices, opinions, and emotions, all of which are valid and essential to include. In the line of what has been previously stated, a truth commission, if validated by the state, could be a possible and meaningful way to hear communities' divergent voices.

Finally, what remains still a challenge is how to 'hear the voice of nature'. We believe that the way to do this is to identify clearly all types of harm caused, as we describe in the following section.

5.3   Recognising the Harm

From both caring and responsibility perspectives, it is the task of the state to recognise the harm that has been done. This involves naming the offences as crimes against nature and biodiversity, naming harm caused to communities, recognising the long-term nature of the damage, identifying cultural and heritage's harm, and, above all, the systemic nature of the harm. It further requires recognising that these harms involve a structural and systemic dimension. Recognising such a multi-layer and complex scenario can only be done in an integrated and interdisciplinary way, integrating culturally informed, scientific, socio-economic, cultural, and international perspectives. In particular, scientists have a very important role when identifying harm to nature. A good example is the impact of lithium extraction at the Atacama Desert. Different biologists have warned that hypersaline lakes are being affected in their biodiversity by extractivist activities. Such biodiversity includes the existence of aquatic crustaceans such as *artemia*,[15] highly relevant for ecological balance (Gajardo & Redón, 2019). Education about what biodiversity is and its implications for everyone's wellbeing might be another crucial way to promote recognition of the harm to nature. We protect what we know.

The challenge then is not for a single person, discipline, or sector. It requires a collegial body in which all these visions can interact. Here again, the truth commission model could contribute towards the realisation of such a task.

5.4   Settling Responsibilities and Accountability

The former steps are worthless if the state does not take action to make individuals, governmental institutions, and companies accountable for actions or omissions that have occurred in the context of environmental harm. This task goes together with enforcing stricter and more exigent laws, in addition to implementing more efficient control and accountability mechanisms. In terms of Daly (2017), these actions would be in the line of securing affected communities' needs of vindication, that is, affirming that the acts that cause environmental harm were both morally and legally wrong. More than punishment, what matters here is constructing a legal framework oriented to protect communities and

---

[15] Artemia salina is a species of brine shrimp—aquatic crustaceans.

ecosystems above other interests and implementing effective monitoring mechanisms, amongst other relevant measures.

5.5   Ensuring Non-repetition

Ensuring non-repetition is perhaps the most difficult but also the most important task of all. Again, seen from the responsibility and caring function of the state, its role is to ensure that citizens and living beings have a future. It has been repeatedly said that Chile will be one of the most affected countries in the context of the climate crisis, if not already. At the moment of writing, several municipalities, in the north, centre, and south of Chile, have declared a state of climatic emergency. Our water resources are dramatically dropping. More than 76 per cent of the Chilean land surface is affected by drought, our glaciers are retreating due to high temperatures, rivers have decreased their levels between 13 and 37 per cent in last 30 years, and it is predicted that rainfalls will decrease 50 per cent by 2030.[16] Making sure that no more harm is caused and that we take care of what we still have is going to be one of the most important tasks of future governments. However, this cannot be achieved by Chile alone. International collaboration will be needed, including rethinking the whole production chain and economic development logic.

6   Final Reflections

In this chapter, we have seen how the Chilean case clearly illustrates that, at least in the Latin American region, environmental harms are embedded in macro-social and economic structures that push developing countries towards implementation of extractivist policies to favour economic growth. The Chilean State not only has followed this policy, it has implemented weak and inadequate regulation, making it possible that where companies cause environmental harm, they can do so without even breaking the law. We have suggested that, in order to address

---

[16] See https://fch.cl/noticias/un-76-de-la-superficie-chilena-esta-afectada-por-sequia-y-suelo-degradado/ (last accessed 23 January 2022).

environmental harm from a restorative justice perspective, the state needs to take a proactive role. First, it must assume its own responsibility for neglecting communities and allowing corporations to act with almost no controls. Second, by accepting as a duty the need to take care of ecosystems and natural resources to guarantee non-repetition of harm. One way we believe the state could act is by promoting and validating the creation of a collegial, international, and interdisciplinary body, similar to a truth commission used in the context of transitional justice, to allow for hearing victims, recognising harms, and setting responsibilities. Chile is a long and thin country with varying climates, a variety of natural resources, and diverse communities. Perhaps this collegial body could draw representatives from different regions of Chile, to enable conflicts to be addressed at the local level and to promote respect for the diversity of Chile's ecosystems and cultures. However, we wonder what impact an initiative like this would have without making changes to the economic system that has been built on top of historical inequalities. Restorative justice has moved forward from addressing ordinary crimes, with the associated attention to victims and offenders from specific communities, to shift its focus onto countries or wide groups of people to address inter-cultural conflicts or mass victimisation in cases of human rights violations. Perhaps now it is the time for restorative justice to take an international perspective and create mechanisms that could help countries and governments to collaboratively establish a more equal and sustainable way of living.

References

Aertsen, I. (2018). Restorative justice for victims of corporate violence. In G. Forti, C. Mazzucato, A. Visconti, & S. Giavazzi (Eds.), *Victims and Corporations. Legal Challenges and Empirical Findings* (pp. 235–258). Milan: Wolters Kluwer-CEDAM.

Barría Meneses, J. (2019). La consulta indígena en la institucionalidad ambiental de Chile: Consecuencias para la minería y las comunidades indígenas Collas de la Región de Atacama. *Investigaciones Geográficas*, (57), 76–93.

Bazemore, G., & Walgrave, L. (1999). Restorative juvenile justice: In search of fundamentals and an outline for systemic reform. In G. Bazemore., & L. Walgrave (Eds.), *Restorative juvenile justice: Repairing the harm of youth crime* (pp. 45–74). New York: Criminal Justice Press.

Beck, M., Bolívar, D. & Vanseveren, B. (2017). Responsibility, care and harm. A reflection on the role of community in cases of child sexual abuse from the experience of the Belgian mediation service Alba. In E. Zinsstag & M. Keenan (Eds.), *Sexual violence and restorative justice: Justice, therapy and collective responsibility* (pp. 229–247) London: Routledge

Bolívar, D. (2014). La mediación víctima-ofensor como alternativa al sistema penal: La perspectiva de las víctimas. *Sistema Penal & Violência*, 6(1), 13–30.

Bolívar, D. (2019). *Restoring harm: A psychosocial approach to victims and restorative justice*. Abingdon Oxon: Routledge.

Braithwaite, J. (2002). In search of restorative jurisprudence. In L. Walgrave (Ed.), *Restorative justice and the law* (pp. 150–167). London: Willan Publishing.

Burchardt, H. J., & Dietz, K. (2014). (Neo-) extractivism: A new challenge for development theory from Latin America. *Third World Quarterly*, 35(3), 468–486.

Chile Sustentable. (2016, July 22). Modelo económico chileno basado en recursos naturales se agota, advirtió la OCDE. Retrieved from: http://www.chilesustentable.net/2016/07/modelo-economico-chileno-basado-en-recursos-naturales-se-agota-advirtio-la-ocde/ (last accessed 24 January 2022).

Choi, J. J., Green, D. L., & Kapp, S. A. (2010). A qualitative study of victim offender mediation: Implications for social work. *Journal of Human Behavior in the Social Environment*, 20(7), 857–874.

Corral, T. H. F. (2004). *Lecciones de responsabilidad civil extracontractual*. Santiago: Editorial Jurídica de Chile.

Daly, K. (2017). Sexual violence and victims' justice interests. In E. Zinsstag & M. Keenan (Eds.), *Sexual violence and restorative justice: Legal, social and therapeutic dimensions* (pp. 108–140). London: Routledge.

Daly, K., & Stubbs, J. (2006). Feminist engagement with restorative justice. *Theoretical Criminology*, 10(1), 9–28.

De Mesmaecker, V. (2013). Victim-offender mediation participants' opinions on the restorative justice values of confidentiality, impartiality and voluntariness. *Restorative Justice*, 1(3), 334–361.

Farthing, L., & Fabricant, N. (2018). Open Veins revisited: Charting the social, economic, and political contours of the new extractivism in Latin America. *Latin American Perspectives, 222*(45), 4–17.

Ferrara, A. (2014). *Assessing the long-term impact of truth commissions: The Chilean Truth and Reconciliation Commission in historical perspective.* London: Routledge.

Forsyth, M., Cleland, D., Tepper, F., Hollingworth, D., Soares, M., Nairn, A., & Wilkinson, C. (2021). A future agenda for environmental restorative justice? *The International Journal of Restorative Justice, 4*(1), 17–40.

Garcés, I., & Alvarez, G. (2020). Water mining and extractivism of the Salar de Atacama, Chile. *WIT Transactions on Ecology and the Environment, 245*, 189–199.

Gajardo, G., & Redón, S. (2019). Andean hypersaline lakes in the Atacama Desert, northern Chile: Between lithium exploitation and unique biodiversity conservation. *Conservation Science and Practice, 1*(9), 1–8.

Global Witness. (2020). Defending tomorrow. The climate crisis and threats against land and environmental defenders. Retrieved from: https://www.globalwitness.org/en/campaigns/environmental-activists/defending-tomorrow/ (last accessed 29 December 2021).

Guerra, L., Frontaura, C., Saieh, C., Astete, B., Frías, N., Ibañez, J., Letelier, M., Martinez, F., Schaeffer, J. (2019). El panel comunitario como alternativa para la solución de conflictos socioambientales. In I. Irarrazaval, E. Piña, M. Jeldes, & M. Letelier (Eds.), *Propuestas para Chile* (pp. 219–253). Santiago: Centro de Políticas Públicas Pontificia Universidad Católica de Chile. Retrieved from: https://politicaspublicas.uc.cl/wp-content//uploads/2020/03/LIBRO-PP-VF.pdf (last accessed 15 December 2021).

Hayner, P. (2011). *Unspeakable truths: Transitional justice and the challenge of truth commissions* (2nd ed.). New York: Routledge.

Höhl, J. (2020). Pueblos indígenas, recursos y gobernanza: Un análisis de la consulta indígena como parte de la Evaluación de Impacto Ambiental del proyecto hidroeléctrico Añihuerraqui, Región de la Araucanía, Chile. *Investigaciones Geográficas, (59)*, 28–40.

Höhl, J., Rodríguez, S., Siemon, J., & Videla, A. (2021). Governance of water in southern Chile: An analysis of the process of Indigenous consultation as a part of environmental impact assessment. *Society & Natural Resources,* 1–20.

Hilson, G. (2012). Corporate social responsibility in the extractive industries: Experiences from developing countries. *Resources Policy, 37*(2), 131–137.

Jerez, B., Garcés, I., & Torres, R. (2021). Lithium extractivism and water injustices in the Salar de Atacama, Chile: The colonial shadow of green electromobility. *Political Geography*, 87, 1–11.

Kochanski, A. (2020). Mandating truth: Patterns and trends in truth commission design. *Human Rights Review*, 21(2), 113–137

Keenan, M., & Zinsstag, E. (2014). Restorative justice and sexual offences. *Monatsschrift für Kriminologie und Strafrechtsreform*, 97(1), 93–106.

Laxminarayan, M. (2014). *Accessibility and initiation of restorative justice*. Leuven: European Forum for Restorative Justice.

Llewellyn J. (2007). Truth commissions and restorative justice. In G. Johnstone & D. Van Ness (Eds.), *Handbook of restorative justice* (pp. 351–371). Cullompton: Willan Publishing.

Magrach, A., & Sanz, M. J. (2020). Environmental and social consequences of the increase in the demand for 'superfoods' world-wide. *People and Nature*, 2(2), 267–278.

McNeish, J. (2012). More than beads and feathers: Resource extractions and the Indigenous challenge in Latin America. In H. Haarstad (Ed.), *New political spaces in Latin American natural resource governance* (pp. 39–60). Basingstoke: Palgrave Macmillan.

Montoya, X. C. (2014). Nuevas estrategias de los movimientos indígenas contra el extractivismo en Chile/New strategies by indigenous movements against extractivism in Chile. *Revista Cidob D'Afers Internacionals*, 141–163.

Mora-Motta, A., Stellmacher, T., Habert, G. P., & Zúñiga, C. H. (2020). Between extractivism and conservation: Tree plantations, forest reserves, and peasant territorialities in Los Ríos, Chile. In F. Fuders & P. Donoso (Eds.), *Ecological economic and socio ecological strategies for forest conservation* (pp. 99–125). Cham: Springer.

Minguet, A. (2021). Environmental justice movements and restorative justice. *The International Journal of Restorative Justice*, 4(1) pp. 60–80.

Nurse, A. (2017). Green criminology: Shining a critical lens on environmental harm. *Palgrave Communications*, 3(1), 1–4.

Ocampo-Melgar, A., Sagaris, L., & Gironas, J. (2019). Experiences of voluntary early participation in environmental impact assessments in Chilean mining. *Environmental Impact Assessment Review*, 74, 43–53

Pali, B., & Aertsen, I. (2021). Inhabiting a vulnerable and wounded earth: Restoring response-ability. *The International Journal of Restorative Justice*, 4(1), 3–16.

Romero-Toledo, H. (2019). Extractivismo en Chile: La producción del territorio minero y las luchas del pueblo aimara en el Norte Grande. *Colombia Internacional*, 98, 3–30.

Sarkin, J. (2018). Redesigning the definition a truth commission, but also designing a forward-looking non-prescriptive definition to make them potentially more successful. *Human Rights Review*, 19, 349–68.

Slack, K. (2012). Mission impossible? Adopting a CSR-based business model for extractive industries in developing countries. *Resources Policy*, 37(2), 179–184.

Schmink, M., & Jouve-Martín, J. (2011). Contemporary debates on ecology, society and culture in Latin America. *Latin America Research Review*, Special Edition, 46(3), 3–10.

Svampa, M. N. (2012). Resource extractivism and alternatives: Latin American perspectives on development. *Journal für Entwicklungspolitik*, 28(3), 43–73.

UN. (2015). United Nations Human Rights Council, Report of the Special Rapporteur on the Rights to Freedom of Peaceful Assembly and of Association, A/HRC/29/25 (2015), 15 January 2017, Retrieved from: http://www.ohchr.org/EN/Issues/AssemblyAssociation/Pages/AnnualReports.aspx (last accessed 29 December 2021).

United Nations Office on Drugs and Crime (UNODC). (2020). Handbook on restorative justice programmes, 2nd edition. Retrieved from: https://www.unodc.org/documents/justice-and-prison-reform/20-01146_Handbook_on_Restorative_Justice_Programmes.pdf (last accessed 29 December 2021).

Walgrave, L. (2008). *Restorative justice, self-interest, and responsible citizenship.* Cullompton: Willan Publishing.

Wood, W., & Suzuki, M. (2020). Are conflicts property? Re-examining the ownership of conflict in restorative justice. *Social and Legal Studies*, 29(6), 903–924.

Zehr, H. (2002). *The little book of restorative justice.* New York: Good Books.

Zehr, H. (2005). *Changing lenses: A new focus for crime and justice* (3rd ed.). Scottsdale: Herald Press, PA.

23

Restorative Justice Conferencing in a New Zealand Environmental Offending Context: Two Models

Mark Hamilton

1 Introduction

It is trite to say that environmental offending, in its many guises, negatively impacts our lives on this planet. Such offending impacts the air we breathe, the water we drink, the land we live upon, and even our spiritual connection with the environment. Environmental offending can also cause intergenerational inequity through the passing of the environment on to future generations in a worse state than we inherited it (Preston, 2011; for an overview of the doctrine of intergenerational equity, see Anstee Wedderburn, 2014). Further, environmental offending can also impact non-human animals, through injury, death, and habitat loss.

Whilst restorative justice has traditionally been used for non-environmental crime, there has been increasing academic and judicial interest in the application to environmental offending (see, e.g., Al-Alosi

M. Hamilton (✉)
Thomas More Law School, Australian Catholic University,
North Sydney, NSW, Australia
e-mail: Mark.Hamilton@acu.edu.au

& Hamilton, 2019, 2021; Hamilton, 2008, 2015b, 2016, 2017, 2019a, b, 2021a, b; Hamilton & Howard, 2020; McDonald, 2008; Preston, 2011; White, 2017), including in New Zealand (Clapshaw, 2009; Fowler, 2016; McElrea, 2004; Sugrue, 2015). In the environmental offending context, restorative justice conferencing has been a common process. It sees relevant stakeholders to offending come face to face before an independent facilitator to talk about the offending and repair the harm arising from it. Relevant stakeholders can include the offender, victim, prosecutor, and experts. Three primary purposes of conferencing are to repair the harm occasioned by offending, make offenders accountable for their offending, and give victims a voice.

Traditionally, such conferencing has occurred in the context of a prosecution which is being heard before a court. Yet more recently, several regional councils in New Zealand, which function as regulatory and prosecutorial authorities, have been holding conferencing to divert offenders away from court proceedings in situations where conferencing has meant that it is no longer in the public interest to continue with prosecution. These two methods of conferencing can be differentiated as the front-end (diversion from court) and back-end (embedded within the court process) models of restorative justice conferencing.

The purpose of this chapter is to explore the use of both the front-end model and back-end model of conferencing in a New Zealand environmental offending context. To achieve this purpose, the chapter is set out in six parts, Part 1 being this introduction. Part 2 will outline the origins of these two models. As will be seen, the genesis of the back-end model is legislation facilitative of restorative justice originally designed for non-environmental offending but which has seen a migration of use to environmental offending. This compares with the genesis of the front-end model which is individual prosecutorial practice and procedure. Part 3 provides an overview of the stages within each model. Whilst there are commonalities between each model, the key differences will become apparent. Three key features of the models will be canvassed in Part 4. The first key feature is the selection criteria used to determine the suitability of conference participants attending conferencing. One selection criterion is an offender's acceptance of responsibility for their offending. There has been recent academic interest in the potential of contrition and remorse evidencing acceptance of responsibility (Al-Alosi & Hamilton, 2019, 2021). Within this content, an offender's motivation in attending conferencing is worthy of exploration. The second key feature is the role of the court vis-à-vis the conference, resulting from the fact that the front-end model is a diversion from prosecution, whereas the back-end model is embedded within the prosecution process with the fact of the restorative justice conference having occurred, and the outcomes from that conference, being considered by the court in sentencing. The third key feature which will be canvassed is the outcomes which have come from both the front-end and back-end conferences. Exploring these three key features in a dedicated part in this chapter will allow for a greater comparative analysis than could be undertaken when outlining the stages of each of the models. Part 5 will provide an outline of some of the benefits and limitations of the two models of conferencing for environmental offending. Part 6 will draw the chapter together in a conclusion.

2 Origins: Legislation Versus Practice and Procedure

The back-end model of conferencing has a longer linage in New Zealand than the front-end model, and therefore it is considered first. The genesis of the back-end model of conferencing is the commencement of the *Sentencing Act* 2002 (NZ) on 30 June 2002 and the *Victims' Rights Act* 2002 (NZ) on 17 December 2002. Whilst these pieces of legislation were not specifically drafted with environmental offending in mind, they do capture such offending, albeit in a limited way.

Section 24A of the *Sentencing Act*, which was subsequently inserted and commenced on 6 December 2014, requires the District Court (the court in which environmental offending is prosecuted in New Zealand) in certain circumstances to adjourn proceedings (a) to enable an assessment of the appropriateness of a restorative justice process in the circumstances of the case, informed by the views of the victims, and (b) to enable the restorative justice process to be undertaken if it is deemed appropriate. Guidance has been provided elsewhere pertaining to assessment criteria to help determine whether a conference, a form of restorative justice process, is appropriate (Al-Alosi & Hamilton, 2019, 2021) and will be explored in greater depth in Part 4 of this chapter.

Section 9 of the *Victims' Rights Act* provides that when a victim expresses the desire to meet with an offender in an attempt to resolve the issues arising from the offending, certain specified individuals employed within the criminal justice system, if satisfied the necessary resources are available, are to refer the request to a suitable person to arrange and facilitate the restorative justice process. A limitation with the application of this section to environmental offending is the fact that the definition of victim is derived from section 4 of the *Victims' Rights Act*, which limits victimhood to currently living humans. Whilst this is reflective of the fact that this Act has a non-environmental offending lineage, it means that human representatives as guardians of the environment or future generations of humans cannot request a restorative justice process through section 9 of the *Victims' Rights Act*. Notwithstanding this limitation, courts have been willing to entertain the prospect of conferencing and have been able to rely on section 24A of the *Sentencing Act*. Before the commencement of section 24A, courts relied on their inherent case management discretion, meaning that conferencing has and still occurs where human guardians represent non-human victims.

Regardless of how a restorative justice process is initiated, a 'court in sentencing or otherwise dealing with an offender…must take into account any outcomes of restorative justice processes that have occurred, or that the court is satisfied are likely to occur, in relation to the particular case' (*Sentencing Act*, s 8(j)). The legislation has facilitated 49 restorative justice conferences in an environmental offending context between the commencement of the *Sentencing Act* on 30 June 2002 and 1 July 2020 (Hamilton, 2021a, p. 131), although establishing the veracity of those figures is difficult for reasons outlined elsewhere (Hamilton, 2021a, p. 132).

Whilst the genesis of back-end restorative justice conferencing in New Zealand is legislation, there is no express legislation establishing the front-end model. Rather its genesis lies in the individual practice and procedure of regulatory authorities such as Environment Canterbury, Environment Southland, and West Coast Regional Council. Environment Canterbury (2012, p. 1) describes developing its front-end model of restorative justice conferencing, which it has named 'Alternative Environmental Justice', 'to fill in an identified gap in the "regulatory toolbox" where an infringement fine does not provide an adequate deterrent, but a prosecution may be overly harsh'. Environment Southland (2017, p. 25) emphasises that its 'Diversion Scheme' 'ensures that only cases which require the full intervention of the criminal law proceed before the Courts'. West Coast Regional Council (2018, p. 13) states that its alternative environmental approach enables it to 'exercise prosecutorial discretion to resolve environmental offending without the offender going to conviction or criminal record'. All these explanations seem based on the rationale that court is not appropriate in some instances of environmental offending and that better outcomes can be achieved outside of court, including through restorative justice processes.

The exact number of conferences held as a diversion from prosecution is unknown; there doesn't appear to be official statistics published in this regard and because the conference is a diversion, the matter doesn't go to trial and therefore there is no published judgement. McLachlan (2014, p. 23) states that Environment Canterbury held 12 front-end conferences between the commencement of its scheme in 2012 and December 2014, but figures beyond that are not readily accessible because there does not appear to be a central repository of information and statistics relating to such conferences.

3 Operation: Diversion Versus Embedding

What follows is a brief outline of the two models, acknowledging that greater detail pertaining to three key features of the models will follow in Part 4 and the benefits and limitations of conferencing will be explored in Part 5.

3.1 Front-End Model

In a front-end model, successful conferencing sees a matter diverted from court adjudication and sentencing. Using Environment Canterbury's

Fig. 23.1   Stages in a front-end model of conferencing

(2012) 'Alternative Environmental Justice', the model can be depicted in the five stages in Fig. 23.1.

Stage 1 is the simultaneous filing of the prosecution with the District Court, assessment of the suitability of the matter for conferencing, and invitation to the offender to attend conferencing. If a conference is to proceed, the prosecution will ask the court to adjourn proceedings to allow the restorative justice conferencing to occur (Stage 2).

The conference itself is Stage 3. This is the cornerstone of Alternative Environmental Justice. Those present at the conference will include Environment Canterbury, through managers, solicitors, and employees involved with the prosecution. Environment Canterbury's role is to present the facts of the offending, contextualise the offence, clarify factual matters in dispute, and evaluate whether conference outcomes 'meet the public interest in resolving offending' (Environment Canterbury, 2012, p. 7); the public interest aspect will be explored in Part 4 because it is a very important aspect of the role of the court in the front-end model. An offender is a participant at the conferencing because they are a stakeholder to the offending. An offender can attend with a support person (such as a family member, friend, community leader, solicitor). The support person may assist the offender 'identify opportunities to put matters right, or suggest conference plan conditions' but their role is not as an advocate or mouthpiece (Environment Canterbury, 2012, p. 7). Victims are also stakeholders to the offence and can attend conferencing. Victims are defined as 'any person or community who has suffered any direct loss, adverse effect, harm, or suffering in any social, economic, aesthetics or cultural dimension' by dint of the offending (Environment Canterbury, 2012, p. 7). Where possible, victims can attend conferencing in person. Where not possible, victims can be represented by an interest group or community group. As participation in conferencing is voluntary, a conference participant may withdraw at any time. The conference will be facilitated by an independent facilitator, who will approach affected victims, conduct a pre-conference meeting with participants, facilitate the conference, produce a conference report recording the events of the conference and the outcomes, and, when agreed, will monitor compliance with conference outcomes (Environment Canterbury, 2012, p. 7). Victims who were present at the conference could also, informally, monitor the offender's compliance with the outcome plan by keeping an eye on the offender's progress with activities or projects agreed at conferencing. Where the victim is concerned with the progress, they could report it to the facilitator or Environment Canterbury (who also plays a role in monitoring compliance).

Although an Alternative Environmental Justice conference 'will not follow a rigid agenda', the following are generally present in an effective conference:

* Outlining of the facts of the offending by Environment Canterbury;
* The reaffirmation by the offender of their acceptance of responsibility for their offending (the importance of such acceptance is explored in Part 4.1);
* The opportunity for the offender to comment on their offending;
* Outlining by an offender how they intend to put matters right and the opportunity to apologise to those present at the conference;
* The development of a conference outcome plan with the assistance of all conference participants and the recording of that outcome plan by the facilitator; and
* Arrangements for the monitoring of the conference outcome plan and explanation by Environment Canterbury of the consequences of non-compliance with the conference outcome plan (Environment Canterbury, 2012, p. 8).

Stage 4 of the front-end model of conferencing is the fulfilment of the outcome plan. It is preferable that the outcome plan obligations be fulfilled within a period of six months and before returning to court and seeking leave to withdraw the charges or that some legal mechanism is in place so that Environment Canterbury has some recourse if the outcome plan is not fulfilled.

Environment Canterbury seeking leave of the court to withdraw the charges is Stage 5 of the front-end model of conferencing and will be guided by the public interest in prosecution which will be fleshed out in Part 4.

3.2   Back-End Model

In a back-end model, the conferencing is embedded in the court process and is a matter that can be considered in sentencing an offender. This is facilitated by the *Sentencing Act* and *Victims' Rights Act* which were explored earlier. Al-Alosi and Hamilton (2021) have outlined the five stages in a back-end model of conferencing, as depicted in Fig. 23.2:

Stage 1 (commencement) involves the regulatory authority, for example, local government, filing the prosecution or originating documents with the District Court.

The establishment component of Stage 2 is concerned with the facts surrounding the offending. Where agreed between the prosecutor and offender, the facts can be reduced to a suitable document and referred to during the court proceedings. Where facts are not agreed, the court will make findings of fact following the hearing of evidence presented by both the prosecutor and offender. The determination component of Stage 2 relates to guilt. That is, acceptance by the court of a guilty plea entered by the offender. Where no guilty plea is entered, it is the role of the court to determine the guilt of the offender beyond a reasonable doubt.

Stage 3 (suitability) involves an assessment as to the suitability of a conference in the circumstances of the case. As a key feature of both the front-end and back-end model alike, this aspect will be explored in depth in Part 4.

Fig. 23.2   Stages in a back-end model of conferencing

In Stage 4, the court proceedings are adjourned for the restorative justice conference to be held which is facilitated by a facilitator who is independent of the court. Conference participants will include offenders (individuals and representatives of an organisational offender), victims (including individuals, representatives of the community, representatives of future generations of humans, and representatives of the environment), and stakeholders (which include representatives of government agencies).

The back-end model of conferencing has been used on at least 49 occasions in a New Zealand environmental offending context, between 30 June 2002 and 1 July 2020 (Hamilton, 2021a, p. 131), for a wide range of offending including discharge of pollution to air, land, and water, and planning breaches. Discharge of offensive odour has come from industrial premises, been associated with coffee roasting, and been the result of burning plastics. Pollution has resulted from the discharge of contaminants onto land (such as copper aluminium chloride, chemicals, aluminium dross, and diesel fuel), discharge of dairy effluent, untreated pig effluent, and human sewage, discharge of chemicals into a stream, operation of an unlawful landfill, and dust nuisance. Planning breaches include the breach of conditions of development consent, destruction, felling, and removal of trees without consent, modification of a Pohutukawa tree (*Metrosideros excelsa*; colloquially known as the Kiwi Christmas tree), contravention of an abatement order, and disturbance of a foreshore through unlawful earth works (Hamilton, 2021a, pp. 133–135).

A wide range of participants attended the back-end restorative justice conferences held in a New Zealand offending context, including local residents, Ōnuku Rūnanga (collective of Maori people, the Indigenous people of New Zealand), Chairperson of a local community board and walkway trust, Chairperson of the Waikato River Enhancement Society, council officers, councillors, and experts (Hamilton, 2021a, p. 135). These participants represented the diversity of victims of environmental offending: individuals, communities, and the environment.

Sentencing is Stage 5 and occurs after the restorative justice conference has been held and the matter returns to court. The court when sentencing the offender must consider the fact that the restorative justice conference has occurred (*Sentencing Act*, s 8(j)).

4   Key Features of the Front-End and Back-End Models

The previous part of this chapter gave a brief overview of the two models of conferencing, reduced to their stages. This part digs a little deeper through a comparative analysis to explore three key features of the models which will reveal where they converge and where they diverge. Those key features are the selection criteria used to establish the suitability of conferencing, the role of the court vis-à-vis the conference, and the outcomes of conferencing.

4.1   Selection Criteria Used to Establish the Suitability of Conferencing

In terms of the front-end model of conferencing, the criteria that Environment Canterbury uses to assess the suitability for conferencing are set out in the *Guidelines for Implementing Alternative Environmental Justice* (Environment Canterbury, 2012). Whilst a series of factors are considered before holding a conference, such as offence factors (prosecutor recommendations, views of victims directly affected, seriousness of the offence), offender factors (compliance history of the offender, any previous convictions, offender's personal factors, and culpability), and public interest factors (deterrence required, whether victims will be assisted by a conference), a 'precondition in every case is that the offender admits their offending', that is, a conference will not occur if an offender 'does not acknowledge their responsibility for the offending' (Environment Canterbury, 2012, pp. 4–5).

With regard to the back-end model of conferencing, section 24A of the *Sentencing Act* and section 9 of the *Victims' Rights Act* which pertains to the assessment of the appropriateness of conferencing does not specify the criteria used for that assessment. Similarly, it is not known what assessment process individual prosecutorial authorities undertake to establish whether a conference is suitable within this model. Notwithstanding, it is suggested that the United Nations Office of Drugs and Crime (UNODC) (2006) critical ingredients for restorative justice success are appropriate criteria (for an overview, see Al-Alosi & Hamilton, 2019, 2021). Those criteria are an identifiable victim, voluntary participation by the victim, an offender who accepts responsibility for their offending, and non-coerced participation of the offender (UNODC, 2006, p. 8). In terms of victims, both the front-end and back-end models of conferencing contemplate the voluntary attendance of victims at conferencing. However, it is not suggested that absence of victims is a reason to not hold conferences in either conferencing model, where it was otherwise appropriate. In terms of an offender's acceptance of responsibility for offending, it is a requirement for both models of conferencing even though the terminology used to refer to that requirement may differ. Non-coercion is interesting, especially in terms of the front-end model as the consequences of not attending a conference, or following an unsuccessful conference, is court prosecution. An offender may well see this as a form of coercion even if it is notionally up to them whether they attend a conference.

Al-Alosi and Hamilton (2019, p. 1483) have equated acceptance of responsibility for offending with an offender's contrition and remorse, which can be demonstrated by an offender taking action to rectify the harm occasioned, voluntarily reporting the offence, action to redress causes of the offence, and genuine regret and future plan to avoid repetition of such offences. It can be seen that actions speak louder than words when assessing contrition and remorse and ergo an offender's acceptance of responsibility for offending. Al-Alosi and Hamilton (2021) subsequently emphasise that an offender who does not readily accept responsibility for offending should not be automatically excluded from conferencing, where they have not denied responsibility either, because there may be genuine reasons why they haven't accepted responsibility for their offending and the conferencing may give them some insight into their offending and thereby a reason to accept responsibility.

Appraising an offender's acceptance of responsibility for offending is different from seeking to uncover an offender's motivation in attending conferencing. This is because an offender's motivation to attend conferencing can be varied, mixed, concealed, can change over time, and could include altruistic, mutually beneficial, and self-serving reasons, all of which are not necessarily adequate predictors of restorative justice

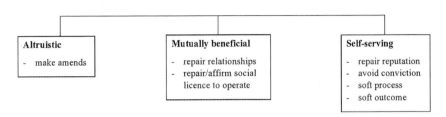

Fig. 23.3 Offender motivation in attending restorative justice conferencing

success. Notwithstanding, some discussion of an offender's motivation in attending conferencing may be of some benefit, even if not as a suitability indicium. An offender's motivation for attending conferencing is depicted graphically in Fig. 23.3:

Additionally, whilst there are many differences between environmental and non-environmental offending in terms of the nature of the offenders (predominantly environmental offenders are government entities, and corporations), victims (there is a wide range of environmental victims, such as individuals, communities, trees, plants, animals), and nature of offending (environmental offending is predominately accidental or negligent, as opposed to deliberate) (for an overview of these factors, see Hamilton, 2021a), there is nothing to suggest that the offender motivations depicted in Fig. 23.3 are unique to environmental offending.

Offender motivation can be differentiated as altruistic (i.e., an offender is concerned with the effect the offending has had on victims), mutually beneficial (in that attending a conference is motivated by mutual benefit to the offender and the victim), and self-serving (in that the offender's motivation in attending conferencing is to derive some benefit to themselves). An offender may wish to attend conferencing to make amends for their offending, with making amends being about healing and putting things right. There are many different ways an offender can heal and put things right, for example, listening to victim stories of harm, explaining the reasons for, and facts surrounding, the offending, offering an apology, paying of restitution and compensation, and making donations, or taking action, towards repairing the physical environment harmed.

As Zehr (2015b, p. 184) reminds us, crime ruptures existing relationships and, where no relationship previously existed, creates relationships of hostility. Therefore, an offender may be motivated to attend conferencing to repair relationships ruptured by offending. Closely associated with relationships, especially in terms of organisational offenders in an environmental offending context, is the desire to repair/affirm a social licence to operate. A social licence to operate 'describes how much community support a project, company or industry has in a region' (Luke, 2018). Good relationships with the community, with whom a social licence to operate exists, will allow an organisation to continue its operation without community resistance. Community resistance could hamper or even affect the viability of an organisation's operations. Good relationships and social licence to operate are not only beneficial for an organisation but may be beneficial for community as well, mainly in terms of the benefits to flow from an organisation's activities, such as investment in infrastructure, and employment opportunities.

An offender may be motivated to attend conferencing to repair their reputation, which may have been damaged through their offending. Reputation may be repaired by an offender taking responsibility for their offending, engaging meaningfully with victims, and devising outcomes to repair any harm that has been occasioned by offending. Avoiding conviction is another offender motivation to attend conferencing. It is more relevant to a front-end rather than back-end conference. Technically, an offender can go through a prosecution, be found guilty or plead guilty, and not have a conviction recorded. Hence, an offender may be motivated to attend a back-end conference for this reason. Yet, in the context of environmental offending, to not have a conviction recorded following prosecution would be rare. However, success in a front-end model conference means diversion from prosecution and therefore conviction. Thus, avoiding conviction may be a significant motivation for many offenders attending a front-end model conference.

There are two primary ways in which a conviction may impact on an offender. The first is indirectly through the effect that having a conviction has on an individual's or organisation's reputation and the flow on effects that a damaged reputation may have for securing future employment or contracts. The second way a conviction may impact on an offender is directly through the operation of legislation. For example, legislation may prevent an individual or corporation from engaging in certain activity following a conviction (for an example, see *Canterbury Regional Council v Bathurst Coal Limited* [2019b] NZDC 23872 ('*Bathurst Coal No 2*')).

An offender may be motivated to attend conferencing because they believe that it is a 'soft' option, either in terms of process or outcome. In terms of process, offenders at conferencing will be expected to give an account of their offending, listen to the hurt and disappointment expressed by the victims, and then work with those victims to come up with solutions to repair the harm that has been occasioned. This may be confronting for offenders and does not appear to be easier than going to court and hiding behind a lawyer. In terms of outcomes, conferencing can result in outcomes which go beyond the traditional fine, which is a court favourite (see Part 4.3). Therefore, conference outcomes may be more onerous on an offender than a traditional court outcome. Additionally, it is within the court's discretion to amend or add to any conference outcome, in terms of a back-end conference, if the court views it as 'soft' or otherwise inadequate. In a front-end model, if the outcome agreement is 'soft' or otherwise inadequate, the court could determine that prosecution was still in the public interest and refuse to grant leave to withdraw the prosecution, an aspect to which this chapter now turns.

4.2   The Role of the Court vis-à-vis the Conference

As would have been obvious from the short outline of the two models in Part 3, the court plays a different function in each model. In a back-end model, the role of the court is to adjourn proceedings to enable a conference to occur (Stage 4; *Sentencing Act*, section 24A) where such a conference is deemed suitable (Stage 3; *Sentencing Act*, s 24A; *Victims' Rights Act*, section 9), and then to take account of any outcomes of the conference during sentencing (Stage 5; *Sentencing Act*, section 8(j)). This approach is designated as a 'take into consideration' approach within the continuum of judicial treatment of restorative justice outcomes, meaning 'the court can endorse the outcomes agreed and make suitable outcomes into court orders' (Hamilton, 2021a, p. 189).

As an example, the offender in *Canterbury Regional Council v Interflow (NZ) Limited* [2015] NZDC 3323, following a water pollution incident, agreed at conferencing to donate $80,000 to the Banks Peninsula Conservation Trust for 'the betterment and improvement of the instream habitat' affected by its pollution and European Settlement ([43]). This donation was made in light of a likely penalty of half that amount, a fact that was known to the offender. The court being satisfied that the agreed donation would be paid, convicted the offender but did not impose any other penalty.

As another example, in *Auckland Council v Andrew Housemovers Ltd* [2016] NZDC 780 following a successful restorative justice conference, the matter came to court for sentencing. The judge held that the 'overall fine that I would have imposed ... would have been very similar to that which has been agreed upon at the restorative justice conference, and for that reason I do not see any benefit whatsoever in imposing a further fine on the company' (*Housemovers*, [14]).

Yet, despite the decisions in *Interflow* and *Housemovers*, 'the court can impose penalties on an offender in addition to any outcome agreement made at conferencing' (Hamilton, 2021a, p. 189). This conclusion is supported by the wording of the legislation and case law. Section 8(j) of the *Sentencing Act* states that the court 'must take into account any outcomes of restorative justice processes that have occurred'. This wording does not dictate how the outcomes must be taken into account and does not say that the court is limited by the outcomes agreed.

In this regard, *Southland Regional Council v Taha Asia Pacific Limited* [2015] NZDC 18010 is illuminating. Despite the offender's attendance at a successful restorative justice conference, and the spending by the offender of $1.2m remediating the polluted site, the court still imposed over $100,000 in fines. A possible explanation for this is the fact that fines have a retributive purpose, whereas restorative justice and remediation fulfil other purposes. Indeed, in a recent decision the court stated that outcomes reached at conferencing do not result in an automatic 'dollar for dollar reduction from the starting point for a fine' (*Canterbury Regional Council v Bathurst Coal Limited* [2019a] NZDC 14416 ('*Bathurst Coal No 1*'), [42]).

In contrast, the role of the court in a front-end model of conferencing is not to sentence an offender. Rather, as a front-end model is a diversion from prosecution, the role of the court is quite different. Firstly, it sits in the background, once the commencing documents have been filed in court (Stage 1) and an adjournment has been granted (Stage 2), functioning as a metaphorical Sword of Damocles ready to adjudicate and sentence if called upon. That is, if the restorative justice conference is unsuccessful. Secondly, the court is to grant leave for the prosecution to withdraw charges (Stage 5) where 'the prosecution has followed a principled and robust process' leading to the conclusion that the prosecution is no longer in the public interest (*Bathurst Coal No 1*, [25]). The notion of the public interest in prosecution derives from the *Solicitor-General's Prosecution Guidelines* (Crown Law Office, 2013), where it forms the second prong of a two-prong test for determining whether a prosecution should be initiated. The first prong of the test, the evidence test, is not relevant in the context of this chapter. Essentially, the public interest test is a weighing of the factors in favour of prosecution in tandem with a weighing of the factors against prosecution (Crown Law Office, 2013, pp. 8–10).

Seriousness of offending, the potential for harm, and likelihood of future offending are three public interest factors. Environment Canterbury, having taken part in a restorative justice conference, can make an argument to the court supporting leave to withdraw changes based on those and other factors. Environment Canterbury could say, for example, that whilst the offending was serious, the conference has given the offender some insight into their offending, and they have made changes to practice and procedure to ensure that such offending does not happen again in the future. Additionally, the offender has agreed to outcomes at conferencing which repair the harm that has been occasioned. This sort of analysis undertaken by the prosecution is what the court means by following a principled and robust process leading to the conclusion that the prosecution is no longer in the public interest. The process is certainly more than a mathematical exercise, such as an argument that the court is likely to impose a fine of $50,000 and the offender has agreed to that value of work or more at a conference and therefore prosecution is no longer in the public interest. The court was critical of Environment Canterbury approaching a leave application in this way, stating that it didn't mention the *Solicitor-General's Prosecution Guidelines* and in particular whether the test for prosecution continued to be met (*Bathurst Coal No 1*, [15]–[16]). Additionally, her Honour stated that a payment made at conferencing does not mean that there will 'be a dollar for dollar reduction from the starting point for a fine' and that in some instances prosecution may still be in the public interest, even if no fine is issued, because of the consequences of an offender facing a conviction being entered (*Bathurst Coal No 1*, [42]). In the circumstances of that case, her Honour declined the application to withdraw the charges (*Bathurst Coal No 1*, [44]), meaning that there was still a public interest in the prosecution. The matter came back for sentencing before a different judge a few months later, who imposed a fine of $18,000 (*Bathurst Coal No 2*, [44]44).

4.3   Outcomes of Conferencing

Three primary purposes of conferencing are to repair the harm occasioned by offending, make offenders accountable for their offending, and give victims a voice, which derive from the central tenets of restorative justice: crime is a violation of people and relationships, responses to crime should be inclusive, and responses to crime should heal and put things right (Hamilton, 2021a, pp. 88–92). These purposes culminate and are reflected in outcome agreements. The 49 uses of conferencing in the back-end model I documented have resulted in a wide range of outcomes.

Apology was a frequent outcome of conferencing and is important because it can give victims a sense of closure and also vindication as recognition that they have done nothing wrong. A conference, because of its interactive nature, will allow a victim to assess the genuineness of any apology, because body language and gesture are interpretative tools (Hamilton, 2017). Despite the benefits of an apology, along with its reciprocal forgiveness, restorative justice success is not measured by their presence (Hamilton, 2015a, p. 175), or indeed, failure measured by their absence. Other outcomes included the publication of a newspaper article educating the community about rural fires and their consequences, a payment towards the education of farmers as to their environmental obligations, undertaking of volunteer work to restore ecological features of the harmed environment, commitments responding to the offending behaviour, payment of reparation to neighbours, and council, the payment of various costs by the offender, and compensation to two local businesses for car-cleaning. Other outcomes relate to the undertaking of work (or the payment for that work) to repair the harm caused by the offending and to stop the harm occurring again in the future and the making of donations (Hamilton, 2021a, pp. 135–137).

It is more difficult to establish the specific conference outcomes of front-end conferences. This is because the conference serves as a diversion from prosecution and therefore outcomes are not depicted in a court judgement. Hence, knowledge of the outcomes of such conferences come from someone associated with the conference itself. Notwithstanding such difficulties, some reported outcomes of Alternative Environmental Justice conferences include:

* A training programme for others in the industry run by a roading contractor who had mishandled contaminated soil;
* Stream care groups and a wetland planting project receiving donations;
* Presentations given by a farmer to Federated Farmers of New Zealand groups relating to his offending, remedies, and how to avoid such offending; and
* Ads placed in a newspaper following a water theft offence (McLachlan, 2014, p. 23).

Outcomes of conferencing can better address the harm occasioned by offending because they involve relevant stakeholder input and can be more tailored to the harm actually caused. Such outcomes are more beneficial than the blunt instrument that are fines.

5   Benefits and Limitations of Restorative Justice Conferencing

Earlier, the primary purposes of conferencing were stated as repair of the harm occasioned by offending, making offenders accountable for their offending, and to give victims a voice. Successful conferencing will achieve these purposes and are a benefit over traditional prosecution which many would argue doesn't repair the harm occasioned, make offenders accountable, and involve victims in any meaningful way. The repairing of harm is not only about the physical projects and activities which form part of the outcome agreement, but also the offender explaining their actions, listening to the victim, and apologising when appropriate. Conferencing makes offenders accountable for their offending, and in a wider sense than being charged, prosecuted, and sentenced, which is the traditional view of accountability (Hamilton, 2021a, p. 68). A wider view of accountability involves facing up, appreciation, and steps to repair (Zehr, 2015a, p. 24, b, p. 47), that is, 'real accountability involves offenders facing up to what they have done, appreciating the harm the offending has caused not only to the victim, but also society, and taking steps to repair that harm, to make things right' (Hamilton, 2021a, p. 84). Giving victims a voice is an important part of conferencing as it allows them to express their hurt and disapproval of the offending and play some active part in its resolution through input into the outcome agreement. This is compared to traditional prosecution, which sees victims as no more than 'footnotes to the crime ... Those who have most directly suffered are not ... part of the resolution of the offence' (Zehr, 2015b, pp. 37–38).

Benefits of conferencing may also be reflected in an offender's motivation to attend conferencing. That is, conferencing has the benefit over traditional prosecution in that it provides a forum for offender and victim dialogue and input which can lead to the making of amends for the offending, can repair relationships, repair/affirm a social licence to operate, and repair an offender's reputation.

Additionally, benefits are found in a front-end model of conferencing where successful conferencing will see an offender avoid prosecution and conviction and simultaneously save court resources through the diversion away from court.

Notwithstanding such benefits, conferencing has limitations. A limitation of conferencing is that it does require a large amount of commitment and energy to succeed. This is a limitation when an offender is not willing to commit to make the conference successful. Indeed, any notion that conferencing is a soft option in terms of process and outcome should be dispelled at pre-conference screening and interviews because it can limit the chances of success.

A further limitation, which relates to the front-end model of conferencing, is the fact that the court loses its sentencing function. In the back-end model, the matter comes back to the court to sentence in light of the conference and conference outcomes, and therefore, the court is not limited by the outcome of the conference. However, in a front-end conference the court's role is to determine whether in light of the conference the public interest in prosecution is met. If the public interest is met, then the court will grant leave to withdraw the charges. In this regard, the court does not have direct control over the sentencing, and therefore, it is possible that conference outcomes may not be representative of proportionate and consistent punishment.

If the conference outcome agreement was deficient, meaning that the public interest in prosecution hasn't been met, the court could refuse to grant leave to withdraw the charges. What would be interesting is a case where a conference outcome agreement is overly onerous on the offender. This may be partly attributable to an offender's desire to avoid conviction. It would be interesting to see what a court would do in this situation, because whilst there is public interest in seeing offenders punished, proportionate and consistent punishment are bedrocks of sentencing. Additionally, there may be public interest in prosecution where an offender appears to be 'buying' their way out of that prosecution and conviction through a conference. It will be interesting to see how the courts deal with these cases at the edge, within a front-end model of conferencing (for further analysis of the benefits and limitations of conferencing following environmental offending, see Al-Alosi & Hamilton, 2019, pp. 1467–1472, Hamilton 2021a, pp. 157–179).

6   Conclusion

It is hoped that the academic, prosecutorial authority and judicial interest in restorative justice conferencing continues in the future. Conferencing is an effective way to repair the harm occasioned by the offending, through devising of outcomes which target the specific harm caused. It can also make offenders accountable for their offending by giving them insight into the consequences of offending, providing a forum for them to explain themselves to relevant stakeholders, and provide input into appropriate outcomes. Conferencing also provides much needed victim input following offending, which is missing from traditional prosecution. Additionally, the front-end model of conferencing provides a way of achieving the public interest in prosecution without burdening the court system with a prosecution which can be resolved more congenially.

Notwithstanding the fact that there may be differing motives for an offender to want to participate in conferencing, it is suggested that not motivation, but an offender's acceptance of responsibility for offending evidenced through contrition and remorse is an appropriate selection criterion to assess whether conferencing is appropriate. This is because motivations to attend conferencing may be mixed, changing, concealed, and not reflective of the potential success of conferencing.

Courts have an important role to play in conferencing, albeit that role is different in a front-end as opposed to back-end model. In the front-end model, the court functions as an ever-present threat; the place where unsuccessful conferences go. The court also has the important function of assessing whether leave to withdraw charges, that is, the prosecution, is granted to a prosecutorial authority. Leave will be granted where the prosecution has followed a robust process leading to the conclusion that the prosecution, because of the fact and outcomes from the conference, is no longer in the public interest. The granting of leave is never guaranteed (*Bathurst Coal No 1*).

While the front-end model of conferencing is a product of prosecutorial practice and procedure, the back-end model is a product of legislation. The role of the court is to allow the conference to occur and then sentence the offender in light of the fact that the conference occurred and the outcomes agreed. That is, the conference in the back-end model is embedded in prosecution, whereas in a front-end model, it functions as a diversion from prosecution.

Whilst there are benefits and limitations of both models of conferencing, further research and practical use of conferencing will determine whether the benefits outweigh the limitations and in which contexts. Evidence to date indicates conferencing is a valuable addition to the resolution of environmental offending.

References

Al-Alosi, H., & Hamilton, M. (2019). The ingredients of success for effective restorative justice conferencing in an environmental offending context. *University of New South Wales Law Journal, 42*(4), 1460–1488.

Al-Alosi, H., & Hamilton, M. (2021). The potential of restorative justice in achieving acceptance of responsibility in the context of environmental crimes. *University of New South Wales Law Journal, 44*(2), 487–512.

Anstee-Wedderburn, J. (2014). Giving a voice to future generations: Intergenerational equity, representation of generations to come, and the challenge of planetary rights. *Australian Journal of Environmental Law, 1*(1), 37–70.

Canterbury Regional Council v Bathurst Coal Limited [2019a] NZDC 14416. Retrieved from: http://www.nzlii.org/nz/cases/NZDC/2019/14416.html (last accessed 03 March 2021).

Canterbury Regional Council v Bathurst Coal Limited [2019b] NZDC 23872. Retrieved from: http://www.nzlii.org/nz/cases/NZDC/2019/23872.html (last accessed 03 March 2021).

Canterbury Regional Council v Interflow (NZ) Limited [2015] NZDC 3323. Retrieved from: http://www.nzlii.org/nz/cases/NZDC/2015/3323.html (last accessed 03 March 2021).

Clapshaw, D. (2009). Restorative justice in resource management prosecutions: A facilitator's perspective. *Resource Management Bulletin, 8*, 53–55.

Crown Law Office (2013). *Solicitor-General's prosecution guidelines.* New Zealand: Crown Law.

Environment Canterbury (2012). *Guidelines for implementing alternative environmental justice.* Christchurch: Environment Canterbury Regional Council Resource Management Act Monitoring and Compliance Section.

Environment Southland (2017). *Compliance policies.* Invercargill: Environment Southland Regional Council.

Fowler, C. (2016). *Environmental prosecution and restorative justice.* Christchurch: Adderley Head.

Hamilton, M. (2008). Restorative justice intervention in an environmental law context: Garrett v Williams, prosecutions under the Resource Management Act 1991 (NZ), and beyond. *Environmental and Planning Law Journal, 25*(4), 263–271.

Hamilton, M. (2015a). Restorative justice intervention in a planning law context: Is the "Amber Light" approach to merit determination restorative? *Environmental and Planning Law Journal, 32*(2), 164–177.

Hamilton, M. (2015b). "Restorative justice activity" orders: Furthering restorative justice intervention in an environmental and planning law context? *Environmental and Planning Law Journal*, 32(6), 548–561.

Hamilton, M. (2016). Restorative justice intervention in an environmental and planning law context: Applicability to civil enforcement proceedings. *Environmental and Planning Law Journal*, 33(5), 487–501.

Hamilton, M. (2017). Restorative justice conferencing in an environmental protection law context: Apology and corporate offending. *Internet Journal of Restorative Justice*. ISSN (online): 2056–2985.

Hamilton, M. (2019a). Restorative justice intervention in an aboriginal cultural heritage protection context: Chief Executive, Office of Environment and Heritage v Clarence Valley Council. *Environmental and Planning Law Journal*, 36(3), 197–211.

Hamilton, M. (2019b). *Restorative justice conferencing in response to pollution offending: A vehicle for the achievement of justice as meaningful involvement.* PhD Dissertation, UNSW.

Hamilton, M. (2021a). *Environmental crime and restorative justice: Justice as meaningful involvement.* Switzerland: Palgrave Macmillan.

Hamilton, M. (2021b). Restorative justice conferencing in Australia and New Zealand: Application and potential in an environmental and Aboriginal cultural heritage protection context. *The International Journal of Restorative Justice*, 4(1), 81–97.

Hamilton, M., & Howard, T. (2020). *Restorative justice in the aftermath of environmental offending: Theory and practice.* Conference Paper, National Judicial College of Australia Sentencing Conference, 1 March.

Luke, H. (2018). Not getting a social licence to operate can be a costly mistake, as coal seam gas firms have found. *The Conversation*. Retrieved from: https://theconversation.com/not-getting-a-social-licence-to-operate-can-be-a-costly-mistake-as-coal-seam-gas-firms-have-found-93718 (last accessed 25 February 2021).

McDonald, J. M. (2008). Restorative justice process in case law. *Alternative Law Journal*, 33(1), 41–44.

McElrea, Judge F. W. M. (2004). The role of restorative justice in RMA prosecutions. *Resource Management Journal*, 3(7), 1–15.

McLachlan, M. (2014). Environmental justice in Canterbury. *Public Sector*, 37(4), 22–23.

Preston, Hon Justice B. J. (2011). The use of restorative justice for environmental crime. *Criminal Law Journal*, 35(3), 136–153.

Sentencing Act 2002 (NZ). Retrieved from: http://www.nzlii.org/nz/legis/consol_act/sa2002121/ (last accessed 03 March 2021).

Sugrue, V. (2015). What happens when values are put to work? A reflection in one outcome from a restorative justice conference in the criminal division of the district court: Environment warranted judge jurisdiction. *Resource Management Journal*, 19–22.

United Nations Office on Drugs and Crime (UNODC) (2006). *Handbook on restorative justice programmes*. Criminal Justice Handbook Series: Vienna: United Nations.

Victims' Rights Act 2002 (NZ). Retrieved from: http://www.nzlii.org/nz/legis/consol_act/vra2002184/ (last accessed 03 March 2021).

West Coast Regional Council (2018). *Compliance and enforcement policy*. Greymouth: West Coast Regional Council.

White, R. (2017). Reparative justice, environmental crime and penalties for the powerful. *Crime, Law and Social Change*, 67, 117–132.

Zehr, H. (2015a). *The little book of restorative justice*. New York: Good Books.

Zehr, H. (2015b). *Changing lenses: Restorative justice for our times*. Virginia and Ontario: Herald Press.

24

Comparing Institutional Responses to the Mining Tailings Dams Collapses in Mariana and Brumadinho (Brazil) from an Environmental Restorative Justice Perspective

Carlos Frederico Braga Da Silva

1   Introduction

In the State of Minas Gerais (translated as 'General Minings'), in the Brazilian countryside, the economic model of extracting large quantities of minerals (mainly gold and iron ore) primarily for international trade has been ongoing since the colonial era. The local population, known in Portuguese as *os mineiros* (the miners), feels that it has always been this way. The dominant business code is driven by pursuit of monetary profits, indifferent to the adverse effects projected beyond the strict contours of extractive processes. Thus, the actual cost-effectiveness of *doing business*

C. F. Braga Da Silva (✉)
Faculty of Sociology, Federal University of Minas Gerais,
Belo Horizonte, Brazil

© The Author(s), under exclusive license to Springer Nature Switzerland AG 2022
B. Pali et al. (eds.), *The Palgrave Handbook of Environmental Restorative Justice*,
https://doi.org/10.1007/978-3-031-04223-2_24

that is harmful to the ecosystem was often hidden.[1] Nonetheless, lack of ecological balance is apparent through indelible scars on the environment left by the bulldozers' giant claws, leaving a multitude of negative impacts on wildlife, domestic and farm animals, and people living near mining sites. The high environmental and human price paid was tragically rendered visible to all after the occurrence of two major environmental and human disasters.

On 5 November 2015, the Fundão mining tailings dam broke in the historic city of Mariana. One of Brazil's most prominent environmental disasters, it resulted in a mud spill that reached the Doce River basin and followed the watercourse to the Atlantic Ocean. It caused a trail of destruction that resulted in challenges for effective restoration and directly killed 19 people. On 25 January 2019, a similar tragedy was repeated, but in the city of Brumadinho. Again, a mining tailings dam collapsed, its contents also leaking. The Paraopeba River, the land and buildings located on the path of the mud 'wave' were heavily affected and 270 people died.

An analysis using a restorative justice lens (Zehr, 1990) would start from the position of the affected people and environments. So, we could ask: Were there opportunities to listen to victims and to recognise and repair their harms? Were the companies' representatives encouraged and keen to participate in reparative processes? Were all the stakeholders given a chance to play a significant role in these processes? In sum, was there a restorative justice approach in the aftermath of the environmental disasters in Brazil's recent past?

This chapter tries to answer some of these questions based on research conducted on the analysis of legal documents. More specifically, the subject of environmental restorative justice (ERJ) is addressed in this chapter by examining the legal agreements reached by the parties during two lawsuits that dealt with the aftermath of the two mining tailings dam disasters. Careful analysis of these documents enables an understanding as to why there was a clear break with the past, and legal professionals

---

[1] Labonne (2016) argues that the mining industry's resistance to change hinders the implementation of efficient disaster protection. Government agencies still lack the capacity to enforce increasingly complex environmental legislation. The result is that the industry engages in self-monitoring, which leads to it cutting corners when confronted with economic downturn.

were able to reach agreements which institutionalised the paradigm of the centrality of victims and legally guaranteed a continuum of reparative actions over time until achieving the integral reparation of the environment.

The chapter posits that argumentative, interpretive, and decision-making processes can evolve from the conventional to the restorative and thus frames restorative justice as an important step of transformative legal reasoning with regard to environmental harms and crimes. By rethinking concepts traditionally adopted by the socio-legal literature, the chapter eventually aims at contributing to the debate around the institutionalisation of restorative justice (Aertsen et al., 2012) at a macro level.[2] I briefly reflect on the transnational aspect of the disasters that took place in Brazil to highlight some of the complexities involved in adequately addressing such cases. I then argue that restorative justice's principles, concepts, and objectives can impact legal operations regarding the discourse on environmental damage. In addition, I reflect on the current impacts of restorative justice on legal debates grounded in national and international environmental law. In the fourth section of the chapter, the legal agreements created in the aftermath of the two environmental disasters will be briefly described, discussed, and compared. In conclusion, the chapter presents some implications that follow from the analysis, resuming the argument that the principles and concepts of restorative justice can drive an innovative method of responding to environmental damage globally.

2 The Transnational Relevance of the Topic in Question

Unfortunately, corporate environmental harms and crimes such as the ones considered in this chapter do not constitute a new form of criminality, having solid links with long-standing extractivist business models. Once inserted in the process of economic globalisation, transnational

---

[2] See Llewellyn and Morrison (2018, p. 343), who want to advance and expand thinking, research, and practice of a restorative approach at the level of institutions and social systems.

mining corporations operate very efficiently to extract mineral resources and make high profits, often causing conflicts. Özkaynak et al. (2015) wrote a report detailing empirical evidence on 346 mining cases from around the world, accessible on EJOLT's website as the Atlas of Environmental Justice, which included interactive discussion with activists and experts on environmental justice activism concerning mining conflicts (see also Zabyelina & Van Uhm, 2020). Reckless practices and aggressive extractive processes do not harmonise with the current concept of sustainable development.

Many consequences of mining activities continue to remain unresolved. For instance, in the State Minas Gerais alone, there are more than 400 mine tailings dams, many abandoned by entrepreneurs who exploited them to the point of exhausting resources. Sometimes, small family mine businesses are passed down from one generation to the next pursuant to successive inheritance chains. In other situations, they integrate into the ownership of large companies abroad. According to the 2020 report of the National Department of Mineral Production (Agência Nacional de Mineração, 2020), numerous tailings dams are paralysed, failing to adopt the minimum measures necessary to ensure safety, even if temporarily. Elsewhere, hazardous waste dam structures have been inadequately maintained.

In 2021, a press report (BBC News Brazil, 2021) detailed that, when carrying out their activities in Brazil, foreign mining companies—mostly from wealthy countries with strict environmental and labour standards—are central to most disputes and complaints involving vulnerable populations. This means that, when operating on Brazilian soil and, consequently, far from direct inspection by the authorities of their respective countries, many companies neither demonstrate adequate compliance mechanisms nor assume commitments to respect the legislation of their homelands. Thus, it is difficult for a single government to hold transnational corporations accountable, clearly revealing that future regulations must consider this international characteristic of mining activities.

To exemplify, the emblematic case of the two Engenho dams, located in Rio Acima, again in the State of Minas Gerais, presents the highest risk in Brazil (Agência Estado, 2019). These structures belong to the bankrupt estate of Mundo Minerals Ltd., an Australian company whose partners are no longer in Brazilian territory. Gold extraction in that site ended in 2011, and the company abandoned the depleted mine, equipment, vehicles, and structures. Chemical residues can still be found in the abandoned site, including arsenic and cyanide moieties, the result of separating gold from other minerals. In the event of a spill, years' worth of improperly stored toxic waste could damage the nearby natural sanctuary of Gandarela National Park, with potential also to reach the Bela Fama drinking water catchment and treatment station, which serves the population of the capital of Minas Gerais. Decommissioning of these dams is now the responsibility of the state water company, with the state treasury having paid the cost of approximately 3 million US dollars. This example is a *drop in the ocean* exemplifying many others, illustrating the routine neglect and indifference to the consequences of exploiting mineral resources elsewhere.

Brazil's situation reflects other countries' experiences throughout the globe, as many legal systems[3] have difficulty holding mining companies (and their boards of directors) accountable for harms and crimes against the ecosystem and the integrity of people's lives. For example, in Italy, on 19 July 1985, the collapse of the Prestavèl fluorite mine tailings dams gave rise to a mudflow, which devastated the Stava valley, causing 268 deaths, and vast material and environmental destruction (Fondazione Stava, 1985, Online, n.d.). The judicial process ended after four trial instances, with the second sentence handed down by the highest Court of Justice in Italy on 22 June 1992. This verdict confirmed the sentences handed down in the first trial, in which ten people were convicted of multiple manslaughter and culpable catastrophe. However, the authorities granted pardons to the defendants and none of the convicted served prison time. Moreover, the compensation—damages of over 132 million Euros allocated to 739 plaintiffs—was only paid nearly 20 years later, in 2004.

---

[3] Fajardo and Fuentes (2014) analysed environmental crimes related to mining activities and the effectiveness of environmental criminal law and European Union liability regimes to prevent and resolve these problems in two Member States. They studied significant cases in Spain and Hungary: the mining accidents in Aznalcollar/ES and Kolontar/HU that brought criminal charges against operators and administrations with different results in both cases. When the authors wrote their article, the criminal charges in Spain had been dismissed, whereas, in Hungary, civil and criminal cases were still ongoing in their courts.

The cases involving mining companies violate the human rights of vulnerable populations.[4] For example, a tailings dam failure occurred in the Mount Polley Mine, British Columbia, Canada, on 4 August 2014, resulting in the loss of about 17 million cubic meters of water and 8 million cubic meters of tailings/materials into Polley Lake, Hazeltine Creek, and Quesnel Lake (British Columbia Government, n.d.). Marshall (2018) made comparative analyses regarding the victims of the disasters of Mount Polley (Canada) and Mariana (Brazil), detailing that, in Canada, those most affected by the collapse were the Xat'sull (Soda Creek) First Nation and the T'exelc (Williams Lake) Band. In Brazil, the disaster affected an Indigenous community of the Krenak located downriver from the mine and several quilombos, Afro-descendant communities of people brought to Brazil as slaves during the 1700s goldrush, in small farming communities adjacent to the mines. By exploring parallels between the two disasters, Marshall (2018, p. 9) highlighted:

> At both mines, the companies' failure to act on warnings and prepare for possible disasters point to an alarming corporate practice of putting production and profit ahead of safety considerations. It also raises serious questions about governments' ability to challenge the power and impunity of the global mining industry, and its willingness to govern in a manner that actively protects the environment and the rights of its citizens—economic, social, cultural, Indigenous and universal human rights.

While unprecedented technological development stimulates corporate activities aiming at more efficient levels of operation, this results in an immense destructive power against the ecosystem, and challenges to humanity reliant on ecosystem services.[5] The two Brazilian dam failures highlighted above had wider international repercussions compared to previous disasters due to the scale of the damage. Moreover, the 2019 disaster is practically a repetition of what happened a few years earlier, in

---

[4] A study about the burst in Mount Polley identified four significant effects on First Nation People's health conditions: environmental dispossession, emotional stress, altered dietary patterns, and changes in physical activity (Shandro et al., 2017).

[5] See the concept of ecocide available at: https://newint.org/features/2016/05/01/make-ecocide-a-crime/ (last accessed 10 July 2021).

2015. The reiteration was a slap in the face and a brutal warning. We were all obliged to ask ourselves: How could we allow an absurd tragedy of this proportion to happen again, in such a short period?

After the second dam-burst, Brazilian authorities, national and international media (Espindola & Guimarães, 2019) were conscious of the scandalised public opinion worldwide.[6] Thus, the Appellate Court of Justice of the State of Minas Gerais published a specific address online[7] to disseminate information through its press office. As one of the main consequences of the dam failures, the Brazilian National Council of Justice established the Environmental Observatory of the Judiciary. The project's general objective is to provide an inter-institutional and international dialogue that can contribute to improving and broadening the range of tools needed for confronting environmental violations. In addition to identifying difficulties and bottlenecks in the Brazilian justice system, the Observatory will identify the judicial units that have worked most with environmental conflict, emphasising positive experiences and proposals for initiatives, projects, and actions by civil society.[8] Innovatively, goal number nine of the National Council of Justice in 2020 provides for integrating the 2030 Sustainable Development Goals Agenda by the Brazilian judiciary. This essentially requires the courts to carry out actions to prevent or divert litigation from the judiciary. The disasters that took place in Brazil have also attracted the attention of restorative justice activists (Wijdekop, 2019), who have turned to healing the harm caused to humans and non-humans by mining-related activities. Next, I examine

---

[6] See Espíndola and Guimarães (2019): 'The Vale disaster in Brumadinho, Minas Gerais, had tremendous repercussions in the press, both nationally and internationally. From the day of the incident, January 25, 2019, until February 25, several spaces were occupied in the media, whether through news, special reports, opinion articles, chronicles, cartoons, interviews, readers' words, among others. The raising of issues by the press was carried out from a broad spectrum of perspectives and had strong shocks, mainly due to the deaths and disappearances. In that first month, *The New York Times* had 16 publications about the Vale disaster; the French newspaper *Le Monde* published 11 articles; in England, the British Broadcasting Corporation—BBC published 24 stories; and the newspaper *El País* published, in Spain, 16 articles about the disaster at Vale. In Brazil, the BBC published 60 and *El País*, 61 articles' (translated from original).

[7] See https://www.tjmg.jus.br/portal-tjmg/noticias/caso-brumadinho/ (last accessed 07 November 2021).

[8] See https://www.cnj.jus.br/observatorio/objetivos-meio-ambiente/ (last accessed 25 December 2020).

whether using restorative justice lenses to address environmental harms and crimes could prove to be an ethical and innovative approach at the universal level.

3   Ethical and Normative Standards That Support an ERJ Approach

This section elaborates on ethical and normative justifications embedded in the legal system. I purposefully analyse agreements and not judicial decisions, starting from the perspective that 'the law is a discourse and a system of rules that contain political and ethical aspects' (Hudson, 2007, p. 63). Likewise, given that Brazil is located within a well-known South American context of populist politicians without major democratic traditions, a not-so-powerful Brazilian judiciary may depend on favourable public opinion to assist it to deal with controversial environmental and socio-economical issues. Thus, it is essential to recognise the central role constitutional systems (not necessarily reduced to justice systems), characterised by political and legal elements, play in responding to environmental law questions.

Apel (2000) has argued that all countries on the planet are at least virtually affected by globalisation, a process more strikingly illustrated by the so-called ecological crisis, a consequence of modern science, technology, and economics. In our time, all human individuals, nations, and cultures are, despite differences, in the same boat, whether they like it or not. Therefore, we need universal ethics, not reduced to the strict contours of interpersonal relationships or the borders of nation-states, to justify actions to deal with environmental damages and crimes, by developing a system of shared responsibility. Applied to environmental restorative justice, Pali and Aertsen (2021) add that essentially, a restorative ethos and praxis to environmental harms calls attention first, to the necessity to repair the damages that have been done to the environment, to its human and other-than-human inhabitants and to communities, their infrastructure, and future generations, and second, to build different relational and ethical systems that prevent future harm.

I criticise the systematic disregard for the illegal and devastating consequences of mining activities worldwide, by pointing to United Nations declarations[9] to describe standards of legal reasoning[10] (Braithwaite, 2002, p. 13). Almost 50 years ago, the 1972 United Nations Conference on the Environment (UN, 1972) took place in Stockholm—the first world conference to make the environment a major relevant topic—resulted in a series of principles for the sound management of the environment, including the Stockholm Declaration. Twenty years later, the 1992 United Nations Conference on Environment and Development (UNCED) (UN, 1993), also known as the Earth Summit, was held in Rio de Janeiro, Brazil. Handl (2012) states that both conferences represented milestones in the evolution of international law and have had an influence on the right to a healthy environment in domestic legal systems. From the 1970s onwards, public acknowledgement of the world's environmental degradation and the inadequacy of state responses motivated constitutional changes and use of the powerful language of human rights. Over time, more than 100 constitutions worldwide have progressively granted constitutional status to the right to a healthy environment. These constitutional provisions consistently influence public policy, legislation, court decisions, and legal agreements. Although no nation has reached the level of ecological sustainability, constitutional protection of the environment represents a transformative step towards that goal (see Boyd, 2012). This express referral to the right to a healthy environment

---

[9] See Preston et al. (2018, p. 35): 'International agreements, developed through political processes often generate proposed governance principles or proposed laws that may become national laws through local adoption and legislation. Many putative principles of "international law", "customary law" or "soft law" that are advanced through scholarship and by political processes however do not become binding laws, though they may have political force. Even when principles are adopted by states, their translation into national law can sometimes emasculate the proposed legal principle; further, safeguards of the integrity of the translation of legal principles from international discourse into practice vary in their effectiveness'.

[10] See Braithwaite (2002, p. 13): 'I have said that fundamental human rights should set legal limits on what restorative processes are allowed to do. But I also suspect that UN human rights instruments can give quite good guidance on the values restorative justice processes ought to observe. Integrating the rights-constraining and values-guiding requirements for restorative justice under the banner of UN human rights instruments might make for simplicity. It also might make for decent and practical global social movement politics for the movement for restorative justice. This is because while no one thinks these UN rights are perfect, they are the distillation of decades of deliberation in which all nations have participated to build a consensual foundation'.

in international or constitutional law allows legal and policy practitioners to identify and use high-level legally justifiable avenues of dealing with ecological damage and crimes committed by transnational corporations.

Nonetheless, there is a need to address international norms and principles by bearing in mind a warning of Teubner (1989), who affirmed that national cultural contexts are still so diverse today that the transplantation of a theory from one context to the other leads to a degree of incomprehensibility that can only be gradually reduced by careful explanation. For example, the Brazilian Constitution (Constitution of the Federative Republic of Brazil, 1988) states that everyone has the right to an ecologically balanced environment, defined as an asset of common use and essential to a healthy quality of life (see Article 225). The Constitution establishes the strict liability of mining corporations which obliges those who exploit mineral resources to restore the degraded environment, in accordance with technical solutions demanded by the competent government body, as provided by law (see § 2°).

Additionally, procedures and activities considered as harmful to the environment subject the offenders (individuals or legal entities) to penal and administrative sanctions, without prejudice to the obligation to repair the damage caused (see § 3°). The Constitution enshrines a substantial right to be protected but does not necessarily establish the necessary procedures for granting effective enforcement. Thus, the legal system refers to itself when providing the preferred way to ensure compliance with the provisions of the superior law (Luhmann, 2004). Brazil's legal framework encourages the implementation of agreements. According to the Brazilian Code of Civil Procedure in force, the state must, whenever possible, encourage the parties to reach a consensual settlement of the dispute (see Brazilian Code of Civil Procedure, 2015, Article 3°, § 2°). Legal actors, for example judges, lawyers, public defenders, and prosecutors, must encourage the use of conciliation, mediation, and other methods of consensual dispute resolution, even during the course of proceedings (see § 3°). Therefore, the primary way of guaranteeing effective law language is to encourage, as much as possible, the resolution of the dispute through consensual means. In the opposite sense, an adversarial language that mirrors old-fashioned dispute resolution within the judiciary must be residual and employed in cases where a coercive sanction is necessary. According to the Brazilian Constitution (see Art. 103-B, Para. 4, I), the National Council of Justice (CNJ, in Portuguese) may issue regulatory acts within its jurisdiction or recommend certain measures. The resolution 125 (CNJ, 2010) instituted the National Judicial Policy for the Adequate Treatment of Conflicts of Interest, aimed at assuring everyone the right to resolve conflicts by means appropriate to their nature and particularity. Thus, before adjudicating a case through sentencing, the judiciary must offer other mechanisms for the settlement of disputes, particularly the so-called consensual means, such as mediation and conciliation, and through providing assistance and guidance to citizens.

So, there are no insuperable prohibitions in dealing with environmental disasters through agreements in the Brazilian law framework. However, maybe some law professionals lack willingness or adequate training in legal culture to use them. Undoubtedly, answers to complex and problematic situations can come by consensus whenever all those interested have seriously established the objective of managing the issue. We all know from experience that restorative justice conferences and circles have proven that agreed solutions come to light from transparent, honest, and clear intersubjective communication. Dialogic procedures need to involve victims, offenders, and the community, often with the support of state or non-state facilitators. Nonetheless, due to the scale of the cases discussed here, legal representatives and public institutions are at the forefront of the agreements signed in their associated class actions. There are always possibilities for non-coercive legal means to manage the problem at hand.

Suggestions for updating the worldwide functioning of legal systems in harmony with restorative justice principles, goals, and concepts are neither recent nor original. Almost 20 years ago, Braithwaite (2002) proposed transforming the legal system into a fairer one through a radical remake of the legal process under the guidance of restorative justice and responsive regulation. The work of the Australian academic seems to have served as inspiration for the terminology used by UN standards. Goal 16 of the sustainable development UN Agenda 2030 calls for a shift in the way justice systems operations should happen:

Goal 16. Promote peaceful and inclusive societies for sustainable development, provide access to justice for all and build effective, accountable and inclusive institutions at all levels.

16.7 Ensure responsive, inclusive, participatory and representative decision-making at all levels.

Recently, the UN General Assembly issued a resolution (2019) which aimed to act in the prevention of environmental disasters as a result of high-impact economic activities, including mining. The UN named the decade between 2021 and 2030 as the 'United Nations Decade for Ecosystem Restoration'.[11] Such international regulation emphasises that ecosystem restoration and conservation contribute to implementing the 2030 Agenda for Sustainable Development.

This UN guidance reveals that Braithwaite's (2002) pioneering ideas do not seem as 'romantic' as they might have seemed at first sight. Significant changes are underway, pushing the legal system towards alternative dispute resolution rather than argumentative competitiveness and traditional adversarial litigation. In the legal context, such novel approaches permit juridical agreements to be made, whereby it is possible for law practitioners who deal with environmental harm to instrumentalise the substantive right to a healthy environment within a reasonable timeframe, following due process and the rule of law. Other professionals, such as social workers, psychologists, NGOs, and activists have gradually changed their attitudes over time towards favouring the restorative justice approach. Hence, there is no logical barrier precluding changes in the actions of legal professionals. Restorative justice principles, concepts, and goals contain epistemological density and have normative support, both nationally and internationally.

---

[11] Available at: https://www.decadeonrestoration.org/about-un-decade (last accessed 22 December 2020).

4 Analysis of Post-Disaster Agreements in Brazil

4.1 The Agreement Regarding Fundão Dam Failure

The failure of Fundão mining tailings dam in Mariana in 2015 resulted in the signing of a Transaction and Conduct Adjustment Agreement (TTAC) (Termo de Transação e Ajustamento de Conduta) on 2 March 2016. It sought a rapid end to litigation in a Federal Court of First Instance. Its premise was that consensual methods for dispute resolution would be a quick and effective way to resolve the dispute without judicially deciding on the assumption of responsibility. The aim was to recover, mitigate, remedy, repair, even indemnify, and, in cases with no possibility of repair, offset the impacts in the socio-environmental and socioeconomic spheres. The TTAC diverted the examination of the lawsuit from judges, instituting instead the Renova Foundation. According to the website:

> The Renova Foundation is the entity responsible for the mobilisation of the reparation of the damages caused by the collapse of the Fundao dam in Mariana (MG). It's a non-profit organisation, the result of a legal commitment called Transaction and Conduct Adjustment Term (TTAC). It defines the scope of action of Renova Foundation, which are the 42 programs that unfold in the many projects that are being implemented in the 670 kilometres of impacted area along the Doce River and its tributaries. The ongoing actions are long-term and expected to take up to 10 years.[12]

The agreement's signatories are Samarco Mineração, supported by its shareholders, Vale and BHP Billiton; the Federal Government of Brazil, in partnership with the national institutes responsible for the environment, conservation of biodiversity, water resources management, mining, and the National Indian Foundation; the states of Minas Gerais and

---

[12] The 'repair' reports and how they were carried out, including the programmes adopted, are available on Renova Foundation's website: https://www.fundacaorenova.org/en/ (last accessed 23 January 2022).

Espírito Santo, together with their respective environmental and water protection bodies.

The agreement ensures social participation in the discussion and follow-up of the actions foreseen as needed to provide transparency to resolving the case and carrying out the repair processes. Furthermore, the TTAC guarantees access to broad, transparent, and public information in accessible, adequate, and comprehensive language for all interested parties. Instead of setting a monetary amount for compensation, the TTAC establishes a regenerative goal for the damaged river and its surroundings. Importantly, the TTAC's 139-page document irrefutably recognises the need for full reparation and does not set a financial limit. Full reparation covers recovery, mitigation, remediation, and repair works, including indemnities, for the socio-environmental and socio-economic impacts caused by the collapse of Fundão dam. In event of an impossibility of recovering, compensatory programmes include measures and actions that aim to compensate for the social and environmental impacts not able to be mitigated or repaired.

The agreement established the need for social assistance to those impacted, to recover productive capacity. In addition, compensatory measures are proportional to irreparable or unmitigated impacts—the objective of remedying environmental and socioeconomic conditions closest to the situation extant prior to the disaster. The parties established the objective of accelerating the recovery process of the Doce River basin, estuarine, coastal and marine regions, particularly the quality and quantity of water from the rivers that flow into it and from the impacted central channel. Therefore, they decided to deal with the environmental and social consequences, by acknowledging that the mud wave reached 680 kilometres of water bodies in the states of Minas Gerais and Espírito Santo. Thus, the agreement is not limited solely to the victims directly affected by the mud and their respective families. The TTAC establishes that Renova Foundation will observe the interests of those who lost goods (cars, herds, real estate, etc.), fishers, small farmers and businesses, and those pursuing activities related to sport and leisure, amongst other economic segments.

Socioeconomic damages resulting from the disaster must be repaired through inspection and supervision by the authorities. TTAC provides for preparation of a survey and registration programme for those affected; compensation and indemnity programme; programme to protect and restore the quality of life of Indigenous peoples; programme to preserve and restore the quality of life of other traditional peoples and communities; social protection programme; communication, participation, dialogue, and social control programme; and animal assistance programme.

Moreover, the condition of adjustment of conduct provides for the reconstruction, recovery, and relocation of people who lived in the localities of Bento Rodrigues, Paracatu de Baixo, and Gesteira. Furthermore, the agreement provides for the recovery of infrastructure and investment in education, culture, and leisure programmes.

A crucial fact is recognising the rights of Indigenous communities or traditional populations and the importance of restoring the historical and cultural heritage of the affected districts. Renova Foundation will carry out a programme to offer specialised care to Indigenous peoples in the territory after establishing permanent communication channels and interaction with society. In addition, it must develop a programme to assist stray and displaced animals, including domestic animals, such as dogs and cats, aimed explicitly at Mariana and Barra Longa.

4.2 The Agreement Regarding the Córrego do Feijão Dam Failure

The lawsuit in a state trial court against Vale began on 25 January 2019 (see Braga Da Silva, 2021), but the settlement to end the litigation only came on 4 February 2021, more than two years after the accident and a year and a half after a state judge held the company responsible for restoring all damages arising from the failure of the mining tailings dam in Córrego do Feijão. This agreement took place before a special Office of the President of the State Court of Justice and his advisers, resulting from a concatenated series of conferences. Additionally, the Governor of the State of Minas Gerais, his secretaries, the Federal and State Prosecutors' Office, the General State Defensor's Office, the State General Attorney's Office, and Vale lawyers, directors, and representatives signed a binding document.

At the beginning of the agreement, the parties emphasised that the mediation sessions in the Court of Appeal, due to the case's complexity, lasted more than three and a half months. They stated that there were 18 joint meetings, several individual sessions and meetings with representatives of those affected, including participation in public hearings in the state and federal legislative houses. The signatories highlighted that more than 100 hours of mediation allowed for the construction of the most substantial agreement presented to the judiciary in the history of Brazil. Those interested highlighted that respect for all those affected by the tragedy guided the conduct of debates.

The agreement consists of a formalised legal document of 130 pages. Its main purpose is to define Vale's responsibilities to fully repair[13] damages, negative impacts, and socio-environmental and socioeconomic losses resulting from the collapse of the mining tailings dam. The solutions and adjustments defined for each specific situation should follow the patterns established by the agreement to achieve the set goals.

The situation prior to the tragedy serves as the reference point for the repair plan as a kind of parameter to achieve in the restoration processes. Comprehensive socio-environmental remediation measures correspond to actions, projects, and work measurable by indicators and not to a monetary limit. Only the so-called environmental compensation defined in the agreement to be delivered to the State Government of Minas Gerais to carry out the works, in the amount of approximately 7 billion US dollars, was defined. In contrast to this compensation, repairing is conceived within a continuous and progressive evaluation process, following Brazilian standards and the guidelines specified in the plan.

The State Government of Minas Gerais views the agreement as the most extensive set of remedial measures in Latin American history (Governo de Minas Gerais, 2021). It foresees the possibility of income transfer to satisfy the urgent and direct demands of those affected and to allow for investments in the river basins damaged by the mud from the tailings dam. The agreement recognises irreparable environmental damage and the possibility that, if new damages are added to the known list

---

[13] See http://www.vale.com/brasil/EN/aboutvale/reports/atualizacoes_brumadinho/Pages/default. aspx (last accessed 22 January 2022).

of damages in future, the parties will adopt new compensatory measures. Moreover, the agreement expressly admitted that losses, negative impacts, and damages resulted from the tragedy. Therefore, the guiding idea is a commitment to the integral repair of the ecosystem, not to the definition of a cash value. Vale undertook to make payments and carry out projects as detailed in the document. All diffuse and collective socioeconomic damage resulting from the failure will be repaired according to the agreement.

The agreement establishes the need to recover the rivers. It foresees the possibility of improving public transportation and essential sanitation services in impacted areas. The municipalities located in the Paraopeba River basin will receive money from the State of Minas Gerais to make recovery investments. In addition, the local population will discuss their specific priorities with the support of Public Prosecutor's and Public Defender's State Offices. The premise governing the activity of socioeconomic repair is respect for the way of life, the autonomy of the people affected and the strengthening of public services available to the population in general.

The agreement provides that part of the money to be paid by Vale must be invested in improving public hospitals in the region and in combatting diseases transmitted by mosquitoes. Restructuring and modernisation of the capacity of action of firefighters, civil defence and the police is also provided for. The subscribers of the agreement planned on building roads, improving the subway, and recovering other public works that were paralysed or destroyed. One of the most critical points is the project of listening to those impacted and their participation in the solutions, resulting from the presence of professionals hired by the company and public servants in the region of the tragedy.

4.3 Comparing the Two Agreement Models

Both agreements have the semantics of restorative justice: they were designed to provide relief, remedies, and offsets for the victims. Furthermore, both agreements are aimed at preventing new disasters and repairing the already degraded environment. As for the attributions of

Renova Foundation and Vale in helping, repairing and caring for those affected, it is possible to say that, in a way, both have tried to comply with their obligations, in the cases of Mariana and Brumadinho. While it is not possible to access the data that shows the amount paid for each person, it is possible to affirm that the agreements followed the parameters of Brazilian jurisprudence on the subject. The parties signed the agreements under the guidance of lawyers and the supervision of the Public Defender's Office and the Public Prosecutor's Office. The judiciary oversaw all deals. For some of those impacted, unfortunately, amounts paid seemed insufficient, while for others, amounts paid and services provided fulfilled the agreed purposes adequately for them. Criticisms about payments or services quality remain common though, especially where controversy exists, causing some people to think justice has not been fully served.

The most significant dimension of the agreements signed following the two tailings dam collapses is simple. This is the first time in Brazil that such agreements contained explicit attentiveness to the obligation to benefit victims and the environment consensually. Moreover, as argued earlier when discussing reactions from the judiciary, press, and public opinion, both agreements were made possible due to the pressure placed on companies from public opinion, NGOs, and legal authorities, allied to an updated law framework. Given the enormous dimensions of the damages in both cases, the limitations of the conventional justice processes were immediately obvious. Immediate healing and responsiveness to the disasters were necessary, and it was absolutely out of the question to wait until the end of conventional legal procedures to define remedies and reparation.

At the same time, there have been and continue to be inspections by the judicial authorities and the appointment of teams to implement the agreements contained in the agreements. Even after starting the repair process, Samarco took about five years to meet the legal conditions established by the authorities as required to resume its legitimate operations. And Vale suffered from and still operates under many restrictions in meeting countless reparative activity requirements. According to many, the companies involved were considered extremely necessary for the economy, too big and important not to continue operating.

As already outlined, only four months after the dam's failure, the parties diverted the Mariana case to create the Renova Foundation. After just a few years, this entity has come under critique. My analysis is that Mariana's agreement—signed just four months after the tragedy of Fundão mining tailings dam—came too soon and ended the first litigation too quickly. Indeed, due to the dimension of the disaster and humans and non-human tragedies, more debate was needed before closing off all legal claims. In early 2021, the State Prosecutor's Office filed a new action requesting dissolution of Renova in a State Court of First Civil Instance. According to the lawsuit, Renova lacks the experience and legal tradition needed to deal with such severe and massive conflicts and has a participatory governance model that is publicly acknowledged as too slow and complicated, causing unjustified delays in delivering the agreed redress within the promised deadlines. Many criticise the high salaries paid to the Foundation's managers, whilst others have mentioned that diverting conflict from judicial oversight did not automatically resolve problematic situations arising from the disasters, even though the private foundation has a substantial budget dedicated exclusively to addressing and solving such situations (Ministério Público do Estado de Minas Gerais, 2021). Reality has proven that more time is needed to deal with the many problems involved.

It is uncontroversial that the extrajudicial procedure for reparation guided by the Renova Foundation is currently being seriously questioned in Brazilian and British courts. Moreover, the existence of lawsuits in Brazil and abroad against Samarco shareholders shows that leaving out the Public Prosecutor's and Public Defender's Offices in the Mariana agreement might have been a significant oversight. These institutional actors are indispensable representatives of diffuse, transcendental, and common people's interests in the Brazilian procedural model. But, unfortunately, in the current moment of the young Brazilian democracy, government representatives do not necessarily speak on behalf of all those affected by the tragedies. And perhaps Renova Foundation still does not have the necessary public recognition and sufficient power to respond, within a reasonable period, to all the criticism that has been addressed to it.

The solution model adopted to address the Brumadinho dam rupture differs from the Mariana agreement. Rather than copying the idea of instituting a private foundation, the legal authorities initially decided to wait for the negotiation process to evolve. As the author has already argued (Braga Da Silva, 2021, p. 106), the legal actors wanted a new model of management of reparation without the 'polluter paying' interference in the process. The objective was to not allow Vale to occupy a leadership position, but instead to keep the company under continuous judicial surveillance until the ultimate goals were reached.

The State and Federal Public Prosecutors and the Public Defender's Office did act on behalf of those affected in this case. In addition, the harmed could be assisted by state-of-the-art programs, as the legal agreements delimit the practices of in-person restorative justice actions at the time of contact between the work teams and the victims. Settlement of the Brumadinho case took place under the authority of public institutions with vast experience in conducting litigation. After a year and a half of valid achievements within the first instance, the trial court judge sent the dossier to second-degree mediation to close the consensual agreement. Thus, the Court of Justice had its organisational chart adapted to implement a consented solution based on the guiding principles of restorative justice. The case reveals an institutional scenario favourable to enabling environmental restorative justice (ERJ).

On 30 April 2019, Vale S/A created a special executive board for recovery and development (a body within the company), focusing on structuring actions, including repairing the damage caused by the dam's failure and coordinating socioeconomic and environmental recovery activities of the affected municipalities. It reports to Vale's CEO and participates in weekly meetings of the regular executive board to account for and discuss progress of the initiatives (Braga Da Silva, 2021). Creating a particular board within Vale to repair the damage changed the corporate culture. Vale hired many professionals with expertise in handling the problematic situations that arose, augmenting significantly the number of employees in its organisational chart and relying on a qualified workforce to develop the task of achieving 'integral reparation'.

Mariana's model has some similarities with the so-called front-end model of justice (Al-Alosi & Hamilton, 2019). A front-end model operates as a diversion to prosecution. The court's oversight is lost because the matter is not brought before the court if the conference is successful and the outcomes are adhered to. In turn, Brumadinho's case relates more to the 'back-end model'. Using that back-end model allows the courts to retain their essential supervisory role. Typically, this model involves the prosecution bringing charges before the court, identifying the utility of holding a restorative justice conference, adjourning proceedings to allow the conference to occur and then returning the matter to court for sentencing. But, as earlier stated, both cases were settled by agreement and not by sentencing.

Sometimes, the traditional model of litigation directs lawyers to pursue clients' interests through all available means, including aggressively. In other countries, alternative suggestions to mitigate the somewhat belligerent manner in which law professionals view environmental damages and crimes are arising, grounded in international parameters. For example, Salyzyna and Simons (2021) mention the background context generated by the United Nations Guiding Principles on Business and Human Rights and the Sustainable Development Goals (UN, 2011), as a means to propose a new way of leading environmental litigation. The suggestion is that, lawyers defending extractive corporations in transnational human rights and environmental cases should adopt a 'moderated resolute advocacy' model. The model would emphasise existing obligations of lawyers not to generate unreasonable costs, create undue delay or advance unfounded legal claims, and instead promote an approach to litigation oriented towards the efficient determination of substantive claims on their merits. The two-pronged approach includes the development of a voluntary litigation code of conduct for defendant extractive corporations alongside legislative action to remove some of the legal obstacles for plaintiffs bringing these cases.

5 Conclusions

In describing problematic situations caused by extractivist mining activities and comparing them with the current scenario of the agreements examined in this study, this chapter shows that in the recent past, whilst abandonment of mining tailings dams has not attracted international attention or activists of the restorative justice movement, the scenario seems to be slowly changing due to growing appreciation of the concept of 'sustainable development'.

Furthermore, the analysis applied to the examined legal documents shows the current predominance of a discourse of ERJ in place of old legalistic ways of approaching environmental harm. This suggests that legal discourses guided by a restorative ethos may be the condition to allow for restorative justice as a direct result of argumentative, interpretive, and decision-making legal activities. Thus, to move from the micro to the macro level, the chapter framed the discourse of ERJ as an innovation in dealing with conflicts related to environmental disasters caused by large corporations. Under a socio-legal approach, the two mining tailings dam failures described in this chapter should be considered as turning points. Therefore, it is possible to say that the Brazilian legal system is beginning to learn something from the principles, concepts, and objectives of restorative justice about environmental protection. This became possible because the legal procedures and discourses described here had incorporated distinct principles, concepts, and objectives, thus institutionalising an ERJ approach consensually. In this way, traditional legal discourses governing sanctioning rules (civil, administrative, and criminal) can be reformulated and gradually replaced by a restorative-oriented law.

Consequently, as old-fashioned legal thinking proved to be insufficient for rapidly dealing with such significant conflicts and keeping the mining activities running according to agreeable terms, innovative approaches appeared, resulting in institutional agreements with normative content. In the aftermath, both legal agreements discussed here were objectively implemented in a reasonable time, under the Brazilian Constitution, in real life.

The disasters that occurred in Brazil produced incisive injuries, some truly irreversible and difficult to assess. Enabling binding legal decisions with a restorative orientation that can significantly influence the institutional discourse; developing principles, concepts, and objectives that can drive the legal debates carried out amongst legal professionals; and fostering dialogue and conciliation as restorative and communicative procedures can possibly benefit all those interested in addressing the harms and responding to the difficult situations that resulted from the disasters. The companies' willingness to address the problematic situations as soon as possible seems to be one of the most relevant elements in handling a restorative approach. Thus, based on the new positive model of coping with problematic situations linked to ecological disasters that is progressively being recognised as legally valid in the current socio-legal context of Minas Gerais, the chapter clearly shows that it is possible to replace old ideas with new ones inspired by ERJ's ethos and perspectives.

References

Aertsen, I., Daems, T., & Robert, L. (2012). *Institutionalising restorative justice.* London: Routledge.

Agência Estado (2019, 02/11). MG tem 400 minas abandonadas ou desativadas; especialistas falam em bomba-relógio. *Correio Braziliense.* Retrieved from: https://www.correiobraziliense.com.br/app/noticia/brasil/2019/02/11/interna-brasil,736713/mg-tem-400-minas-abandonadas-ou-desativadas-especialistas-bomba.shtml (last accessed 08 December 2021).

Agência Nacional de Mineração (2020). *Relatório Anual de Segurança das Barragens de Mineração.* Retrieved from: https://www.gov.br/anm/pt-br/assuntos/barragens/relatorios-anuais-de-seguranca-da-barragens-de-mineracao-1/RelatorioAnual2020Final.pdf (last accessed 12 December 2021).

Al-Alosi, H. & Hamilton, M. (2019). The ingredients of success for effective restorative justice conferencing in an environmental offending context. *University of New South Wales Law Journal,* 42(4), 1460–1488.

Apel, K.-O. (2000). Globalisation and the need for universal ethics. *European Journal of Social Theory.* https://doi.org/10.1177/13684310022224732.

BBC News Brazil (2021, 03/09). *Mineradoras estrangeiras são campeãs de denúncias e conflitos no Brasil.* Retrieved from: https://www.bbc.com/portuguese/58377635 (last accessed 12 December 2021).

Boyd, R. (2012). *The environmental rights revolution: A global study of constitutions, human rights, and the environment* (Law and Society). Vancouver: UBC Press.

Braithwaite, J. (2002). *Restorative justice and responsive regulation.* Oxford: Oxford University Press.

Brazilian Code of Civil Procedure (2015). Retrieved from: http://www.politi-caeprocesso.ufpr.br/index.php/codigo-de-processo-civil-english-version/ (last accessed 07 November 2021).

British Columbia Government (n.d.) *Mount Polley Mine Tailing Dam Breach.* Retrieved from: https://www2.gov.bc.ca/gov/content/environment/air-land-water/spills-environmental-emergencies/spill-incidents/past-spill-incidents/mt-polley (last accessed 12 December 2021).

CNJ (National Council of Justice) (2010). Resolução n° 125, 29 November 2010. Retrieved from: https://atos.cnj.jus.br/files/compilado 18553820210820611ffaaaa2655.pdf (last accessed 08 November 2021).

Constitution of the Federative Republic of Brazil (1988). Retrieved from: https://wipolex-res.wipo.int/edocs/lexdocs/laws/en/br/br325en.pdf   (last accessed 06 November 2021).

Espindola, H. & Guimarães, D. (2019). História ambiental dos desastres: Uma agenda necessária [Debate]. *Revista Tempo e Argumento*, Florianópolis, 11 (26), 560–573.

Fajardo, T., & Fuentes, J. (2014). *The Aznalcollar and the Kolontar Mining Accidents: A case study on mining accidents and the criminal responsibility of operators and administrations. A study compiled as part of the EFFACE project.* Granada: University of Granada and University of Jaen. Retrieved from: https://efface.eu/case-studies/index.html (last accessed 12 December 2021).

Fondazione Stava 1985 Online (n.d.). *Legal liability.* Retrieved from: https://www.stava1985.it/legal-liability/?lang=en (last accessed 18 June 2021).

Governo de Minas Gerais (2021). *Acordo Judicial.* Retrieved from: https://www.mg.gov.br/conteudo/pro-brumadinho/acordo-judicial (last accessed 08 December 2021).

Handl, G. (2012). *Declaration of the United Nations Conference on the Human Environment, Stockholm, 16 June 1972, Rio Declaration on Environment and Development, Rio de Janeiro, 14 June 1992.* Audiovisual Library of International Law. Retrieved from: https://legal.un.org/avl/ha/dunche/dunche.html (last accessed 05 November 2021).

Hudson, B. (2007). The institutionalisation of restorative justice: Justice and the ethics of discourse. *Acta Juridica*, 2007(1), 56–72.

Labonne, B. (2016). Mining dam failure: Business as usual? *The Extractive Industries and Society*, 3 (3), 651–652.

Llewellyn, J., & Morrison, B. (2018). Deepening the relational ecology of restorative justice. *The International Journal of Restorative Justice*, 1(3), 343–355.

Luhmann, N. (2004). *Law as a social system*. Oxford: Oxford University Press.

Marshall, J. (2018). *Tailings mam spills at Mount Polley and Mariana: Chronicles of disasters foretold*. Retrieved from: https://www.policyalternatives.ca/sites/default/files/uploads/-publications/BC%20Office-/2018/08/CCPA-BC_TailingsDamSpills.pdf (last accessed 12 December 2021).

Ministério Público do Estado de Minas Gerais (2021). MPMG pede na justiça extinção da Fundação Renova. Retrieved from: https://www.mpmg.mp.br/portal/menu/comunicacao/noticias/mpmg-pede-na-justica-extincao-da-fundacao-renova.shtml (last accessed 08 December 2021).

Özkaynak, B., Rodriguez-Labajos, B., Aydın, C.İ., Yanez, I., & Garibay, C. (2015). Towards environmental justice success in mining conflicts: An empirical investigation. EJOLT Report No.14, April 2015. Retrieved from: http://www.ejolt.org/wordpress/wp-content/uploads/2015/04/EJOLT_14_Towards-EJ-success-mining-low.pdf (last accessed 22 January 2022).

Pali, B., & Aertsen, Ivo. (2021). Editorial: Inhabiting a vulnerable and wounded earth: restoring response-ability. *The International Journal of Restorative Justice*, 4(1), 3–16.

Preston, J., Martin, P., & Kennedy, A. (2018). Bridging the gap between aspiration and outcomes: the role of the court in ensuring ecologically sustainable development. In C. Voigt, & Z. Makuch, Z. (Eds.), *Courts and the environment* (pp. 35–58). Cheltenham, UK: Edward Elgar Publishing.

Braga Da Silva, C. F. (2021). A maximalist approach of restorative justice to address environmental harms and crimes. Analysing the Brumadinho dam collapse in Brazil. *The International Journal of Restorative Justice*, 4(1), 98–122.

Shandro, J., Jokinen, L., Stockwell, A., Mazzei, F., & Winkler, M. (2017). Risks and impacts to First Nation health and the Mount Polley mine tailings dam failure. *International Journal of Indigenous Health*, 12 (2), 84–102.

Teubner, G. (1989). How the law thinks: Toward a constructivist epistemology of law. *Law & Society Review*, 23 (5), 727–758.

Salyzyna, S., & Simons, P. (2021). Professional responsibility and the defence of extractive corporations in transnational human rights and environmental litigation in Canadian courts [Unpublished manuscript]. Faculty of Law, Common Law Section, University of Ottawa.

UN General Assembly (1972). *United Nations Conference on the Human Environment*, 15 December 1972, A/RES/2994. Retrieved from: https://www.un.org/ga/search/view_doc.asp?symbol=A/CONF.48/14/- REV.1 (last accessed 29 October 2021).

United Nations (2011). *Guiding principles on business and human rights: Implementing the United Nations "Protect, Respect and Remedy" framework.* Retrieved from: https://www.ohchr.org/documents/publications/guiding-principlesbusinesshr_en.pdf (last accessed 12 December 2021).

United Nations Conference on Environment and Development Rio de Janeiro, Brazil (1993). *Agenda 21: programme of action for sustainable development; Rio Declaration on Environment and Development; Statement of Forest Principles: The final text of agreements negotiated by governments at the United Nations Conference on Environment and Development (UNCED), 3–14 June 1992, Rio de Janeiro, Brazil.* United Nations Dept. of Public Information.

Zabyelina, Y. & Van Uhm, D. (Eds.) (2020). *Illegal mining: Organised crime, corruption, and ecocide in a resource-scarce world.* Cham: Springer International Publishing AG.

Zehr, H. (1990). *Changing lenses: A new focus for crime and justice.* Scottdale, PA: Herald Press.

Wijdekop, F. (2019). Interview with Dominic Barter. In E. Biffi & B. Pali (Eds.), *Environmental justice: Restoring the future* (pp. 54–60). Leuven: European Forum for Restorative Justice.

25

Restorative Environmental Justice with Transnational Corporations

Martin Wright and Ulrike Tabbert

1   Introduction

According to a commonly used definition, restorative justice is a voluntary process that engages those responsible for and harmed by a criminal offence in constructive dialogue about the harm caused and what can be done to set things right. Even though it is often applied on a small scale and in cases of individual victims and offenders, restorative justice is, in theory, applicable to complex cases as well. Cases are considered to be complex if they involve corporate wrongdoers and/or multiple victims/survivors as in the case of the poison gas disaster in the Union

M. Wright (✉)
London, UK
e-mail: info@actionforbhopal.org

U. Tabbert
University of Huddersfield, Huddersfield, UK
e-mail: ulritab@googlemail.com

Carbide pesticide factory in Bhopal, India, in 1984. People who suffered harm to their body and soul, as in the case of Bhopal, often want the same things in the aftermath of such disasters: medical treatment and compensation for themselves and action to reduce the likelihood that others may suffer in the same way in future. They may see deterrence through punishment as a way to do this or as a method of achieving 'justice', which may mean official condemnation of the harmful deed, punishment of the wrongdoer, or both. Some of them say that they would like an explanation of why the harmer committed the harmful act, and possibly an apology, which is more problematic when the harmer is a corporation, especially a transnational one.

This chapter will consider the applicability of restorative justice in cases where a corporation has caused environmental damage in addition to harming people: possible advantages, problems that arise, and whether restorative principles have to be modified. These issues will be considered, taking Bhopal's 1984 disaster as an example. We begin with a brief outline of the historical background of the disaster and the reactions from the corporations and the survivors in its aftermath; then, with that as the basis, we review possible advantages and potential difficulties that come with a restorative justice approach on this scale, followed by considerations of responsibilities of shareholders and corporations in the light of principles of corporate social responsibility (CSR) and environmental, social, and governance (ESG).

2   Bhopal's Disaster and Its Aftermath

2.1   Historical Background

Almost four decades have passed since 'the world's worst industrial disaster' (Edwards, 2015, p. 1) at the pesticide plant of Union Carbide India Ltd. (UCIL) in Bhopal on the night of 2/3 December 1984. An investigation into the causes for the release of deadly methyl isocyanate (MIC) clouds into the town of Bhopal found, among other unsafe practices, that 'a large volume of water had been introduced into a MIC storage tank, causing a chemical reaction that forced the pressure release valve to open and allowed the gas to leak'.[1] The Bhopal Medical Appeal (BMA) reports that within the first two weeks, some 10,000 people had died, and those exposed to the gas (up to 500,000 people) had suffered severe health damage. Later, congenital malformations were found in 9 per cent of babies of gas-exposed mothers, compared to 1.3 per cent of unexposed mothers (BMA, 2019). Almost four decades later, the former plant site has still not been cleaned up and deadly poisons, such as mercury, lead, cadmium, arsenic, and more, continue to contaminate the soil and ground water.

2.2   UCC's and Dow's Corporate Response

UCIL's stock at the time of the disaster was majority owned (50.9 per cent) by American-based Union Carbide Corporation (UCC); the remainder was in the hands of the Indian government and private investors. In the immediate aftermath of the fatal gas leak, UCC tried to conceal the composition of the MIC gas and its extreme toxicity, even to the extent of preventing the use of an antidote, sodium thiosulphate, because its effectiveness could have given a clue to this information (BMA, 2012, pp. 22–25).

Another twist to the story is that UCC over the course of the past three and a half decades has sold its controlling shareholding in UCIL and then further changed its face: firstly in 2001 by becoming a wholly owned subsidiary of Dow Chemical Company which, in a second step, merged with DuPont to form DowDuPont in 2017. In 2019, finally, DowDuPont split into three companies: Dow, DuPont, and Corteva Agriscience.

As has been argued, all of these changes 'not only made it the biggest chemical conglomerate in the world but obscured the chain of corporate responsibility' (Kershen, 2021, p. 157). After Dow took over UCC, it did not clean up the toxic waste which UCC had left on the abandoned site, thus choosing not to follow the generally accepted principles in such cases: 'the polluter pays' and 'successor liability'. In fact, a statement by

---

[1] See http://www.bhopal.com/Cause-of-Bhopal-Tragedy (last accessed 25 March 2019). For a more detailed account of the events leading to the disaster, see Weick (2010, pp. 538–40).

The Dow Chemical Company[2] (TDCC) states that responsibility for the clean-up of the Bhopal site lies with the Madhya Pradesh State government, because in 1998, several years before UCC became a subsidiary of TDCC, the State Government took over the facility and assumed all accountability for the site.[3]

UCC had a poor safety record even before 1984, despite its slogan 'Production at a cost, safety at any cost' (Pearce, 1990, p. 420). Several examples can be found in the *Bhopal Reader* (Hanna et al., 2005), for instance, 'Bhopal sitting at the edge of a volcano', by investigative journalist Rajkumar Keswani, reprinted from *Rapat Weekly*, 1 October 1982, which says that '[a]ccidents began in the plant from the time it started' and gives examples such as a leak of phosgene gas in 1981 and 1982.

Dow has also had safety problems elsewhere. For example, in 2013 there was a fire at a Dow Chemical electronic materials facility in North Andover, MA, USA (Kemsley, 2013), in 2016 an explosion at the Dow Chemical Co. in North Andover, MA, USA,[4] and in 2020 an industrial accident at a Dow Chemical facility in Mount Meigs, AL, USA.[5] These relatively recent incidents were reported in the news and in each case cost lives and damaged the surrounding environment. They show that the Dow Group repeatedly has difficulties to ensure the safety of their workers and protect the environment surrounding their factories, a phenomenon which besets the chemical industry sector as a whole. Dow acknowledges this (see Nesmith et al., 2013, pp. 170–174):

> Major process safety incidents in chemical manufacturing are infrequent but when they do occur, the consequences can be severe. Corporate performance goals are often measured in such a way that a single plant could operate for many years without a major process accident and still be operating below the required performance standard. Process safety near-misses reporting is intended to be a more sensitive indicator of actual process

---

[2] See https://corporate.dow.com/en-us/about/legal/issues/bhopal (last accessed 25 March 2019).

[3] See https://corporate.dow.com/en-us/about/legal/issues/bhopal/site-remediation.html (last accessed 27 March 2021).

[4] See https://www.ehstoday.com/safety/article/21917594/explosion-rocks-dow-chemical-co--facility-four-seriously-injured (last accessed 27 March 2021).

[5] See https://www.powderbulksolids.com/safety-compliance/worker-dies-accident-dow-chemical--site-alabama (last accessed 9 December 2021).

safety performance. Simply put, a near miss is "an event that presents an opportunity to learn valuable information that may prevent future accidents." Since the wholesale implementation of the process safety near miss program in Dow Chemical the number of process safety incidents has declined by 80%. While there are many contributors to that remarkable result, the process safety near miss program has been a major one.

The above illuminates the dilemma in which Dow finds itself, in addition to the usual pressure to maximise profits: on the one hand, the *zeitgeist* demands that Dow, like other companies, presents itself as an eco-friendly and socially responsible company, in order to appear attractive to investors and trade partners, while on the other hand, processes in the chemical industry repeatedly lead to incidents with life- and environment-threatening consequences, and Dow faces demands from the victims of the Bhopal gas disaster and their legal representatives.

2.3   Survivors' Response

There have been several survivors' organisations[6] and voluntary groups asking for justice for victims, such as the International Campaign for Justice in Bhopal, Bhopal Medical Appeal, and Action for Bhopal. Survivors of the Bhopal disaster have made heroic efforts to provide medical treatment for fellow survivors and to secure redress. Survivors' medical needs were treated in a clinic established in 1996, named Sambhavna (meaning 'compassion' in Hindi). In 2004, two leading campaigners, Rashida Bee and Champadevi Shukla, were awarded the prestigious Goldman Prize for their work for the environment, and they used the money to establish a second clinic, Chingari (meaning 'spark'), for children born damaged by the gas and poisoned water (BMA, 2012, p. 134).

To secure financial redress, at first the civil law was used by survivors. After the disaster, American lawyers acted for the victims and the Indian government passed the Bhopal Gas Leak Act (1985). Many individual

---

[6] The Bhopal Gas Peedit Mahila Stationery Karmchari Sangh, the Bhopal Gas Peedit Mahila Purush Sangharsh Morcha, the Bhopal Gas Peedit Nirashrit Pension Bhogi Sangharsh Morcha, the Bhopal Group for Information and Action, and Children Against Dow Carbide.

lawsuits were also filed at various courts in the federal court system of the USA. They were eventually transferred and centralised at the southern district court of New York (where UCC is based), presided over by Judge Keenan (Jaising & Sathyamala, 1995, p. 175). There, UCC's lawyers successfully initiated a transfer of proceedings to India. The Indian government, ostensibly acting on behalf of the victims, took over the lawsuits. In February 1989, after the Indian Supreme Court urged the parties involved to reach a settlement, UCC agreed to pay US$ 470 million for damages caused in the Bhopal disaster. This settlement, although immediately criticised for the extremely low sum given the extent and long-term effects of the damage, was confirmed by the Indian Supreme Court in 1991 (Image 25.1).

Afterwards further lawsuits were filed both in India and the USA, aiming at re-negotiating the sum to be paid and cleaning up of the contaminated site, but none of them has been successful so far. Dow is a named respondent both in a curative petition (pending at the time of writing) in India's Supreme Court to correct inadequacies of the 1989 settlement

Image 25.1   Solar evaporation pond, leaking toxic chemicals. (Photo © Annie Murray)

and in Madhya Pradesh High Court,[7] seeking remediation of the abandoned Union Carbide factory site. The petitioners state that the number of people eligible for compensation and the amounts they should receive have both been seriously underestimated.

The criminal law has also been invoked. At present, UCC is wanted for criminal charges of 'culpable homicide not amounting to murder' by the court of the Chief Judicial Magistrate in Bhopal; UCC has, however, repeatedly refused to attend the court and is therefore regarded as an absconder from justice. The legal battle is still pending.

As legal action has largely failed to deliver a satisfactory outcome, the survivors mounted numerous public protests—non-violent, to their great credit (BMA, 2012, p. 157). For example, survivors Bee and Shukla visited America in 2003. They held a 12-day hunger strike in Wall Street, and then stood outside the Indian embassy in Washington, to try to shame the Indian government for making no effort to secure the extradition of Warren Anderson, the CEO of UCC, who was wanted in India on criminal charges.

They also went to Dow's shareholder meeting and demonstrated outside the building. As described in the report *The Bhopal Marathon* (BMA, 2012, p. 113), Dow officials invited them to come in and talk, but they refused, saying that they would continue their protests until Dow accepted responsibility towards the victims. Advocates of problem-solving dialogue may speculate whether at this point a meeting with an experienced facilitator might have led to an acceptable outcome; those with many years' experience of dealings with UCC and Dow would probably take a more pessimistic view (reasons for their distrust are considered later in this chapter). Apart from other considerations, harmers often demand that settlements of this kind be conditional upon their not accepting responsibility, while the survivors strongly feel that justice requires full and public accountability.

Back in India, survivors also tried to gain the support of the Indian government. In 1989, a group of 75 gas-affected women, together with 30 children and 12 men, walked 500 miles to Delhi to meet the Prime

---

[7] See   https://www.bhopal.org/the-dow-chemical-companys-bhopal-related-legal-liabilities/   (last accessed 08 April 2019).

Minister. They were told that he was out of town, but they met their regional Member of Parliament who made promises, which were not kept (BMA, 2012, pp. 50–53). In 2006, a second walk to Delhi was undertaken. Prime Minister Manmohan Singh at first refused to see the protesters, but after they held a five-day hunger strike, he met them and agreed to four of their six demands, but refused to endorse any trade bans or agree to prosecute Dow as the successor of Union Carbide (BMA, 2012, pp. 146–155). In 2008, a third walk again found the Prime Minister too busy to meet the delegation. According to the BMA, '[n]ews reached them that the PM and his ministers were looking for a way to assist Dow Chemical in its bid to dispute successor liability [while] Dow was dangling the carrot of a $1 billion investment. 'The demonstrators were beaten up and hustled away by police' (BMA, 2012, pp. 154–156).

There were numerous protests at the London Olympics in 2012, where Dow Chemical sponsored a £7 million fabric wrap for the Olympic stadium, despite numerous protests. The BMA and the Bhopal Group for Information and Action (BGIA) produced a report, *The Bhopal Marathon* (2012), on which a large part of the information of this chapter is based.

3   A Restorative Approach for Bhopal?

3.1   Possible Advantages of a Restorative Approach

In this chapter, we consider a restorative approach and outline the potential advantages and possible difficulties accompanying it. A central question is what could a restorative approach offer in the case of Bhopal in the aftermath and in addition to the legal action and public protests mentioned above? Numerous attempts have been made to reach a new outcome that would help the survivors of the disaster and would allow for a clean-up of the still-contaminated plant site by entering into a dialogue with UCC's successors, aiming for a re-negotiation of the earlier settlement. Until now, none of these attempts has succeeded. Suggestions that a dialogue could be enabled and guided by restorative justice principles (see Wright, 2019 among others) were ignored by Dow but were also not accepted by those representing the victims because in their opinion Dow could get off too lightly and other potential wrongdoers would not be sufficiently deterred.

It is understandable then that these representatives, like many victims of wrongdoings, wanted there to be 'no more Bhopals'. If anything at all positive could be said to have followed from this disaster, it is the hope that it 'caused a shift in the way that the industry viewed process safety' (Macleod, 2014, p. 8). Unfortunately a gas leak at the LG polymer plant at Visakhapatnam, Andhra Pradesh, on 7 May 2020 suggests that lessons have not been learned (BBC, 2020; India Today, 2020); another example is the deadly and devastating explosion of ammonium nitrate, stored in the Beirut harbour, on 4 August 2020 (Sky News, 2020). Inquiries into the causes of disasters like these should focus on future preventive standards, not merely on allocating blame.

One of the key authors on restorative justice, John Braithwaite (2002), has provided the inspiration for the idea of tackling the situation in Bhopal in a restorative way (described in detail elsewhere by Wright, 2019). This approach is also recommended in a British report on the enforcement of sanctions called *Regulatory Justice*, by Professor of Law at University College London Richard Macrory (2006a, pp. 69–72), which mentions restorative justice as a means of securing compliance. There are, however, three basic problems with restorative justice (outlined in detail in the following section): firstly, the fact that the harmers may refuse to engage with the process; secondly, the reluctance of victims to engage with it; and thirdly, the fact that 'pure' restorative principles, such as voluntariness, may not work with corporate wrongdoers.

We nevertheless argue that potential advantages are comparable with those of restorative justice applied in individual cases and are briefly summarised as follows: compensation for survivors, a remediated environment, and the possibility of employment if and when the company resumes operation. For the company, it can repair its credentials as an organisation which maintains sustainable principles and shows concern for the wellbeing of its employees and the local community by observing ESG principles. This admittedly entails the cost of the remedial measures, which will be recovered when other companies are willing to do business again. Furthermore, it will help to satisfy survivors who regard impunity as injustice: they wish to see the company pay a price for the harm it caused.

3.2   Potential Difficulties with a Restorative Approach

Making contact: The first hurdle encountered by any restorative justice advocate wishing to initiate the process is to make contact or, as Braithwaite put it in a personal conversation, 'getting to hello' (modifying the title of Fisher and Ury's (1991) famous book *Getting to Yes*). In the case of Bhopal, a first contact with Dow was initiated after an appeal including full-page advertisements in newspapers drawing attention to the situation of victims in Bhopal, the Bhopal Medical Appeal. This appeal was launched in 1994 to raise money for a clinic to treat survivors.[8] The BMA's campaign not only made possible the creation of the Sambhavna Clinic in 1996 (mentioned above), but also drew attention to the dire situation, including the fact that toxic contamination of the water supply was still causing further illness.

Some readers felt that the work of the clinic should be supported by preventive efforts: to remediate the contaminated site, so as to prevent more and more people from becoming ill. Two peace-promoting organisations agreed to sign letters, although not to take on a full-scale campaign. The initial approach, in 2015, was to contact Mr. Matt Davis, the Dow Chemical Company's vice-president of global government and public affairs at the time, asking for a meeting with the assistance of a professional facilitator (who had offered his services pro bono). The reply was signed by Mr. Scot Wheeler, Issues Management and Crisis Communications department, and repeated the company's official line: that it had nothing to do with the 1984 tragedy, the $470 million settlement had been approved by the Supreme Court of India, Dow never owned nor operated the UCC plant, and when UCC became a subsidiary of Dow in 2001, Dow did not assume UCC's liabilities. UCC (it

---

[8] See BMA website: https://www.bhopal.org/ (last accessed 22 January 2022).

added) 'continues to exist today' [2015] as a separate entity—although this has been questioned by the BMA (2012, pp. 96–100). Further letters were sent to individual directors but remained unanswered.

Respect for survivors' wishes: However, well-wishers from afar owe respect to the views and feelings of those directly on the receiving end of the disaster (Wright, 2019). Survivors may wish for what they regard as 'justice', in some form of official retribution. They may also wish action to be taken to reduce the possibility that others will suffer as they have, and they may believe that deterrent punishment will help to achieve this. Furthermore, they may be sceptical of the benefits of restorative justice, and it is not surprising that there is deep distrust of outsiders. For example, in 2006, after the takeover of UCC by Dow, an American firm called Cherokee Investment Partners produced a business plan for Bhopal, supported by an officer of the Madhya Pradesh Pollution Control Board, who was suspected of having a conflict of interest. According to this business plan, the land would be donated to Cherokee, which would remediate it and sell it at a profit. Cherokee would indemnify stakeholders against future environmental liability, to which survivors objected because it was viewed as letting Dow escape its obligations. What further added to this feeling of distrust is that Cherokee in a remediation project in California had broken basic safety standards. For these and other reasons, survivors rejected the deal[9] (Tremblay, 2007).

Modifying restorative principles: As to the restorative justice process itself, it usually takes place voluntarily between an individual victim and an individual wrongdoer, giving them an opportunity to meet on a human level, to express feelings, ask questions, and perhaps agree on some form of reparation or apology. The voluntary principle may however have to be stretched by 'building them a golden bridge', in other words encouraging the 'difficult people' to cross from their old position to a new one, as recommended by William Ury in his book *Getting Past No: Negotiating with Difficult People* (1991). First, there may be practical difficulties: in a commercial context, people who want to offer to mediate

---

[9] See https://www.bhopal.net/dubious-dubey-and-cherokee-bosses-in-bhopal-stew/.

https://www.bhopal.net/dubey-and-dow-front-cherokee-lambasted-in-bhopal/ (last accessed 4 October 2020).

may be dealing with a board of directors and a group of victims who may be in different countries. Face-to-face meetings are more difficult to arrange but can have a potential for encouraging empathy; however, this is reduced if the encounters have to be electronic. Letters, and perhaps photos, are easier to send (if the directors' contact details can be found) but have the disadvantage that they do not carry much impact and are easier to ignore.

It may be necessary to modify the voluntary principle further: John Braithwaite quotes the old saying 'Speak softly but carry a big stick'. What might the 'big stick' be? It could be the law, if there is a relevant law, as in the cases of the polluting Pacific Gas and Electric Company, described in the film *Erin Brockovich*, and also of the chemical giant Ciba-Geigy, which polluted the water in the New Jersey town Toms River. In each of these cases, the company settled out of court when faced with a major lawsuit [spurred on by facilitators Eric Green and Jan Schlichtmann in the case of Toms River (Fagin, 2013)]. Those cases, however, took place within American jurisdiction. In the case of Bhopal or, more generally, when the pollution has been caused in another country, the company can sidestep such an approach by the simple method of refusing to attend the court in that country, as the Dow Chemical Company and UCC have done in India.

An alternative to the use of force or rather the threat of the law is the use of so-called soft power. An interesting use of this 'soft power' has been described by Sarah Corbett (2019). A group called the 'Craftivists' was contacted by the non-governmental organisation (NGO) ShareAction, who wanted to persuade a large retail company to pay a real living wage to its workers. The CEO had refused to meet them. Therefore, 24 'craftivists' bought shares and identified company directors, chief investment officers of their biggest shareholders, and some others. They found out a few facts about them, their interests such as music or gardening, where they used to work, and so on, and embroidered a handkerchief for each of them reflecting their interests. For each of them there was also a hand-written letter, with a positive message, but also expressing shock at discovering that the company paid less than a living wage. They hand-delivered the gift-wrapped presents, and four meetings resulted. Before the next shareholders' meeting, the firm announced a living wage for 50,000 staff.

Clearly this type of approach will only be suitable in a limited number of cases, but it shows that an approach based on persuasion, without the 'big stick', can work—although we do not know how often it has been tried aside from this case.

Mindful investments and shareholder pressure: Another way of applying pressure is through share value. Efforts have been made to persuade pension funds, asset managers, charitable organisations which hold cash reserves, and so on, either to disinvest or to practise what is called impact investing: they inform the company that they will not invest in it, or will sell their shares, unless it remediates its contaminated site.

With regard to shareholder pressure, campaigners have considered asking for the support of investors and their advisers who wish to apply ethical principles by disinvesting, but it appears that the chemical industry as a whole is already off-limits to those with ethical concerns, so that no additional pressure could be applied in this way. Alternatively, campaigners may buy at least one share so that they can attend shareholder meetings and present their case (either in the meeting itself or informally afterwards). The NGO ShareAction, mentioned above, specialises in this method. Shareholder resolutions can be put forward, but it is difficult to get the support of a sufficient number of shareholders (a 'big enough stick'). Dow, as mentioned, uses an additional tactic: it argues that Dow was not running the company at the time of the poisonous gas leak and denies the principle of 'successor liability'. This ignores the fact that Dow is now the owner of UCC which undoubtedly was responsible for pollution in Bhopal in (and even before) 1984, and, furthermore, contamination of water by discarded chemicals has continued after Dow took over UCC in 2001.

At the time of writing, news headlines reported that small-scale investors united through social media and invested in GameStop stocks (and other stocks) against at least two large hedge funds (using 'shorts', meaning they had bet on falling prices). This led the company's value to increase by billions, at least on paper, just within a few days, instead of falling as expected. Although the intention might not entirely be of a Robin Hood nature (robbing from the rich to give to the poor), this example nevertheless underlines the power of (united) shareholder action.[10]

---

[10] See https://www.nytimes.com/2021/01/28/business/gamestop-stock-market.html (last accessed 26 November 2021).

4   How to Mitigate the Difficulties— Enforcement or Compliance

In this section, we discuss ways to mitigate the difficulties with a restorative justice approach in the case of Bhopal which we outlined in the previous section. The first to mention is the call for an international law on ecocide which, in the eyes of experts, is needed to tackle the ongoing climate and ecological crisis. Such law could become the 'fifth offence' prosecuted by the International Criminal Court (ICC) 'alongside war crimes, crimes against humanity, genocide and the crime of aggression'.[11] A draft law was unveiled in June 2021 that defines ecocide as 'unlawful or wanton acts committed with knowledge that there is a substantial likelihood of severe and widespread or long-term damage to the environment being caused by those acts' (Stop Ecocide Foundation, 2021). Such a law would, however, not be without problems of enforcement (Kershen, 2019; Mehta, 2019) both at trial stage and in the enforcement of a sentence after the verdict. In the case of Bhopal, it is, however, the harm done not just to the environment but also and foremost to the people (survivors and their offspring) that call for a restorative approach.

On a smaller scale, one international initiative does provide a method of encouraging compliance. The United Nations Global Compact (UNGC)[12] invites corporations and other organisations to join by signing up to a set of ten basic principles relating to human rights, labour, anti-corruption, and the environmental Sustainable Development Goals. Some of these are broad, others refer to relatively specific problems such as seafarers stranded far from home by coronavirus restrictions (the 'crew change crisis'). This approach raises hope that the Compact could cover the specific case of Bhopal or at least the globally widespread dumping of toxic waste. As regards compliance, the Compact has no 'big stick', except companies' unwillingness to be seen as hypocritical or as failing to provide a regular Communication on Progress, which is required of signatories to the Compact. Under the Integrity Measures, it offers a platform to

---

[11] See https://www.theguardian.com/environment/2021/jun/22/legal-experts-worldwide-draw-up--historic-definition-of-ecocide?CMP=Share_iOSApp_Other (last accessed 10 December 2021).

[12] See https://www.unglobalcompact.org/ (last accessed 10 December 2021).

facilitate dialogue when allegations of abuse of the Ten Principles are raised by a credible third party about any participant. If the allegations are deemed in scope of the dialogue facilitation process, a company will be required to respond to the concerns raised within two months of receipt. If they fail to respond, the company can be listed as non-communicating on the UNGC website and de-listed after one year—a hit to their reputation which survivors can exploit. A UK-based group called Action for Bhopal,[13] of which the first author is a founding member, has written to Dow through the Compact and has raised concern over the handling of the Bhopal disaster by Dow.

Another way of making contact is provided by the Office for Economic Co-operation and Development (OECD). An NGO can complain to its country's National Contact Point if it alleges, with evidence, that a corporation is breaching OECD guidelines on issues such as human rights and the environment. The NGO is required to propose what should be done to resolve the problem, and there is provision for facilitation or mediation. For example, the Global Legal Action Network has brought forward complaints relating to a coal mine in Colombia.

5 The Trend Towards Corporate Responsibility: Doing the Right Thing

The recent trend towards corporate responsibility could in our view aid compliance or even pave the way towards a restorative solution of the Bhopal conflict. The first argument in favour of Dow's corporate responsibility is that Dow is the successor to UCC which did not act in accordance with safety measures in operation in Bhopal and India at that time (as discussed above), and secondly, because Dow is now responsible for the site and fails to safeguard the water supply by cleaning it up. (UCC, however, claims that Madhya Pradesh has taken over accountability for it).[14] This refusal continues to cause harm to people and the environment

---

[13] See https://actionforbhopal.org/ (last accessed 10 December 2021).
[14] See Science & Sustainability, https://corporate.dow.com/en-us/science-and-sustainability.html (last accessed 5 February 2021).

to the present date. Although Dow has managed to get away with denying its legal obligations following from their corporate legacy, they cannot successfully suppress their moral obligations and are confronted with them on a daily basis. One option for Dow is, of course, to continue their strategy of denial which has been successful—at least in legal terms—for decades. However, it is argued that an increasing awareness of the damage mankind has caused to people and nature has brought Dow to a 'tipping point' and a restorative approach might offer them a 'way out' towards acting in accordance with their liability as successor of UCC and towards 'shaming' in a reintegrative way (Braithwaite, 1989).

As we have seen, campaigners often insist that the corporation accepts responsibility or liability, which company lawyers tend to resist. A solution from a linguistic point of view could be that both sides might accept replacing these legal terms with moral ones, such as 'duty', during a restorative justice process.

An approach based on enlightened self-interest is proposed by William Ury's (1991) 'golden bridge', as mentioned above: in other words, offer a resolution which does not involve losing face or helps to rebuild Dow's reputation (see also Fisher & Ury, 1991). Those who want to see retribution may not like this, but they may consider, firstly, that it represents the best chance of persuading the company to meet its obligations; secondly, that retribution can take the form of condemnation, administered separately through court action or the media; and thirdly, that the cost of remediating the damage will be substantial and can therefore be regarded as a form of 'punishment', for which a term such as 'taking the consequences' might be an acceptable replacement. An important point in a case such as Bhopal is that the resolution should include not only responsible actions in future (avoiding pollution to reduce climate change and so on) but remedying the contamination caused in the past.

There are grounds for hope in the increasing interest in assessing a company's performance not only by its profit but by ethical criteria, variously known as CSR or ESG principles. An American example is the Statement on the Purpose of a Corporation (Business Roundtable, 2019), by which companies undertake not only to generate value for shareholders (the sole aim proclaimed by neoliberal economists like Milton Friedman) but also to treat employees and suppliers fairly and to support 'the communities in which we work' by respecting people, protecting the environment, and 'embracing sustainable practices across our businesses'. One critic, however, has commented on the Roundtable Statement that '[t]o the extent that it provided a fig leaf that enabled CEOs to pursue business as usual—well, it was probably worse than doing nothing at all' (Makower, 2019). The Statement appears to include no mechanism for ensuring compliance.

There are indications that ESG companies are recommended by financial journalists and advisers and listed in indexes such as Standard & Poor. The criteria for ESG compliance need to be spelt out, for example, by including remediation of past damage, especially by clearing up toxic waste, as well as avoiding it in future. If and when a law of ecocide will be introduced, conforming with it should also be an ESG requirement. As Frank Pearce (1990, p. 425) has proposed,

> A responsible corporation would accept that it bears responsibility for the dangerous side effects of its productive processes, its products and the waste that it generates.

This of course extends the issue far beyond Bhopal and would be a desirable addition to the world-wide campaign to reverse the damage inflicted on the planet in this anthropocene era. It could also involve organisations campaigning for human rights as well as the environment, because a degraded environment also degrades people's lives. The NGO Pure Earth (2016) has shown that the illnesses and deaths caused by toxic waste inflict millions of disability-adjusted life years (DALYs).

Campaigners' instinct may be to appeal to the humanitarian feelings of directors, especially if they can be approached individually and not as a group and without the intervention of their lawyers. They may also hope that directors will calculate that philanthropy could be good for their reputation (Macrory, 2006b, p. 84). The less optimistic view is that 'they simply don't care', as evidenced by the reported remark of one Dow official that '$500 [compensation] is plenty good for an Indian', and one director's statement that to do anything philanthropic would be 'unfair to shareholders'—overlooking the unfairness to former employees and other residents of Bhopal (BMA, 2012, p. 104).

Almost four decades after the deadly disaster in Bhopal, times have changed, as have business principles, in the face of demands for eco-friendly and socially responsible investments. The Dow Group does not remain untouched by these developments. In fact, Jim Fitterling, CEO of Dow Inc., is most conscious of these new conditions, as a random sample of his tweets (@JimFitterling) amply illustrates:

13/08/2020: @DowNewsroom is the leading customer of clean energy in our sector; and we will continue to lead the way to a low-carbon future. @euniceheath5

07/08/2020: Advancing a circular [recycling] economy also advances inclusion. This project, which exchanges plastic bottle caps for wheelchairs, helps people with accessibility challenges get an education, gain employment and become more self-reliant.

06/08/2020: We need continual, uninterruptible power sources—even a millisecond interruption can be problematic. The more abundant, dispatchable and competitive non-carbon energy sources, the more we can reduce carbon emissions.

Dow is thus seeking to construct both an eco-friendly and a socially responsible image for itself to increase its attractiveness for investors who not only seek a reliable provider of chemical goods or an investment with a potential of growth but also pay attention to the company's ethical approach to doing business.

This trend for renewed attention to ethics, sustainability, and eco-friendliness in business does not happen in isolation nor is it limited to the fields of industry, business, and (financial) investment; it is a consequence of social re-thinking in times where resources are no longer perceived as 'endless' and environmental degradation leads to climate change, the consequences of which are painfully perceived all over the world. There is renewed awareness of environmental issues (particular by millennials and their flagship movement #FridaysForFuture) and the way in which humans have endangered and damaged the earth and its atmosphere. This is mirrored in the notion of the Anthropocene, a geological epoch in which human impact dominates the earth's geology and ecosystem. The Anthropocene in geological terms covers an epoch from the early years of the appearance of the human species on this planet up until the present day and is thus by no means a recent or new phenomenon. What is recent is the growing awareness of the damage done by humans to our planet and its ecosystem during the Anthropocene, given that the gravest and probably irreversible damage has been done in recent centuries, maybe starting with the industrial revolution. Even more recently, the Anthropocene is interpreted by Gibbons (2019, p. 282)

> not as the de facto zeitgeist but as a contributing component within a more complex configuration of contemporary crises—of environment, of financial instability, of social inequality, of political conflict, of terrorism, of technology, and others—that together, relatedly, engender a new cultural sensibility.

Part of this 'new cultural sensibility' which Gibbons recognises is, most certainly, the aforementioned trend to invest in companies which adhere to eco-friendly and socially responsible business principles, to which Dow vividly pays attention both in its internet performance and in its daily business operations. Here is an extract from their webpage.[15]

> *Shared value for all*
> How can science help create a more sustainable world? It takes collaboration and innovation. At Dow, we're working to deliver a sustainable future for the world by connecting and collaborating to find new options for materials that make life better for everyone.
> *Our focus areas*
> We identified three focus areas where we believe we can make the biggest difference and drive industry-wide change. These global priorities represent areas where we are using our science, size and global relationships across our value chains to seek and create shared opportunity for Dow and society.

The areas where Dow wishes to contribute (according to their website) are climate protection, circular economy (designing products to be easier

---

[15] See Science & Sustainability, https://corporate.dow.com/en-us/science-and-sustainability.html (accessed February 2021).

to recycle), and safer materials. To be consistent with these ideals, it would be commercially good practice for Dow to accept the duty of ensuring that the poisonous chemicals left on the site by its predecessor UCC are cleaned up, so that the water supply is no longer contaminated and causing further harm to the people in the city of Bhopal and their children.

6   The Role of Government

One last factor we wish to consider in the aftermath of a disaster of such large dimensions in Bhopal is the local or national Indian government. Governments have the power to initiate and pass laws and to make sure that laws are obeyed. Good regulations, enforced by fair and firm inspection, are not a bureaucratic burden but can help companies to improve their reputation and their prosperity by doing the right thing. To give one example, Unilever has announced that it will insist that its suppliers also pay a living wage (Unilever, 2021).

In our view, the Indian government, however, may have adopted a policy of encouraging foreign companies to invest by keeping regulations to a minimum or by not enforcing them, for example, by refusing to prosecute those responsible even for serious offences. In the case of Bhopal, as we have outlined, it was also alleged that the Indian Prime Minister and his cabinet ministers were looking for a way to assist Dow (BMA, 2012, p. 155). The international policy adviser Simon Anholt argues on the basis of wide experience (2020, p. 228) that

> Nothing improves a country's prosperity more than a powerful and positive national image, and nothing improves a country's image more than working internationally and contributing to the international community, tackling the 'grand challenges' in partnership with other nations and organisations.

Seen in this light, there might be a way forward if the Indian government used its power to put pressure on Dow not to make a profit by exploiting people and environment in India but to adhere to CSR and ESG principles and thereby improve the image of both India and Dow.

7   Conclusion

There is movement in the right direction, although obviously much remains to be done. The need for urgent steps to preserve the planet is beginning to be recognised and put into practice. Further, there is growing recognition that commercial enterprises should aspire to social and environmental criteria for success as well as financial ones.

The international community should recognise that the 'environmental' agenda must give higher priority to eliminating toxic waste, including left-over pollution from the past, and that this is not only an environmental issue but affects human rights to the essentials of life: clean air and clean water. Although experience with the law of genocide gives grounds for doubt whether criminal prosecution is the most effective enforcement mechanism, the creation of an international law of ecocide is an urgent necessity and might provide the impetus to reaching a mediated out-of-court settlement, such as the one that was reached under domestic law at Toms River (Fagin, 2013). The polluter should pay both by compensation and by bearing the cost of cleaning up. In cases where a company has been taken over or restructured, the principle of successor liability should apply. As part of its reparation a company should co-operate with investigations into the cause of the damage, with a view to preventing similar disasters, for example, through safety measures, staffing levels, training, and company culture. The threat of a penalty could be a disincentive to such co-operation: it has been said that 'punishment is the enemy of truth', but reparative consequences benefit everyone. Commercial companies are beginning to see the advantages of declaring adherence to principles of benefitting the world and its inhabitants, and governments and NGOs have a vital part to play in monitoring the companies' compliance with these ideals.

References

Anholt, S. (2020). *The good country equation: How we can repair the world in one generation.* Oakland, CA: Berrett-Koehler Publishers, Inc.

BBC (2020). *India gas leak: At least 11 dead after Visakhapatnam incident.* Retrieved from: https://www.bbc.com/news/world-asia-india-52569636 (last accessed 5 February 2021)

Bhopal Medical Appeal (BMA) (2012). *The Bhopal marathon*. Brighton. Retrieved from: https://www.bhopal.org/wp-content/uploads/2015/03/Bhopal-Marathon-email.pdf. (last accessed 14 October 2021).

Bhopal Medical Appeal (BMA) (2019). *Suppressed study finds birth defects seven times higher with gas-exposed parents*. Retrieved from: https://www.bhopal.org/suppressed-study-finds-birth-defects-seven-times-higher-with-gas-exposed-parents/ (last accessed 5 February 2021).

Braithwaite, J. (1989). *Crime, shame and reintegration*. Cambridge: Cambridge University Press.

Braithwaite, J. (2002). *Restorative justice and responsive regulation*. New York: Oxford University Press.

Business Roundtable (2019). *Statement on the purpose of a corporation*. Retrieved from: https://s03.s3c.es/imag/doc/2019-08-19/Business-Roundtable.pdf (last accessed 5 February 2021).

Corbett, S. (2019). The art of gentle protest. In J. A. Huxley (Ed.), *Generation Y, spirituality and social change* (pp. 98–106). London/Philadelphia: Jessica Kingsley Publishers.

Edwards, D. (2015). Editorial. *Process Safety and Environmental Protection*, 97, 1–2.

Fagin, D. (2013). *Toms River: The story of science and salvation*. New York: The Random House Publishing Group.

Fisher, R., & Ury, W. (1991). *Getting to yes: Negotiating an agreement without giving in*. London: Random House Business Books.

Gibbons, A. (2019). Entropology and the end of nature in Lance Olsen's theories of forgetting. *Textual Practice*, 33(2), 280–299.

Hanna, B., Morehouse, W., & Sarangi, S. (Eds.). (2005). *The Bhopal reader: Remembering twenty years of the world's worst industrial disaster*. Goa/New York: Other India Press/The Apex Press.

India Today (2020). *Vizag gas leak: How events unfolded & why it happened | All you need to know*. Retrieved from: https://www.indiatoday.in/india/story/vizag-gas-leak-tragedy-visakhapatnam-lg-polymers-all-you-need-to-know-1675509-2020-05-07 (last accessed February 2021).

Jaising, I., & Sathyamala, C. (1995). Legal rights ... and wrongs: Internationalising Bhopal. In J. Kirkby, P. O'Keefe, & L.Timberlake (Eds), *The earthscan reader in sustainable development* (pp. 174–181). London: Earthscan Publ.

Kemsley, J. (2013). FATALITY Dow Chemical worker dies from burns sustained in fire at electronic materials facility. *Chemical & Engineering News*, 91(42), 7.

Kershen, L. (2019). Implementing restorative justice to environmental harm. In E. Biffi & B. Pali (Eds.), *Environmental justice: Restoring the future* (pp. 46–53). Leuven: European Forum for Restorative Justice.

Kershen, L. (2021). Restorative approaches to environmental harm: Shifting the levers of power. *The International Journal of Restorative Justice*, 4(1), 157–165.

Macleod, F. (2014) Impressions of Bhopal. *Inst. Chem. E Loss Prevention Bulletin.* 240, December, 3–9.

Macrory, R. (2006a). *Regulatory justice: Making sanctions effective. Final report.* London: Cabinet Office. Better Regulation Executive.

Macrory, R. (Ed.) (2006b). *Reflections on 30 years of EU environmental law: A high level of protection?* Groningen: Europa Law Publishing.

Makower, J. (2019). *The Business Roundtable's statement of purpose, one year on.* Retrieved from: https://sciencemetro.com/energy/the-business-roundtables-statement-of-purpose-one-year-on/ (last accessed 5 February 2021).

Mehta, J. (2019). Restorative justice for nature: A tribute to Polly Higgins and the power of a law of ecocide. In E. Biffi & B. Pali (Eds.), *Environmental justice: Restoring the future* (pp. 26–28). Leuven: European Forum for Restorative Justice.

Nesmith, G., Keating, J. T., & Zacharias, L. A. (2013). Investigating process safety near misses to improve performance. *Process Safety Progress*, 32(2), 170–174.

Pearce, F. (1990). Responsible corporations and regulatory agencies. *Political Quarterly,* 415–430.

Pure Earth (2016). *World's worst pollution problems: The toxics beneath our feet.* Retrieved from: https://www.worstpolluted.org/2016-report.html (last accessed 5 February 2021).

Sky News (2020). *What happened in Beirut: The explosion that shocked the world.* Retrieved from: https://www.msn.com/en-gb/news/world/what-happened-in-beirut/ar-BB17DDRL (last accessed 5 February 2021).

Stop Ecocide Foundation (2021). Independent expert panel for the legal definition of ecocide: Commentary and core text. Retrieved from https://static1.squarespace.com/static/5ca2608ab914493c64ef1f6d/t/60d1e6e604fae2201d03407f/1624368879048/SE+Foundation+Commentary+and+core+text+rev+6.pdf (last accessed 18 January 2022)

Tremblay, J.-F. (2007). An offer rejected: Prominent U.S. company gets nowhere with its offer to clean up Bhopal. *Chemical and Engineering News*, 85(6). Retrieved from: https://cen.acs.org/articles/85/i6/Offer-Rejected.html (last accessed 14 October 2021).

Unilever (2021). *Unilever commits to help build a more inclusive society.* Retrieved from: https://www.unilever.com/news/press-releases/2021/unilever-commits-to-help-build-a-more-inclusive-society.html (last accessed February 2021).

Union Carbide Corporation. (n.d.). *Union Carbide Corporation's (UCC) response efforts to the tragedy and the settlement.* Retrieved from: https://www.bhopal.com/ucc-tragedy-response-efforts.html (last accessed 5 February 2021).

Ury, W. (1991). *Getting past No: Negotiating with difficult people.* New York: Bantam Books.

Weick, K. E. (2010). Reflections on enacted sensemaking in the Bhopal Disaster. *Journal of Management Studies*, 47(3), 537–550.

Wright, M. (2019). Restorative justice with corporations: The idea and the practicality. In B. Pali, K. Lauwaert, & S. Pleysier (Eds.), *The praxis of justice: liber amicorum Ivo Aertsen* (pp. 281–292). The Hague: Eleven International Publishing.

26

Environmental Restorative Justice: Activating Synergies

Ivo Aertsen

The global only exists by the generosity of the local

1   Introduction

I came across the above quote painted on the fence of Labiomista last summer. Labiomista is the name of an arts centre in the city of Genk, Belgium, designed and developed by the Flemish artist Koen Vanmechelen and built on the ground of a former coal mine (Vanmechelen & Magiels, 2019). This (literally translated) 'mix of life' initiative is meant to be more than just a typical arts centre. It provides a space for the cross-pollination of arts, local community, and city council and accentuates and explores the transition from culture to nature. Prominent in the Labiomista is the theme of the chicken, a forest bird long ago domesticated by humans. In reflecting on the ongoing process of the chicken's evolution through

I. Aertsen (✉)
KU Leuven, Leuven, Belgium
e-mail: ivo.aertsen@kuleuven.be

cross-breeding to sustain diversity in healthy ecosystems, these artistic depictions of the chicken are a metaphor for a living, organic, and unfolding cross-fertilisation of ideas—a symbol standing for a 'cosmopolitan renaissance':

> We are in a revival, a cosmopolitan renaissance where we make the combination again between imagination and knowledge. (...) We have to be brave to see the real role played by the human animal and to seek out solutions far beyond our familiar surroundings. In other cultures, on other continents, with other people. Diversity is a resource to be tapped on a global level; the local level has to get a global extension, and vice versa. For without the generosity of the local level, the global level shrivels. Renewal is generated by the connection between both.[1]

I begin with this short detour to a Flemish arts centre because, for me, it reveals a common thread of relationality and willingness to test out the unfamiliar, to braid together the known into something new, that runs throughout many chapters of this book on environmental restorative justice (ERJ), with its multitude of innovative ideas and approaches presented and discussed.

In this concluding chapter, I will not reiterate all the promising ideas, of which—it must be admitted—some are still in an embryonic phase or remain somewhat abstract. At the same time, various chapters report on very concrete steps that have been made already, be it at the level of local practice or more general policymaking. This is an important observation indeed: ERJ is not just a vague idea, nor an idealistic excursion or *spielerei* (frivolity). The new field shows the contours of a starting practice, theoretically and conceptually underpinned by existing notions and insights on 'restorative justice', gradually developing internationally in an ongoing relationship between practice and research. One of the main opportunities is that this exercise of reflecting on, and testing, the applicability of restorative justice to the complex field of environmental harm and crime invites us to further develop—even to 'rethink'—some of our core notions of restorative justice.

---

[1] Retrieved from https://www.labiomista.be/en/cosmopolitan-renaissance (last accessed 11 February 2020).

In what follows, I discuss some of the main themes I consider important without aiming for completeness and certainly without claiming to reach final conclusions. I join the authors of this book in a common, and hopefully effective, learning process and hope to contribute to what might become a future agenda for ERJ. My reflections go back, not only to the wealth of ideas and experiences presented in the chapters of this book, but also to several other initiatives organised on the topic, such as the inspirational discussions held at the international Oñati workshop organised by Brunilda Pali, Miranda Forsyth, and Gema Varona in June 2021; a first exploratory seminar on ERJ that we organised at KU Leuven in April 2019; the creation of a special issue on ERJ for *The International Journal of Restorative Justice* (2021, issue 1);[2] discussions held in the Working Group on Environmental Restorative Justice and resulting publications of the European Forum for Restorative Justice (EFRJ); and networking with colleagues who are involved in this topic in different regions of the world. In addition, the EU-funded research project on Victims and Corporations, coordinated by our colleagues from the Università Cattolica del Sacro Cuore, Milan (Forti et al., 2018), helped me considerably to structure some nascent ideas on how to reconceptualise and reorient restorative justice towards the environmental field.

2   What Brings Us Here?

What unites the authors of this edited volume is the exploration of 'A new justice framework for preventing and addressing environmental harms', as the title of this handbook instructs us. Whereas many of us come from the field of restorative justice, others—environmental scholars, activists, and artists—have joined the collective. As argued by various authors in this volume and implicitly assumed by all, the field of restorative justice needs a considerable extension of its scope and a truly interdisciplinary approach. It is good for restorative justice scholars and practitioners to leave their cocoon, to become aware of the relatively restricted reach of restorative justice programmes in the Western world

---

[2] https://www.elevenjournals.com/tijdschrift/TIJRJ/2021/1.

and its limited potential to fundamentally reorient our justice systems. Many working in the field of restorative justice feel that a more encompassing framework is needed, one in which there is room for the exploration of new dimensions.

A first dimension relates to restorative justice's extended field of application. During its first three decades, until the beginning of the 2000s, restorative justice was associated with conventional types of crime that have easily identifiable individual victims and offenders who are supported by some members of their immediate communities of care. More recently, the search for restorative justice models to be applied to more serious crime, to forms of collective and political violence, to cases of terrorism and extremism, to hate crime and racism, to various types of (historical) violence and abuse within institutional contexts, and to corporate crime—to name the most important areas—has started. A strategic choice has supported this expansion: not only with the intention to demonstrate the wide applicability and relevance of restorative justice, but also in order to incite the interest and involvement of larger groups of citizens who are expected to be concerned by these types of crimes and harms. In short, developing new areas of application has been a strategic lever for the restorative justice movement.

A second dimension has to do with the deliberate objective to reach out beyond the field of criminal law. We are now talking about human-caused 'harms' and 'injustices'. This extension, again, has been defended vigorously by many authors in this volume and for understandable reasons: (1) many forms of harmful environmental behaviour are not—and will never be—subject to legal criminalisation, so sticking within a criminal justice context would imply that many or even most of these human-made harms would never be dealt with by restorative justice responses and would therefore be devoid of an integral and balanced approach with respect to the needs of all stakeholders; (2) environmental matters typically form a field that in most countries is increasingly regulated by civil and administrative law; (3) operating exclusively within the field of criminal law would run the risk of uncritically reinforcing legal understandings of justice that confirm or even strengthen social and economic power constellations that promote current hegemonic arrangements enabling widespread environmental harm (see also, in a more general way, the critical comments by Hydle & Henriksen in this volume on juridification and judicialisation as possible manifestations of a trend to 'colonise the lifeworld', in Habermas' terms). Some of these arguments may be even more decisive for countries in the Global South (indeed, this has been touched upon for several Latin American, Asian, and African countries). In other words, the search for 'justice frameworks' and for 'doing justice' mechanisms can no longer be restricted to the field of criminal law.

A third dimension that indicates the need for a broader framework concerns the preventative approach. Several authors in this volume enlarged their analysis into this direction, whereas restorative justice traditionally reflects a more reactive approach. This preventative orientation goes hand in hand with the—broadly formulated—relational approach that is so strongly emphasised in many of the book chapters. I will come back to this dimension later on—it suffices here to remind the reader of the 'ongoing relational process' character in time and space that analytically shapes the interactive and 'healing' worlds of human and other-than-human beings and ecosystems as a whole.

Reviewing these three (new) dimensions as sketched out in the previous paragraphs, it is indeed restorative justice that brings us together here, but restorative justice not so much in the sense of a practical framework or tool for intervention, but much more conceived as an all-encompassing approach, *an ethos*, to look at the most diverse forms of ecological harm and threat, including human alienation from nature and problematics as broad as climate change. Whereas this broad approach might be attractive, fundamental, and inspirational for many, it might also cause increasingly a blurring of definitions, concepts, boundaries, and therefore has potential to impact a whole field of scientific study and practice (Walgrave, 2021, pp. 287–333). As such, care must be taken to ensure that while the ethos informs and helps push the boundaries of our 'cosmopolitan renaissance', both conceptual clarity and practical outcomes and tools for intervening should remain at the forefront of ERJ.

3   Understanding Harm

Environmental *crime*, as we are most familiar with, refers to unlawful acts such as the illegal taking or trading of non-human species (flora and fauna), pollution offences, and the transportation of banned or toxic substances (radioactive or hazardous material). Environmental *harm*, caused by different types of human behaviour, negligence, or omissions, is a far more complex and multi-layered phenomenon. The subject of environmental harm can be wide and diverse: natural resources (private or public propriety, communal property such as air, water, and forests, and unowned property such as light), public infrastructure, heritage, environmental meaning (sense and use of the environment by a community), and impact on future generations.

An important feature of environmental crimes and harms relates to their (oftentimes) *invisibility* in terms of effect in time and causal relationships, the diffuse victimisation of human and non-human beings, the lack of clarity about a particular agent's accountability or cross-border character of their operations, and the underlying structural and cultural issues (Varona, 2020). The multi-dimensional character of environmental harm is abundantly explained in many chapters of this volume. Besides harm at the physical, material, financial, psychological, and social levels for both individuals and communities, there is the harm to nature: to other-than-human beings and the whole ecosystem. Harm to people and the environment can stay *silent* as well during many years until the passage of time reveals its seriousness (Bolivar et al., this volume); the latter was found to be the case often also for victims of corporate violence—for example, in the pharmaceutical industry or in the production of asbestos—because of the long latency period—up to 30 or 40 years—before physical consequences appear (Aertsen, 2018). Harms are also not always visible or tangible, which complicates, delays, or impedes adequate responses.

Green criminology and green victimology have largely contributed to a better understanding of the heterogeneous and sometimes diffuse nature of environmental crime and victimisation, including so-called victimless crimes or faceless victimisation (see, amongst others, in this volume the chapters by Di Ronco & Chiramonte, Varona, and White). Environmental victims are not always aware of their victimisation, not only because of the factors mentioned above, but also because of the frequently amorphous character of the harming behaviour or its repeat and normalising manifestations. Finally, environmental crime often affects large groups of victims (Hall, 2014).

The occurrence of harm in the case of environmental crime is often obscured by structural and cultural issues. Harm appears within a context of systemic injustices, extreme power imbalances, and high victim vulnerability. Harm is often presented by those in power as 'inevitable' or as 'collateral damage' in the pursuit of economic development and progress. Harm also appears indirectly, in the form of a kind of secondary victimisation: when there is a lack of adequate reaction, harmed people begin to lose trust in government, the justice system, and corporations, while local communities can become divided due to differing opinions, a lack of solidarity, or dependence on the victimising employer (Bolivar et al., this volume).

An important theme throughout this book (and elsewhere) concerns the warning against taking a one-sided anthropocentric approach when looking at environmental harm and preferred responses. In this respect, a consensus seems to exist, namely, that we have to adopt a fundamental change of mindset: from an anthropocentric to a bio-centric and eco-centric approach (White, 2018). We should stop considering the harmed environment 'as a disrupted resource' in function of our human needs. Harm, and the devastating effects of environmental crime, must be understood more completely, taking into account the consequences for other-than-human beings, the ecosystem in general and future generations. Nature has, indeed, its own rights.

Many of the foregoing book chapters discuss the legal recognition of the Rights of Nature at the constitutional level or otherwise, where 'legal systems would regulate the interests and relationships between humans and all other beings within the Earth Community' (Wessels & Wijdekop, this volume). The international Rights of Nature movement is significant in this respect.[3] Applied to climate change, Jones (this volume) speaks of

---

[3] See Global Alliance for the Rights of Nature: https://www.garn.org/.

'wrongs' or 'climate injustices' when disproportionate emissions go together with the exploitation of people and natural resources, misinformation about climate science, and failure to attend to voices or needs of those most affected. In still another context, harms of 'minimalisation' and repression, including the criminalisation of activists, on top of the original ecological harm, are interpreted as 'harm to knowledge' by Di Ronco and Chiaramonte (this volume).

'Injustices' are also interpreted by Varona (this volume) when she refers to the cultural justification of harm against animals for utilitarian reasons, where 'harm might be tacitly acknowledged, but not its injustice'. She argues for a de-objectivation of animal life and animals' full recognition as 'sentient beings'. But Varona is also aware that we as human beings will always adopt an irreducible anthropocentric view when talking about harm and repair. It seems impossible to make abstraction of our human constitution and being, we can only try to connect as best as possible. To face this challenge, 'imaginative thinking' will be needed to feel and understand the distant suffering of—in this case—animals. We can do so through dialogical processes or forms of inter-species communication, for which Varona proposes a circle process in three phases.

A way to understand and to address environmental injustices in individual cases and also at the structural level—and in a departure from traditional crime and justice models—is to look at them through the lens of 'harm landscapes' or 'harmscapes' as elaborated in literature and used by Dore and colleagues for wildlife offences in South Africa (this volume). Contemporary 'harmscapes' are characterised by both radical uncertainty and unpredictability and they reflect the complex nature of some types of environmental harm, being 'multifaceted, inter- and intra-generational, even circular' (Dore, Hübschle & Batley, this volume).

In sum, when speaking about 'environmental harm' and recognising its diverse, multifaceted, and often unclear character, 'it can be very tricky to identify what harm clearly and definitively can and should be repaired, and who are the victims, offenders, regulators and broader community actors, when these roles are blurred and shifting across time and space' (Amparo et al., this volume). This broad understanding of harm poses enormous challenges to restorative justice to come up with appropriate answers. As we will see below, there are no one size-fits-all models available. A uniform model is probably not desirable either, as long as the perspective of looking at and addressing environmental harm in concrete cases remains open and fluid, and suitable procedures are being adopted in a creative way in order to involve all types of 'stakeholders'.

4 Doing Justice

The previous chapters in this book not only informed us about what is 'harm' in the case of environmental injustices, but they also brought to the fore new concepts and understandings of what is—or can be—'justice'. How can justice be conceived in this respect, and how do endeavours towards justice based on legally established frameworks relate to justice processes as experienced in the lifeworld of human and other-than-human stakeholders?

It will not be a surprise—as referred to above—that 'environmental justice' is still often formulated from an anthropocentric perspective and that, in other words, we are never sure that full justice—be it always an open, unfinished aspiration—has been done. In that sense, Forsyth et al. (2021, p. 23) define environmental justice as 'a normative ideal [that] is achieved when communities have the right and capacity to participate meaningfully in environmental decisions affecting them, and when no individual or group is disproportionately impacted by the outcomes of such decisions, including future generations'. Furthermore, they stress the greater emphasis that environmental justice, as compared to traditional restorative justice, places on 'structural causes and consequences of uneven distribution of harms across time, space and demographics' (ibid., p. 24).

While various authors in this volume focus their attention to justice processes provided by judicial and other regulatory institutions for the sake of human beings' communities, others explore new ideas of justice beyond the human atmosphere. Bosselman (this volume) thinks of 'justice' in function of a sustainable environmental future and therefore elaborates on three dimensions of 'social justice': '*space* to include all humans ('global' or 'intragenerational justice'), *time* to include humans living in the future ('intergenerational justice') and the human community as part of the larger community of life ('inter-species justice')'. To attain real sustainability, Bosselman stresses that we should not 'overlook our belonging to the community of life. We are not separate from nature as conventional Western philosophy and jurisprudence assumes, nor are we in any way superior to non-human species. We are just part and parcel of a single ecological system, called Earth'. For this reason indeed, Bosselman prefers the notion of 'ecological justice' to the notion of 'environmental justice' as the former 'adds the concern for inter-species justice in reflection of ethical ecocentrism'. What therefore is at stake in a restorative justice approach is 'to bring non-human victims—fauna, flora and eco-systems—into the conversation'. This brings Bosselman (see also Wessels and Wessels & Wijdekop, this volume) to speak of 'Earth Jurisprudence' and to describe 'Earth restorative justice as a social movement concerned with restoring the broken human-nature relationship'.

The question 'who is justice for?' is also a vital one for White (this volume), 'particularly in relation to non-human environmental entities such as animals, plants, rivers and mountains'. Conceptions of justice dealing with humans, with ecosystems and biospheres, and with non-human animals and plants form together a system of eco-justice—a system of justice that has also to find its place in relation to criminal law. Whereas in 'environmental justice', which according to White refers to 'the equitable distribution of environments among people in terms of access to, and use of, specific natural resources in defined geographical areas', humans are at the centre of the analysis, 'ecological justice' refers to the relationship of humans generally to the rest of the natural world and includes concerns relating to the health of ecosystems, and the plants and animals that also inhabit these systems. To answer the question 'what non-human rights might consist of', White—referring to literature—identifies four key dimensions of justice as including distribution, recognition, participation (basically involving human advocacy), and capability. 'Doing justice', then, is seen by White in function of desired outcomes: 'the protection of the needs, rights and interests of non-human environmental entities such as rivers, birds, trees and mountains, as well as humans'. However, such a system of active, participatory justice will be confronted at a certain moment with conflicting rights and interests, involving humans, ecosystems, and species and therefore—according to White—will require a weighing within an institutional context of criminal justice and environmental regulatory systems.

The inherently *intergenerational* nature of environmental justice was formally recognised in law in the Philippines as the first country to do so. According to Amparo and colleagues (this volume), citing from national jurisprudence in 1993: 'every generation has a responsibility to the next to preserve that rhythm and harmony [of nature] for the full enjoyment of a balanced and healthful ecology'. In other words, future generations are given 'standing' or the right to sue. The intergenerational dimension of environmental justice has been emphasised—for self-evident reasons—in the chapters on climate change (Almassi and Jones, respectively). For Almassi (this volume), 'climate injustices' consist of 'moral damages to our relationships', at the intercultural, inter-species, and intergenerational level. Therefore, 'climate justice' implicates an 'asynchronous' and 'relational process of moral repair'. This process should not be reduced to compensation or restitution but should entail full restorative or reparative mechanisms: these offer 'a better way to acknowledge and ameliorate the injustices of anthropogenic climate change than compensation for "loss and damage"'. 'Ameliorative responsibility' means being accountable and 'making amends to repair and renew the conditions upon which cross-generational relationships can be rebuilt'.

This intergenerational relational ethic is also sensible to *epistemic injustices*, which occur 'when people are wronged in their capacities as knowers' (Almassi, this volume). The latter happens to historically marginalised and oppressed (Indigenous) people, when their 'lived experiences and perspectives [are] dismissed, devalued, misunderstood, misattributed, and appropriated by dominant knowers and knowledge systems' (see also the notion of 'cognitive justice', Rodriguez, this volume). Epistemic repair, then, is not just a personal process but should be considered as 'the restoration of conditions for trustful, collaborative knowledge-making', which requires a multi-agential process actively involving not only the directly affected stakeholders but us all.

For Jones (this volume), climate injustice also refers to the unjust position, across the world, of those least responsible for climate change but suffering its severest impacts and having the least access to decision-making power. Climate justice requires—now that the 'connection between proximity and responsibility' is being lifted—global action. In this regard and more fundamentally, Jones refers to the work of Margaret Urban Walker, who argues for an 'expressive-collaborative' approach to ethics, 'focusing on responsibilities, including those where factors such as colonialism and exploitation create relationships between distant actors'. For Walker (2006), restorative justice has to address these historic injustices, by correcting the 'profound distortion of relationship' that has prevented 'white communities from recognising their own complicity and from acknowledging the need for apology'. Restoring this type of relationship aims at achieving a certain level of 'moral adequacy', not only at the level of personal emotional connections, but foremost by 'resetting a moral compass' more generally. One of Walker's principles of restorative justice is then 'doing active justice', which entails elements of transformative justice, 'conscious of its responsibility to seek fundamental change to societal structures, processes, values and relationships' (Jones, this volume).

ERJ should be able to (better) engage with Indigenous understandings of justice, as Killean (this volume) argues when speaking about the role of transitional justice in a country such as Colombia. Transitional justice, she observes, does not usually engage much with environmental issues, let alone with harms to other-than-human beings. By accepting the 'dominance of legalism' in general and the role of international criminal law in particular, transitional justice conforms—once again—to an anthropocentric approach. Killean argues for the involvement of inter-species communication in transitional contexts, for example, by 'involving victimised communities with a "special relationship" to land and other-than-human victims'.

In short, it becomes clear that restorative justice, applied to environmental crime, goes far beyond the level of a one-off reaction to a particular harmful event. If it really aspires at 'doing justice', restorative justice has to be situated in a broader time and space context and must broadly address the relational, cultural, social, economic, and political dimensions of the problem.

5   The Role of Law

Obviously, the type(s) of justice as discussed above largely exceed the capacities of any (human-made) legal system, and this is recognised by environmental justice movements (Minguet, 2021). Still, to keep both feet on the ground, it is worthwhile to examine how existing legal systems deal with environmental injustices so far (and—see next section—whether and where restorative justice can find a place in this).

Various authors in this volume analyse the role of law and in particular the role of jurisprudence and the functioning of the courts. As there are important differences between legal systems worldwide (White, this volume), the possibilities for legally dealing with environmental harm can be very different and sometimes hard to compare. Where we still might be able to compare our criminal law systems (as we do in order to understand the position of restorative justice in various countries), other legal frameworks related to regulatory provisions in civil and/or administrative law are less addressed by restorative justice scholars. An important exception to the latter is offered by the Australian colleagues in various chapters of this volume. In particular, the chapter by Forsyth on how restorative justice can be 'braided' systematically into environmental regulation in a broad sense is instructive in this respect. She explores the different domains of environmental regulation (approvals and licensing, inspections, and responding to environmental harm) and explains—theoretically and practically—how practice and policymaking in these domains can be improved by making use of an iterative process of recurrent phases (which had me thinking of the phases followed in participatory action research), based on restorative justice values and principles.

But as said, most analyses remain in the—for restorative justice practitioners—safe haven of existing civil law and criminal law procedures. The chapter by Braga Da Silva clarifies how attempts have been made in Brazil, where court procedures have been dealing with two recent mining tailings dam disasters, to integrate restorative elements in legal reasoning, procedures and agreements, and how both legal and non-state actors have been operating within this context. Interestingly, a civil law context clearly seemed to offer more possibilities for a restorative justice approach than criminal law procedures (Braga Da Silva, 2021). This seems also to apply to the Bhopal case (Wright & Tabbert, this volume), where civil litigation results in financial compensation, but in criminal procedures 'the legal battle is still pending'.

How restorative justice can interact with environmental criminal law is further explored by Perini (this volume). Where criminal law is based on a principle of 'absolute prohibition', administrative and civil law procedures establish (only) 'burdens' on the responsible actors based on the principle of 'pricing' (putting a price on environmental offending). Therefore, criminal law and civil law offer qualitatively different contexts, the former focusing much more on important social values (such as the value of environment). For Perini, a synergy of restorative justice and environmental criminal law can support the protection of relationships between the individual and their environment and between community members. Restorative justice in a criminal law context may help to understand collective victimhood and to develop 'regenerative pathways' to restore relationships. Such a restorative approach seems to be consistent with the 'environment personalistic conception' as this appears in the European legal space, based on European Union and Council of Europe regulation and (also human rights) jurisprudence.

This relational approach to law is also discussed in other chapters of this volume, including the one by Wessels as applied to Earth jurisprudence. Based on constitutional and other legislation in countries such as Ecuador, an eco-centric approach to litigation becomes possible, showing 'the potential to influence and strengthen relationships between humans and Nature when overarching policies and executive agendas of governments fail'. Here, the harm 'is not remedied by placing an emphasis on either retribution or compensation'. However, Wessels argues, as most national legal systems do not recognise the rights of nature, this approach is less likely to change legal relationships.

This brings us to the critical issue of criminalising behaviour that is harmful to the environment. The 'dominance of legalism', as referred to above, often coincides with the dominance of a 'crime-driven lens', hence the role of punishment comes to the fore (Killean, this volume). What is, then, from a restorative justice perspective, the added value of criminalisation, taking into account the—*de iure* and *de facto*—limitations of criminal law as discussed above and the prevailing absence of legal recognition of the Rights of Nature as a value or an entity on its own? How realistic is it to expect a real impact of criminalisation at the level of human and social relations and at a level that transcends a predominantly anthropocentric approach? Here, the discussion on the notion of 'ecocide' appears, and whether and how it could be meaningful to include the crime of ecocide in the Rome Statute as a fifth crime against peace, and thus whether the crime of ecocide could be brought under the jurisdiction of the International Criminal Court (ICC) in The Hague. Whereas this type of international criminalisation at the highest level is being promoted strongly from an environmental activist perspective, it poses fundamental questions for restorative justice advocates.

These issues were considered in depth when the EFRJ's Working Group on Environmental Restorative Justice provided feedback to the European Commission on the occasion of a public consultation in spring 2021 with respect to a possible revision of Directive 2008/99/EC of the European Parliament and of the Council of 19 November 2008 on the protection of the environment through criminal law. One of the survey questions was about 'measures to foster a more deterrent criminal sanctioning system with regard to environmental crime', referring to the need for more 'effective, proportionate and dissuasive' sanctions. The Working Group's point was that the identification of such sanctions 'should be supported by scientific evidence, in order to better understand what kind of sanctions and what kind of procedures have the potential to meet these expectations. The assumption that criminal sanctions always entail a deterrent effect, should be critically approached'. Reference was made to possible harmful side-effects of punishment on the convicted entity and/or on his/her environment. Moreover: 'Criminal justice procedures alone do not provide adequate tools to confront the perpetrator with the consequences of his/her harmful behaviour and to appeal to his/her responsibility. The criminal justice context often allows, or even encourages, perpetrators to adopt a strong defensive, minimalising or rationalising attitude. The victim's needs are usually not at the centre of a criminal procedure, and little consideration is given to how to repair the harm created'. And finally: 'Criminalisation of harmful behaviour may convey a message of public disapproval, but the realisation of its potential to initiate a sense of responsibility to perpetrators will be enhanced by incorporating participatory and restorative elements in its procedures'.

With respect to the issue of criminalisation and the debate on the inclusion of ecocide in the Rome Statute, we should have a clearer idea in advance on what is precisely hoped for in terms of impact of these (ICC and other) procedures and how such goals can realistically be achieved. Public censure might be an important function of criminal law indeed, but the element of retribution or punishment in the sense of intentional pain infliction and additional suffering is dysfunctional and unnecessary. Proposing criminalisation in and of itself is not enough, as it cannot go without a reflection on the contents of proposed penal sanctions and their effects. Moreover, results of international criminal court procedures towards the reparation of harms to victims are rather mixed, and therefore, we should not take it for granted that human and non-human victims of environmental crime will benefit automatically from this type of jurisprudence. These elements should be part of impact assessment analyses by bodies such as the European Commission when preparing (binding) EU legislation.[4] Be that as it may, two of the conclusions of the accompanying paper by the European Forum's Working Group with their feedback to the European Commission were the following:[5]

* Criminal law provisions related to environmental crime should not operate in an isolated way but must be conceptually integrated in a more encompassing regulatory and sanctioning framework entailing a broader range of formal and informal justice mechanisms. Restorative justice has to be situated at the bridge between different types of justice mechanisms.

---

[4] See, for example, European Commission Impact Assessment Report Brussels with the proposal for a new Directive on environmental criminal law (Brussels, 15.12.2021, SWD(2021) 465 final/2). Retrieved from https://eur-lex.europa.eu/legal-content/EN/TXT/PDF/?uri=CELEX:52021 SC0465R(01)&from=EN (last accessed 9 February 2022).

[5] European Forum for Restorative Justice Comments on the EU Directive 2008/99/EC of the European Parliament and of the Council of 19 November 2008 on Improving Environmental Protection through Criminal Law. Retrieved from https://www.euforumrj.org/sites/default/files/2021-08/EFRJ%20contribution%20EU%20Directive%20Environmental%20Criminal%20Law%203.05.2021%20%281%29.pdf (last accessed 9 February 2022).

* Constructing a legal basis for the use of restorative justice in processes dealing with environmental crime is essential. Legislation should primarily facilitate (not restrict) the use of restorative justice in these cases by providing a safe space (legally, socially, psychologically) where genuine dialogue is encouraged and supported and effective reparation can take place.

6   Realising Restorative Justice

The development of ERJ does not only require an encompassing regulatory and sanctioning framework, but also a broad policy framework at various levels allowing for multi-actor forms of government (see chapters by Amparo and Bosselman in this volume). Also, Vasilescu (this volume) warns of the dangers, on the basis of a comparison of the goals, values, and outcomes of participatory governance versus restorative justice, of a 'context blind process'. She says that when developing restorative justice in the field of environmental crimes and conflicts (or in general), '[f]ailure to take into consideration the features of the context of the process may put at risk its effectiveness and reinforce social inequalities'. What is needed to ensure 'an effective *and* just ecological transition' is hearing and understanding the justice perspectives of all involved, not just the concept of justice as defined by a single actor. Restorative justice, in particular, can help us to come to a shared vision of justice. Other lessons can be learned from the blending of participatory governance and restorative justice, including the potential to educate citizens through participatory processes and to make them feel responsible for, and to engage with, policy-related problems of a more general nature. This would provide a positive model on how to 'involve the public in combatting environmental crime' as is increasingly termed in official policy papers.

Now, turning to the practical applicability of restorative justice to a variety of cases of environmental harm and crime, it is clear that some of our familiar conceptualisations related to actors and processes in

I. Aertsen

restorative justice have to be reconsidered or adapted.[6] This is so with respect to the definitions and positions of the victim, the offender, and the community and the process of participation and desired outcome of restoration. Before I elaborate, it should nevertheless be mentioned that classic forms of alternative dispute resolution, such as 'environmental mediation', may apply and have been applied for many decades: 'Environmental mediation is a process in which representatives of environmental groups, business groups and governmental agencies sit down together with a neutral mediator to negotiate a binding solution to a particular environmental dispute' (Amy, 1983). But indeed, to make restorative justice fit the complexity of environmental crime as discussed above and to respond to an eco-centric rather than an anthropocentric approach, notions and practices will have to be modified in a flexible and creative way.

Let us first have a look at the stakeholders. In the case of corporate environmental crime, there might be many 'victims', directly or indirectly affected by an incident: persons internal to the company (employees and their families, shareholders, and investors) and external (harmed people with no relationship to the company, but also other corporations, institutions, and governmental or public bodies). Moreover, future generations, other-than-human beings, and the whole ecosystem also have to be included as possible 'victims'. As referred to above, the latter forms a challenge as it often concerns harms that are not easily definable or tangible and extending over a longer period of time (with uncertain impact on the future). The unclear causal relationship between criminal behaviour (or a crime of omission or negligence) and harm makes it often hard or impossible to collect evidence, and judicial action can be impeded because of legal prescription.

As a principle, all categories of victims and types of victimisation should at least be acknowledged in restorative justice responses. In many cases, additional expert knowledge will have to be called in to define and circumscribe the harm in all its facets. Human victims should explicitly

---

[6] The following paragraphs are based on what we have elaborated in the European Forum for Restorative Justice Comments on the EU Directive 2008/99/EC (see previous footnote) and on the topic of restorative justice for victims of corporate violence (Aertsen, 2018).

receive recognition for the harm suffered, information on the causes, circumstances, and responsibilities around the harmful events, and financial compensation or restitution *in natura* for material and non-material harm. A central need for victims is to be heard and to tell their story. Sometimes a personal approach is required, sometimes a group approach, sometimes it will be an ecosystem approach. For victimless crimes or other-than-human victimisation, delegates of environmental groups, public bodies, community representatives, or scientific experts can be involved to represent the victim dimension.

The 'offender' dimension may constitute an even bigger challenge. This requires the creation of a space for open, non-defensive communication so that the disconnection between harmer and harmed is lifted, and the corporation's representatives can begin considering their responsibilities as a corporation. Why should a company come to the table to listen and talk to victims and community representatives? Would that not legally imply an admission of guilt? The latter can be avoided by special legislation—as exists for more common forms of restorative justice—ensuring that participation in a restorative process does not imply guilt in a legal sense and thus offering legal protection to the offending entity. However, often some social, political, or judicial pressure will be needed to engage the offending entity into a process of dialogue. Self-interest plays a role also in two ways: (1) obtaining a financial deal with (groups of) victims through a process of negotiation, mediation, or arbitration is often more beneficial for a company in terms of financial costs and time; (2) a most important incentive for a company seems to be the will to avoid a process of being labelled as an enemy of the environment, hence it is a matter of reputation and public image. Probably a mixture of extrinsic and intrinsic motivations is at play, as nowadays also a certain degree of socio-ethical responsibility can be expected from profit-making companies (see also Hamilton, this volume, arguing that a company's motivations can be varied, mixed, and changing over time). Here, it is useful to remember the role of shame and shaming processes as effective deterrents (Braithwaite, 1989, 2002; Barnard, 1999; Gabbay, 2007). The active involvement of peers can evoke a powerful process of 'constructive' (i.e., *not* stigmatising) shame to the offending entity and as a result facilitate reacceptance in society. This resonates with the option of not just labelling offending companies as being always the 'bad guys'.

The 'community' in responses to environmental crime is often embodied by representatives of communities or interest groups, by surrogate victims or governmental representatives and independent experts explaining the community impact in a given case. 'Sentencing circles' and 'community impact panels' have been described as potential practical tools for active community participation (Boyd, 2008). However, possible risks have been pointed out as well, such as power imbalances in an informal group, a biased or incomplete assessment of the environmental impact by individual community members, or a lack of uniformity in sanctioning practices, together with the risk of double jeopardy. Moreover, communities are usually not homogenous groups and should not be idealised. However, active involvement of the community in restorative justice processes can shed a special light and can clarify the concrete impact and ripple effect of environmental crime. Restorative justice can also offer special opportunities for society, in particular for developing its social capital and civic interconnectedness; a space is provided for ongoing norm clarification and democratic, political debate (Dodge, 2009; Dzur, 2011), and for citizens to explore and to challenge, for example, a culture of extractivism (Bolivar et al., this volume) and, more generally, 'the morality of commerce, or socioeconomic inequity, or the temptations of great wealth, or the responsibilities of the powerful, or what "represents the law of the land", in a purposeful and meaningful way' (Chiste, 2008, pp. 99–100).

How can the restorative justice principle of 'participation' be effectuated more concretely in cases of environmental crime, given its plurality of stakeholders? Depending on the specific characteristics of a case, a 'forum' will have to be created where participation of multiple stakeholders can take place in a structured and safe way. Models of large-group dialogue and 'community processing' can offer guidance in this respect. Inspiration can also be found in the model of Truth and Reconciliation Commissions (Boyd, 2008). Strengths of the latter are its innovative public character, transparency of the process, the possibility to exercise censure and public condemnation (and shaming), the ability to deal with large numbers of victims and offenders and to include peers and members of the community, and the possibility of providing different types of restoration depending on the needs identified (e.g., social housing or educational services). Storytelling and truth-telling are usually seen as indispensable elements of reconciliation and healing: truth not only to be considered in its legal and factual sense, but also in its narrative, relational, and dialogical dimension. Dealing with the issue of climate injustice at the international level, the idea of a Global Truth and Reconciliation Commission has been referred to (Jones, this volume). A recurrent theme in the debate concerns the issue of representation of other-than-human beings and the whole ecosystem in restorative justice processes. While some chapters remain tacit in this respect, and others warn 'things get very complicated then' (White, in this volume), still others come with concrete proposals on how, for example, elements of nature such as trees and rivers can be given a voice in conferences (see, amongst others, Hamilton and Wessels & Wijdekop, this volume; see also the Ecocide Mock Trial at the UK Supreme Court, Kershen, 2019; and see Pali et al., this volume).

Finally, the element of 'restoration'. Victims of crime, in general, value reparation or restitution by the offender him/herself, and, if possible, on a voluntary basis or through a process of persuasion. In various jurisdictions, proposals have been made to include reparation as an alternative to, or as a purpose of, the sentencing process in cases of environmental crime (see, in this respect, the chapters by White and others in this volume). As practice in New South Wales (Australia) and elsewhere demonstrates, court orders may include the obligation for the offending company to publicise the offence and, its consequences, to carry out specified projects for restoration or the enhancement of the environment, to pay a specified amount to the Environmental Trust, or to organise a training course for its employees. Hamilton (this volume; see also Hamilton, 2021a, 2021b; Pain et al., 2016) concretely explains how restorative conferences have been carried out in New Zealand for a wide range of environmental cases and with participants representing individuals, communities, and the environment. Notwithstanding its legal basis (in New Zealand), the practice has not been implemented broadly. One of its limitations seems to relate to the high level of commitment and energy needing to be mobilised each time. However, seen from an eco-centric restorative justice perspective, the model is most promising. Furthermore, in Scotland, Croall (2017, p. 7) advocates for the involvement of 'appropriate community and environmental groups along with enforcers and prosecutors', where the outcomes of such restorative justice processes may be that firms breaching health and safety regulations are asked to conduct research into safety issues as well as improving safety procedures. A Community Environmental Justice Forum has been applied in Canada (British Columbia) for some crimes under certain conditions, consisting of a two- to five-hour circle process led by an impartial facilitator with voluntary participation by the company, the community, and the enforcement agency (Wijdekop, 2019).

7   Conclusion

The reference to 'a mix of life' in the beginning of this concluding chapter stands for the broad relational approach adopted by many authors of this book when envisaging 'a new justice framework for preventing and addressing environmental harms'. The focus is overwhelmingly on developing new, encompassing, and more just types of relationship: between humans reciprocally and with their institutions and between humans and nature, taking into account the interrelation between local and global spaces and between different time frames. This new, common endeavour is summarised in some way by Tepper (this volume) when conceiving ERJ as a principle to be included in the United Nations Decade on Ecosystem Restoration, stressing the central objective of restoring relationships between human and nature: ERJ 'can contribute towards creating and strengthening relational, knowledge-sharing and collaborative approaches to repair and care for the ecosystems upon which we all depend'. She proposes three channels for restorative justice to contribute to this ongoing relational process: helping to sustain a social-ecological, relational ethos; through education and dialogue (providing support to conflict resolution, to top-down organised participatory events and to bottom-up informal processes); and by full involvement of all concerned.

Extending the field of restorative justice in this way may be challenging—and confusing—for many. We permanently oscillate between, on the one hand, the (unlimited) exploration of a better and more complete framework for eco-justice, including a variety of new stakeholders and a diversity of new types of relationship, and on the other hand, the development of an integrated, innovative conceptual framework focusing on practical outcomes. The former exploration carries with it the risk of completely blurring the boundaries of the field and lapsing into an esoteric world of vague intuition if we fail to keep an eye on conceptual clarity and practical application. The latter, developing a robust—but not rigid—conceptual framework, will allow for innovative, realistic practices, programmes, and policies aimed at enhancing restorative justice thoroughly. In short, expanding the field of restorative justice to the broad phenomenon of environmental harm can bring a promise or can reveal a weakness of the movement. We have to broaden the field of application and the legal context for sure, but pragmatism and realism seem to be imperative too. The domain of environmental harm opens up great opportunities for restorative justice to develop new understandings and practices of doing justice. Pilot projects and test cases are crucial in order to build *ad hoc* expertise, which should be subject to sound evaluation and which can be followed by setting up training programmes for facilitators and other stakeholders. Once more, practice and theory should go hand in hand.

References

Aertsen, I. (2018). Restorative justice for victims of corporate violence. In G. Forti, C. Mazzucato, A. Visconti, & S. Giavazzi (Eds.), *Victims and corporations: Legal challenges and empirical findings* (pp. 235–258). Milan: Wolters Kluwer-CEDAM.

Amy, D.J. (1983). The politics of environmental mediation. *Ecology Law Quarterly*, 11(1), 1–19.

Barnard, J.W. (1999). Reintegrative shaming in corporate sentencing. *Southern California Law Review*, 72, 959–1007.

Boyd, C.C. (2008). Expanding the arsenal for sentencing environmental crimes: would therapeutic jurisprudence and restorative justice work? *William & Mary Environmental Law and Policy Review*, 32(2), 483–512.

Braga Da Silva, C.F. (2021). A maximalist approach of restorative justice to address environmental harms and crimes: Analysing the Brumadinho dam collapse in Brazil. *The International Journal of Restorative Justice*, 4(1), 98–122.

Braithwaite, J. (1989). *Crime, shame and reintegration*. Cambridge: Cambridge University Press.

Braithwaite, J. (2002). *Restorative justice and responsive regulation*. Oxford: Oxford University Press.

Chiste, K.B. (2008). Retribution, restoration and white-collar crime. *The Dalhousie Law Journal*, 31(1), 85–121.

Croall, H. (2017). Corporate responsibility? *Scottish Justice Matters*, 5(1), 6–7.

Dodge, J. (2009). Environmental justice and deliberative democracy: How social change organisations respond to power in the deliberative system. *Policy and Society*, 28(3), 225–239.

Dzur, A. (2011). Restorative justice and democracy: Fostering public accountability for criminal justice. *Contemporary Justice Review*, 14(4), 367–381.

Forsyth, M., Cleland, D., Tepper, F., Hollingworth, D., Soares, M., Nairn, A., & Wilkinson, C. (2021). A future agenda for environmental restorative justice? *The International Journal of Restorative Justice*, 4(1), 17–40.

Forti, G., Mazzucato, C., Visconti, A., & Giavazzi, S. (Eds.), (2018). *Victims and corporations: Legal challenges and empirical findings*. Milan: Wolters Kluwer-CEDAM.

Gabbay, Z.D. (2007). Exploring the limits of the restorative justice paradigm: restorative justice and whitecollar crime. *Cardozo Journal of Conflict Resolution*, 8, 421–485.

Hall, M. (2014). Environmental harm and environmental victims: Scoping out a 'green victimology'. *International Review of Victimology*, 20(1), 129–143.

Hamilton, M. (2021a). Restorative justice conferencing in Australia and New Zealand: application and potential in an environmental and Aboriginal cultural heritage protection context. *The International Journal of Restorative Justice*, 4(1), 81–97.

Hamilton, M. (2021b). *Environmental crime and restorative justice: Justice as meaningful involvement*. Cham: Palgrave Macmillan.

Kershen, L. (2019). Implementing restorative justice to environmental harm. In E. Biffi & B. Pali (Eds.), *Environmental justice: Restoring the future* (pp. 40–53). Leuven: European Forum for Restorative Justice.

Minguet, A. (2021). Environmental justice movements and restorative justice. *The International Journal of Restorative Justice*, 4(1), 60–80.

Pain, N., Pepper, R., McCreath, M., & Zorzetto, J. (2016). *Restorative justice for environmental crime: An antipodean experience.* Conference paper. International Union for Conservation of Nature Academy of Environmental Law Colloquium, Oslo Norway.

Vanmechelen, K. & Magiels, G. (2019). *The global only exists by the generosity of the local.* Tielt: LannooCampus.

Varona, G. (2020). Restorative pathways after mass environmental victimisation: Walking in the landscapes of past ecocides. *Oñati Socio-Legal Series,* 10(3), 664–685.

Walgrave, L. (2021). *Being consequential about restorative justice.* The Hague: Eleven.

Walker, M.U. (2006) Restorative justice and reparations. *Journal of Social Philosophy,* 37(3), 377–395.

White, R. (2018). Ecocentrism and criminal justice. *Theoretical Criminology,* 22(3), 342–362.

Wijdekop, F. (2019). *Restorative justice responses to environmental harm. An IUCN report.* Retrieved from www.restorativejustice.nl/user/file/rapportiucnnl.pdf (last accessed 9 February 2022).

Index[1]

---

[1] Note: Page numbers followed by 'n' refer to notes.

Printed in the United States
by Baker & Taylor Publisher Services